Handbook of UNMANNED AERIAL VEHICLES

Handbook Of UNMANNED AERIAL VEHICLES

Franco Victor Bacha
Sok Chea Am

Kruger Brentt
Publishers
2025

Kruger Brentt Publishers UK. LTD.
Company Number 9728962

Regd. Office: 68 St Margarets Road, Edgware, Middlesex HA8 9UU

ISBN: 978-1-78715-370-7

Disclaimer:

Every possible effort has been made to ensure that the information contained in this book is accurate at the time of going to press, and the publisher and author cannot accept responsibility for any errors or omissions, however caused. No responsibility for loss or damage occasioned to any person acting, or refraining from action, as a result of the material in this publication can be accepted by the editor, the publisher or the author. The Publisher is not associated with any product or vendor mentioned in the book. The contents of this work are intended to further general scientific research, understanding and discussion only. Readers should consult with a specialist where appropriate.

Every effort has been made to trace the owners of copyright material used in this book, if any. The author and the publisher will be grateful for any omission brought to their notice for acknowledgement in the future editions of the book.

For information on all our publications visit our website at http://krugerbrentt.com/

PREFACE

This book provides a comprehensive introduction to the world of aerial vehicles, offering readers fundamental knowledge about various types of aircraft and unmanned aerial vehicles (UAVs). This handbook serves as an essential resource for enthusiasts, students, hobbyists, and professionals interested in understanding the basic principles, technologies, and applications of aerial vehicles.

Aerial vehicles have revolutionized transportation, surveillance, photography, and numerous other fields. From traditional airplanes and helicopters to cutting-edge drones and quadcopters, these vehicles offer unique capabilities for exploration, monitoring, and transportation in both civilian and military contexts.

The primary objective of this handbook is to provide readers with a foundational understanding of aerial vehicles, covering key topics such as aerodynamics, propulsion systems, flight control, navigation, safety considerations, and regulations. Each chapter offers insights, practical tips, and illustrative examples to help readers grasp the fundamental concepts and operational principles of aerial vehicles.

Contributions from experts in the field explore various types of aerial vehicles, including fixed-wing aircraft, rotary-wing aircraft, hybrid aircraft, and UAVs. Special emphasis is placed on emerging trends and technologies, such as electric propulsion, autonomous flight, sense-and-avoid systems, and applications in fields like agriculture, filmmaking, environmental monitoring, and disaster response.

As aerial vehicles continue to evolve and find new applications, it is essential for enthusiasts and professionals alike to stay informed about the latest developments and best practices in aerial vehicle operation and management. By fostering a deeper understanding of aerial vehicles and their capabilities, we can unlock new opportunities for innovation, exploration, and problem-solving in diverse domains.

We extend our sincere gratitude to the contributors whose expertise and dedication have enriched this handbook. It is our hope that "Handbook of Unmanned Aerial Vehicles" will serve as a valuable resource and inspiration for readers interested in exploring the fascinating world of aerial vehicles.

Franco Victor Bacha

Sok Chea Am

CONTENTS

PREFACE V

1. DESIGNING AND BUILDING A FLY-BY-WIRELESS UAV PLATFORM 1

1. Introduction: An Overview 1

2. Wireless Network Technologies 2

3 Global Electronics Architecture 4

4. On-board Wireless System 4

5. Results 9

6. Concise summary 11

References 12

2. INTEGRATING OCCUPANCY GRIDS WITH A POLYGONAL OBSTACLE WORLD MODEL FOR AUTONOMOUS AERIAL NAVIGATION 13

1. Introduction 13

2. Flight Testbed 14

3. Model Overview 16

4. Occupancy Grid Mapping 18

5. Extracting Objects from the Grid Map 21

6. Estimating Ground Planes 26

7. Results 27

8. Concise Summary 28

References 29

3. MICRO-AIR VEHICLES USE FIELD PROGRAMMABLE GATE ARRAYS (FPGA) FOR BIO-INSPIRED VOLTAGE CONTROL SYSTEMS 31

1. Introduction: An Overview 31

2. Methodology for FPGA 34

3. LORA III autopilot 37

4. Execution of visual feedback mechanisms 43

5. Architecture 46

6. Hardware co-simulation 50

7. Concise Summary 55
References 56

4. PLANNING AND GUIDANCE FOR ADVANCED UAV TRAJECTORY GENERATION 61

1. Introduction: An Overview 61
2. Motion-planning Methodologies 63
3. Trajectory Generation and Guidance Module - TG2M 66
4. TG2M Simulation Results 84
5. Observations 88
References 89

5. PROTOTYPING OF UNMANNED HELICOPTER MODELING AND CONTROL .. 91

1. Introduction: An Overview 91
2. Helicopters flight principles 92
3. Specification of the Model 94
4. Identification Results 104
5. Control Prototyping 107
6. Control Results 113
7. Concise summary 114
References 115

6. STABILIZATION OF SCALE MODELS OF VERTICAL TAKE-OFF AND LANDING VEHICLES WITHOUT MEASURING THEIR SPEEDS 117

1. Introduction: An Overview 117
2. Notations and Mathematical Background 119
3. UAV Model and Control Strategy 120
4. Position Controller 124
5. Attitude Controller 127
6. Stability Analysis 131
7. Simulation Results 132
8. Concise summary 134
References 134

7. OPTIMIZATION OF FLIGHT CONTROL SYSTEM DESIGN USING GENETIC PROGRAMMING 137

1. Introduction: An Overview 137
2. Aircraft flight control 137
3. Evolutionary design(ED) 150

4. *Controller synthesis and testing* 156
5. *Controller testing* 164
6 *Concise summary* 169
References 171

8. FLY-THE-CAMERA PERSPECTIVE: MANAGEMENT OF A REMOTELY OPERATED QUADROTOR UAV AND CAMERA SYSTEM 175

1. *Introduction: An Overview* 175
2. *System Modeling* 181
3. *Control Method* 188
4. *Stability Analysis* 193
5. *Simulation system* 195
6. *Concise summary* 198
7. *Appendix* 199
References 201

9. DRAGONFLIES CAN TEACH US ABOUT HOW TO FLY INTELLIGENT AERIAL VEHICLES 203

1. *Introduction: An Overview* 203
2. *A robotic dragonfly model* 205
3. *Flight pattern of a typical dragonfly* 207
4. *Force measurements and computations* 208
5. *Aerodynamic force for anti-phase hovering flight* 209
6. *Impact of phase difference on hovering flight* 210
7. *Impact of phase disparity during forward flight* 211
8. *Dragonfly-inspired M.A.V* 213
9. *Concise summary* 215
References 215

10. DC SUPPLY SYSTEM DETECTOR OF UNMANNED AERIAL VEHICLE (UAV) 217

1. *Introduction: An Overview* 217
2. *Hardware Circuit and Measurement Principle* 218
3. *Software Flowchart and Methods of Measurement* 220
4. *Result Analysis and Methods of Improving Precision* 221
5. *Concise summary* 223
References 223

11. FORMATION FLIGHT OF UNMANNED AERIAL VEHICLES UTILIZING SLIDING MODE DISTURBANCE OBSERVERS 225

1. Introduction: An Overview 225
2. The vehicle: X-Shaped Rotorcraft 226
3. Formation Strategies 229
4. Controller Architecture 231
5. Sliding Mode 232
6. Controller Design 236
7. Simulation Results 241
8. Concise summary 247
9. List of Symbols 248
References 249

12. DESIGN AND EXPERIMENTS OF AUTONOMOUS FORMATION FLIGHT 251

1. Introduction: An Overview 251
2. Formation Controller Design 252
3. Test-Bed Development 257
4. Modeling and Parameter Identification 261
5. Controller Implementations and Simulation 267
6. Flight-Testing Of Formation Control Laws 270
7. Concise summary 275
References 275

13. VIBRATION-INDUCED PARTICULATE MATTER OSCILLATOR NOISE AND ITS MITIGATION 279

1. Introduction: An Overview 279
2. Defining Acceleration Sensitivity 280
3. Measurement Techniques 284
4. Results - Acceleration Sensitivity of Non-oscillatory Devices 290
5. Results - Acceleration Sensitivity of Different Classes of Oscillators 296
6. Acceleration Sensitivity Reduction 302
7. Concise summary 305
References 306

14. LOCALIZATION OF QUADROTOR UAV FORMATIONS UTILIZING NEURAL NETWORK CONTROL AND WIRELESS SENSOR NETWORKS 311

1. Introduction: An Overview 311
2. Background 312

3. Leader-Follower Formation Tracking Control 315
4. Optimized energy-delay sub-network routing (OEDSR) protocol for UAV Localization and Discovery 327
5. Simulation Results 332
6. Concise summary 334
References 334

15. ASYMMETRIC HOVERING FLAPPING FLIGHT: A COMPUTATIONAL ANALYSIS 337

1. Introduction: An Overview 337
2. Flapping kinematics and parameters 339
3. Investigation tools 340
4. Results and discussion 343
5. Concise summary 354
References 355

16. PATH PLANNING FOR UNMANNED AERIAL VEHICLES (UAVS) IN SEARCH OPERATIONS 357

1. Introduction: An Overview 357
2. Problem Definition 358
3. Related Work 359
4. Our Approach 360
5. Sequential Monte Carlo Methods 361
6. Applying SMC methods to UAV Search 363
7. Implementation and Test of a Scenario 365
8. Concise Summary 369
References 370

17. OPTIMAL CIRCULAR FLIGHT OF MULTIPLE UNMANNED AERIAL VEHICLES FOR TARGET SURVEILLANCE IN URBAN ENVIRONMENTS 373

1. Introduction: An Overview 373
2. UAV Path Planning for STTP 374
3. Numerical Tests 381
4. Concise summary 384
5. At a Glance 384
References 385

18. STIFFNESS-FORCE FEEDBACK IN UNMANNED AERIAL VEHICLE (UAV) TELEOPERATION 387

1. Introduction: An Overview 387

2. Haptic feedback 388

3. Tuning of the stiffness feedback 395

4. Experiment 396

5. Concise Summary 402

References 403

INDEX 405

Chapter 1 DESIGNING AND BUILDING A FLY-BY-WIRELESS UAV PLATFORM

1. INTRODUCTION: AN OVERVIEW

Within the past few years, study into unmanned aerial vehicles (UAVs) has sped up, and some very interesting devices have been made. UAVs are useful for many things, like keeping an eye on forests and finding fires, keeping an eye on the coast and economic exclusive zones, finding pollution in watersheds, and even military tasks. This chapter talks about work that is part of a bigger project called AIVA. It involves designing and building an aerial platform with instruments, communications, flight control, and artificial vision systems so that it can take off, fly, and land itself. The chapter is mostly about the on-board wireless distributed data acquisition and control system, which is one of the most novel parts of the project. Unmanned aerial vehicles (UAVs) usually have one main, complicated unit with one or more CPUs. Instrumentation devices are wired to this unit. They also have big mechanical links. The standard single processing unit is swapped out for several simpler units (wireless nodes) that are spread out across the aircraft. This method is called "fly-by-wireless." So, the nodes are put close to the sensors and controlled surfaces, making a network of nodes that can collect data, process it, and act on it. This suggested fly-by-wireless platform has some benefits over traditional systems. For example, it is more flexible and modular, and it's easier to set up because there are no connecting cables. But it also brings about a number of problems. The wireless network that powers the distributed data acquisition and control system on board needs to meet strict quality of service (QoS) standards, such as steady throughput, limited delay, and reliable packet delivery. In order to make the system more self-sufficient, it is also important to make sure that the battery-powered wifi nodes don't use too much power. There are a lot of different types of wireless network systems on the market right now. Section 2 gives an overview of the most important tools and talks about how well they meet the needs listed above. Because of this study, we decided to use the Bluetooth wireless network technology to build a prototype of the fly-by-wireless system. It was decided that commercial off-the-shelf parts would be the best way to build the system while still keeping development costs low and performance high. As the system was being built, other goals were also pursued. These included creating a framework that lets nodes communicate effectively regardless of the technology used, coming up with a way to model the embedded system, and creating an application-oriented operating system with a modular structure. The following parts are set up in this way: As the needs of the application are taken into account, Section 2 gives an overview of the different wireless network technologies that are available. Section 3 talks about the overall electronics architecture of the UAV platform, and Section 4 talks about the onboard wireless system that was developed. Section 5 shows the performance results that were achieved with this system in experiments, and Section 6 talks about the conclusions and what needs to be done next.

2. WIRELESS NETWORK TECHNOLOGIES

These are the main types of wireless networks technologies used today: satellite networks, mobile cellular networks, broadband wireless access, wireless local networks (WLAN), and wireless personal networks (WPAN). The first three are very different from the last two. This is different because the user does not own the network equipment. Instead, the network operator does. The user pays the network operator for the services they provide. They also cover a larger area, which is another difference. WLAN and WPAN, on the other hand, are short-range technologies where the user usually owns all of the transmission gear. Because these features work better for the purpose, the rest of this part will only talk about wireless network technologies in these two groups. A lot of people use IEEE 802.11 (IEEE, 2007), which is also known as WiFi, as their WLAN these days. These networks come in a number of different physical forms and radio bands. All of these versions, though, use the same MAC algorithm, which is a CSMA/CA (Carrier Sense Multiple Access/Collision Avoidance) system called DCF. Due to the chance of collisions, this protocol is not good enough to provide the QoS promises needed by the onboard wireless data acquisition and control system because it is based on statistics. In order to handle real-time traffic, the 802.11 standard defines a different MAC protocol called PCF (Point Coordination Function). It is based on a polling scheme and can guarantee QoS. While the DCF protocol must be used, the PCF protocol does not have to be, and devices that support it are not easy to find. A newer standard, IEEE 802.11e (IEEE, 2007), was released not long ago. It's meant to make 802.11 networks more efficient and better support Quality of Service (QoS), but it's not easy to find on the market either. Concurrently to the development of the 802.11, the European Telecommunications Standards Institute (ETSI) has developed another WLAN standard: HIPERLAN/2 (ETSI, 2002). HIPERLAN/2 networks are designed to operate at the 5 GHz band using OFDM (Orthogonal Frequency Division Modulation). Its physical layer is similar to the one used by the IEEE 802.11a due to agreements made by the two standard bodies. On the other hand, the MAC protocols used by these networks are radically different. HIPERLAN/2 uses a demand based dynamic TDMA (Time Division Multiple

Access) protocol, which is able to provide extensive support of QoS to multiple types of traffic, including those generated by data acquisition and control systems (Afonso & Neves, 2005). However, the 802.11 standar dwon the battle for the wireless LAN market and as such no available HIPERLAN/2 products are known at the moment.

Due to its design characteristics, 802.11 and HIPERLAN/2 modules present relatively high power consumption. Although these networks can be suitable to interconnect devices like computers, there is an enormous potential market to provide wireless communication capabilities to smaller and cheaper devices running on batteries without the need of frequent recharging. Such devices include computer peripherals, biomedical monitoring appliances, surveillance units and many other sensing and actuation devices. To provide communication capabilities to such devices, various low cost short range networks, known collectively by the term wireless personal area network (WPAN), are being developed.

At the IEEE, the task of standardization of WPAN networks is under the scope of the IEEE802.15 group. One of these standards, the IEEE 802.15.4 (IEEE, 2006) defines the physical and MAC layer of ZigBee (ZigBee, 2006), which aims to provide low power and low bit rate WPANs with the main purpose of enabling wireless sensor network applications. At the physical layer, the IEEE 802.15.4 relies on direct sequence spread spectrum (DSSS) to enhance the robustness against interference, and provides gross data rates of 20/40 kbps, at the 868/915 band, and 250 kbps, at the 2.4 GHz band. As in 802.11 networks, the basic ZigBee MAC protocol is a contention based CSMA/CA mechanism. A complementary mechanism defined in the 802.15.4 standard, the guaranteed time slot (GTS), enables the provision of some QoS guarantees to real-time traffic.

Bluetooth (Bluetooth, 2003) is another WPAN technology. It operates in the 2.4 GHz band using frequency hopping spread spectrum (FSSS) and provides a gross data rate of 1 Mbps. A star structure, or piconet, is used by Bluetooth. It is made up of one master and up to seven active slaves. No matter what kind of gadget it is, transmissions can go up to 10 or 100 m. At the MAC layer, Bluetooth devices use a protocol built on polling that can handle both real-time and asynchronous traffic. For the purpose we're talking about, Bluetooth has better general features than the other networks we've talked about so far. It requires a lot less power than 802.11 and HIPERLAN/2, has a faster gross data rate than ZigBee, and uses a MAC system that can handle real-time traffic. The Bluetooth spread spectrum has a bandwidth of 79 MHz, while ZigBee works in a band of less than 5 MHz. This makes Bluetooth better at blocking interference. Bluetooth also has a feature called "adaptive frequency hopping" that lets it dodge frequency bands that are affected by interference. Because it had these qualities and was already out there when the prototype was being made, Bluetooth was chosen as the wireless network technology that would help power the system that will be explained in the next few parts.

3 GLOBAL ELECTRONICS ARCHITECTURE

Figure 1 shows a big picture of the general design of the AIVA fly-by-wireless UAV platform's computer and transmission system that is built in. It's a system with multiple tasks and processors that are linked by an asynchronous local bus that lets the speeds of the tasks and processors change. For a Bluetooth piconet master node, the system architecture allows one processing unit. There is also one flight controller unit, one data logger and earth link, and one embedded vision system (EVS). Many important processes are always running in each of these nodes.

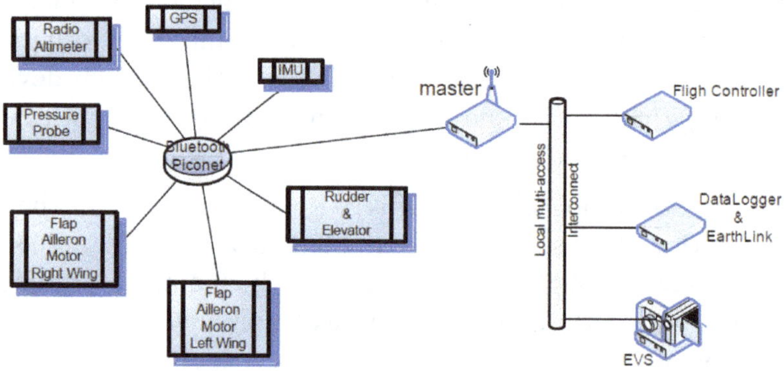

Figure 1. Global electronics architecture of the AIVA UAV platform

This design makes it simple to add or remove processing units from the platform. For example, new cameras or vision units can be added. For the first case, a new module needs to be linked to the Bluetooth piconet. For the second case, the new module needs to be linked to the local multi-access bus.

4. ON-BOARD WIRELESS SYSTEM

Figure 1 shows the Bluetooth piconet, which is part of the AIVA UAV platform's onboard wireless distributed data collection and control system. It is based on Bluetooth (BT) wireless network technology.

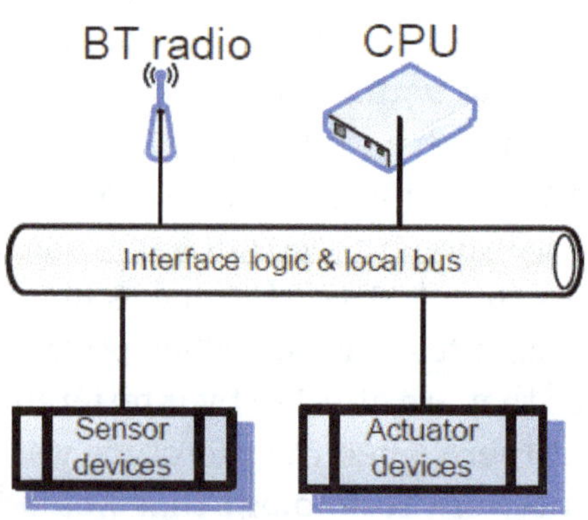

Figure 2. Architecture of a Bluetooth wireless node

Figure 2 shows how a wireless node is put together in general. There is a market-ready Bluetooth module with the radio electronics inside each node, a microcontroller that runs the code that controls the node's behavior, a local bus that lets the node's parts talk to each other, and sensors and/or actuators that are specific to the node's purpose.

4.1 Physical Architecture

The physical part of the platform is built around a low power Texas Instruments MSP430 microcontroller, a Von-Neumann 16 bit RISC architecture with mixed program, data and I/O in a 64Kbytes address space. Besides its low power profile, which uses about 280 µA when operating at 1 MHz @ 2.2 Vdc, MSP430 offers some interesting features, like single cycle register operations, direct memory-to-memory transfers and a CPU independent hardware multiplication unit. From the flexibility perspective, a flexible I/O structure capable of independently dealing with different I/O bits, in terms of data direction, interrupt programming, and edge triggering selection; two USARTs supporting SPI or UART protocols; an onboard 12 bit SAR ADC with 200 kHz rate; and PWM capable timers, are all relevant features.

The Bluetooth modules chosen for the implementation of the wireless nodes are OEM serial adapter devices manufactured by connectBlue. The master node uses an OEMSPA33i module and the slave nodes use OEMSPA13i modules (connectBlue, 2003). These modules include integrated antennas; nevertheless, we plan to replace them with modules with external antennas in future versions of the platform, to be able to shield the modules in order to increase the reliability of the system against electromagnetic interference.

While the module used on the master (OEMSPA33i) allows up to seven simultaneous connections, the module used on the slaves (OEMSPA13i) has a limitation of only three simultaneous connections. However, this limitation does not represent a constraint to the system because the slaves only need to establish one connection (to the master).

The connectBlue modules implement a virtual machine (VM) that enables the provision of a serial interface abstraction to the microcontroller, so Bluetooth stack details can be ignored and focus can be directed to the application. The manufacturer's virtual machine implements a wireless multidrop access scheme where the master receives all frames sent by the slaves and all slaves can listen to the frames sent by the master, in a point-to-multipoint topology.

The AIVA onboard wireless system is composed by one Bluetooth piconet containing seven nodes: one master (MM - Master Module) and six slaves (SAM - Sensing & Actuation Modules). The nodes are spread over the aircraft structure, as shown in Figure 3. The master node (MM) is placed at the fuselage body, and acts as the network and flight controller, onboard data logger, and communications controller for the link with the ground station. On each wing, there is a SAM node for an electrical propulsion motor and for control surfaces (ailerons and flaps). These wing nodes are responsible for motor speed control and operating temperature monitoring, as well as control surfaces actuation and position feedback.

In the fuselage body, there are other two SAM nodes, one for a GPS module and other for an inertial measurement unit (IMU), which provide information assessment for navigational purposes. At the tail, there is another SAM node for elevator and rudder control, and position feedback. Finally, at the nose there is a The SAM node is linked to a nose probe that has a special design and six separate pressure sensors that send useful digital data for flight control. There is also a sound probe in this node that helps the automatic takeoff and landing system by giving it information. Figure 4 shows how the nose node is physically laid out. In the bottom left corner is the Bluetooth module. On the left side of the board is the microprocessor, and on the right side is the sensor hardware.

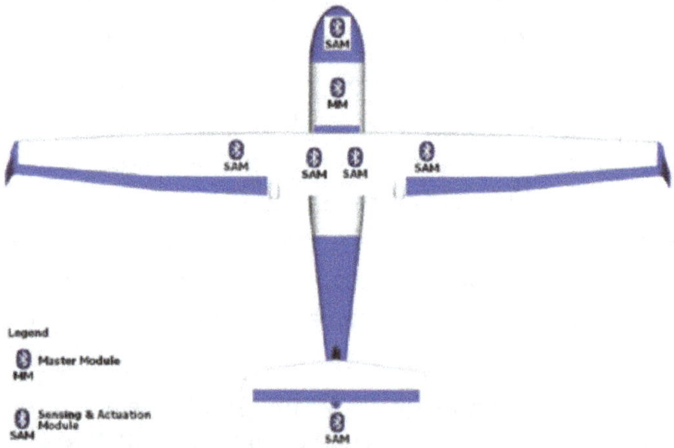

Figure 3. Node distribution on the aircraft structure

Figure 4. Physical layout of the nose node

4.2. Logical Architecture

A transport layer and an application layer make up the two-layered state machine design that was used to build the system. The transportlayer, which is run by the master node, offers a packet delivery service that can send data packets across the network without any problems. The transport layer doesn't depend on the application and talks to the top level application layer through a buffering data space and a set of signaling control bits that keep the two layers in sync. Figure 5 shows the structure and how signals move between the two layers.

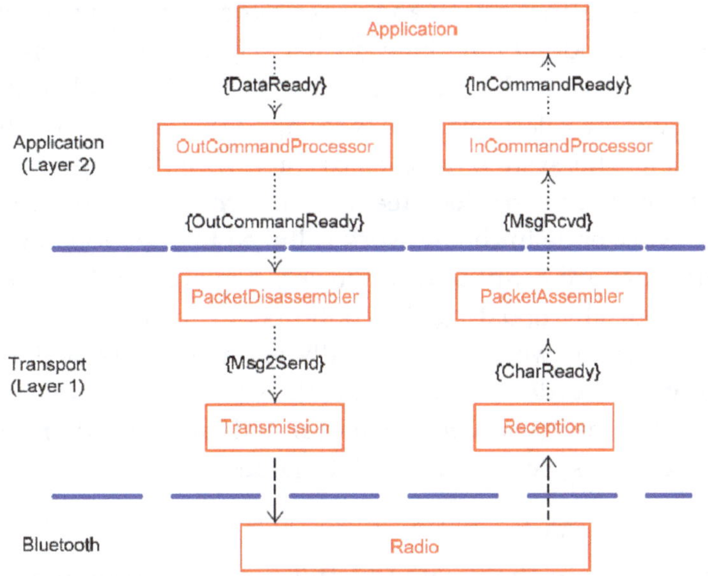

Figure 5. Hierarchy and signalling at the logical level of the platform

The asynchronous reception process delivers characters to upper processes. Analysing the hierarchy from the lower level to the upper level, Char Ready condition goes TRUE every time a new character arrives to the interface. The next process in the hierarchy is Packet Assembler, a state machine that performs packet re-assembly, reconstructing the original packet from a group of segments, and delivers packets for the next process in the hierarchy. When Msg Rcvd (message received) goes TRUE, a new message is ready for processing. Thus, for incoming data, the model at layer 1 receives characters and delivers ready-to-process messages to the application layer. When the application layer understands that the message is ready to process, a command processor for incoming messages is activated in order to decode the embedded command and semantics contained in the message, to eventually execute some action, and to pass relevant information for the final application.

For outgoing data, the resident application eventually makes available some data to transmit to the master, signaling this event with a Data Ready signal. This causes the output command processor to execute its cycle, preparing one message to be sent. When the message is ready, Out Command Ready goes TRUE, signaling to the lower layer that there is a message to send to the network. At this phase, frame segmentation starts (if needed) by means of a state machine for packet disassembly. This state machine breaks the original message in smaller segments prepared to be serialized. Each time a segment is ready to be sent, Msg2Send goes TRUE and serialization is triggered. So, for outgoing data, the transport layer receives messages from application layer, and sends segments to the radio module in order to be sent over the wireless medium.

Layer 2 depends on the application and doesn't know anything about how the lower layer works on the inside; it only knows what services and interfaces are available from it. In other words, its logical design can be used for other things. The main purpose of the fly-by-wireless program is to copy a system table to all network nodes as often as possible. This system table keeps track of all the important system values, including the state parameters, loop feedback information, and the different sensors and actuators. Every node in the network is linked to a section of a table that holds all the variables that are connected to sensing, actuators, and meters. This layer is in charge of periodically updating the contents of each table based on the current situation in the area, as well as periodically acting on data sent from the master node (flight controller orders). In this way, the system is seen as a two-dimensional array that is permanently placed at the master node and has different copies that are spread out and kept in sync among the slave nodes.

4.3 Other Design Issues

All Bluetooth modules in the created platform are set to non-discoverable mode, enhancing the system's security. The Bluetooth node discovery procedure is relatively slow, taking several seconds; however, this is not an issue as the master retains the addresses of all slave nodes intended to participate in the piconet, thereby circumventing this process. The establishment of the piconet occurs on the ground before to takeoff, hence the resultant delay is not an issue. Bluetooth technology restricts piconet operation to a maximum of seven active slaves; however, this limitation is not a significant issue for the created system, as it utilizes only six slaves, which may present constraints if the number of nodes were to increase. The quantity of slaves within the network may be augmented by interlinking many piconets to create a scatternet. A device engaged in two pico networks could facilitate traffic relay between them. Nonetheless, this architecture is likely to adversely affect network performance, complicating the provision of QoS guarantees to the application. Furthermore, there are now limited practical implementations of scatternets. Considering that free space propagation loss is proportional to the square of the distance, it is anticipated that the on-board wireless network will neither experience nor generate interference with other networks operating within the same frequency band, such as the extensively utilized WiFi networks, as the former predominantly operates in the sky, whereas the latter typically functions on the ground.

5. RESULTS

The efficacy of the designed wireless system was assessed in a laboratory setting. The experimental configuration utilized to get the results delineated in this part comprises six slaves transmitting data periodically to the master (uplink direction) at a consistent predetermined sampling rate. Each sampling packet measures 15 octets, the maximum length permitted by the transport layer due to constraints imposed by the virtual machine utilized by the Bluetooth module. Figure 6 illustrates the cumulative uplink throughput received by the master node as a function of the sampling rate employed by the six slaves. Bluetooth employs a contention-free MAC protocol, resulting in uplink transmissions that are unaffected by collisions. Consequently, network throughput increases linearly with the offered load until it reaches saturation, which occurs when the slaves transmit data at sampling rates exceeding 200 Hz. The statistic indicates that the application's maximum throughput is approximately 160 kbps, considerably lower than Bluetooth's gross data rate of 1 Mbps. The disparity can be attributed to the overhead imposed by the Bluetooth protocol and the virtual machine, encompassing the inter-packet gaps, Forward Error Correction (FEC), Automatic Repeat reQuest (ARQ) mechanisms, packet headers, and the additional overhead from control packets, such as the POLL packet, which the master transmits to authorize slaves to send data.

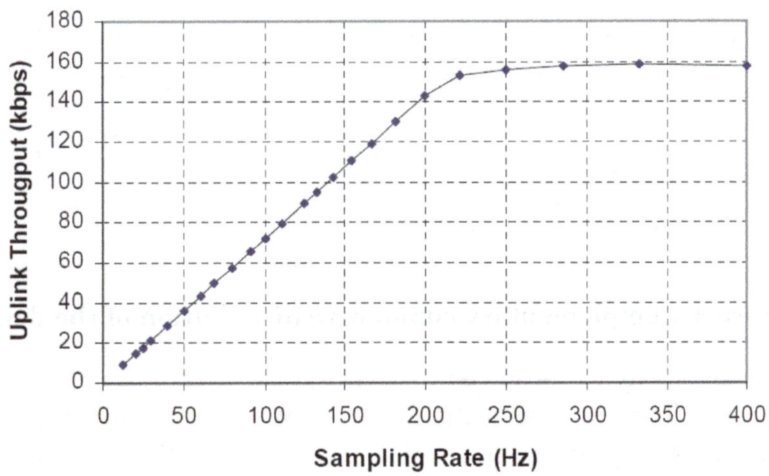

Figure 6. Uplink throughput as a function of the sampling rate

Figure 7 presents the packet loss ratio (PLR) averaged over the 6 slaves as a function of the sampling rate. As the figure shows, the PLR is limited to less than 0.5 % in the region where the network is not congested, but increases rapidly after the saturation point. The flight control application should be able to tolerate such small losses; otherwise a change in the supporting wireless technology should be made in the attempt to obtain higher link reliability.

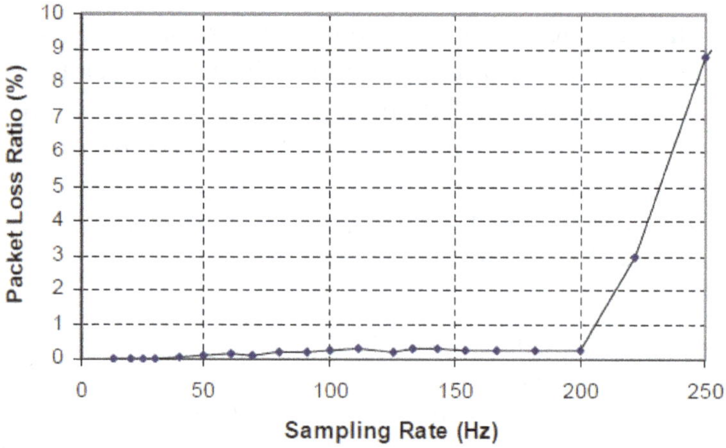

Figure 7. Packet loss ratio as a function of the sampling rate

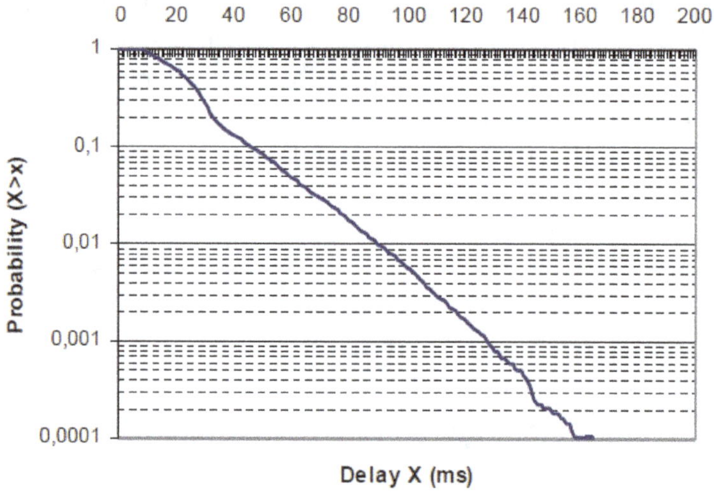

Figure 8. Complementary cumulative distribution of the delay

Concerning to the delay experienced by the packets as they travel from the slaves to the master, results showed that the delay is not adversely affected by the rise in the offered load, as long as the network operates below the saturation point. For sampling rates up to 200 Hz, the registered average delay was 27 ms and the standard deviation was 16 ms.

Figure 8 presents the complementary cumulative distribution (P{X>x}) for the random variable X, which represents the delay for all the samples collected using sampling rates in the range from 0 to 200 Hz. With this chart, it is possible to see the probability that the delay exceeds a given delay bound, which is an important metric for real-time applications such as the one considered in this chapter. The chart shows, for instance, that less than 1 % of the sample packets suffer a delay higher than 90 ms, while less than 0.1 % of the packets suffer a delay higher than 120 ms.

Experimental testing were conducted with a different number of slaves in the piconet (ranging from 1 to 6), in both the uplink and downlink directions. The average latency in the downlink direction (from the master to the slaves) was marginally greater than that recorded in the uplink direction, however remained below 40 ms for measurements including up to 4 slaves. The average master-to-slave delay increased to 600 ms with 5 slaves in the network, and further deteriorated to 1000 ms with 6 slaves.

6. CONCISE SUMMARY

This chapter presented the design and development of a fly-by-wireless UAV platform built on top of Bluetooth wireless technology. The developed onboard wireless system is composed by one master node, connected to the flight controller and six slave nodes spread along the aircraft structure and connected to several sensors and actuators.

In order to assess the suitability of the developed system, several performance evaluation tests were carried out. The experimental results showed that, for the slave-to-master direction, the system prototype is able to support a sampling rate of up to 200 Hz for each of the 6 slaves simultaneously without significant performance degradation in terms of throughput, loss or delay. On the other hand, although the master-to-slave delay with 1 to 4 slaves in the network is low, its value increases significantly with 5 and 6 slaves, which is unacceptable given the real-time requirements of the desired application. This problem is caused by implementation issues related to the proprietary embedded virtual machine provided by the manufacturer of the Bluetooth module that is used in the master node of the prototype.

This chapter detailed the design and development of a fly-by-wireless UAV platform utilizing Bluetooth wireless technology. The onboard wireless system has one master node linked to the flight controller and six slave nodes distributed throughout the aircraft structure, interfacing with various sensors and actuators. To evaluate the system's appropriateness, multiple performance assessment tests were conducted. The experimental results indicated that, in the slave-to-master path, the system prototype can sustain a sampling rate of up to 200 Hz for each of the 6 slaves concurrently without notable performance deterioration regarding throughput, loss, or latency. Conversely, while the master-to-slave delay with 1 to 4 slaves in the network is minimal, it escalates considerably with 5 and 6 slaves, rendering it unsatisfactory for the real-time demands of the intended application. The issue arises from implementation challenges associated with the proprietary embedded virtual machine supplied by the manufacturer of the Bluetooth module utilized in the prototype's master node. The strategy of utilizing the manufacturer's virtual machine, which conceals the Bluetooth protocol stack capability, enabled the development emphasis to shift towards the application, hence decreasing development expenses. The drawback, however, is the absence of control over the system's behavior at the Bluetooth stack level, which hinders performance optimization and the resolution of issues such as the confirmed master-to-slave delay. The resolution to the identified issue may involve either substituting the Bluetooth module with a newer one from the same manufacturer or directly interfacing with the Bluetooth stack, circumventing the virtual machine. Notwithstanding the constraints of the existing prototype, the aggregate outcomes provided by the experimental assessments are commendable. However, additional testing are required to assess the system›s performance under more severe interference situations and in a real-world aircraft environment.

REFERENCES

1. Afonso, J. A. & Neves, J. E. (2005), Fast Retransmission of Real-Time Traffic in HIPERLAN/2 Systems, *Proceedings of Advanced Industrial Conference on Telecommunications (AICT2005)*, pp. 34-38, ISBN 0-7695-2388-9, Lisbon, Portugal, July 2005, IEEE Computer Society.
2. Bluetooth SIG (2003), Specification of the Bluetooth system. Available at http://www.bluetooth.org/.
3. Connect Blue (2003), Serial Port Adapter - 2nd Generation, *User Manual*. Available at http://www.connectblue.com/.
4. ETSI TR 101 683 V1.1.1 (2000), Broadband Radio Access Networks (BRAN)—HIPERLAN Type 2—Data Link Control (DLC) Layer—Part 1: Basic Data Transport Functions.
5. IEEE Std 802.11 (2007), IEEE Standard for Information Technology—Telecommunications and information exchange between systems—Local and metropolitan area networks—Specific requirements—Part11: Wireless LAN Medium Access Control (MAC) and Physical Layer (PHY) Specifications.
6. IEEE Std 802.15.4 (2006), IEEE Standard for Information Technology—Telecommunications and information exchange between systems—Local and metropolitan area networks—Specific requirements—Part 15.4: Wireless Medium Access Control (MAC) and Physical Layer (PHY) Specifications for Low-Rate Wireless Personal Area Networks (WPANs).
7. ZigBee Standards Organization (2006), ZigBee Specification. Available at http://www.zigbee.org/.

Chapter 2

INTEGRATING OCCUPANCY GRIDS WITH A POLYGONAL OBSTACLE WORLD MODEL FOR AUTONOMOUS AERIAL NAVIGATION

1. INTRODUCTION

1.1 Overview

This chapter delineates a mapping procedure applicable to autonomous systems for the purposes of obstacle avoidance and trajectory planning. It represents an advancement over widely utilized obstacle mapping methodologies, such as occupancy grids. Challenges faced in extensive outdoor environments are addressed, resulting in the creation of a compressed map suitable for transmission over low-bandwidth networks. The methodology is capable of real-time operation and functions within whole three-dimensional environments. The efficacy of the suggested method is illustrated under actual operational settings on an unmanned aerial vehicle employing stereo vision for distance measurement.

1.2 The Challenge of Mapping and Obstacle Representation

For cars to achieve autonomy, they must comprehend the environment in which they operate. Similar to humans or animals, they must perceive, comprehend, and retain at least the aspects of the surrounding environment that are nearby. Only then can the cars function autonomously within their environment, devoid of manual remote control. Previous presentations have showcased successful outcomes of autonomous flights in urban environments utilizing unmanned aerial vehicles. (Hrabar et al., 2005; Zufferey & Floreano, 2005; Griffiths et al., 2007; Scherer et al., 2007). Beside that, the majority of obstacle detection, mapping, and avoidance research are carried out with ground vehicles and many approaches used in flight applications are based on practices derived from that wealth of knowledge.

A main requirement for autonomous vehicles is to detect obstacles and to generate environ- mental maps from sensor data and a large number of approaches have been developed over the years (Thrun, 2002). One approach to represent the environment is the use of grid-based maps (Moravec & Elfes, 1985; Konolige, 1997). They allow an easy fusion of data from different sensors; including noise reduction and simultaneous pose estimation, but they have large memory requirements. Further, they do not separate single objects. A second approach, called feature-based maps, focuses on individual objects. An early work (Chatila & Laumond, 1985) uses lines to represent the world in 2-D. Later approaches use planar (e.g. Hähnel et al., 2003) or rectangular (Martin & Thrun, 2002) surfaces for 3-D modeling - but mostly to rebuild the world with details and possible texture mapping. A suitable model for autonomous behavior is the velocity obstacle paradigm (Fiorini & Shiller, 1998) that can be added with the introduced specifications on how to measure the obstacles.

These map kinds, along with others not addressed here, possess distinct advantages and disadvantages. Consequently, various map types are required for distinct robotic tasks (Kuipers, 2000). Utilizing grid-based maps for sensor fusion and feature-based polygonal metric maps for local planning is often beneficial, particularly for obstacle avoidance. (Fulgenzi et al., 2007). Additionally, non-metric topological maps are most suitable for global search tasks like route planning. In a complex scenario, a robot must deal with all of these different maps and keep them updated. These tasks need the usual information exchange.

To generate maps from sensor data that are applicable to autonomous applications, it is a straightforward procedure to generate a grid map from sensor data and extract out the features. Outdoor scenarios, however, can be too large to store the whole scene in a data array with reasonably accurate resolution. Additionally, the area boundaries may be unknown before mapping.

This methodology integrates grid maps with polygonal obstacle representations and addresses the challenge of extensive environments by employing compact grid maps that encompass just critical areas for sensor fusion. Distinctive characteristics are identified, their geometries are computed, and incorporated into a global map that is memory-efficient and readily extendable. This map is not confined to the sensor environment and is utilized for path planning and various applications.

2. FLIGHT TESTBED

2.1 ARTIS – A Flying Robot

The presented mapping and world modeling approach is developed within the ARTIS (Autonomous Rotorcraft Testbed for Intelligent Systems) research project that deals with mid-sized unmanned helicopters (Dittrich et al., 2003). One of the helicopters is shown in figure 1. It has a main rotor diameter of 3 meters and a total weight of up to 25 kg. Flights of more than 30 minutes are possible.

Figure 1. The unmanned helicopter ARTIS

The 5 kW turbine engine possesses the capacity to transport over six kilograms of experimental payload, alongside the avionics system and power supply. The current setup consists of a dedicated image processing computer and a stereo camera system, weighting 2 kg, allowing for the integration of additional sensors such as multiple cameras or laser scanners in future applications.

2.2 Sense and Avoid Setup

A vision system, apart from the flight controller, is mounted on the helicopter for applications requiring environmental sensing capabilities. The current design employs solely cameras due to their lightweight nature, passive operation, and minimal power usage. A dedicated computer processes visual data. This leads to enhanced speed without affecting the real-time performance of the flight control computer. Data exchange for interaction between image-based results and flight control is facilitated using a local network. A mission planning and automated control system is included into the flight control computer (Dittrich et al., 2008). It computes trajectories around obstacles, accounts for modifications resulting from real-time image-based map updates, and directs the helicopter to navigate these routes. The autopilot guarantees steady hovering, enabling the vehicle to operate entirely autonomously.

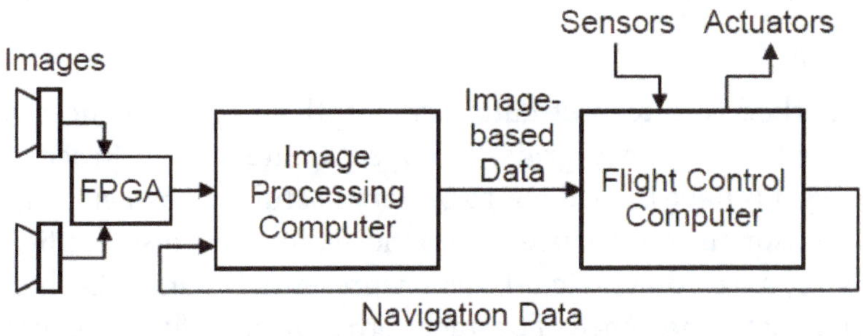

Figure 2. Overview of the on-board hardware for vision applications

Figure 2 illustrates the connection between vision hardware and flight controller. Since obstacle mapping and other image-based algorithms require flight information, a navigation solution provides the global position and attitude of the helicopter.

A stereo camera (Videre Design STOC, fig. 3) with a baseline of 30 cm and a field of view of approximately 51° x 40° is used. It creates images with 640 x 480 pixels and has an inbuilt FPGA processor that calculates a depth image out of the two input images in real-time with 30 Hz. This image is a result of a complex processing step where regions of the two camera images are matched (e.g. Scharstein & Szeliski, 2002), and it acts as a depth sensor in the mapping process. Basic pre-processing to enhance the depth image quality is already done by the camera. In the depth images shown in the figures of this article, near distances are represented by light colors and farther distances by darker colors. White space indicates that depth values are missing or have been filtered due to bad image quality, e.g. low texturing.

Figure 3. Stereo camera mounted at the helicopter (left), left on-board camera image (centre), depth image (right)

The helicopter's position and attitude is provided in six degrees of freedom by the flight control computer using a differential GPS sensor, a magnetometer and an inertial measurement unit. The raw data of these sensors are integrated by an Extended Kalman filter (Koch et al., 2006) to provide an accurate solution in all six degrees of freedom. Filtered navigation data is sent with a rate of 100 Hz to the vision computer. All computer clocks are synchronized so that a fitting recording pose of an image with a given timestamp can be obtained.

3. MODEL OVERVIEW

3.1 Grid Maps

The basic method to interpret data from depth image sequences follows classical approaches with occupancy grids (Moravec & Elfes, 1985; Raschke & Borenstein, 1990). These grids have turned out to be very useful for obstacle mapping since they allow easy sensor fusion, reduce sensor noise, and are also applicable to multiple vehicles. Here, a world-centric 3-D grid represents the map. Each cell consists of a value describing the presence of obstacles: the log odd of the occupancy probability. Higher values refer to a higher probability of the cell being occupied. The map is created incrementally by starting with an empty grid and writing the actual sensor information with each new depth image.

3.2 Feature Maps

As already denoted, occupancy grid maps are advantageous for sensor fusion and will be used to interpret data from depth image sequences. In addition to that, feature maps are built out of these occupancy grids to store global obstacle information in a compressed way and to be an input for applications that use the map.

For this reason, it is a primary requirement to determine the required level of detailing to represent objects in a feature map. For obstacle avoidance applications, the important criteria are:

1. Small details are not needed,
2. an identification of planes for texture projection is not needed, and
3. real-time capabilities are more important than centimeter accuracy.

These criteria imply that it is sufficient to mark a box around an object and simply avoid this area.

The simplest way of modeling potentially danger areas is the usage of cuboidal bounding boxes that are aligned with the coordinate axes. Unfortunately, they are too rough for object modeling, e.g. a set of objects like houses with a rectangular base shape and an arbitrary angle to the coordinate axes. Aligned bounding boxes will be larger than the houses and narrow paths between them will not be found because they exist inside the area marked by the occupied box.

In a real scenario like a city, objects can have any ground shape from the top view, but it is likely that they have vertical walls. This assumption is made in a lot of mapping approaches (e.g. Iocchi et al., 2000). For this reason, prism shapes are used here. They do not require an axis alignment like the bounding boxes. Additionally, they can deal with any ground shape without dramatically increasing the complexity. If walls are not vertical like roofs, the prism's volume will be much larger than the real object. To improve the modeling of objects, they can be subdivided vertically and represented by multiple prisms. In other words, the world is split into layers in different heights, and each layer has its own polygonal 2-D obstacle map. This is sufficient for a lot of applications like flight trajectory planning in urban scenarios (Dittrich et al., 2007).

Nevertheless, the prism model can only handle complex shapes in 2-D as seen from the top view. A gabled roof or a tunnel with an up- or downhill road must be modeled by multiple obstacle prisms. These scenarios are more complex than a simplified urban canyon with a horizontal ground plane and obstacles with vertical walls. Avoiding roofs that are represented by bounding prisms will be only a small restriction and a tunnel, however, will be an unusual case, at least with respect for flying robots.

3.3 The Modelling Process

As already mentioned, advantages of both grid and feature maps are combined, see figure 4. The mapping process works as follows:

1. Create an occupancy grid around the vehicle's position if not existing in the map.
2. If a new grid is allocated or the grid is extended, check whether previously stored features can be inserted.
3. Insert the actual sensor data information to the grid.
4. Find clusters of occupied grid cells and mark them as single obstacle features.
5. Find out which obstacle features are new and which are updates of objects that have already been identified in the previous loop cycle. Preexisting objects may also be removed.
6. Calculate the shape of each new or updated feature.
7. To insert the next sensor data, go back to step 1.

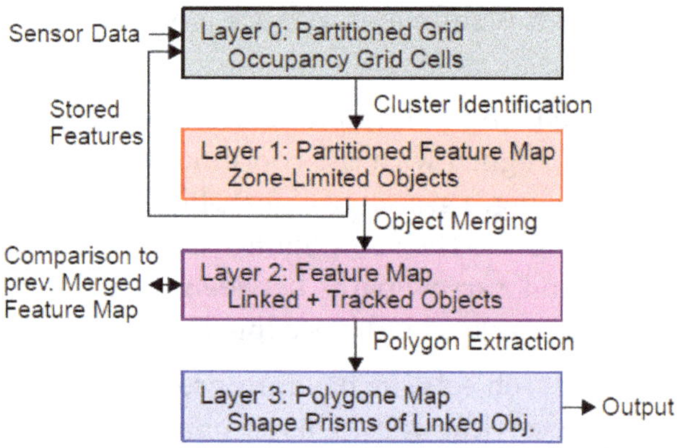

Figure 4. The general mapping process with different map layers

This approach uses different map types as illustrated. The occupancy grid map for sensor inputs (layer 0) is partitioned into zones and each zone is searched for obstacles separately (layer 1). This method is described in the following section. Next, a map with separate obstacles, but without zone separation is generated (layer 2), and finally, the prism shapes are extracted out of them (layer 3). Grid resolution and zone sizes are user-defined but may not change over time when processing image series.

4. OCCUPANCY GRID MAPPING

4.1 Sensor Interpretation and Local Mapping

To iteratively construct an occupancy grid from sequences of depth image data, a local grid map is generated. This local map is world-centric, akin to global output maps, but incorporates solely the occupancy data from a single depth image captured from its respective camera position and orientation. The local map represents the analysis of an individual depth image as follows: Employing projective geometry and the pinhole camera paradigm, each pixel corresponds to an object coordinate in the world. Assuming the absence of transparent items in front of the object, a clear path exists along a ray from the camera center to the object coordinate. No information is accessible beyond the opaque barrier. If the distance represented by a pixel beyond a threshold, this image point indicates free space. A line is delineated for each pixel, as depicted in figure 5, which illustrates the interpretation of an ideal sensor (left) or accounts for measurement noise (right). The uncertainty of stereo-based depth measurement escalates quadratically with increased distance values. The visible region is covered, and the impediments are depicted.

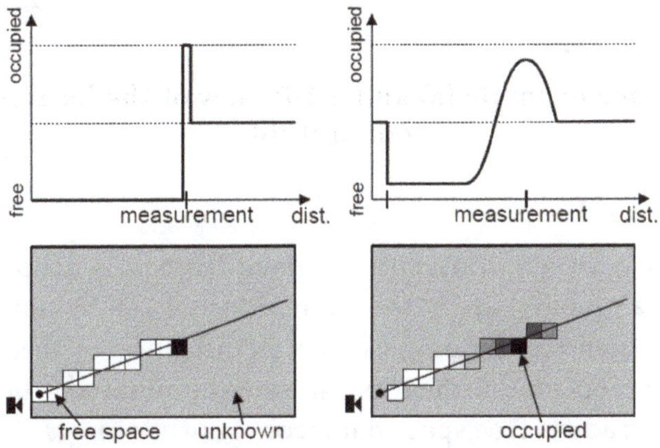

Figure 5. Interpretation of an ideal depth image pixel (left) and a more realistic model that considers depth uncertainty (right)

The result of processing all rays of a single image is illustrated in figure 6. Obstacles and free areas of the image are also visible in the corresponding map. The thickness of an obstacle cannot be derived from a single image so it depends on the depth uncertainty first. Information about the true size and the area behind obstacles are included when the vehicle moves there.

The size of the local grid is determined by considering that the data fusion of the empty local map with one image will only affect grid cells inside the sensor range, i.e. a small environment of the actual position. There is no need to update cell values outside this area. Hence, it is satisfactory to have only an occupancy grid representation of the map inside the sensor range. Further, the local map definition is independent to the outer camera orientation angles to get a fast implementation. With that, the camera is set to the center of the local map, with maximal half a grid cell deviation to keep the grid cells aligned to a global rasterization. To give an example, a sensor range of 30m implies zones with a size of each 60 x 60 x 60m to cover all attitude angles without truncating far distances. In applications where looking and moving straight upwards or downwards is restricted, smaller zones with a height of e.g. 15m are satisfactory. If the grid resolution is set to 0.5m, grid arrays with 120 x 120 x 30 cells are created to represent a local map.

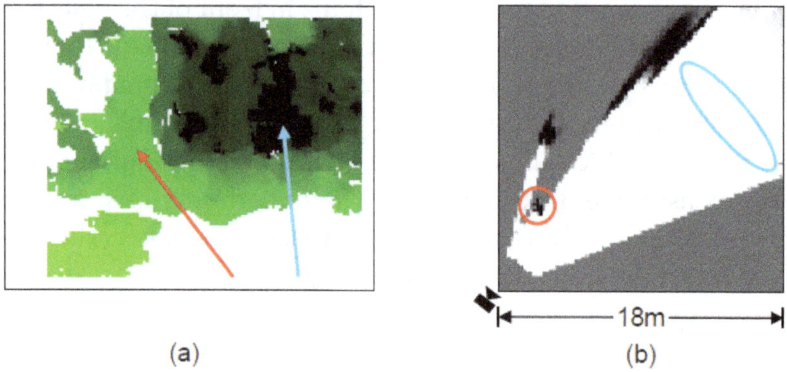

Figure 6. Depth image example (a) and a 2-D view of the local map that was built out of it (b)

4.2 Creating Global Maps

The fusion of data between consecutive image frames is achieved by combining the local grid with a global map by the summation of cell values at identical global coordinates. Like other methodologies, the significance of free or occupied cells increases with repeated measurements. Noise is eliminated when occupancy data from one image contradicts free space data from another image. Consequently, static items can be readily identified. The map values are truncated to a predetermined range to facilitate the use of integers for the cell array. The range must be sufficiently extensive to guarantee resilience against failures.

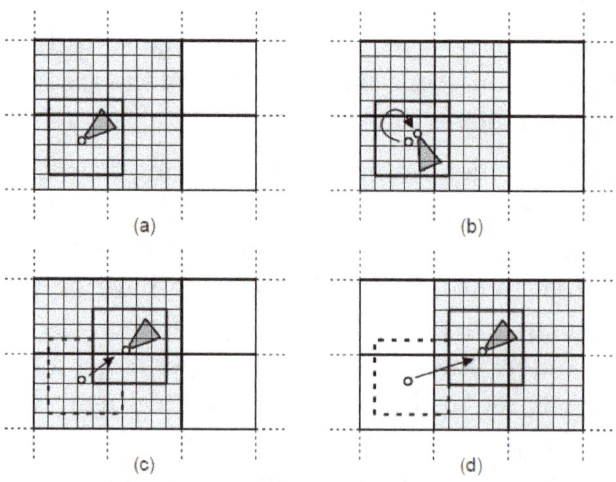

Figure 7. 2-D view of the global map that is divided into zones. For zones around the actual camera position, an occupancy grid representation exists

Since grids are usually stored as continuous data blocks in the computer memory, the boundaries of the map must be known a priori. If the vehicle moves outside, extensive reallocation or shifting methods must be applied because the map boundaries change.

To avoid shifting a large number of map cells when moving, the implementation divides the global map into cuboidal zones. Their position is fixed. The size of each zone is equal to the local map so each zone overlaps a maximum of eight zones in the global 3-D map. The environmental cuboid moves through the global map with the movement of the camera, i.e. with the helicopter. Only those global zones overlapping with the local grid are represented by an occupancy grid array.

Figure 7 illustrates how the zone partitioning works. The actual camera and local map position is shown in figure 7a, the other graphics show possible effects on the map, caused by the next measurement. Often, rotations (7b) or movements (7c) will not have an effect on the zone boundaries. But if the movement is larger so that boundaries are crossed (7d), new memory is allocated for these zones. Grid information is discarded for zones that fall outside the immediate vicinity.

5. EXTRACTING OBJECTS FROM THE GRID MAP

5.1 Determining Separate Objects and Bounding Boxes

Object features are detected by segmenting the global map into occupied and free areas applying a threshold. A single object is a cluster of connected occupied cells of the 3-D array. These objects are recognized with a flood fill algorithm. By saving the minimal and maximal values of the coordinates of cells belonging to the object, the bounding box is calculated and put into the global feature map as it is illustrated in figure 8. Unlike the cell array, these boxes need much less memory and it is easy to make them available to further applications, independent of the existence of a grid. For each object, the binary cell shape is stored in addition to the bounding box so that this shape can be re-inserted. Compression with an oct- tree structure is applicable here. The polygonal shape is calculated later.

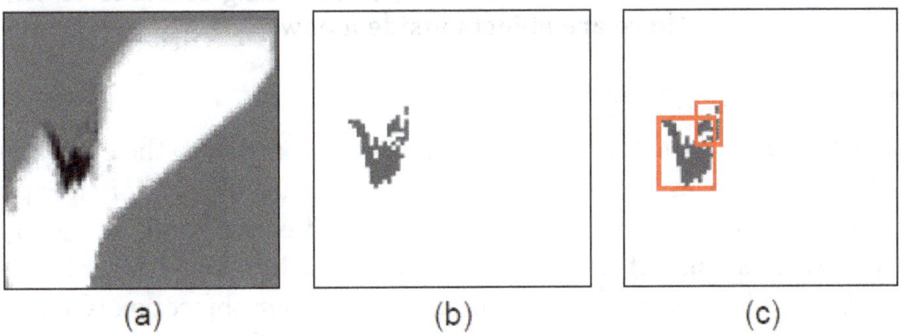

Figure 8. Extracting features from an occupancy grid (a) with thresholding (b) and bounding box calculation (c). Bounding boxes can overlap

5.2 Integration Between Features and Grid

In contrast to the occupancy grid, the feature map is not confined to the vicinity of the vehicle's current location. Objects are stored autonomously, irrespective of the existence of a grid. Should any grid data be eliminated, the associated features will persist (fig. 9). Conversely, things may be incorporated into the grid (fig. 10). A-priori knowledge regarding impediments can be incorporated here.

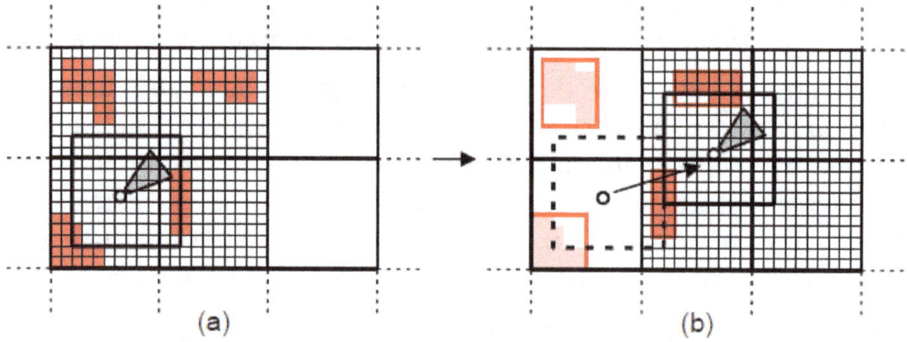

Figure 9. Vehicle movement from (a) to (b). Features are stored when the grid data is discarded due to helicopter movement

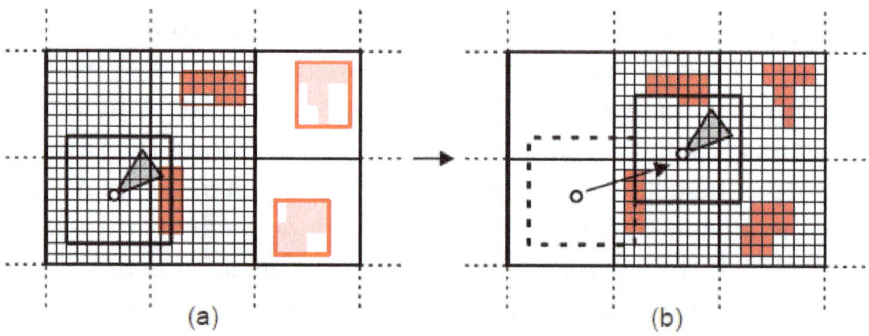

Figure 10. Vehicle movement from (a) to (b). Inserting features to the grid if there are objects inside a new zone

5.3 Merging and Tracking Obstacles

Single objects found in the temporary grid zones are limited to the zone boundaries since each zone is processed separately. To build an output map for the application that has no zone partitioning, the features of different zones are merged if a connection exists as illustrated in figure 11. First, it is checked whether the bounding box of an object is located at a zone boundary. If another object box is located at the same boundary from a different zone, the cell shapes of both objects are tested for connection. Connected objects are linked together, and the linkage is not limited to only two objects.

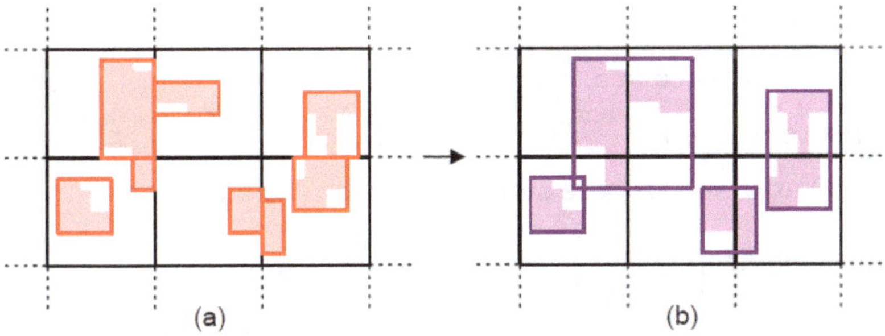

Figure 11. Merging tangent objects in the partitioned map (a) to linked objects (b) that are not partitioned into zones. Simplified 2-D view

After an update step where sensor data is inserted, the zones are checked for obstacles again without reference to the previous state. Since it is useful to know which specific object has been updated with the actual sensor information, the linked objects are tracked. This is done as follows:

- Calculate the center of mass of each linked object.
- For each linked object, try to find a matching object in the map with linked objects of the previous state. This is the object with the nearest center of mass determined by the Euclidean distance (Prassler et al., 2000). To avoid the matching of objects that are too far away, distances above a threshold will not result in a match.
- If no match exists for an object of the actual state, give it a new unique ID. Otherwise, copy the ID from the matched object.

With object tracking, each linked obstacle in the world gets a unique ID that remains constant over time, if the tracking is successful. The merging and tracking step can be skipped for linked objects whose coordinates are completely outside the zones where a grid exists since there will be no update.

Due to sensor updates it is possible that a linked object becomes split. Usually, one of the new objects gets the old ID, the others receive the new IDs. If multiple linked objects are merged together over time, one ID is kept and the others removed.

5.4 Horizontal Shape Slicing

The approach presented in this paper models the shape of each linked object with a prism (see section 5.5) with horizontal bases. Since a bounding prism of complex shapes may be too rough, the cell-based shape $s(x, y, z)$ is partitioned into horizontal slices with the height of one cell. Similar slices are merged and a polygonal shape prism is calculated for each multi-slice.

A single slice in height z is denoted as $sz(x, y)$ with $sz(x, y) = s(x, y, z)$. Without loss of generality, z is valid from 1 to n. Similar consecutive slices are put together to multi-slices S using the following algorithm:

- Set $i = 1$, set $z = 1$.
- Create a multi-slice S_i: $S_i = \{sz\}$.
- If $z = n$, break.
- If the slices sz and $sz+1$ are similar given by the equation

$$\sum_x \sum_y | s_z(x,y) - s_{z+1}(x,y) | < t \tag{1}$$

with a similarity threshold t,

set $Si = Si \cup \{sz+1\}$,

set $z = z + 1$ and go to step 3.

Otherwise,

set $i = i + 1$;

set $z = z + 1$ and go to step 2.

The result is a set of multi-slices $\{S1, ..., Sm\}$. The parameter t of equation 1 controls the degree of similarity consecutive single slices must have in order to be unified. The extreme case $t = 0$ puts every single slice into a separate multi-slice and $t = \infty$ merges all single slices together. As illustrated in figure 12, grid-based object shapes are splitted into parts that have approximatively vertical walls. They can be approximated by their 2-D shape from top view and easily modeled by prisms without forfeiting large irregularities in the shape.

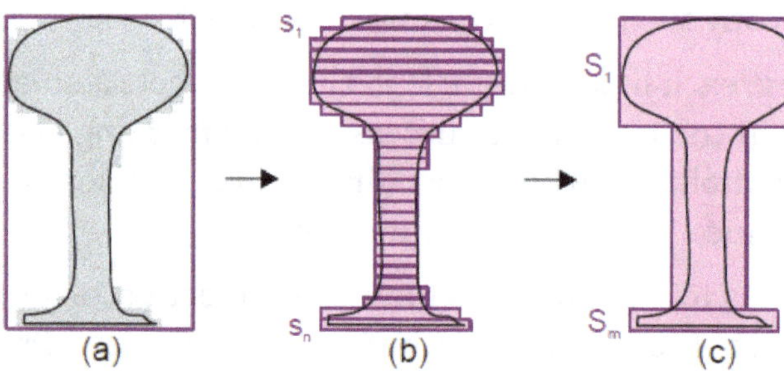

Figure 12. Grid shape and bounding box of linked feature (a), single slices (b) and merged multi-slices (c). Side view of a tree as an example

In the next step, the 2-D shape is calculated for each multi-slice Si ($1 \leq i \leq m$). Occupied cells of a multi-slice are calculated through the logical disjunction of all single slices sz inside Si. It is

$$S_i(x,y) = \begin{cases} 0, \text{ if } \sum_{s_z \in S_i} s_z(x,y) = 0; \\ 1, \text{ otherwise.} \end{cases} \tag{2}$$

5.5 Extraction and Approximation of the Polygonal Shape

Now, algorithms developed for binary images can be applied to a multi-slice *Si* since its shape is represented by a binary 2-D array. Figure 13 illustrates the process. The polygonal extraction is done in two steps. First, the contour is calculated with a tracing algorithm (Ren et al., 2002). The primary objective is to initiate at a single edge pixel and systematically seek the subsequent edge pixel until the entire contour is encompassed. The output consists of a list of pixels arranged in a counterclockwise manner for outer contours and in a clockwise manner for inner contours. Numerous lists are feasible. Each contour pixel can be regarded as a vertex of a polygon by utilizing the coordinates of the matching grid cell center.

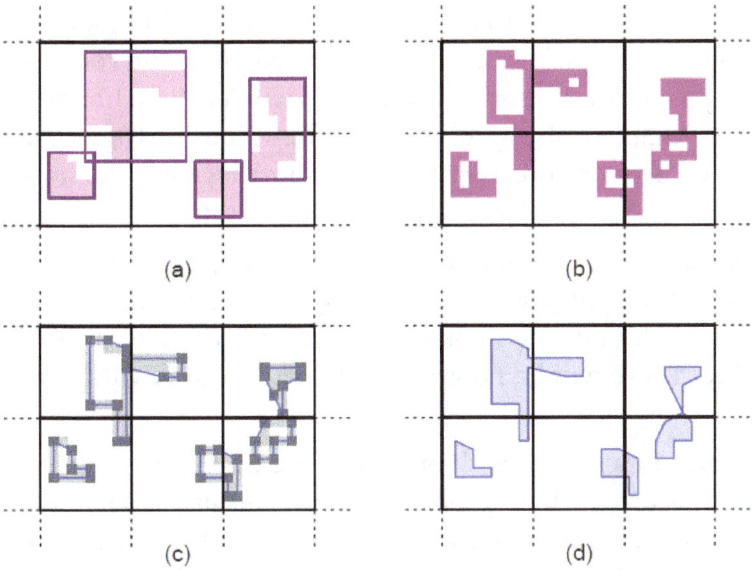

Figure 13. Shape Extraction, top view. Cell-based shapes (a), contours (b), approximated polygons with its vertices (c) and final ground shapes (d)

The second step is for data compression and to accelerate later applications. A polygon with a lower number of vertices is calculated here with the approximation algorithm presented by Ramer (1972). It is parameterized by a maximal distance value that specifies the accuracy of the algorithm. Every point of the original shape has this maximal distance to the lines defining the approximated shape.

This 2-D polygon acts as the ground shape of the right prism that is calculated for each *Si*. The prism's height and vertical position is determined by the span of single slices $sz \in Si$. As seen in figure 13, the prism shapes can be smaller than the grid-based shapes since the center coordinates of the grid cells are used for the shape vertices and an approximation is performed. In obstacle avoidance applications, a safety distance to the objects is added; it must be larger than half a cell size plus the approximation accuracy to ensure that the grid cells of each object are completely covered by the polygon hull.

6. ESTIMATING GROUND PLANES

6.1 The Height Histogram

In addition to the prism-based shape detection of obstacles, the floor plane is extracted out of the sensor data. This simplifies the resulting world model. The floor will not be a shape prism that has to be tracked. Furthermore, obstacles and the floor are not merged which helps to identify objects that hit the ground.

The floor plane detection is similar to classical Hough Transform based approaches (e.g. Okada et al., 2001) with the limitation that only the horizontal planes are searched.

The actual sensor data leads to a cloud of points (x, y, z) in global coordinates. The vehicle's position must be known and a calibrated pitch and roll angle measurement is assumed to ensure that a horizontal ground will lead to a horizontal plane in map coordinates. A sampled histogram $n(z)$ of all z-values of these points is built. The sampling should be more precise than the occupancy grid. Points are inserted with blurring considering their depth- dependent uncertainty.

Since all measurements of a horizontal plane have approximately the same z-value, planes lead to significant peaks in the histogram and can be detected there. If more than one plane is found, the floor is the one with the lowest height, i.e. with the largest z-value. There will be no ground plane if no peak with a high confidence is found.

It can be useful to increase the floor plane height, e.g. to force a minimal distance to the ground in urban flight scenarios. First, the peak maximum $n_{max} = n(z_{max})$ is not taken. Rather, z is decreased while $n(z) > t \cdot n_{max}$, starting at $z = z_{max}$. The result is denoted as zground. In the experiments, t is set to 0.5. Second, an offset $z_{offset} < 0$ can be added to this height.

Figure 14. Depth image (left), depth image with marked floor pixels (center) and original camera image with floor pixels for comparison (right)

Figure 14 shows an example of the ground estimation. Pixels that lead to a larger z-value are marked in the depth image and for comparison in the original camera image. As seen in the right subfigure, the floor is marked contrary to the house at the right side. It is not necessary to insert the marked measurements into the obstacle map.

6.2 A Floor for each Zone

To combine the data of multiple images, the ground plane estimation is done separately for map zones. The map partitioning as described in section 4.2 is used but without the vertical separation. This leads to rectangular 2-D zones from the top view and each zone can have a floor plane height. A height histogram is calculated for each zone by accumulating these sensor data elements that lead to object points in that zone. The histogram is built incrementally over time so that multiple sensor updates can be stored in one zone if the vehicle stays there. The result is a kind of an elevation map with an estimated height value for each zone.

7. RESULTS

A first experiment tests the mapping algorithm in a simulation environment where two views are generated and captured with cameras (fig. 15). The grid resolution is 0.25m and each map zone has a size (*x*, *y*, height) of 128 x 128 x 32 cells. Figures 15 and 16 show the result of the simulation experiment where two obstacle arches of 6 x 6m size are placed with a distance of 50m.

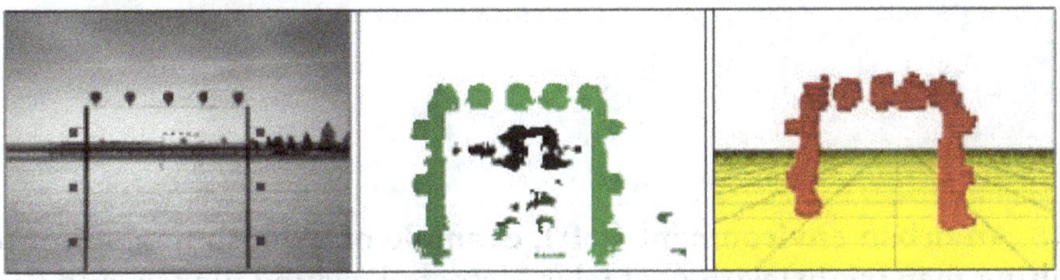

Figure 15. Original image from left camera (left), depth image (center) and extracted occupancy grid (right)

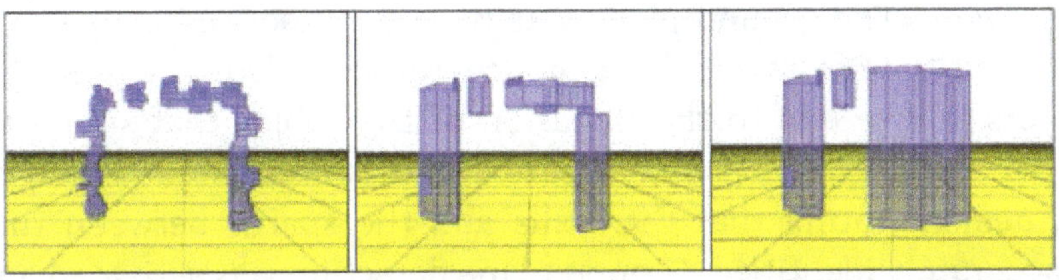

Figure 16. Prism shapes calculated from occupancy grid of figure 15 with different similarity thresholds for shape slice merging: zero (left), 100 (center) and infinite (right)

As shown in the image, the first arch is mapped while the second arch in the background is filtered due to its far distance from the viewpoint. The subfigures include the occupancy grid map (fig. 15, right) and the resulting shape prisms (fig. 16). As shown in the figures, the final map representation is only a small set of simple shapes. The similarity threshold for horizontal shapes should not be too large to allow multiple prisms for features generated from grid cell clusters in order to represent non-vertical walls. For a better illustration, floor planes marked with lines every 8 meters are shown in the images. The computation speed on the 3 GHz test computer was generally in the interval from 15 to 20 frames per second.

In a second test series, helicopter flights are performed outdoors. The grid array size is the same as in the simulation and its resolution is 0.5m, so that the map size of each zone is doubled. The helicopter is manually directed through obstacle posts and in an urban environment near house walls.

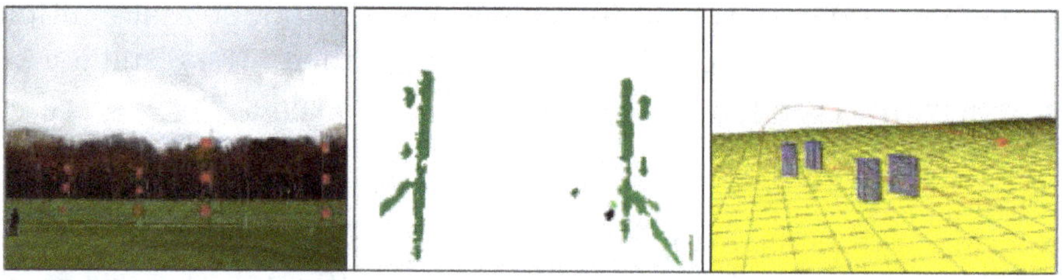

Figure 17. Obstacle posts on a flight field (left), example of a depth image (center), and the resulting map with obstacles and flight path (right)

Figure 18. An urban environment (left), example of a depth image (center), and the resulting map (right) with obstacles, detected ground planes, and flight path

Figure 17 illustrates the detection of obstacle posts, with each post represented by a singular polygon. The ground plane was not identified; the image solely depicts a plane at the altitude from which the helicopter ascended. The prisms exceed the actual barriers in size due to the inclusion of the cables that support the posts. The flight path is indicated in red. Figure 18 illustrates the outcome of mapping residences, demonstrating that the walls and the spaces between the buildings have been accurately identified. Ground plane detection was activated and achieved success.

8. CONCISE SUMMARY

This study delineates a methodology for extracting obstacles from sensor data for autonomous applications. The emphasis is on developing a concise representation of the bounding forms necessary for obstacle avoidance, rather than detailed object description. Occupancy grid mapping serves as the foundation for shape extraction in sensor data fusion. A shape delineated by one or more right prisms is computed for each cluster of occupied grid cells. The shapes are computed in real-time with each new sensor data input. Furthermore, ground planes are recognized. This approach produces compact map representations of barriers, enabling the transmission of actual map updates via low-bandwidth networks for use in external applications. To validate the methodology, experiments were conducted in a simulated laboratory environment as well as in real-world conditions, where a helicopter navigated past obstacle posts and urban settings. The findings indicate that obstacles can be mapped using GPS/INS-based positioning and stereo vision, with feature extraction ensuring the resultant map is appropriate for obstacle avoidance.

REFERENCES

1. Chatila, R. &Laumond, J.-P. (1985). Position referencing and consistent world modeling for mobile robots. In *IEEE International Conference on Robotics and Automation*, pp. 138-145.
2. Dittrich, J.; Bernatz, A. &Thielecke, F. (2003). Intelligent systems research using a small autonomous rotorcraft testbed. In *2nd AIAA Unmanned Unlimited Conference, Workshop and Exhibit*, paper 6561, San Diego.
3. Dittrich, J.; Adolf, F.; Langer, A. &Thielecke, F. (2007). Mission planning for small VTOL UAV systems in unknown environments. In *AHS International Specialists' Meeting on Unmanned Rotorcraft*, Chandler.
4. Dittrich, J.; Andert, F. & Adolf, F. (2008). An obstacle avoidance concept for small unmanned rotorcraft in urban environments using stereo vision. In *64th Annual Forum of the American Helicopter Society*, Montréal.
5. Fiorini, P & Shiller, Z. (1998). Robot motion planning in dynamic environments using velocity obstacles. *International Journal of Robotics Research*, Vol. 17, No. 7, pp. 760- 772. ISSN 0278-3649.
6. Fulgenzi, C.; Spalanzani, A.; Laugier, C. (2007). Dynamic obstacle avoidance in uncertain environment combining PVOs and occupancy grid. In *IEEE International Conference on Robotics and Automation*, pp. 1610-1616, Roma.
7. Griffiths, J.; Saunders, A.; Barber, B.; McLain, T.; Beard, R. (2007). Obstacle and terrain avoidance for miniature aerial vehicles. Valavanis, K. P. (ed.), *Advances in Unmanned Aerial Vehicles*, pp. 213-244. ISBN 978-1-4020-6113-4.
8. Hähnel, D.; Burgard, W. &Thrun, S. (2003). Learning compact 3D models of indoor and outdoor environments with a mobile robot. *Robotics and Autonomous Systems*, Vol. 44, No. 1, pp. 15-27. ISSN 0921-8890.
9. Hrabar, S.; Corke, P.; Sukhatme, G.; Usher, K. & Roberts, J. (2005). Combined optic flow and stereo-based navigation of urban canyons for a UAV. In *IEEE/RSJ International Conference on Intelligent Robots and Systems*, pp. 302-309, Edmonton.
10. Iocchi, L.; Konolige, K. &Bajracharya, M. (2000). Visually realistic mapping of a planar environment with stereo. In *Seventh International Symposium on Experimental Robotics*, Waikiki.

11. Koch, A.; Wittich, H. &Thielecke, F. (2006). A vision-based navigation algorithm for a VTOL UAV. In *AIAA Guidance, Navigation and Control Conference and Exhibit*, paper 6546, Keystone.
12. Konolige, K. (1997). Improved occupancy grids for map building. *Autonomous Robots*, Vol. 4, pp. 351-367. ISSN 0929-5593.
13. Kuipers, B. J. (2000). The spatial semantic hierarchy. *Artificial Intelligence*, Vol. 119, pp. 191- 233. ISSN 0004-3702.
14. Martin, C. &Thrun, S. (2002). Real-time acquisition of compact volumetric 3D maps with mobile robots. In *IEEE International Conference on Robotics and Automation*, pp. 311- 316, Washington D.C.
15. Moravec, H. P. &Elfes, A. (1985). High resolution maps from wide angle sonar. In *IEEE International Conference on Robotics and Automation*, pp. 116-121.
16. Okada, K.; Kagami, S.; Inaba, M. & Inoue, H. (2001). Plane segment finder: Algorithm, implementation and applications. In *IEEE International Conference on Robotics and Automation*, pp. 2120-2125, Seoul.
17. Prassler, E.; Scholz, J. &Elfes, A. (2000). Tracking multiple moving objects for real-time robot navigation. *Autonomous Robots*, Vol. 8, No. 2, pp. 105-116. ISSN 0929-5593.
18. Ramer, U. (1972). An iterative procedure for the polygonal approximation of plane curves.
19. *Computer Graphics and Image Processing*, Vol. 1, No. 3, pp. 244-256. ISSN 0146-664X.
20. Raschke, U. &Borenstein, J. (1990). A comparison of grid-type map-building techniques by index of performance. In *IEEE International Conference on Robotics and Automation*, pp. 1828-1832, Cincinnati.
21. Ren, M.; Yang, J. & Sun, H. (2002). Tracing boundary contours in a binary image. *Image and Vision Computing*, Vol. 20, pp. 125-131. ISSN 0262-8856.
22. Scharstein, D. &Szeliski, R. (2002). A taxonomy and evaluation of dense two-frame stereo correspondence algorithms. *International Journal of Computer Vision*, Vol. 47, pp. 7-42. ISSN 0920-5691.
23. Scherer, S.; Singh, S.; Chamberlain, L. &Saripalli, S. (2007). Flying fast and low among obstacles. In *IEEE International Conference on Robotics and Automation*, pp. 2023-2029, Roma.
24. Thrun, S. (2002). *Robotic mapping: A survey*. Tech. Rep., Carnegie Mellon University. No. CMU-CS-02-111.
25. Zufferey, J.-C. &Floreano, D. (2005). Toward 30-gram autonomous indoor aircraft: Vision- based obstacle avoidance and altitude control. In *IEEE International Conference on Robotics and Automation*, pp. 2594-2599, Barcelona.

3 Chapter

MICRO-AIR VEHICLES USE FIELD PROGRAMMABLE GATE ARRAYS (FPGA) FOR BIO-INSPIRED VOLTAGE CONTROL SYSTEMS

1. INTRODUCTION: AN OVERVIEW

Innovative systems for the control of Micro-Air Vehicles (MAVs) can be inspired by the many instances of clever sensors and systems found in nature that support animal behavior. Despite their visual system's low resolution, winged insects exhibit intimidating behaviors. They have been using optic flow (OF) signals to navigate in strange 3D surroundings for over 100 million years. Insects are equipped with sophisticated sensors known as Elementary Motion Detectors (EMDs), which can function as optic flow sensors, allowing them to see and respond to the flowing image of their surroundings (figure 1). Our bioinspired optic flow sensors work on the basis of research we have done on the electromyography (EMD) of common houseflies. We have done this by applying optical microstimuli to two single photoreceptor cells within a single ommatidium and recording the electrophysiological activity of a single neuron (Franceschini, 1985, Franceschini et al. 1989). According to Koenderink (1986), the OF-field provides the angular velocity (in rad/s) at which each contrasting item in the surrounding environment is passing the eye. Insects can navigate quickly through even the most uncertain situations without the aid of velocimeters or rangefinders, which is one thing we can learn from them. For example, insects use optic flow to avoid collisions (Collett, 1980; Wagner, 1982; Tammero and Dickinson, 2002), follow a corridor (Kirchner and Srinivasan, 1989; Baird et al., 2005; Ruffier et al., 2007; Serres et al., 2008); follow the terrain (William, 1965; Srygley and Oliveira, 2001); fly against the wind (Kennedy 1939, Kennedy 1951), and cruise and land (Srinivasan et al., 1996). It's interesting to note that during cruising and landing, insects appear to keep a steady visual flow with respect to their surroundings (Kennedy, 1951; David, 1978, Srinivasan et al. 1996, 2000). Recent construction of a number of MAV autopilots demonstrates how insects could accomplish this by utilizing an optic flow regulator feedback control system (Ruffier and Franceschini, 2003, 2005, Serres et al. 2008). It may be necessary for future MAVs' visual guidance systems to include optic flow sensors that cover different regions of the visual field, much like insects do (Figure 2). We have built multiple terrestrial and airborne robots based on OF sensing thanks to the biorobotic methodology our lab has developed over the past 20 years (Pichon et al., 1989, Franceschini et al., 1992, 1997; Mura and Franceschini, 1996; Netter and Franceschini., 2002, Ruffelier et al., 2004). Initially, Fly, also known as "le robot-mouche," was a tiny, fully autonomous robot that operated on Earth and had 114 optical flow sensors installed. At a rather fast pace of 50 cm/s, this robot was able to navigate its way to a goal through an unidentified field of obstacles (Franceschini et al. 1992, 1997). In the past ten years, we have created small aerial demonstrators (mass less than 1 kg) with limited degrees of freedom, which we have named FANIA, OSCAR, OCTAVE, and LORA, respectively (Viollet and Franceschini, 1999, 2001 ; Netter and Franceschini, 2002; Ruffier and Franceschini, 2003, 2005; Kerhuel et al., 2007; Serres et al., 2008).

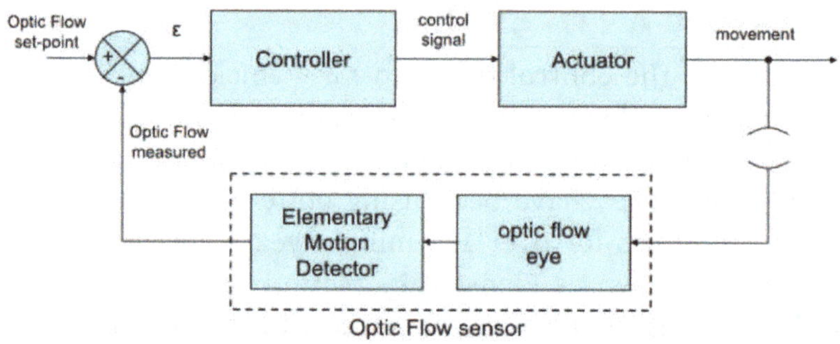

Figure 1. Principle of an optic flow regulator. The feedback control loop aims at maintaining the OF measured constant and equal to an optic flow set point

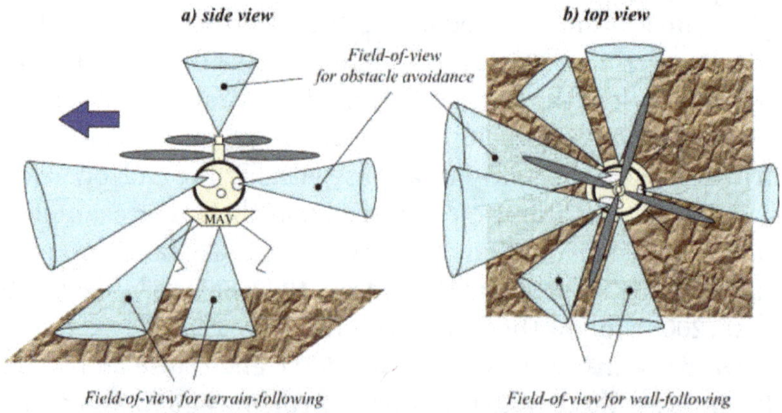

Figure 2. Several optic flow sensors covering various, strategic fields of view for an insect- like visual guidance of Micro-Air Vehicles (MAV)

In addition to visuo-motor control loops, inner control loops are necessary (i) to lock the micro-aircraft in certain desired attitudes, and (ii) to improve the control performances and robustness.

- Attitude control systems laid out in parallel with the *visuo-motor control loops* may use inertial and/or magnetic angular sensors (Ruffier et al., 2005; Viollet et al., 2001; Serres et al., 2008)
- Control speed and accuracy can be improved with advanced actuators and more classical control loops based on proprioceptive sensors.

MAVs suffer, however, from stringent constraints on avionic payload, which requires highly miniaturized sensors, actuators and control systems (Viollet et al., 2008, Ahmad and Taib, 2003, Van Nieuwstadt and Morris, 1995).

Some of the aerial robots we developed thus far (FANIA and LORA) embedded only the optic flow sensors, the electromechanical actuators, and some internal stabilizing loops. The *visuo-motor control* system therefore operated off-board, yet real time experimentation was made possible by the joint use of *Simulink* (The Mathworks) and *Dspace* (Dspace) softwares. This permanent exchange between simulation (from the high level models) and experimental tests (from the robot) is appealing because it permits quick implementation of any new visuo-motor control systems onto a physical demonstrator, in addition to easy monitoring, and validation and tuning procedures. In this approach, the robot's behavior may be limited by the wire umbilical, however, unless a wireless link is established between the robot and the real-time board.

In our quest to achieve robot's complete - computational and energetic - autonomy, we now designed a complete digital system that integrates both optic flow sensors and robot's control systems within the same target: a 0.5 gram Field Programmable Gate Array (FPGA). An FPGA offers significantly more computational capabilities-per-gram than an embedded microprocessor or Digital-Signal Processor (DSP) because it supports custom, application- specific logic functions that may accelerate processing.

The following sections detail our FPGA-based autopilot project that satisfies three requirements: (i) total autonomy; (ii) real-time processing computational requirements; and (iii) light weight required due to the extremely constrained avionic payload (Chalimbaud and Berry, 2007). Our research involves the FPGA implementation of a visuo-motor control system designed for a miniature hovercraft (part 3). The system is termed LORA, which stands for Lateral Optic flow Regulation Autopilot. utilizing Intellectual Property (IP) cores and VHDL descriptions imported in the Simulink high-level graphical environment (The Mathworks) and the System Generator software interface (Xilinx), the FPGA integration work was carried out utilizing a top-down design technique (section 2). Next, digital Xilinx blocks are used in place of the high-level Simulink blocks from the developed autopilot. Behavioral studies were used to develop the digital specifications for a number of functions, including sampling time, fixed-point binary formats (section 4), and architectures (section 5). Using System Generator, only models constructed from Xilinx blocks were converted into hardware logic devices. Ultimately, a hardware/software simulation based on the FPGA (via a JTAG protocol) and the Simulink environment (on a PC) was used to study the overall behavior of the combined systems. We were able to examine the hovercraft's behavior in a variety of visual settings thanks to this co-simulation (section 6).

2. METHODOLOGY FOR FPGA

2.1 FPGA general description

A field programmable gate array (FPGA) is a versatile integrated circuit programmed by the designer instead of the manufacturer. In contrast to an application-specific integrated circuit (ASIC), which executes a comparable function within an electronic system, a field-programmable gate array (FPGA) can be reprogrammed post-deployment. A FPGA is configured by uploading a bitstream into the static on-chip random-access memory (RAM). Similar to the object code for a microprocessor, this bitstream results from compilation tools that convert high-level abstractions created by a designer into an equivalent logic gate representation. A platform FPGA is engineered from low-density to high-density configurations utilizing IP cores and bespoke modules. These devices are user-programmable gate arrays featuring several customizable components. The programmable device consists of input/output blocks and internal adjustable logic blocks. Programmable I/O blocks serve as the interface between package pins and the internal customizable logic. The internal customizable logic comprises elements arranged in a systematic array: Configurable logic blocks (CLBs) offer functional components for combinatorial and synchronous logic, encompassing fundamental storage elements; memory modules deliver extensive storage capabilities, either single or dual-port; multiplier blocks incorporate adder and accumulator functionalities (MAC); digital clock managers furnish self-calibrating, entirely digital solutions for clock distribution delay compensation, clock multiplication and division, as well as coarse- and fine-grained clock phase adjustment. Numerous FPGAs additionally include embedded system capabilities, including high-speed serial transceivers, one or more hard embedded CPUs, Ethernet media access control cores, or other integrated high-functionality modules. A general routing matrix offers a configuration of routing switches connecting each component. Each programmable component is connected to a switch matrix, facilitating several connections to the overarching routing matrix. All programmable components, including routing resources, are governed by values retained in memory cells. The values are stored in memory cells during configuration and can be reloaded to modify the functions of the programmable parts. FPGAs are particularly utilized in domains or algorithms that can leverage the extensive parallelism inherent in their architecture. They increasingly assume bigger functions, to the extent that some are currently sold as complete systems on chips (SoC). In this regard, they can fulfill the computational requirements for real-time processing on autonomous MAVs.

2.2 FPGA design process

There are many design entry tools used for FPGA design. The easiest and most intuitive is the schematic entry. In this case, the required functionality is drawn using a set of library components. These one include the components primitives available on the FPGA as well as some higher level functions. Intellectual Property cores (IP) can be configured and placed in the schematic as a black box.

A trend in the digital hardware design world is the migration from graphical design entries to Hardware Description Languages (HDLs). They allow specifying the function of a circuit using a specific language such as VHDL or Verilog. These languages are specially made to describe the inherently parallel nature of digital circuits (behavioural and data flow models) and to wire different components together (structural models). In addition, HDLs are well adapted for designs with synchronous Finite State Machine (FSM) (Golson, 1994, Chambers, 1997). State machines impose a strict order and timing for operations making them similar to programs for CPUs.

However, HDL descriptions are difficult to design in the case of complex systems using digital signal processing functions or control applications. Indeed, the high level mathematical modelling tools are often used to validate a system model and to make a floating or fixed point model. Developments in the simulation capabilities of high-level mathematical modeling tools have opened new design flow possibilities. These tools are includes in methodologies that help to hand complex design efficiently, minimize design time, eliminate many sources of errors, reduce the manpower required to complete the design and to produce optimal solutions. Nowadays, we can integrate complex system using IP cores and HDL descriptions imported in a high-level graphical environment. A software interface converts a fixed point model into hardware using traditional low level design implementation tools. Our design approach is based on this latter method.

In our project the high-level environment is *Matlab/Simulink* (The Mathworks), and the used interface tool is *System Generator for DSP* (Xilinx). *Matlab* is interactive software for doing numerical computations that simplifies the implementation of linear algebra routines. Powerful operations can be performed by utilizing the provided *Matlab* commands. *Simulink* is an additional toolbox of the *Matlab* software that provides a graphical environment for modeling, simulating and analyzing dynamic systems. *System Generator for DSP* toolbox was specifically designed to allow the fast development of complex system requiring powerful digital signal operations. It is presented in the form of a 'toolbox' added to the *Simulink* graphic environment, which allows the interfacing of certain functions fulfilled with the others 'toolbox' dedicated to *Simulink* (Turney, 1999, Ownby, 2003).

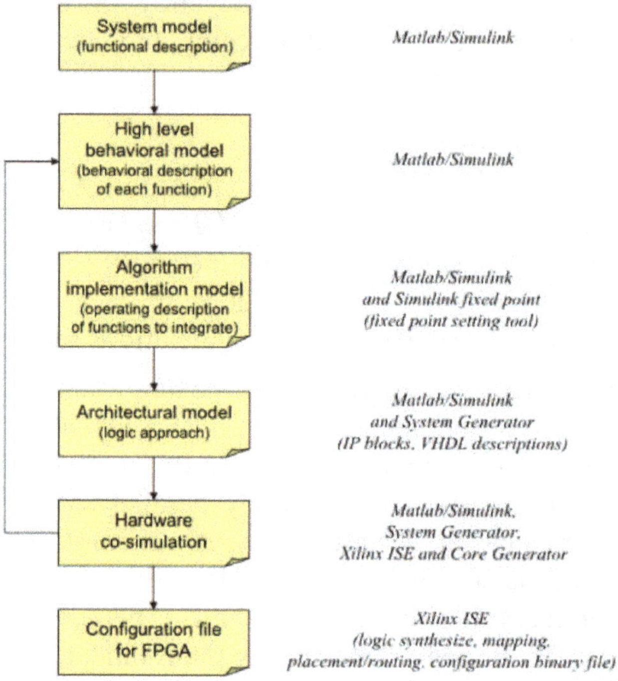

Figure 3. Top-down design methodology with a view to designing, simulating and implementing 'visuo-motor control loops' in FPGA (from Aubépart and Franceschini, 2007)

The design flow is presented in Figure 3. This methodology requires design various stages as well as the linking framework. This top-down methodology simplifies the integration problems by using detailed descriptions (Browny et al., 1997; Aubépart & Franceschini, 2005, 2007, Murthy et al., 2008).

Firstly, we start to create a '*system model*' as well as one or more simulated environment configurations for validating its principle. In this state, a functional approach involves dividing the system into elementary function blocks. The elements used in constructing the '*system model*' are all capable of operating on real (double precision, floating point) or integer (quantized, fixed point binary) data types. When the model is first entered, simulation is typically performed using floating data types to verify that its theoretical performance is as desired.

Secondly, we define the '*high-level behavioral model*' which describes the computation sequences and timing data (sampling frequency, delays, etc.). For example, mathematical operators that occur within an algebraic feedback loop may have an associated delay to avoid the system instability. The fixed time step, insert delays and/or rate changes need to be also considered to ensure a stability of a system based on a feedback loop.

Thirdly, an '*algorithm implementation model*' must be defined for each function to integrate. This model identifies the data type and the procedures used in each function. It takes account some factors, such as the binary format, in order to optimize the digital calculations. The internal data types are then converted to the bit true representations that will be used in the hardware implementation, and the model is re-simulated to verify its performance with quantized coefficient values and limited data bit widths, which can lead to overflow, saturation and scaling problems.

Fourthly, the hardware constraints are study in the '*architectural model*' that defines one or several implementation architectures for each function to integrate. They are replaced by IP blocks available in the *System Generator* toolbox. We can also define black boxes, such as VHDL descriptions, which can be incorporated into the model and the elaboration process. The importation of VHDL descriptions will be limited to complex synchronous descriptions, such as Finite State Machines.

Fifthly, the final verification will be completed by implementing the hardware co-simulation of the *System Generator* '*architectural models*'. The co-simulation process uses *Xilinx ISE* and *core generator* to synthesize and generate and FPGA programming bit file. A new *Simulink* library was created containing the hardware co-simulation blocks. These blocks were copied into the *Simulink* project file for replacing all the *System Generator* '*architectural models*'. The port names, types and rates are matching the original design. The hardware implementation is then executed by connecting the FPGA board to the computer and using a standard JTAG connection. When the simulation is run, stimuli are send to FPGA and output signals are receive by *Matlab/Simulink* environment, closing the loop.

The designer can ultimately utilize the netlister and test-bench generator provided by System Generator. The netlister produces a hierarchical VHDL representation of the model's structure, annotated with all element parameters and signal data types, which will be integrated into the FPGA using traditional low-level tools included in Xilinx ISE: logic synthesis, mapping, placement, routing, and programming file generation.

3. LORA III AUTOPILOT

This section introduces the 'system model' and the execution of a comprehensive control system utilizing the System Generator toolkit. This study incorporates a comprehensive visuo-motor control system known as LORA III (LORA signifies Lateral Optic flow Regulation Autopilot, Mark 3), specifically engineered for a distinct type of aerial vehicle: a fully actuated miniature hovercraft capable of "flying" mere millimeters above the ground while autonomously regulating both its velocity and proximity to surrounding walls—without any measurements of speed or distance (Serres et al., 2008).

3.1 Robot position in travel environment

The hovercraft is modeled to navigate an unfamiliar textured corridor at a ground speed vector on a level surface, as illustrated in Figure 4. The walls include vertical stripes with varying spatial frequencies and contrasts that emulate a richly textured setting (Iida, 2001; Ruffier & Franceschini, 2005). The corridors are categorized into two types: straight or tapering.

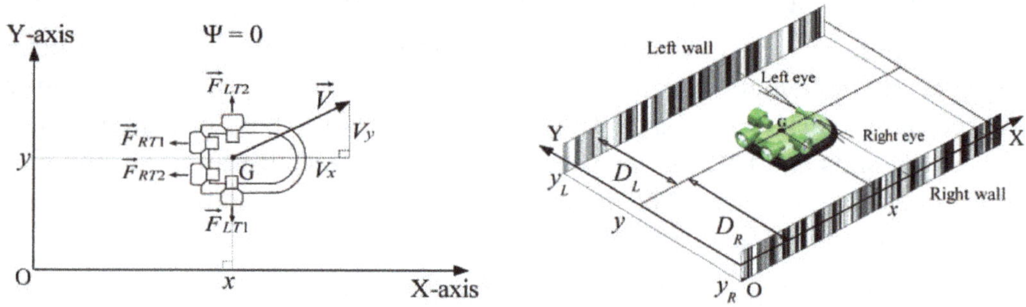

Figure 4. Miniature hovercraft moving through an unknown textured corridor: the groundspeed vector $\vec{V}$ is projected onto the corridor-fixed coordinate frame. Four thrusters (two rear thrusters and two lateral thrusters) allow the hovercraft to be fully actuated in the plane. (Adapted from Serres et al., 2008)

The robot's position (x, y) is computed in including side forward speed V_x and the side speed V_y. The hovercraft is fully actuated because in addition to the pair of rear thrusters providing forward motion (surge axis) and heading control (yaw axis), the vehicle is equipped with a pair of lateral thrusters generating independent side-slip motion (sway). The angle ψ defines the hovercraft's yaw angle with respect to the corridor axis, which is kept at a reference value equal to zero (ψ = 0). In this study, the hovercraft's heading ψ is assumed to be stabilized along the X-axis of the corridor.

The hovercraft's motion is defined by dynamic equations involving the forward thrust ($F_{Fwd} = F_{RT1} + F_{RT2}$) produced by the rear thrusters (left: RT1, right: RT2) and the lateral thrust ($F_{Side} = F_{LT2} - F_{LT1}$) produced by the lateral thrusters (left: LT1, right: LT2). In the simulations, the maximum forward speed is 2m/s and the maximum side speed is 0.5m/s. At such low speeds, the drag-versus-speed function can be linearized. The following equations referred to the center of gravity G define the dynamics of the simulated hovercraft (Figure 5):

$$\begin{aligned} m.\frac{dV_x}{dt} + \xi_x.V_x &= F_{RT1} + F_{RT2} = K_T.(U_{RT1} + U_{RT2}) \\ m.\frac{dV_y}{dt} + \xi_y.V_y &= F_{LT2} - F_{LT1} = K_T.(U_{LT2} - U_{LT1}) \end{aligned} \quad (1)$$

Where m = 0,73 kg is the total mass of the hovercraft, and ξ_x and ξ_y are translational viscous friction coefficients along the X-axis and Y-axis, respectively. KT (0.10 N/V) is a simple gain that relates the thrust to the applied voltage: U^{RT1} and U^{RT2} are the forward control signals received by the rear thrusters, U^{LT2} and ULT1 are the side control signals received by the lateral thrusters.

3.2 Dual optic-flow regulator

The solution concurrently tackles the concerns of automatic speed regulation and lateral wall avoidance for the small hovercraft. LORA is a dual optic flow regulator including two interdependent control loops: a forward control loop and a lateral control loop. Figure 5 illustrates the block diagram comprising several processing phases. This scheme consists of two components. Initially, all functions designated for integration into the FPGA (blue, green, and red functions). Secondly, the dynamics of hovercraft, the lens/photoreceptor system, and visual surroundings replicate the trajectories produced by the LORA dual regulator scheme (Cyan and cyan hatching functions). These 'high-level behavioral models' will be employed when defining digital, temporal, and architectural specifications for the functions to be integrated.

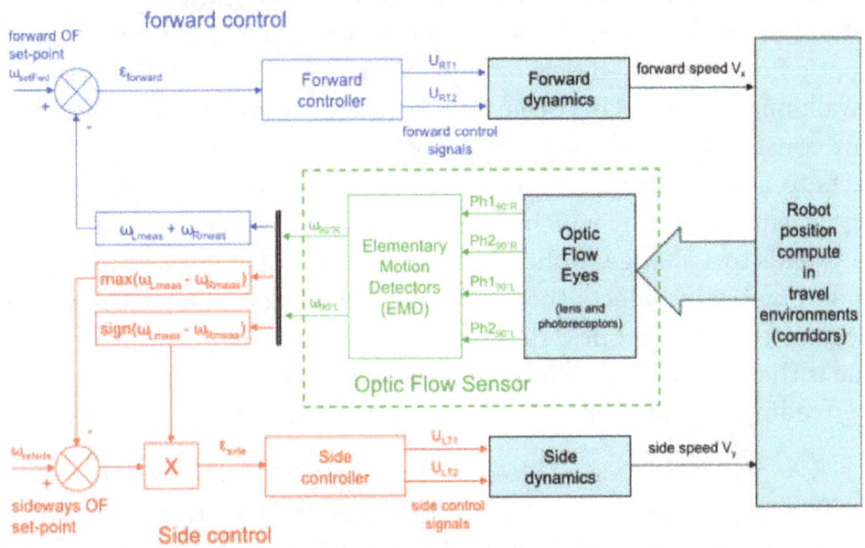

Figure 5. LORA III functional block diagram. LORA III autopilot is based on two interdependent visual feedback loops working in parallel with their own optic flow set- point (the forward control system is the upper one, and the side control system is the bottom one). Optic flow sensors measure the right and left optic flows in accordance with the hovercraft's speed and position in the environment (adapted from Serres et al., 2008a)

In Serres et al. (2008), the full control system is described in detail. We summarize here the principal aspects of them. That the hovercraft is fully actuated means that each groundspeed component V_x and V_y can be controlled independently. LORA III regulates (i.e., maintains constant) the lateral OF by side and forward controls, according to the following principles:

- The first *lateral OF regulator* adjusts the air vehicle's lateral thrust (which determines the lateral speed V_y, i.e., the sway speed) so as to keep the lateral optic flow equal to the *sideways OF set-point*. The outcome is that the distance to the wall becomes proportional to the vehicle's forward speed V_x which is defined in (ii): the faster the air vehicle travels, the further away from the walls it will be. The clearance from the walls will depend directly on the *sideways OF set-point*.
- The second *lateral OF regulator* adjusts the air vehicle's forward thrust (which determines the forward speed Vx, i.e., the surge speed) so as to maintain the sum of the two (right and left) optic flows equal to the *forward OF set-point*. The outcome is that the air-vehicle travels all the faster as the environment is less cluttered. The forward speed attained by the vehicle will depend directly on the *forward OF set-point and will become proportional to the local width of the corridor*.

The first lateral optic flow regulator is based on a feedback signal that takes into account the left or right optic flow measured. The feedback is simply the larger of the two optic flows measured (left or right): $\max(\omega_{Lmeas}, \omega_{Rmeas})$: it corresponds to the nearest wall: min(DL, DR). This optic flow regulator was designed to keep the lateral optic flow constantly equal to the side optic flow set-point $\omega_{SetSide}$. The hovercraft then reacts to any deviation in the lateral optic flow (left or right) from $\omega_{SetSide}$ by adjusting its lateral thrust, which determines the hovercraft's side speed Vy: this eventually leads to a change in the distance to the left (DL) or right (DR) wall. A sign function automatically selects the wall to be followed, and a maximum criterion is used to select the higher optic flow value measured between ωRmeas and ω_{Lmeas}. This value is then compared with the sideways optic flow set-point $\omega_{SetSide}$. The error signal ε_{side} feeding the side controller is therefore calculated as follows:

$$\varepsilon_{side} = \text{sign}(\omega_{Lmeas} - \omega_{Rmeas}) \times (\omega_{SetSide} - \max(\omega_{Lmeas}, \omega_{Rmeas})) \quad (2)$$

The identified transfer function of the side dynamics Gy(s) relating the hovercraft's ordinate *y* to the control signal approximates a first-order low-pass filter (with a time constant of 0.5s) in series with an integrator:

$$G_y(s) = \frac{Y(s)}{(U_{LT1} - U_{LT2})(s)} = \frac{1}{s} \times \frac{\frac{K_T}{\xi_y}}{1 + \frac{m}{\xi_y}s} = \frac{1}{s} \times \frac{0.1}{1 + 0.5s} \quad (3)$$

A lead controller Cy(s) was introduced into this feedback loop to increase the damping, thus improving the stability and enhancing the response dynamics. The lead controller Cy(s) (Eq.3) is tuned to reach a phase margin of 45° and a crossover frequency of 4rad/s (0.64Hz):

$$C_y(s)=\frac{(U_{LT1}-U_{LT2})(s)}{\varepsilon_{side}(s)}=10\times\frac{1+1.5s}{1+0.5s} \tag{4}$$

The second optic flow regulator is the forward control system. It is intended to keep the sum of the two lateral optic flows measured ($\omega_{Rmeas}+\omega_{Lmeas}$) constant and equal to a *forward optic flow set-point* ω_{SetFwd} by adjusting the forward thrust, which will determine the hovercraft's forward speed V_x. At a given corridor width, any increase in the sum of the two lateral optic flows is assumed here to result from the hovercraft's acceleration. This control scheme thus automatically ensures a 'safe forward speed' that is commensurate with the local corridor width. The sum of the two optic flows measured is compared with a *forward OF set-point* ω_{SetFwd}, and the error signal ω_{Fwd} (the input to the forward controller) is calculated as follows:

$$\varepsilon_{Fwd} = \omega_{SetFwd} - (\omega_{Rmeas} + \omega_{Lmeas}) \tag{5}$$

The model GVx(s) for the dynamics of our hovercraft is described by a first order low-pass filter with a time constant of 0.5s:

$$G_{Vx}(s)=\frac{V_x(s)}{(U_{RT1}+U_{RT2})(s)}=\frac{\frac{K_T}{\xi_x}}{1+\frac{m}{\xi_x}s}=\frac{0.1}{1+0.5s} \tag{6}$$

A proportional-integral (PI) controller $C_{Vx}(s)$ (Eq. 7) is tuned to cancel the dominant (aeromechanical) pole of the hovercraft and to reduce the forward time constant computed in the closed-loop by a factor of 1.57. The integral action is introduced to cancel the steady state error:

$$C_{Vx}(s)=\frac{(U_{RT1}+U_{RT2})(s)}{\varepsilon_{Fwd}(s)}=10\times\frac{1+0.5s}{s} \tag{7}$$

3.3 Bio-inspired visual system

The visual system of the hovercraft consists of two optic flow sensors. Each optic flow sensor is designed by a lens/photoreceptor assembly (the eye) including at least two photoreceptors (i.e. two pixels) driving an Elementary Motion Detector (EMD) circuit.

Each eye consists of two photoreceptors mounted slightly defocused behind a lens, which creates a bell-shaped Angular Sensitivity Function (ASF) for each of them (Hardie, 1985). The ASF, which is often modelled in the form of a truncated Gaussian curve (Netter & Franceschini, 2002) and characterized by the 'acceptance angle' $\Delta\rho=4°$ (i.e. the angular width at half height). The ASF plays an important role in the visual processing chain, because it serves as an effective low pass anti-aliasing spatial filter. The visual axes of each pixel are separated by an inter-receptor angle $\Delta\phi = 4°$.

The EMD principle was originally based on the results of experiments in which a combined electrophysiological and micro-optical approach was used. The activity of a large field motion detecting neuron in the housefly's eye was recorded with a microelectrode while applying optical microstimuli to a single pair of photoreceptor cells located behind a single facet (Franceschini, 1985, Franceschini et al., 1989). Based on the results of these experiments, a principle was drawn up for designing an artificial EMD capable of measuring the angular speed ω of a contrasting object (Franceschini et al., 1986, Blanes, 1986).

Thus, in each of EMDs, the lens/photoreceptor combination transforms the motion of a contrasting object into two successive photoreceptor signals separated by a delay Δt:

$$\Delta t = \frac{\Delta \varphi}{\omega} \tag{8}$$

Where Δϕ is the inter-receptor angle and ω is the relative angular speed (the optic flow). An electronic device based on some linear and nonlinear functions estimates the angular speed ωEMD:

$$\omega_{EMD} = e^{-\left(\frac{\Delta t}{\tau}\right)} \Leftrightarrow K.\omega = \frac{K.\Delta \varphi}{\Delta t} = \frac{K'}{\Delta t} \tag{9}$$

Our original EMD functional scheme (Franceschini et al., 1986, Blanes 1986, Aubépart & Franceschini, 2007) consists of five processing steps giving ω_{EMD} (Figure 6):

- A first-order high-pass temporal filter (fc = 20Hz) produces a transient response whenever a contrasting border crosses the photoreceptors' visual field. This filter enhances the contrast information while eliminating the DC components of the photoreceptor signals.
- A higher order low-pass temporal filter (fc = 30Hz) attenuates any high frequency noise, as well as any interferences brought about by the artificial indoor lighting (100Hz) used.
- A thresholding device/step normalizes the signals in each channel.
- A time delay circuit is triggered by one channel and stopped by the neighboring channel. This function measures the time Δt elapsing between similar transitions occurring in two adjacent photoreceptors.
- A converter translates the delay Δt measured into a monotonic function that will approximate the angular speed ω_{EMD}. A simple inverse exponential function makes for a relatively large dynamic range (eq. 9).

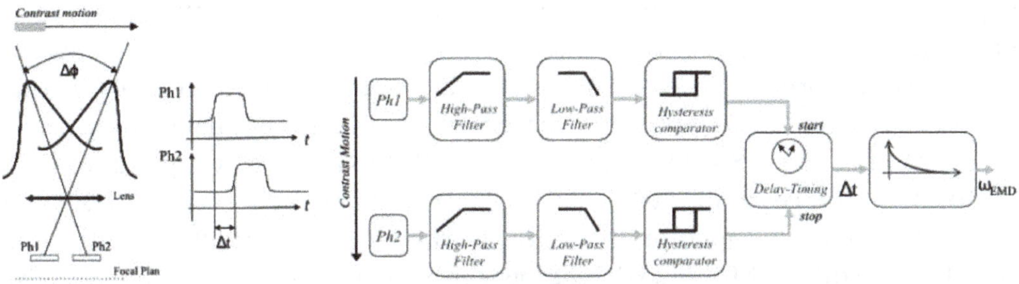

Figure 6. Principle scheme of Elementary Motion Detector (EMD). (Aubépart & Franceschini, 2007)

4. EXECUTION OF VISUAL FEEDBACK MECHANISMS

4.1 Sampling time consideration

In aerial robotic applications, the sampling time must comply with the requirements imposed by the digital control system so that the MAV can be controlled throughout its safe flight envelope. The maximum sampling time TSMAX will depend on the minimum delay Δt_{min} encountered by the robot's EMDs during the fastest manoeuvres in the most critical applications. Indeed, in an extreme case the smallest delay Δt_{min} which could be determined by the EMD will be that corresponding to one only counter clock period, whose clock signal corresponds to the sampling time.

One example of a fast maneuver is automatic wall-following, which is performed by measuring the side optic flow in the side direction. When the eye-bearing hovercraft is in pure translation at speed Vx and follows a textured wall at distance D, the image of the wall slips at an angular speed ω that depends on both V_x and D:

$$\omega = \frac{V_x}{D} = \frac{\Delta\varphi}{\Delta t} \tag{10}$$

If we take an extreme case where the hovercraft is allowed to follow the wall at the minimum distance D = 0.05m at the maximum speed V_x = 2m/s, equations 8 and 10 show that the 90° oriented EMD onboard the hovercraft, with its inter-receptor angle $\Delta\phi$ = 4°, will be subject to a minimum delay $\Delta t_{min} \approx$ 1.8ms. Accordingly, the sampling frequency fS_{min} will have to be set at values of at least 500Hz. When considering a MAV flying over an unknown terrain, we selected a minimum sampling frequency of 1kHz (Aubépart & Franceschini, 2007).

The maximum sampling frequency fS_{max} is less constraining to choose. It must be in keeping with the timing specifications to which the lens/photoreceptors devices are subject, especially in the case photoreceptors using the current-integrator mode (Kramer et al., 1997). On the other hand, the maximum sampling frequency fS_{MAX}, is limited by the lower end of the illuminance range over which the sensor is intended to operate, because at low illuminance levels, the integration of the photoreceptor signal takes a relatively long time and the sampling procedure will have to wait for this integration process to be completed (Aubépart & Franceschini, 2007). Taking the range [100Lux–2000Lux] to be a reasonable working illuminance range for the hovercraft, this gives fS = 2.5kHz, which we call the 'nominal sampling frequency' (Aubépart & Franceschini, 2005). At twice this sampling frequency (5kHz), the hovercraft would still operate efficiently in the [200Lux - 2000Lux] range, but it would then be difficult for it to detect low contrasts under artificial indoor lighting conditions. In addition, we avoided CMOS cameras equipped with digital outputs, which have not high frame rates, which still need to scan every pixel internally (Yamada et al., 2003, Zufferey et al., 2003).

As regards the forward and side controllers or others functions ('sign', 'add' and 'max' functions) present in the visual feedback loops, the sampling time should be adapted to the maximum frequency of the processed signals. Generally, the highest frequency corresponds to the smallest time-constant of control loops. In our loops, it is equal to 0.5-second that corresponds to a cut-off pulsation of 2rad/s. In discrete time we may consider that a sampling frequency 100 times higher than the system cut-off frequency would give results similar to continuous time. An equivalent sampling frequency of about 32 Hz could be sufficient. Since this value is in the lower part of the minimal value necessary to EMDs, we chose 2.5kHz as the nominal sampling.

4.2 Digital specifications

In feedback control loops, the digital specifications were mainly defined for EMDs design (i) and controllers design (ii). The secondary functions, such as 'sign', 'add' and 'max' functions, are easily adapted to the digital choices (binary format, implementation, etc.) without modifying the control loops operation.

In EMDs (i), the digital specifications were found during the filter design. Due to the low values of the high-pass and low-pass filter corner frequencies (f_{CHP} = 20Hz, f_{CLP} = 30Hz) in comparison with the sampling frequency (fS = 2,5kHz), it was not possible to obtain a digital band-pass filter meeting the Bode specifications. The high-pass filter section and low-pass filter section were therefore designed separately and cascaded.

Infinite Impulse Response (IIR) filters were synthesized because they require far fewer coefficients than Finite Impulse Response (FIR) filters, given the low cut- off frequencies and short sampling times involved:

$$y(n) = \sum_{i=1}^{n} b(i) \cdot x(i) - \sum_{i=1}^{n-1} a(i) \cdot y(i) \tag{11}$$

A transposed Direct-Form II structure was used because this structure reduces the number of delay-cells and decreases the quantization errors. Ripples on the low-pass filter temporal response were prevented by using a 4th-order Butterworth Filter, the phase of which was linearized over the frequency range of interest. The filters require 17 coefficients in all (4 coefficients for the 1st-order high-pass section, 12 for the 4th-order low-pass section, and 1 for the adjustment between the two filters). Three Direct-Form II filters suffice in fact to perform all the filtering, including that carried out by the two cascaded 2nd order low-pass filters.

A specific binary format was developed and used to prevent offset and stabilization problems. A two-complement fixed-point binary format, denoted $[s,m_I,m_D]$, was defined. The bit number of integer parts, m_I, and the decimal part, m_D, were defined so as to ensure maximum accuracy and to eliminate overflow from the filter calculations. Based on the results of a study carried out with *Filter Design and Analysis* and *Fixed-point Blockset* of the Mathworks tools, 6 bits were selected for the integer part m_I and 29 bits for the decimal part m_D. The large m_D bit number is due to the low value required to make the coefficients in the low-pass filter section comply with a Bode template characterized by a low cut-off frequency at high sampling frequencies.

Other digital specifications were defined in EMDs as regards the bit number of the counter output giving the delay time Δt, and the inverse exponential function giving the angular speed ω_{EMD}.

The delay time Δt is measured in terms of a count number at a given clock period. The minimum delay to be measured determines the minimum clock period (400μs for fS = 2.5kHz). The maximum delay to be measured is taken to be ~100ms, which is compatible with the wide range of angular speed values encountered by the hovercraft eyes (eq.4): ω ~40°/s to ~10000°/s, for Δφ = 4°. Using an 8-bit counter at fS = 2.5kHz gives an delay of 102.4ms.

The measured angular speed ω is a hyperbolic function of Δt (eq. 8), but we used a function that decreases more slowly: an inverse exponential function with a time constant τ = 30ms. A Look-Up Table (LUT) was used to convert the delay Δt into an output that decreases monotonically (exponentially) with the delay and therefore approximately reflects the angular speed ω_{EMD} (eq. 9). The Look-Up Table features an 8-bit input resolution (at fS = 2.5kHz) and a 12-bit output resolution for memorizing the results of the conversion.

The digital correctors design (ii) is easy in using 'c2d' Matlab function (discretize continuous-time system). The correctors' continuous-time models have been converted to discrete-time models with sampling time TS. The discretization method used was the bilinear Tustin approximation. Table 1 presents the discretized polynomial transfer functions.

Table 1. Correctors' discretization using Tustin bilinear approximation with fS = 2.5kHz

Correctors	Continous-time	Discrete-time
side	$C_y(s)=10\times\frac{1+1.5s}{1+0.5s}$	$C_y(z)=\frac{29.99z-29.98}{z-0.9992}$
forward	$C_{Vx}(s)=10\times\frac{1+0.5s}{s}$	$C_{Vx}(z)=\frac{5.002z+4.998}{z-1}$

As during EMD design, a specific binary format was defined to avoid offset and stability problems. Once again, we used the '*Filter Design and Analysis*' tool as well as the '*Fixed-point Blockset*' tool. A first approximation gave two signed fixed-point binary formats: 9 bit and 8 bit for the integer part m_I of the forward and side correctors, respectively, 10 bit for their decimal parts m_D. However, simulations of the full feedback control system showed that accuracy was too low. The decimal parts m_D were then increased to 13 bit for best accuracy in computation.

5. ARCHITECTURE

5.1 Optic flow sensor

Figure 7 shows the "EMD *architecture*". This architecture has several important features, such as the optimization of the digital filters, the simplicity of design owing to the use of Intellectual Property (IP) cores, and the added flexibility in the circuit design. Special care was taken to restrict the space taken by the digital filter implementation. Instead of implementing eight IIR digital filters in parallel - which would each require a high number of mathematical operators (adders, subtracters and multipliers) - a single structure called the "*Filter Compute Unit*" was developed, with which high speed sequential processing can be performed, as shown in figure 8. This unit consists of only one multiplier, one adder, one Read Only Memory (ROM), one Random Access Memory (RAM), two multiplexers, three registers and two binary transformation functions. The ROM contains the 17 filter coefficients obtained at a sampling frequency fS = 2.5kHz. The RAM is used to store the intermediate values computed. The multiplexers minimize the number of operators.

In this unit, each photoreceptor signal is processed during the sampling time TS. A Finite State Machine (FSM) was written in VHDL language and imported into the *Simulink* environment. The FSM specifies the filtering sequence for each photoreceptor channel. Even though this solution somehow complicates the design (due to the mixing of VHDL description and System Generator IP blocks), it has the advantage to minimize the number of logical gates necessary for integration in the FPGA.

Once the filtering process is completed, the delay Δt between the excitations of two neighboring photoreceptors starts being measured. A comparator determines the instant at which the band-pass filtered signal from each photoreceptor channel reaches the threshold value. The resulting logical signal is used to trigger the measurement of the delay Δt corresponding to each of the two eyes (an eye consists of a lens and only two photoreceptors). Specifically, the logical signal delivered by a photoreceptor starts a counter that will be stopped by the logical signal delivered by the neighboring photoreceptor.

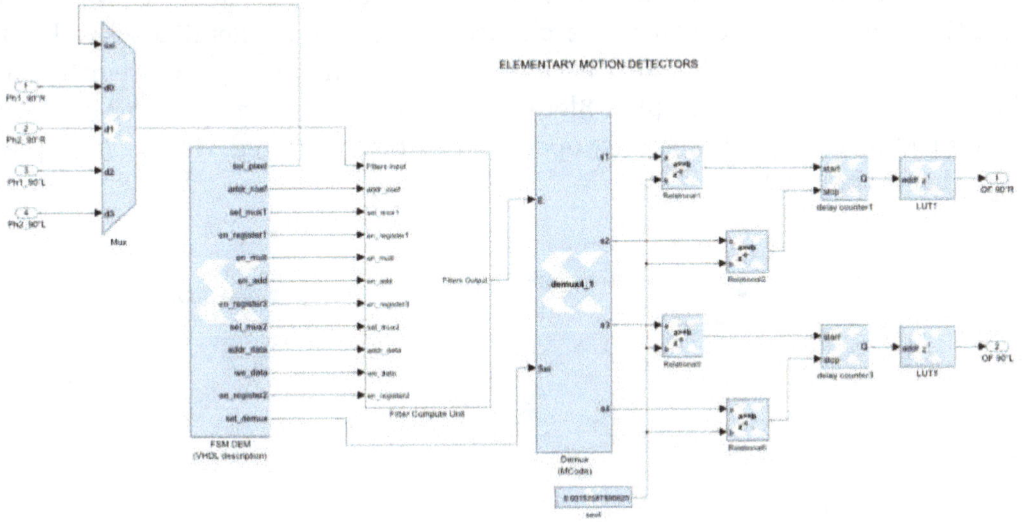

Figure 7. Elementary Motion Detector Architecture

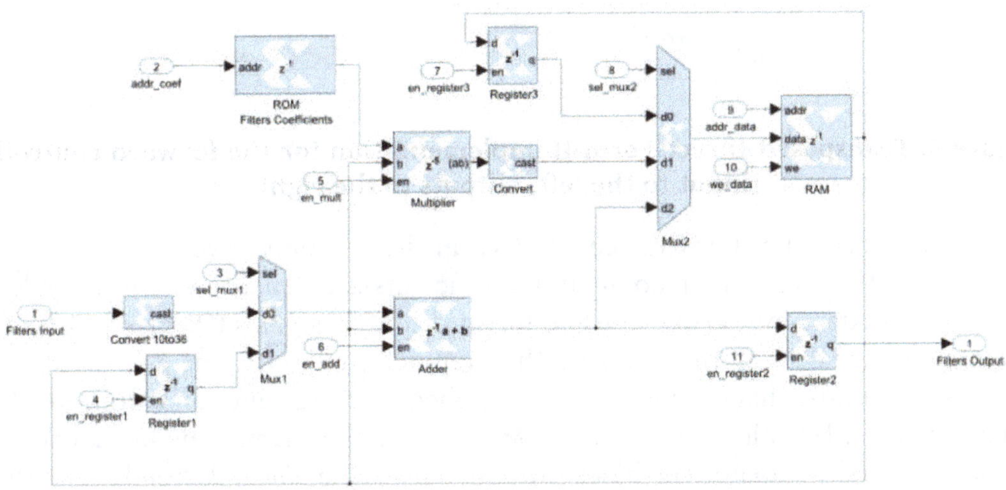

Figure 8. "Filter Compute Unit" architecture. The Filters' input is at port Nb 1. Others inputs (from 2 to 11) are control signals from a Finite State Machine that specifies the sequence of the filtering process for each photoreceptor channel

The final piece of the architecture is the inverse exponential function with a time constant τ = 30ms, which allows for a wide range of delays ranging up to 102.4ms. This component was implemented in the form of Look-Up Tables (ROM) to facilitate the conversion of each delay data into an estimated angular speed ωEMD (i.e., the optic flow).

5.2 Forward and Side correctors

The forward and side controllers (see Figure 5) have discretized polynomial transfer functions that can be designed as first-order IIR filters. For their realization, we chose the transposed-Direct Form-II architecture. *System generator* IP blocks allow for graphical integration of IIR filters because they use only multipliers, registers and adders. Using the Xilinx Multiplier Block called "Mult" (instead of the "CMult Xilinx Constant Multiplier") allowed this operation to be integrated via the multipliers embedded in the FPGA.

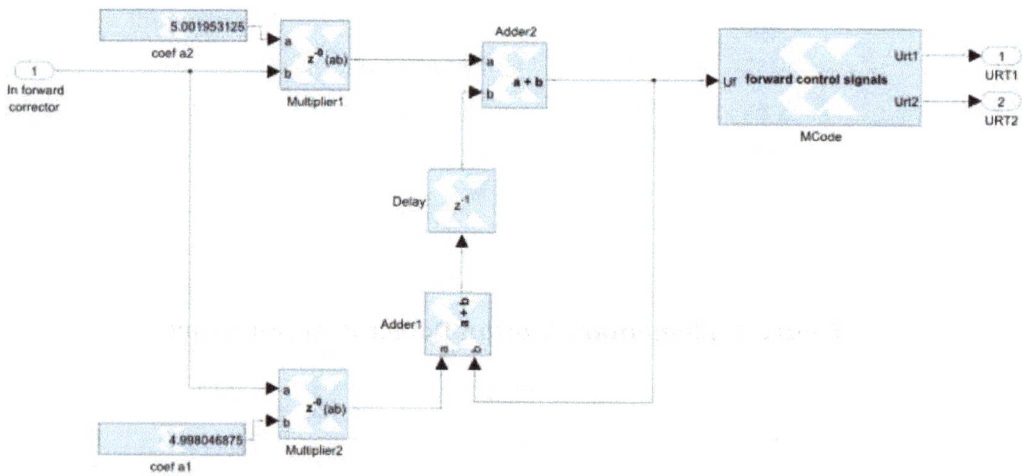

Figure 9. Transposed-Direct-Form-II implementation for the forward controller (input to the left, outputs to the right)

Figure 9 shows the forward controller, in which only two multipliers are implemented because the denominator coefficients are equal to one (see Table 1). A Xilinx "MCode" block allows the two forward control signals (URT1 + URT2) to be determined, for driving the two rear thrusters RT1 and RT2. The Xilinx "MCode" block is a container that executes a user- supplied MATLAB function within *Simulink*. A parameter of this block specifies the M-code function name. The block executes the M-code and calculates the block outputs during the *Simulink* simulation. When hardware is generated, the same code is translated into equivalent behavioral VHDL in a straightforward manner.

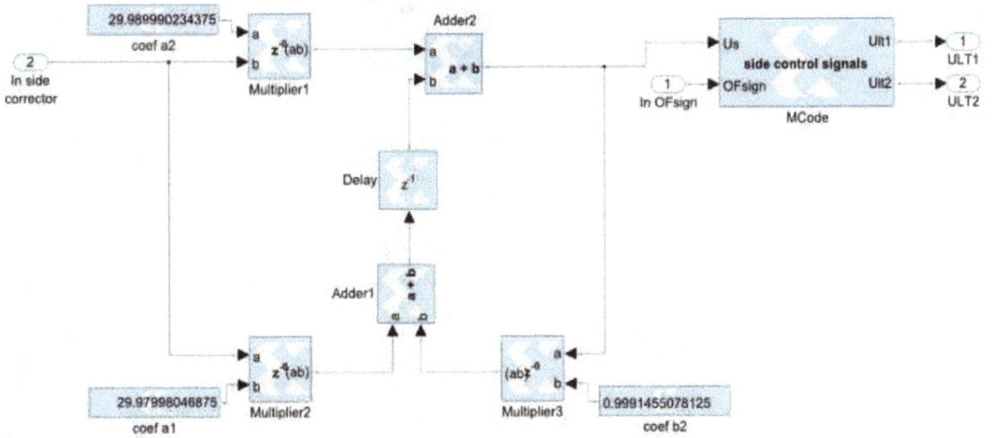

Figure 10. Transposed-Direct-Form-II implementation for the side controller (input to the left, outputs to the right)

The *side controller* was designed in the same way as the *forward controller*, figure 10. An additional multiplier was needed to perform calculation with the denominator coefficient b2. The two side control signals (U_{LT1} - U_{LT2}) controlling the two lateral thrusters LT1 and LT2 were again obtained by a Xilinx "MCode" block in which the sign of the difference between the right and left lateral optic flows measured was used.

5.3 Complete visuo-motor control system

The complete control system based on *System Generator* is presented in Figure 11. The secondary functions such as the determination of the larger value (OF max) and sign between right and left optic flows measured are easily realized with Xilinx "MCode" blocks. An adder and two subtracters are put in to close the loops. Xilinx "Gateway Out" blocks and "Gateway In" blocks limit the I/O from the Xilinx portion of the user's *Simulink* design. "Gateway Out" blocks convert the *System Generator* fixed point data type into *Simulink* double format while "Gateway In" blocks convert *Simulink* integer, double and fixed point data types into the *System Generator* fixed point type. Each block will define a top-level I/O port in the HDL design generated by *System Generator*.

Forward dynamics, side dynamics and "robot position compute" are defined by high level *Simulink* models linked to Matlab files that set several constants before simulation. "Lens/photoreceptors model" is designed with S-functions. An S-function is a computer language description of a *Simulink* block and is used to add our own blocks to *Simulink* models. Generally, such functions make it possible to accelerate simulations.

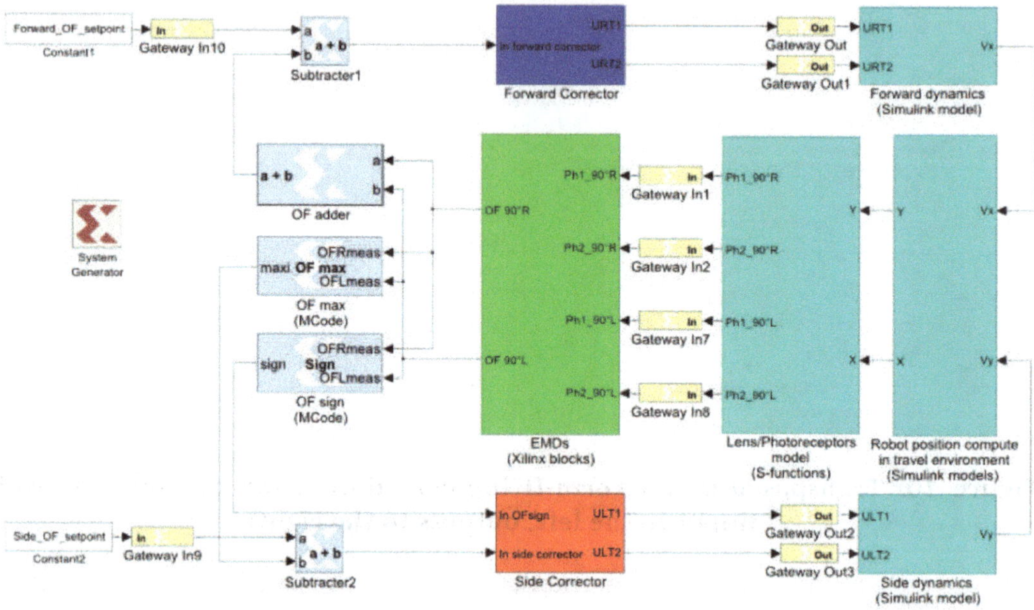

Figure 11. Complete control system based on System Generator. Xilinx "Gateway Out" blocks and "Gateway In" blocks limit the I/O from the Xilinx portion of the user's Simulink design

The *System Generator* block provides for control of system and simulation parameters, and is used to execute the code generator. Every *Simulink* model containing any element from the Xilinx Blockset must contain at least one *System Generator* block. Once a *System Generator* block is added to a model, it is possible to specify how code generation and simulation should be handled.

6. HARDWARE CO-SIMULATION

The final tests were achieved by making a hardware co-simulation of the system. The various functions (the EMDs and the two complete visuo-motor control loops) were implemented on a small Virtex-4 FPGA (type XC4FX12, size 17mm X 17mm, mass 0.5gram) whose digital processing capacities are largely sufficient. We designed a specific electronic board, figure 12, embedding the FPGA to validate the various digital properties of the whole system in co-simulation. This board was realized with an aim to install it on-board the miniature hovercraft and therefore also included the required interface components (ADC, DAC, etc.). The complete board weighs only 17.3grams and measures 90mm x 50mm and it is well suited to an embedded technological solution.

Figure 12. Specific electronic board based on a small FPGA Virtex-4: (i) right hand side: components for the power supply; (ii) middle part: FPGA XC4FX12 (top) and the Flash- memory to configure it (bottom); (iii) left hand side: electronic front-end related to the photosensors and several Analog-Digital and Digital-Analog Converters

The FPGA power consumption was evaluated using Xilinx "XPower" tool. The consumption is estimated at 149mW for a 2.5 kHz sampling frequency of the visuo-motor control loops and for the FPGA running at a clock frequency of 4MHz. The total power consumption of the electronic board, figure 12, is estimated at ~500mW

6.1 Hardware co-simulation results

A *Simulink* library was created from *System Generator* and copied into the *Simulink* project file replacing all the Xilinx System Generator blocks (i.e. between "Gateway In" and "Gateway Out" blocks, figure 11). The simulated robot is equipped with two lateral eyes oriented at ±90° to the walls (the inter-receptor angle is $\Delta\phi$=4° and the acceptance angle is $\Delta\rho$=4°). At 2.5kHz sampling frequency, the computing temporal step is Δt = 400µs and the simulation spatial accuracy is $\delta\theta$ = 0.005°. The two OF set-points were chosen according to the results of behavioural studies on honeybees that were video-filmed when flying through a straight corridor (Serres et al., 2008b). The forward OF set-point was set to ω_{setFwd} = 314°/s and the side OF set-point was set to $\omega_{setSide}$ = 238°/s.

6.2 Automatic speed control and lateral positioning in a straight corridor

The simulated visual environment is a 3-meter long, 1-meter wide straight corridor with textured walls. The right and left walls are lined with a random pattern of various grey vertical stripes covering a 1-decade contrast range (from 4% to 38%) and a 1.1-decade angular frequency range (from 0.068 c/° to 0.87 c/° reading from the corridor midline).

In figure 13, the hovercraft can be seen to follow either the right (red or black trajectory) or the left wall (blue trajectory), depending on the sign of the error signal εside, (Eq. 1). The robot can be seen to generate a steady state clearance of 0.24m from either wall (figure 13a), while reaching a steady state forward speed of V_x = 1m/s (figure 13b). Thus, the hovercraft adopts a wall-following behavior in much the same way as honeybees do in a similar situation (Serres et al., 2008a).

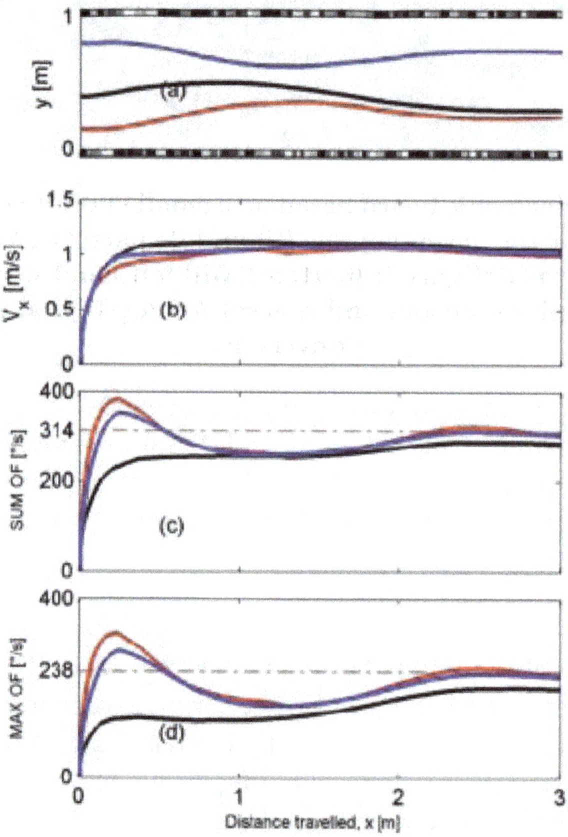

Figure 13. Hovercraft automatic wall-following behavior: (a) Simulated trajectories starting from three different initial positions y0 (red: y0 = 0.15m; black: y0 = 0.40m; blue: y0 = 0.80m). (b) Forward speed profiles. In the steady state, the forward speed can be seen to have reached V_x = 1m/s in the three cases. (c) Sum of the two lateral optic flows ($\omega_R + \omega_L$). (d) Larger value of the two lateral optic flows, right and left

6.2.1 Automatic response in a tapered corridor

For the dual OF regulator, a tapered corridor acts like a *non-constant OF disturbance*. The forward control system adjusts the forward speed proportionally to the local corridor width (the width varies from 1.24m to 0.50m). The lateral control system controls the distance to the right wall in proportion to the forward speed at all times.

This simulation experiment shows that the dual OF regulator is able to cope with the major disturbance caused by a tapered corridor, by making the robot decelerate or accelerate appropriately, figure 14.

The hardware co-simulation results presented here closely match the *Simulink* results presented by Serres et al. (2008a).

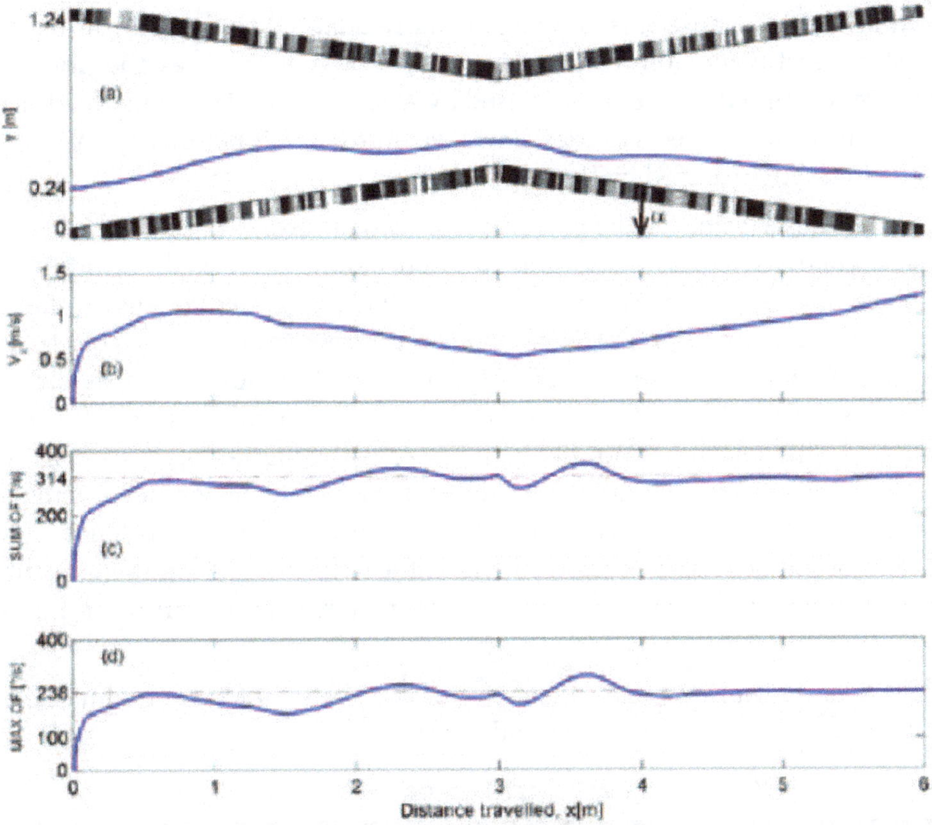

Figure 14. Automatic response in a tapered corridor: (a) Simulated trajectory of the hovercraft moving to the right in a tapered corridor starting at the initial position y0 = 0.24m. This trajectory shows that the hovercraft automatically slows down when the local corridor width decreases and accelerates again when it widens again after the constriction. (b) Corresponding forward speed profile V_x. The forward speed turns out to be a quasi linear function of the distance travelled x, and hence of the local corridor width. (c) Sum of the two optic flow ($\omega_R + \omega_L$). The speed control system succeeds to keep the sum of the optic flows measured virtually constant. (d) Larger value ("max") between the two optic flows measured. The side control system succeeds to keep the larger value of the two optic flows measured virtually constant

6.3 Hardware integration

In order to make the interface between the peripheral components and the FPGA-integrated system, several pilots have being developed and tested: driver ADC108S102, driver ADC101S101, driver DAC1101S101, PWM generator. These drivers link the components or the corresponding boards to the integrated *visuo-motor control loops* (generated by *System Generator*). They are integrated in the Virtex-4 FPGA and designed in VHDL because this language is more adapted to the realization of both the PWM generator and the communications protocols, such as the SPI protocol (*Standard Peripherical Interface*) that is used by ADC and DAC. For example, it is easier to manage the timing aspects of these low level hardware functions by a VHDL Finite State Machine synchronized by the FPGA clock.

The complete description can be performed in two different ways: (i) In *Simulink*, the *visuo-motor control system* is associated to import some drivers' descriptions (using Xilinx black box in the *System Generator* library); (ii) in Xilinx ISE environment, the VHDL generated description of the *visuo-motor control system* is connected to drivers descriptions (using ISE schematic description). The first solution (i) is difficult to apply because it requires the addition of peripheral components or functions high level models. This solution is not likely to improve the digital integration of control loops (validated in co- simulation) or drivers (validated in low level simulations and/or tested with peripheral components). The second solution does not allow any simulations of the closed loop system, but facilitates the integration stages (synthesis, mapping, placement and routing) and the generation of FPGA configuration binary file.

Table 2 shows the working characteristics of the Virtex-4 FPGA obtained after the integration of the LORA III autopilot and drivers.

Table 2. Working characteristics of the XC_4FX_{12} $Virtex^{-4}$ FPGA

DSP48	15 out of 32	42%
Slices	2569 out of 5472	47%
Block RAM 18kb	9 out of 36	25%

The DSP48 is a DSP-oriented component. The DSP48 is basically a multiplier followed by an adder with several optional registers on the ports and between the multiplier and adder. The multiplier takes two 18-bit signed signals and multiplies them, giving a 36-bit result. This is then sign extended to 48 bits and can be fed into the adder or routed directly to the outputs of the DSP48. The adder, which can be configured either as an adder or a subtractor, can accept the sign-extended output of the multiplier and 48-bit input to the DSP48. In addition, the adder can also accept itself as an input, to form an accumulator.

A slice is a basic element of Configurable Logic Block (CLB). A CLB element contains four interconnected slices. These slices are grouped in pairs. The elements common to both slice pairs are two logic-function generators (or look-up tables), two storage elements, wide-function multiplexers, carry logic, and arithmetic gates. The slices are used to provide logic, arithmetic, and ROM functions. Virtex-4 device features a large number of 18 Kb block RAM memories. True Dual-Port RAM offers fast blocks of memory in the device. Block RAMs are placed in columns, and the 18 Kb blocks are cascadable to enable a deeper and wider memory implementation, with a minimal timing penalty.

7. CONCISE SUMMARY

We have implemented a digital integration of an autopilot named LORA III (Serres et al. 2008a) onto a Virtex-4 FPGA. The autopilot comprises two interdependent visuo-motor control loops designed to regulate the visual steering of a small hovercraft inside a corridor, without utilizing measurements of speed or distance from the walls. A hierarchical approach was employed to construct and simulate the whole visuo-motor control system. The latter was analyzed utilizing both a sophisticated graphical interface: Simulink from Mathworks, and the System Generator for DSP toolkit from Xilinx. The exceptional analytical capability arises from System Generator's ability to construct the planned sensory-motor control loop within the Simulink environment. The design approach has streamlined integration issues by employing multiple levels of abstraction, each validated at every development stage. This methodology optimized the digital specifications and topologies of each control loop for the LORA III autopilot. Furthermore, conclusive experiments were conducted utilizing hardware co-simulation. Consequently, we successfully tested the final descriptions in FPGA using a JTAG connection from the Matlab/Simulink environment. Integrated architectures were validated by examining the hovercraft's operation in straight or tapered corridors. The integration of the visuo-motor control loops necessitated the fabrication of a specialized electronic board utilizing a Virtex-4 FPGA (Figure 12). Integrating the control system with external components (ADC, DAC, motor control) necessitated the design of many drivers, which were contained within the same 0.5 gram FPGA. Future endeavors will involve the installation of the FPGA-based sensory-motor control board into the miniature hovercraft for which it was designed. Subsequent tests analogous to those conducted in co-simulation will be performed. Further enhancements are planned to augment the resilience of the LORA III control system and enable the robot to navigate more complex pathways. The passive OF sensors and the basic processing system outlined below are especially appropriate for Micro-Air Vehicles (MAVs), where strict limitations are placed on allowable avionic payload and onboard energy resources. Recent advancements in FPGA implementation have been proposed for both the visuo-motor control of micro-air aircraft and the automated visual guidance and retrorocket control of forthcoming planetary landers.

REFERENCES

Ahmad, Z. & Taib, M.N. (2003). A study on the dc motor speed control by using back-emf voltage, *Proceeding of IEEE Asian conference on Sensors (AsiaSense)*, pp. 359-364

Aubépart, F. & Franceschini, N. (2005). Optic flow sensors for robots: Elementary Motion Detectors based on FPGA, *Proceedings of IEEE International Workshop on Signal Processing Systems*, pp.182-187, Athens, Greece, 2-4 November, 2005

Aubépart, F. & Franceschini, N. (2007). Bio-inspired optic flow sensors based on FPGA: Application to Micro-Air-Vehicles. *Journal of Microprocessors and Microsystems*, Vol. 31, No.6, (2007) (408-419) , ISSN°0141-9331

Baird, E.; Srinivasan, M. V.; Zhang, S. & Cowling, A. (2005). Visual control of flight speed in honeybees. *The Journal of experimental Biology,* Vol.208, No.20, (3895-3905), ISSN°0022-0949

Blanes, C. (1986). Appareil visuel élémentaire pour la navigation à vue d'un robot mobile autonome, DEA thesis (in french), Aix-Marseille University

Browy, C.; Gullikson, G & Indovina, M. (1997). A Top-Down approach to IC design, www.indovina.us/~mai/a_top_down_approach_to_ic_design.pdf

Chambers, P. (1997). The ten commandments of excellent design: VHDL code examples, *Electronic design*, Vol. 45, No.8, (123-126)

Chalimbaud, P. & Berry, P. (2007). Embedded active vision system based on an FPGA architecture. *EURASIP Journal on embedded systems*, Vol.2007, No.1 (January 2007) (12p.), ISSN°1687-3955

Collett, T.S. (1980). Some operating rules for the optomotor system of a hoverfly during voluntary flight. *Journal of Comparative Physiology A*, Vol.138, No.3, (September 1980) (271-282), ISSN°0340-7594.

David, C. (1978). The relationship between body angle and flight speed in free-flying *Drosophila. Physiological Entomology*, Vol.3, No.3, (191-195), ISSN°0307-6962 Franceschini, N. (1985). Early processing of color and motion in a mosaic visual system, *Neurosciences. Res. Suppl.* 2, (17-49)

Franceschini, N.; Blanes, C & Oufar, L. (1986). Passive non-contact optical velocity sensor, *Dossier technique N° 51549* (in French), Agence Nationale pour la Valorisation de la Recherche/Direction de la Valorisation, Paris

Franceschini, N.; Riehle, A. & Le Nestour, A. (1989). Directionally Selective Motion Detection by insect neurons, In: *Facets of Vision*, Stavenga D.G. & Hardie R.C. (Eds.), (360-390), Springer, ISBN°0387503064, Berlin

Franceschini, N.; Pichon, J-M. & Blanes, C. (1992). From insect vision to robot vision, *Philosophical transactions of the Royal Society of London. Series B, Biological sciences* Vol. 337, No.1281, (283-294), ISSN°0080-4622

Franceschini, N.; Pichon, J-M. & Blanes, C. (1997). Bionics of visuomotor control, *In Evolutionary robotics: from intelligent robots to artificial life, AAAI books,* : T. Gomi (Ed.), (49-67), Ottawa

Franceschini, N. (1998). Combined optical, neuroanatomical, electrophysiological and behavioural studies on signal processing in the fly compound eye, In: *Biocybernetics of vision: Integrative Mechanisms and Cognitive Processes,* C. Taddei-Ferretti, (Ed.), (341-361), World Scientific, London

Golson, S. (1994). State Machine design techniques for Verilog and VHDL, *Synopsys Journal of High-Level Design,* (September 1994), (1-48)

Hardie, R.C. (1985). Functional organisation of fly retina, In: *Progress in Sensory Physiology,* D. Ottoson (Ed.), Vol.5 (1-79), Springer, Berlin Iida, F. (2001). Goal-Directed Navigation of an Autonomous Flying Robot Using Biologically Inspired Cheap Vision, Proceedings of the 32nd International Symposium on Robotics (ISR), pp.1404-1409, 19-21 April 2001

Kennedy, J.S. (1939). Visual responses of flying mosquitoes. *Proceedings of Zoological Society of London,* Vol.109, 221-242.

Kennedy, J.S. (1951). The migration of the desert locust (Schistocerca gregaria Forsk.) I. The behaviour of swarms. *Philosophical Transactions of the Royal Society of London. Series B, Biological Sciences,*. Vol.235, No.625 (May 31, 1951), (163-290), ISSN°02610523.

Kerhuel, L.; Viollet, S. & Franceschini, N. (2007). A sighted aerial robot with fast gaze and heading stabilization, *Proceedings of the IEEE International Conference on Intelligent Robots and Systems (IROS),* pp.2634-2641, ISBN°978-1-4244-0912-9, San Diego, USA, October 29-November 2 2007

Kirchner, W.H. & Srinivasan, M.V. (1989). Freely flying honeybees use image motion to estimate object distance. *Naturwissenschaften,* Vol.76, No.6, (281-282), ISSN 0028- 1042.

Koenderink, J.J. (1986) Optic flow, *Vision Research,* Vol.26, No.1, (161-179), ISSN°0042-6989 Kramer, J.; Sarpeshkar, R. & Kock, C. (1997). Pulse-Based Analog VLSI Velocity Sensors, *IEEE Transactions on Circuits and Systems* II, Vol.44, No.2 (February 1997) (86-101), ISSN° 1057-7130

Liu, S.C. & Usseglio-Viretta, A. (2001). Fly-like visuomotor responses of a robot using a VLSI motion-sensitive chips, *Biological Cybernetics,* Vol.85, No.6 (december 2001) (449-457), ISSN°0340-1200

Mura, F. & Franceschini, N. (1996). Obstacle avoidance in a terrestrial mobile robot provided with a scanning retina, In: *Intelligent Vehicles Vol.II,* M. Aoki and I. Masaki (Eds.), (47-52), MIT Press, Cambridge.

Murthy, S. N.; Alvis, W.; Shirodkar, R.; Valavanis, K. & Moreno, W. (2008). Methodology for implementation of Unmanned vehicle control on FPGA using System generator, *Proceedings of the 7th international Caribbean conference on devices, circuits and systems (ICCDCS)*, pp.1-6, ISBN°978-1-4244-1956-2, Mexico, April 28-30, 2008

Netter, T. & Franceschini, N. (1999). Neuromorphic optical flow sensing for nap-of-the- earth flight, *Proceeding of SPIE Conference on Mobile Robots XIV*, Vol.3838, pp.208- 216, Boston, USA, 20 September 1999

Netter, T. & Franceschini, N. (2002). A Robotic aircraft that follows terrain using a neuromorphic eye, *Proceeding IEEE International Conference on Robots and Systems (IROS)*, pp.129-134, ISBN 0-7803-7398-7, Lausanne, Swiss, 30 September - 4 October 2002.

Ownby, M. & Mahmoud W. H. (2003). A design methodology for implementing DSP with Xilinx System Generator for Matlab, *Proceedings of the 35th Southeastern Symposium on System Theory*, pp. 404-408, 16-18 March 2003

Pichon, J-M.; Blanes, C. & Franceschini, N. (1989). Visual guidance of a mobile robot equipped with a network of self-motion sensors, *Proceeding of SPIE Mobile RobotsIV.*, Vol.1195, (44-53), ISBN°9780819402349

Ruffier, F. & Franceschini, N. (2003). OCTAVE, a bioinspired visuo-motor control system for the guidance of Micro-Air-Vehicles, *Proceeding of SPIE Conference on Bioengineered and Bioinspired Systems*, Vol.5119, pp.1-12, Bellingham, USA

Ruffier, F.; Viollet, S. & Franceschini, N. (2004). Visual control of two aerial micro-robots by insect-based autopilots, *Advanced Robotics*, Vol.18, No.8, (771-786)

Ruffier, F. & Franceschini, N. (2005) Optic flow regulation: the key to aircraft automatic guidance, *journal of Robotics and Autonomous Systems*, Vol.50, No.4, (177-194), ISSN° 0921-8890

Ruffier, F., Serres J., Masson, G.P. and Franceschini N. (2007). A bee in the corridor: regulating the optic flow on one side. *Proceedings of the 7th Meeting of the German Neuroscience Society - 31st Göttingen Neurobiology Conference*, Göttingen, Germany, Abstract T14-7B.

Serres, J., Ruffier, F. & Franceschini, N. (2006). Two optic flow regulators for speed control and obstacle avoidance, *Proceedings of the first IEEE International Conference on Biomedical and Biomechatronics (Biorob)*, pp.750-757, Pisa, Italy.

Serres, J., Ruffier, F., Masson, G.P. & Franceschini N. (2007). A bee in the corridor: centring or wall-following? In *proceedings of the 7th Meeting of the German Neuroscience Society - 31st Göttingen Neurobiology Conference*, Göttingen, Germany, Abstract T14-8B.

Serres, J., Dray, D., Ruffier, & Franceschini N. (2008a). A vision-based autopilot for a miniature air vehicule: joint speed control and lateral obstacle avoidance. *Autonomous Robots*, Vol.25, No.1-2 (August 2008) (103-122). ISSN° 0929-5593

Serres, J., Masson, G.M., Ruffier, F. & Franceschini, N. (2008b). A bee in the corridor: centering and wall-following. Naturwissenchaften. (in press)

Srinivasan, M.V., Zhang, S.W., Lehrer, M. and Collett T.S. (1996). Honeybee navigation *en route* to the goal: visual flight control and odometry. *Journal of Experimental Biology*, Vol.199, (237-244)., ISSN°0022-0949.

Srinivasan, M.V., Zhang, S.W., Chahl, J.S., Barth, E., Venkatesh, S. (2000) How honeybees make grazing landings on flat surface. Biological Cybernetics, Vol.83, No.3, (171- 183), ISSN°0340-1200

Srygley, R.B., and Oliveira, E.G. (2001). Orientation mechanisms and migration strategies within the flight boundary layer. In Insect Movements: Mechanisms and Consequences, T.P. Woiwod, D.R. Reynolds, and C.D. Thomas, (Eds). (183–206), Wallingford, Oxon, UK: CABI Publishing, CAB International), ISBN 0851994563.

Tammero, L.F. and Dickinson, M.H. (2002). The influence of visual landscape on the free flight behavior of the fruit fly *drosophila melanogaster*. *Journal of Experimental Biology*, Vol.205, No3 (327-343), ISSN°0022-0949

Turney, R. D.; Dick, C.; Parlour, D. P. & Hwang, J. (1999). Modeling and implementation of DSP FPGA Solutions, *Proceedings of International Conference on Signal Processing Applications and Technology (ICSPAT)*, November 1-4, 1999, Orlando, USA (http://www.xilinx.com/products/logicore/dsp/matlab_final.pdf)

Van Nieuwstadt, M. & Morris, J.C. (1995). Control of rotor speed for a model helicopter: a design cycle, *Proceedings of American Control Conference*, Vol.1, pp. 668-672

Viollet, S. & Franceschini, N. (2001). Aerial Minirobot that stabilizes and tracks with a bio- inspired visual scanning sensor, In: *Biorobotics: Methods and applications*, B. Webb and T. Consi (Eds.), (67-83), MIT Press, ISBN°026273141X, Cambridge

Viollet, S.; Kerhuel, L. & Francechini, N. (2008). A 1-gram dual sensorless spedd governor for Micro-Air-Vehicles, *Proceedings of 16th Mediterranean Conference on Control and Automation*, pp.1270-1275, ISBN°978-1-4244-2504-4, Ajaccio, France, June 25-27, 2008

Wagner, H. (1982). Flow-field variables trigger landing in flies. *Nature*, Vol.297, (147-148). William C. (1965). *Insect migration*, Second edition ed. Collins editor, London.

Yamada, H.; Tominaga, T. & Ichikawa, M. (2003). An autonomous flying object navigated by real-time optical flow and visual target detection, *Proceedings of the IEEE International Conference on Field Programmable Technology*, pp.222-227, ISBN°0- 7803-8320-6, Tokyo, 15-17 December 2003

Zufferey, J.C.; Beyeler, A. & Floreano, D. (2003). Vision-based Navigation from Wheels to Wings, *Proceedings of IEEE International Conference on intelligent Robots and Systems*, pp.2968-2973, ISBN°0-7803-7860-1, Las-Vegas, USA, 27-31 October 2003

Chapter 4 PLANNING AND GUIDANCE FOR ADVANCED UAV TRAJECTORY GENERATION

1. INTRODUCTION: AN OVERVIEW

As technology and legislation advance (JAA & Eurocontrol, 2004), remotely controlled, semi-autonomous, or autonomous Unmanned Aerial Systems (UAS) will significantly contribute to service provision and the enhancement of safety and security for both military and civilian sectors (e.g., surveillance and monitoring) (Coifman et al., 2004). The prospective market for UAVs extends far beyond mere surveillance. Unmanned Aerial Vehicles (UAVs) are optimal for risk evaluation and mitigation in hazardous environments, including combat zones and places affected by disasters such as volcanic eruptions, wildfires, floods, and terrorist incidents. As UAVs gain autonomy, they will assume further responsibilities, including air-to-air combat and planetary science exploration (Held et al., 2005).

As the operational capabilities of UAVs are developed there is a perceived need for a significant increase in their level of autonomy, performance, reliability and integration with a controlled airspace full of manned vehicles (military and civilian). As a consequence researchers working with advanced UAVs have moved their focus from system modeling and low-level control to mission planning, supervision and collision avoidance, going from *vehicle* constraints to *mission* constraints (Barrientos et al., 2006). This mission-based approach is most useful for commercial applications where the vehicle must accomplish tasks with a high level of performance and maneuverability. These tasks require flexible and powerful trajectory-generation and guidance capabilities, features lacking in many of the current commercial UAS. For this reason, the purpose of this work is to extend the capabilities of commercially available autopilots for UAVs. Civil systems typically use basic trajectory-generation algorithms, capable only of linear waypoint navigation (Rysdyk, 2003), with a minimum or non-existent control over the trajectory. These systems are highly constrained when maneuverability is a mission requirement. On the other hand, military researchers have developed algorithms for high-performance 3D path planning and obstacle avoidance (Price, 2006), but these are highly proprietary technologies that operate with different mission constraints (target acquisition, threat avoidance and situational awareness) so they cannot be used in civil scenarios.

This chapter presents a robust Trajectory Generation and Guidance Module (TG2M), a software tool capable of generating complex six-degrees-of-freedom trajectories in 3D space with velocity, acceleration, orientation and time constraints. The TG2M is an extension module to the Aerial Vehicle Control Language (AVCL), a software architecture and interpreted language specification that address the issues of mission definition, testing, validation and supervision for UAVs (Barrientos et al., 2006). The AVCL platform incorporates a 3D visual simulator environment that uses a Geographic Information System (GIS) as the world's data model. The GIS backend means all objects are geo-referenced and that several official and commercial databases may be used for mission planning (roads, airports, power lines, crop fields, etc.). The language specification contains a wide set of instructions that allow the off/on-line supervision and the creation of vehicle-independent missions.

The TG^2M is designed to model 3D cartesian paths with parametric constraints. The system uses two types of mathematical descriptors for trajectories: analytical functions and polynomial interpolation. The two main contributions of the module are its geometrical representation of the trajectory and its parametric definition. Simple maneuvers like lines and circumference arcs are created with analytical functions that constrain the geometry of the desired path; then the parametric constraints are *applied*. These constraints are typically kinematic: constant velocity and acceleration, trapezoidal curves to accelerate or stop, etc. More complex maneuvers are described with polynomial interpolation and fitted to the critical control path points, meeting desired position, time and velocity constraints. These polynomial functions are based on third and fourth order splines with fixed boundary conditions (for example initial and final velocities), which join all control points with a continuous and smooth path (Jaramillo-Botero et al., 2004).

Section 2 of this chapter delineates contemporary approaches derived from military aerial survey applications, showcasing sophisticated mission-planning systems adept at accommodating mission-specific limitations. This section provides a concise overview of the AVCL architecture, detailing its components, modules, and primary features, while presenting an innovative method to human-mission planning definition and testing. Section 3 presents the AVCL integrated TG2M framework, including the many approaches employed for trajectory planning and guidance of UAVs, along with the mathematical analysis of these methods (analytical functions and polynomial interpolation). Conversely, simulation-based outcomes (refer to section 4) utilizing a mini-helicopter simulator integrated into the AVCL environment will demonstrate the capabilities of the TG2M during both aggressive and straightforward operations. Lastly, the "final observations" in section 5 encompass remarks regarding the TG2M framework and the forthcoming enhancements to the TG2M now in development.

2. MOTION-PLANNING METHODOLOGIES

A planning algorithm should provide feasible and fiyable optimal trajectories that connect starting with target points, which should be compared and valued using specific criteria. These criteria are generally connected to the following major concerns, which arise during a plan generation procedure: feasibility and optimality. The first concern asks for the production of a plan to safely "move" the UAV to its target state, without taking into account the quality of the produced plan. The second concern asks for the production of optimal, yet feasible, paths, with optimality defined in various ways according to the problem under consideration (LaValle, 2006). Even in simple problems searching for optimality is not a trivial task and in most cases results in excessive computation time, not always available in real-world applications. Therefore, in most cases we search for suboptimal or just feasible solutions. The simplest way to model an UAV path is by using straight-line segments that connect a number of waypoints, either in 2D or 3D space (Moitra et al., 2003, Zheng et al., 2005). This approach takes into account the fact that in typical UAV missions the shortest paths tend to resemble straight lines that connect waypoints with starting and target points and the vertices of obstacle polygons. Although waypoints can be efficiently used for navigating a flying vehicle, straight-line segments connecting the corresponding waypoints cannot efficiently represent the real path that will be followed by the vehicle due to the own kinematics of the traced path. As a result, these simplified paths cannot be used for an accurate simulation of the movement of the UAV in an optimization procedure, unless a large number of waypoints is adopted. In that case the number of design variables in the optimization procedure explodes, along with the computation time. This is why this section presents some background based on the state-of-the-art mission/path planning for UAVs. The purpose (apart from the state-of-the-art survey) is to compare to the AVCL architecture in order to observe how its TG2M embedded framework (see section 3 for details) is a complement of some lacking approaches found in this specialized literature.

2.1 Background

Researchers at top research centers and universities (e.g. JPL at Caltech, Carnegie Mellon, NASA, ESA, among others) are dealing with the development of new strategies that allow high-level mission UAV planning within the desired flight performance. As examples of these approaches we present the NASA Ames Research Center, the Georgia Institute of Technology, and the University of Illinois.

For the past decade NASA has focused on developing an efficient and robust solution to Obstacle Field Navigation (OFN), allowing a fast planning of smooth and flyable paths around both known and unknown obstacles (Herwitz, 2007). Evolutionary algorithms have been used as a viable candidate to solve path-planning problems effectively, providing feasible solutions within a short time. Given a number of UAVs that are launched from different and known initial locations, the algorithm must create 2-D trajectories with a smooth velocity distribution along each trajectory and with the goal of reaching a predetermined target location, while ensuring collision avoidance and satisfying specific route and coordination constraints and objectives. B-Splines curves are used in order to model both the 2-D trajectories and the velocity distribution along each fight path. A flying laboratory for autonomous systems, based on the Yamaha RMAX helicopter is being developed, which it incorporates the OFN planner and one all-digital camera system with a state-of-the-art tracking and passive ranging capabilities. Machine stereo-vision is being used to determine safe landing areas and monocular vision is used to track the landing location without access to GPS. A Ground Control Station (GCS) is being integrated into the simulation environment to investigate the issues related to high altitude and long endurance flights.

As an alternative approach other researchers do not focus on generating complex trajectories profiles; their aim is to develop robust mission management, capable of successfully integrating the available onboard hardware (e.g. cameras, sensors, etc). The Georgia Tech's UAV Lab developed the GTMax architecture (Alison et al., 2003) based on a Yamaha R-Max mini-helicopter, is an example of this alternative approach. The GTMax system is capable of fully autonomous flight with decentralized software modules for trajectory generation, offline simulation, supervision, guidance and control. The trajectory module generates lineal waypoints between targets with initial and final velocity parameterization. Commands to the helicopter take the form of different types of waypoints. All trajectories generated are assumed to be physically feasible by the helicopter, the kinematic model used for the trajectory generation uses specifiable limits on the maximum speed and acceleration the aircraft may have during a maneuver. From a high-level mission management perspective the GTMax architecture allows the UAV to autonomously locate targets with an identifying symbol within a designated search area. Once the correct target is identified the mission coordination and fight path generation is done at a centralized location on the ground, and the commands are transmitted to the UAV via a wireless datalink. The GTMax GCS interfaces with the primary fight computer and displays vehicle status, the object tracker information and the fight plans generated by the mission planner. The Vision Monitoring Station receives streaming video from the camera and the output of the image processor. This allows the operator to monitor the efficiency of the image processing as well as visually document the results of the search in the final phase of the mission.

For other projects the main goal is the unification of high-level mission commands and development of generic languages that support online UAV mission planning without constraining the system to a single vehicle. The University of Illinois at Urbana-Champaign (Frazzoli, 2002) presents a new framework for UAV motion planning and coordination. This framework is based on the definition of a modelling language for UAV trajectories, based on the interconnection of a finite number of suitably defined motion primitives, or maneuvers. Once the language is defined, algorithms for the efficient computation of feasible and optimal motion plans are established, allowing the UAV to fulfill the desired path.

If we analyze the UAS described above, almost all of them share one important limitation: their software architecture is tightly coupled to one vehicle and the capabilities of its low- level controller. Civil applications require open and extendable software architectures capable of *talking* to vehicles from different suppliers. The AVCL addresses those limitations, allowing to model different vehicles into a single common language (vehicle- independent missions). In the same fashion the described vehicles show that complex and simple maneuvers could be a suitable solution depending on the kind of mission to fulfill. For this reason the AVCL is extended with the TG2M framework, capable of generating simple and complex 3D paths with the necessary vehicle constrains. The next sub-section introduces the AVCL architecture and some of the features provisioned.

2.2 The Aerial Vehicle Control Language - AVCL

The AVCL is not just a language capable of describing the missions and capabilities of an heterogeneous group of vehicles, it is part of a bigger framework that includes its interpreter, a definition of a base-vehicle, and a Mission Planner that uses GIS as the data- model for the world (Barrientos et al., 2006). The Mission Planner (MP) is not tied to a particular set of vehicles, sensors or commands. At any given time new functionality can be loaded and displayed to the human operator as new options and commands. This means that the MP tool is to be extended through Vehicle and Command Libraries without recompiling, and those new capabilities and better vehicles can be added easily. The Mission Planner is a great tool for simulation and direct comparison of various trajectory trackers, UAV models and controllers, because it can display *N* missions at the same time.

When considered just as a language the AVCL concept is the abstraction layer that allows the human supervisor to create missions that are vehicle and payload independent, promoting code reuse. At the same time the AVCL statements and commands hide device specific software/hardware, and serve as mission definition and storage. As an example of the code used to define operations within a mission:

uav.Sensors(0) = parser.loadObject ('camera.lib')

uav.Sensors(0).LookAt (p^1)

uav.Sensors(1) = parser.loadObject ('laser.lib')

uav.Sensors(1).TurnOn()

uav.doLine (way_points = {p^1, p^2, p^3, p^4}, vel = 0.9 ms)

Compared to previous vehicle *programming* languages the AVCL and its interpreter provide several advantages: intuitive handling of different systems of units; its use of the object- oriented-programming paradigm; facilities for inter-vehicle communications; run-time definition of relations between vehicles, sensors and other equipment; it may be extended easily through C, C++ or C# code; the interpreter is a light-weight application written in C++, therefore it may be deployed in many SW/HW architectures. Before the development of the TG_2M module the AVCL framework relied on a simpler guidance module to connect waypoints with straight-line segments, and while the language could describe complex maneuvers and mission constraints the framework lacked the capacity to *fly* a vehicle through complex paths.

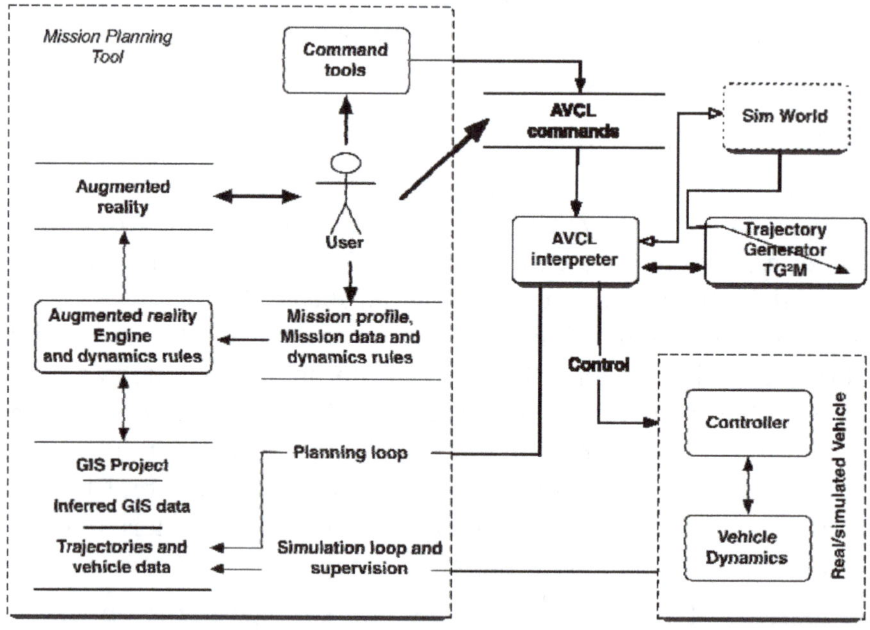

Figure 1. The AVCL simplified diagram for mission planning

3. TRAJECTORY GENERATION AND GUIDANCE MODULE - TG2M

The TG2M is designed to model 3D cartesian paths with parametric constraints. The system uses two types of mathematical descriptors for trajectories: polynomial interpolation and analytical functions. Complex maneuvers are described with polynomial interpolation based on third and fourth order splines with fixed boundary conditions (e.g. initial and final velocities user definition), which join all control points with a continuous and smooth path. Likewise, simple maneuvers like lines and circumference are created with analytical functions that constrain the geometry of the desired path. These constraints are typically kinematic: constant velocity and acceleration, trapezoidal curves to accelerate or stop, etc.

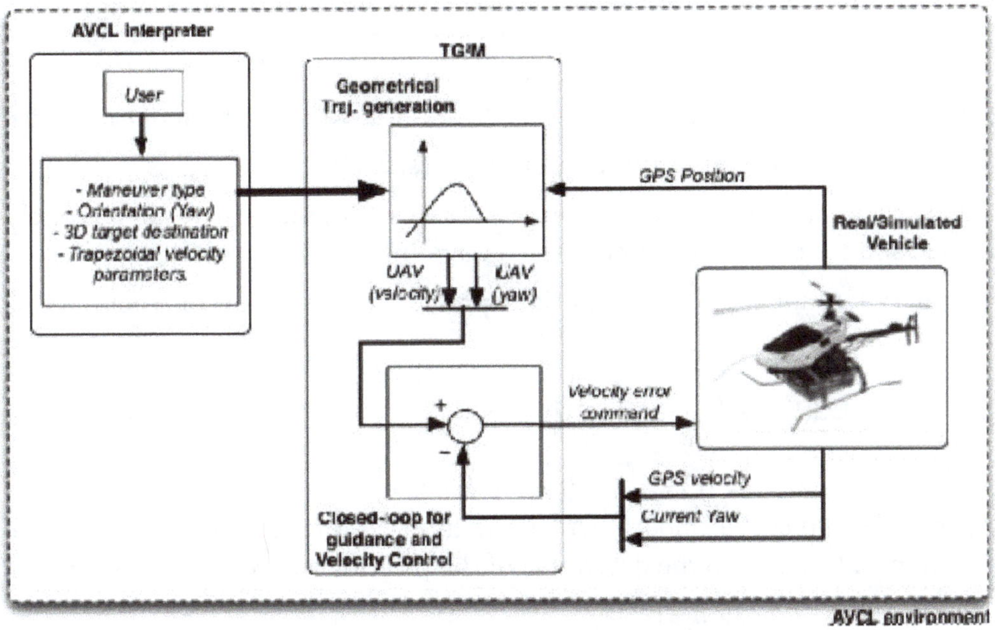

Figure 2. The TG2M framework

As shown in Fig. 2, two main modules compose the TG2M framework: the geometrical trajectory generation and the online closed-loop guidance. All the parameters and fundamental data (e.g. the desired maximum speed, etc) are provided by the user via the AVCL interpreter.

Polynomial interpolation using splines

The fundamental idea behind the spline interpolation is based on the definition of smooth curves through a number of points. These curves are represented by a polynomial function (normally of third-grade) defined by some coefficients that determine the spline used to interpolate the numerical data points. These coefficients *bend* the line so that it passes through each of the data points without any erratic behavior or breaks in continuity. The essential idea is to define a third-degree polynomial function $S(t)$ of the form:

$$S\,(t)= a_i t^3 + b_i t^2 + c_i t + d_i \qquad (1)$$

This third-degree polynomial needs to conform to the following conditions in order to interpolate the desired knot-points as depicted in Fig. 3:

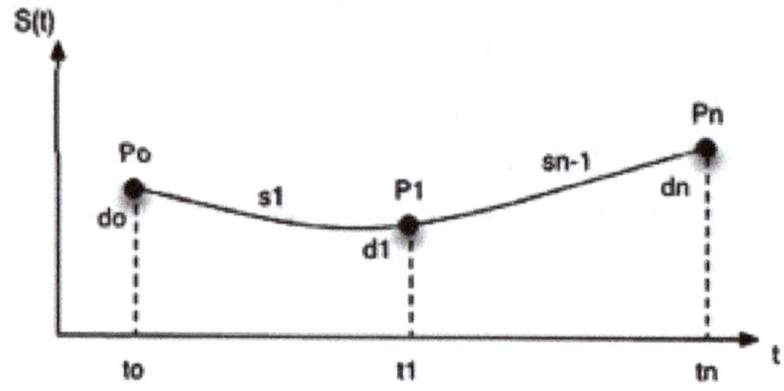

Figure 3. Knot-points to interpolate using 3D splines (with free boundary conditions) $S(t)$, $S'(t)$and $S''(t)$will be continuous on the interval $[t_0, t_n]$.

Since the curve $S(t)$must be continuous across its entire interval, it can be concluded that each sub-function must joint at the data points, so: $s_i(t_i) = s_i{-}1(t_i)$.

Taking into account the set of conditions previously described, for each $i = 1, 2.....n - 1$,

$t_i \in [t_i, t_i{+}1]$, $S(t_i) = s_i(t_i)$, and letting $h_i = \sum_{i-1}^{n-1} t_{i+1} - t_1$, Eq. (1) is re-writing as:

$$s_i(t_i) = a_i h_i^3 + b_i h_i^2 + c_i h_i + d_i \tag{2}$$

Also, to make the curve smooth across the interval, the first derivative of the $S(t)$ function must be equal to the derivative of the position reference points; this yields:

$$Si'(t_i) = 3a_i h_i^2 + 2b_i h_i + c_i$$

$$Si'(t_i) = s'_{t-1}(t_i)$$

$$Si'(t_i) = c_i = 3a_{i-1} h_i^2 + 2b_{i-1} h_i + c_{i-1} \tag{3}$$

Applying the same approach for the second derivative:

$$S_i''(t_i) = 6a_i h_i + 2b_i$$

$$S_i''(t_i) = s_{i+1}''(t_i)$$

$$S_i''(t_i) = 2b_{i+1} = 6a_i h_i + 2b_i \tag{4}$$

For the solution of the polynomial coefficients in Eq. (1), the system in Eq. (5) must be solved as:

$$A \cdot Y = f \tag{5}$$

where the matrix $A \in \Re^{nxn}$ (n is the number of knot-points to interpolate) corresponds to:

$$A=\begin{bmatrix} 1 & 0 & 0 & \cdots & \cdots & \cdots & 0 \\ h_0 & 2(h_0+h_1) & h_1 & \ddots & & & \vdots \\ 0 & h_0 & 2(h_1+h_2) & h_2 & \ddots & \ddots & \vdots \\ \vdots & \ddots & \ddots & \ddots & \ddots & \ddots & \vdots \\ \vdots & & \ddots & \ddots & \ddots & \ddots & \vdots \\ \vdots & & & \ddots & h_{n-2} & 2(h_{n-2}+h_{n-1}) & h_{n-1} \\ 0 & \cdots & \cdots & \cdots & 0 & 0 & 1 \end{bmatrix} \tag{6}$$

The term $h \in \Re^{n-1}$ in Eq. (6) is the time vector defined for each point (P_0, P_1, ... P_n). The b_i coefficients in Eq. (1) are stacked in $Y \in \Re^{nx3}$, which yields the term $f \in \Re^{nx3}$ as:

$$f=\begin{bmatrix} 0 \\ \frac{3}{h_1}(d_2-d_1)-\frac{3}{h_0}(d_1-d_0) \\ \vdots \\ \frac{3}{h_{n-1}}(d_n-d_{n-1})-\frac{3}{h_{n-2}}(d_{n-1}-d_{n-2}) \\ 0 \end{bmatrix} \tag{7}$$

From Eq. (2) and (3), the *ai* and *ci* coefficients are respectively obtained as:

$$a_i=\frac{b_{i+1}-b_i}{3h_i}$$

$$c_i=\frac{1}{h_i}(d_{i+1}-d_i)-\frac{h_i}{3}(2b_i-b_{i+1})$$

$$d_i=S(t_i) \tag{8}$$

Example 1. *Simple UAV trajectory planning using 3D splines with fixed boundary condition. Let's define three knot-points as:* P_0 = [0,0,0], P_1 = [5,10,10], P_2 = [0,10, 20] *with the following time condition for each point:* t = [0,10, 20]*(note that the time vector components are given in seconds). Calculate a 3D spline considering zero initial and final velocities.*

The natural 3D splines with free boundary conditions may generate smooth paths but without control over the velocities over the knot-points. Because of this, Eq. (6) and (7) must be complemented in order to generate a 3D spline with the fixed boundary conditions, in this case, with zero initial and final velocities. Re-writing those equations yield:

$$A=\begin{bmatrix} 2h_0 & h_0 & 0 & \cdots & \cdots & \cdots & 0 \\ h_0 & 2(h_0+h_1) & h_1 & \ddots & & & \vdots \\ 0 & h_0 & 2(h_1+h_2) & h_2 & \ddots & \ddots & \vdots \\ \vdots & \ddots & \ddots & \ddots & \ddots & \ddots & \vdots \\ \vdots & & \ddots & \ddots & \ddots & \ddots & \vdots \\ \vdots & & & \ddots & h_{n-2} & 2(h_{n-2}+h_{n-1}) & h_{n-1} \\ 0 & \cdots & \cdots & \cdots & 0 & h_{n-1} & 2h_{n-1} \end{bmatrix} \tag{9}$$

The same system $A \cdot Y = f$ must be solved with the new $A \in \Re^{nxn}$ from Eq. (9) and $f \in \Re^{nx3}$ defined in Eq. (10). Note that the first derivative of the $S(t)$function has been added in the first and last position of the f vector, allowing the control of the initial and final velocities of the curve.

The first procedure is to obtain the time vector as follows: $h=\sum_{i=1}^{n-1} t_{i+1}-ti$ =[10 10], then the system: $Y = A{-}1\, f$ must be solved to obtain the polynomial coefficients as:

$$A=\begin{bmatrix} 20 & 10 & 0 \\ 10 & 40 & 10 \\ 0 & 10 & 20 \end{bmatrix}, \quad Y_x=\begin{bmatrix} b_{0x}=0.15 \\ b_{1x}=-0.15 \\ b_{2x}=0.15 \end{bmatrix}, \quad Y_y=\begin{bmatrix} b_{0y}=0.15 \\ b_{1y}=0 \\ b_{2y}=-0.15 \end{bmatrix}, \quad Y_z=\begin{bmatrix} b_{0z}=0.1125 \\ b_{1z}=0.0750 \\ b_{2z}=-0.2625 \end{bmatrix}$$

$$f_x=\begin{bmatrix} 1.5 \\ -3 \\ 1.5 \end{bmatrix} \qquad f_y=\begin{bmatrix} 3 \\ 0 \\ -3 \end{bmatrix} \qquad f_z=\begin{bmatrix} 3 \\ 1.5 \\ -4.5 \end{bmatrix} \tag{10}$$

Finally, the two polynomials for each (x, y, z) component are evaluated from the time [0, 10] to [10, 20]. Figure 4 shows the results.

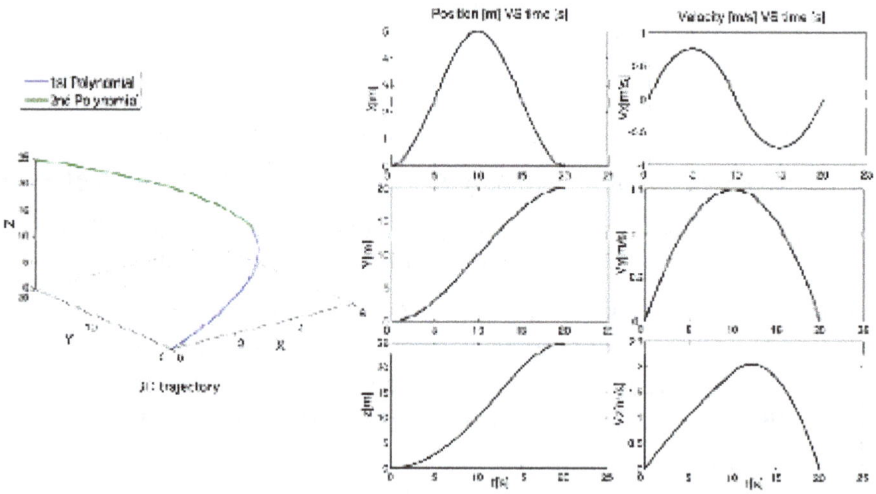

Figure 4. 3D spline with fixed boundary condition

The end of Example 1: Simple UAV trajectory planning using 3D splines

In the previous example, the 3D splines with fixed boundary conditions allowed to define a smooth curve across the knot-points. Nevertheless, two basic problems must be taken into account: the 3D spline only allows the user to establish the initial and the final velocities of the whole trajectory, limiting the user to have total control over the other points. The second problem is the smoothness of the acceleration curves (linear). To solve these problems 4D splines must be used. These address the user total control over the velocity profile across the whole trajectory and its results in additional *smoothness* for position, velocity and even acceleration curves. This could be more effective for UAVs tasks where the mission requires a strong control of the vehicle acceleration and velocities at each defined knot-point.

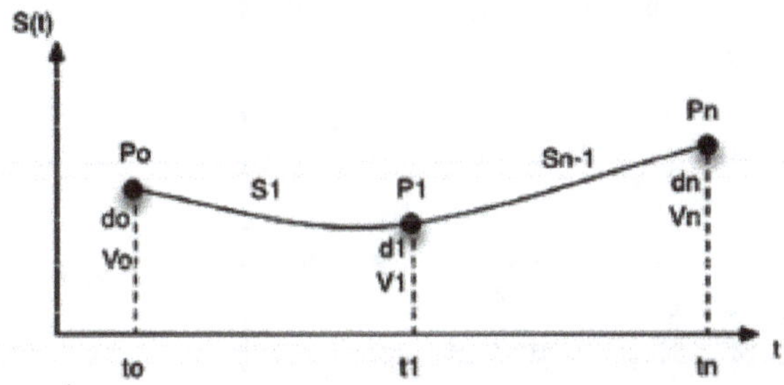

Figure 5. Knot-points to interpolate with 4D splines

The fourth-degree polynomial is defined by:

$$S(t)= e_i t^4 + a_i t^3 + b_i t^2 + c_i t + d_i \tag{11}$$

The same 3D-spline conditions previously described also apply to the 4D-spline. To obtain a generalized solution of the system to solve ($A \cdot Y = f$), we start from the three-point case as depicted in Fig. 5. The polynomials for each trajectory segment as a function of time *t* are:

$$\begin{aligned}
S_n(t_0) &= e_0 t_0^4 + a_0 t_0^3 + b_0 t_0^2 + c_0 t_0 + d_0 = f(t_0) \\
S_n(t_1) &= e_0 t_1^4 + a_0 t_1^3 + b_0 t_1^2 + c_0 t^1 + d_0 = f(t_1) \\
S_{n-1}(t_1) &= e_1 t_1^4 + a_1 t_1^3 + b_1 t_1^2 + c_1 t_1 + d_1 = f(t_1) \\
S_{n-1}(t_1) &= e_1 t_n^4 + a_1 t_n^3 + b_1 t_n^2 + c_1 t_n + d_1 = f(t_n)
\end{aligned} \tag{12}$$

Taking the first and second derivatives (velocities and accelerations), we obtain:

$$\begin{aligned}
S_1'(t_0) &= 4e_0 t_0^3 + 3a_0 t_0^2 + 2b_0 t_0 + c_0 = V_0 \\
S_1'(t_1) &= 4e_0 t_1^3 + 3a_0 t_1^2 + 2b_0 t_1 + c_0 = V_1 \\
S'_{n-1}(t_1) &= 4e_1 t_1^3 + 3a_1 t_1^2 + 2b_1 t^1 + c_1 = V_1 \\
S'_{n-1}(t_1) &= 4e_1 t_n^3 + 3a_1 t_n^2 + 2b_1 t_n + c_1 = V_1
\end{aligned} \tag{13}$$

The second derivatives of Eq. (12) yield a set of accelerations. Equaling the acceleration functions for the intermediate points ($t1$ for each case) and setting to zero the initial and final acceleration of the path segment yields:

$$\begin{aligned} S_0''(t_0) &= 12e_0t_0^2+6a_0t_0+2b_0 = s'_{n-1}{}''(t_1) = 12e_1t_1^2+6a_1t_1+2b_1 \\ S_0''(t_0) &= 12e_0t_1^2+6a_0t_1+2b_0 = s'_{n-1}{}''(t_1) = 12e_1t_n^2+6a_1t_1+2b_1 \end{aligned} \quad (14)$$

Equations (12), (13) and (14) conform the complete system in order to obtain the ten polynomial poefficients. Solving $A \cdot Y = f$, the matrix $A\ Y \in \Re 5^{(n-1)\times 5(n-1)}$ corresponds to:

$$A = \begin{bmatrix} 1 & t_0 & t_0^2 & t_0^3 & t_0^4 & 0 & 0 & 0 & 0 & 0 \\ 1 & t_1 & t_1^2 & t_1^3 & t_1^4 & 0 & 0 & 0 & 0 & 0 \\ 0 & 0 & 0 & 0 & 0 & 1 & t_1 & t_1^2 & t_1^3 & t_1^4 \\ 0 & 0 & 0 & 0 & 0 & 1 & t_2 & t_2^2 & t_2^3 & t_2^4 \\ 0 & 1 & 2t_0 & 3t_0^2 & 4t_0^3 & 0 & 0 & 0 & 0 & 0 \\ 0 & 1 & 2t_1 & 3t_1^2 & 4t_1^3 & 0 & 0 & 0 & 0 & 0 \\ 0 & 0 & 0 & 0 & 0 & 0 & 1 & 2t_1 & 3t_1^2 & 4t_1^3 \\ 0 & 0 & 0 & 0 & 0 & 0 & 1 & 2t_2 & 3t_2^2 & 4t_2^3 \\ 0 & 0 & 2 & 6t_1 & 12t_1^2 & 0 & 0 & -2 & -6t_1 & -12t_1^2 \\ 0 & 0 & 2 & 6t_0 & 12t_0^2 & 0 & 0 & -2 & -6t_2 & -12t_2^2 \end{bmatrix} \quad (15)$$

The polynomial coefficients are stacked into the $Y \in \Re^{5(n-1)}$ vector, and with the $f \in \Re^{5(n-1)}$ term, the total system is defined from Eq. (14) and (15):

$$\begin{aligned} Y &= \begin{bmatrix} d_0 & c_0 & b_0 & a_0 & e_0 & d_1 & c_1 & b_1 & a_1 & e_1 \end{bmatrix}^T \\ f &= \begin{bmatrix} f(t_0) & f(t_1) & f(t_1) & f(t_n) & V_0 & V_1 & V_1 & V_n & 0 & 0 \end{bmatrix}^T \end{aligned} \quad (16)$$

Example 2. *UAV trajectory planning using 4D splines with total velocity control*

Lets define three knot-points as: P0 = [0,0,0], P1 = [5,7, 5], P2 = [4 , 2,10] *in the time* t = [0, 4, 8] *(note that the time vector components are given in seconds). Calculate the 4D spline considering the following velocity profile:* V0 = [0,0,0], V1 = [0.5,0.8,1], V2 = [1,1.5,1].

Solving the system: $Y = A{-}1\, f$, we obtain the following numerical data:

$$A = \begin{bmatrix} 1 & 0 & 0 & 0 & 0 & 0 & 0 & 0 & 0 & 0 \\ 1 & 4 & 16 & 64 & 256 & 0 & 0 & 0 & 0 & 0 \\ 0 & 0 & 0 & 0 & 0 & 1 & 4 & 16 & 64 & 256 \\ 0 & 0 & 0 & 0 & 0 & 1 & 8 & 64 & 512 & 4096 \\ 0 & 1 & 0 & 0 & 0 & 0 & 0 & 0 & 0 & 0 \\ 0 & 1 & 8 & 48 & 256 & 0 & 0 & 0 & 0 & 0 \\ 0 & 0 & 0 & 0 & 0 & 0 & 1 & 8 & 48 & 256 \\ 0 & 0 & 0 & 0 & 0 & 0 & 1 & 16 & 192 & 2048 \\ 0 & 0 & 2 & 24 & 192 & 0 & 0 & -2 & -24 & -192 \\ 0 & 0 & 2 & 0 & 0 & 0 & 0 & -2 & 0 & 0 \end{bmatrix}, \quad Y_{x,y,z} = \begin{bmatrix} d_{0x} & d_{0y} & d_{0z} \\ c_{0x} & c_{0y} & c_{0z} \\ b_{0x} & b_{0y} & b_{0z} \\ a_{0x} & a_{0y} & a_{0z} \\ e_{0x} & e_{0y} & e_{0z} \\ d_{1x} & d_{1y} & d_{1z} \\ c_{1x} & c_{1y} & c_{1z} \\ b_{1x} & b_{1y} & b_{1z} \\ a_{1x} & a_{1y} & a_{1z} \\ e_{1x} & e_{1y} & e_{1z} \end{bmatrix} \quad f_{x,y,z} = \begin{bmatrix} 0 & 0 & 0 \\ 5 & 7 & 5 \\ 5 & 7 & 5 \\ 4 & 2 & 10 \\ 0 & 0 & 0 \\ 0 & 0 & 0 \\ 0 & 0 & 0 \\ 0 & 0 & 0 \\ 0 & 0 & 0 \\ 0 & 0 & 0 \end{bmatrix} \quad f_{x,y,z} = \begin{bmatrix} 0 & 0 & 0 \\ 5 & 7 & 5 \\ 5 & 7 & 5 \\ 4 & 2 & 10 \\ 0 & 0 & 0 \\ 0 & 0 & 0 \\ 0 & 0 & 0 \\ 0 & 0 & 0 \\ 0 & 0 & 0 \\ 0 & 0 & 0 \end{bmatrix}$$

Finally, the two polynomials for each (x, y, z) component are evaluated from the time [0, 4] to [4, 8] seconds. Figure 6 shows the results:

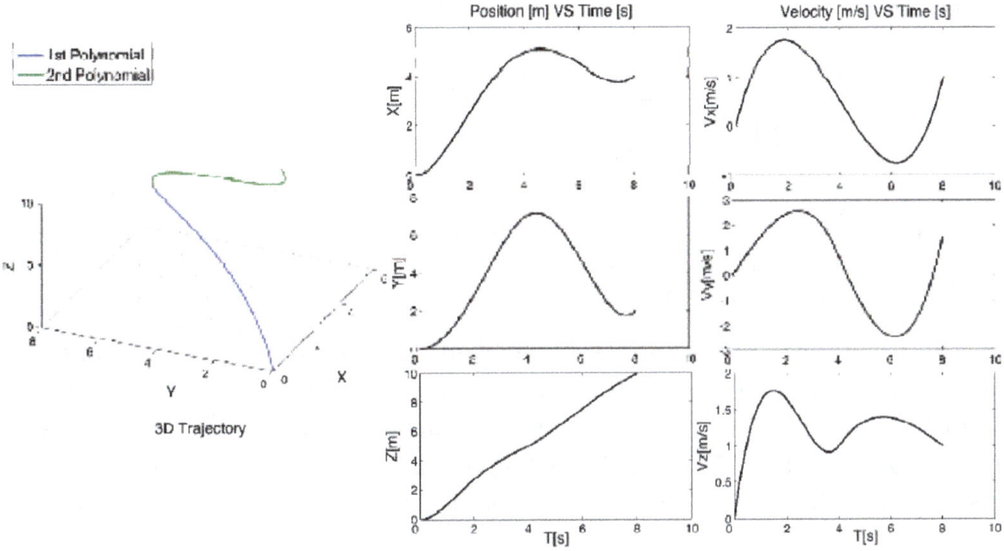

Figure 6. 4D spline with total control of the knot-points velocities

The end of Example 2: UAV trajectory planning using 4D splines with total velocity control

3.2 Simple Maneuvers with Analytical Functions

The TG2M framework facilitates the specification of straight lines and circles through analytical functions that delineate the geometry of the intended trajectory for straightforward movements. These limitations are predominantly kinematic: constant velocity and acceleration, trapezoidal profiles for acceleration or deceleration, etc. This parameterization is advantageous when the mission necessitates that the UAV effectively halts at the intended endpoint of the trajectory, owing to user control over the acceleration slope tilt level. The required set of velocities for both lines and circumferences must adhere to the specified trapezoidal velocity profile:

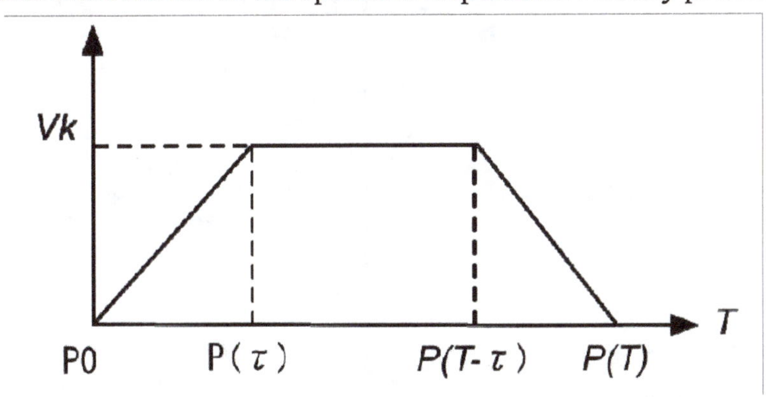

Figure 7. Trapezoidal velocity profile used for simple UAV maneuvers

From Fig. 7, three fundamental segments compose the total function to define the straight- line. The two intermediate points of the trapezoidal curve: $P(\tau)$, $P(T-\tau)$are calculated as [1]:

$$P(\tau)_{x,y,z} = \frac{1}{2\left\|P(T)_{x,y,z} - P_{0x,y,z}\right\|}\frac{V_k}{t}t^2\left(P(T)_{x,y,z} - P_{0x,y,z}\right) + P_{0x,y,z}$$

$$P(T-\tau)_{x,y,z} = \frac{V_k(T-2t)}{\left\|P(T)_{x,y,z} - P_{0x,y,z}\right\|}\left(P(T)_{x,y,z} - P_{0x,y,z}\right) + P(\tau)_{x,y,z}$$

(17)

Once that the intermediate points have been defined, the three segments that compose the total function are defined by Eq. (18). In the first section (from P0 to $P(\tau)$) the initial velocity is set $V_i = 0$ and progresses toward a final velocity V_k. The second segment (from $P(\tau)$ to $P(T-\tau)$) is traced at constant maximum velocity V_k. Finally the last segment (from $P(T-\tau)$to $P(T)$) drives the vehicle from V_k to zero velocity.

$$f(t) = \begin{cases} \frac{V_k}{2\tau(P(\tau)-P_0(}t^2(P(\tau)-P_0)+P_0 & 0 \le t \le \tau \\ \frac{V_k}{(P(T-\tau)-P(\tau)(}(t-\tau)(P(T-\tau)-P(\tau))+P(\tau) & \tau < t \le T-\tau \\ \frac{(t-T+\tau)}{(P(T)-P(T-\tau)(}\left(-\frac{V_k}{2\tau}(t-T+\tau)+V_k\right)(P(T)-P(T-\tau))+P(T-\tau) & T-\tau < t \le T \end{cases}$$

(18)

The same approach is applied to generate circumferences with the trapezoidal profile used for the straight-lines. Three knot-points define the circumference path and the objective is to find the trace angle across the trajectory.

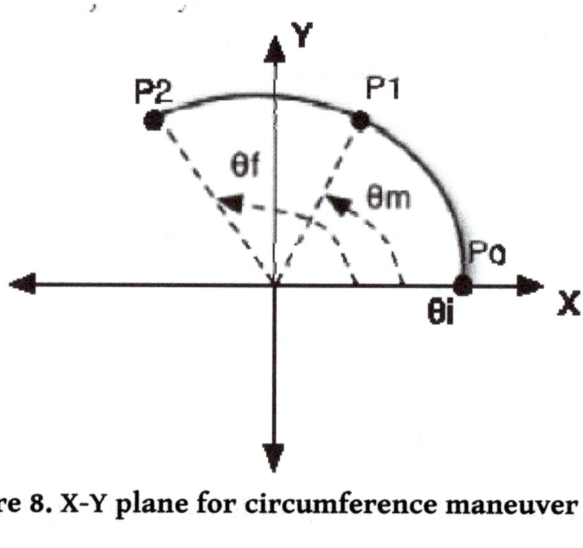

Figure 8. X-Y plane for circumference maneuver

The first step is to determine the center of the circle that joins the three knot-points. The equation of the trajectory plane is defined by the cross product between the vectors formed by points $P0$, $P1$ and $P2$, as shown in Eq. (19).

$$\overrightarrow{P_0 P_1} \times \overrightarrow{P_0 P_2} = N(A,B,C)$$
$$Ax + By + Cz = cte \tag{19}$$

The center of the circle $c(x_c, y_c, z_c)$ is always equidistant to any point over the circle[2], then:

$$(x_{0,1,2} - x_c)^2 + (y_{0,1,2} - y_c)^2 + (z_{0,1,2} - z_c)^2 = K_2$$
$$Ax_c + By_c + Cz_c = cte \tag{20}$$

To obtain the relation between the angle motion (θ), the angular velocity (ω) and the tangential velocity (vt), constant velocity is assumed across the path, yielding:

$$\omega = \frac{d\theta}{dt} = \frac{v_t}{r}$$
$$d\theta = \frac{v_t}{r} dt$$
$$\int d\theta = \int \frac{v_t}{r} dt$$
$$\theta(t) = \frac{v_t}{r} t + \theta_i \quad \text{where} \quad t \in [0,T] \tag{21}$$

Likewise, the relation between the angle motion (θ) and the angular acceleration of the curve is given by:

$$at + v_0 = r\frac{d\theta}{dt}$$
$$d\theta = \frac{at + v_0}{r} dt$$
$$\int d\theta = \int \frac{at + v_0}{r} dt$$
$$\theta(t) = \frac{a}{2r} t^2 + \frac{v_0}{r} t + \theta_i \quad \text{where} \quad t \in [0,T] \tag{22}$$

If the known parameter is the total motion time (T) of the trajectory, we set the acceleration term to the left-side of the equation, yielding:

$$\theta(t) = \theta_f = \frac{a}{2r} t^2 + \frac{v_0}{r} t + \theta_i$$
$$a = \frac{2\left[(\theta_f - \theta_i) - v_0 T\right]}{T^2} \tag{23}$$

Otherwise, if the constrained parameters are the initial and final velocity, the acceleration function is given by:

$$a = \frac{v_{ft} - v_{0t}}{T}$$

$$T = \frac{2(\theta_f - \theta_i)}{v_{0t} + v_{ft}} \tag{24}$$

Example 3. UAV straight-line trajectory with trapezoidal velocity profile

Lets defined a straight-line trajectory starting from P_0 = [0,0, 20] *to* $P(T)$= [10,0,20] *with 10 meters of displacement in the X coordinate at a constant altitude of 20 meters. Define the 3D cartesian trajectory with a maximum velocity of 1m/s taking into account a trapezoidal velocity profile with 30% of acceleration time.*

The distance vector to trace is: $dist = P(T) - P_0$ = [10,0,0].

The total velocity is obtained from the norm of the vector *dist*: $V_T = \|dist\| = 10.$

The total motion time is calculated as: $T = \frac{VT}{V_k\ (1 - pt)} = 7$ where *Vk* is the maximum velocity at 1m/s and *pt* is the percentage of acceleration (30%=0.3).

The acceleration motion is also obtained as: $a = \frac{V_k}{T * pt} = 0.2381$.

The number of the total points to generate is: n = 28 (the user defines this parameter depending on the UAV controller data rate). In the same way, the number of points in the first and last segment of the trapezoidal profile is calculated as: $k = 2(pt \cdot n) = 16$, and for the constant velocity segment: $kc = n - 2k = 12$. Once that the preliminary data have been calculated, Eq. (16) is used to obtain the intermediate points (see Fig. 7):

$$Px(\tau) = \frac{1}{2V_t} at^2 (dist)_x + P_{0x} = 2.10 \qquad Px(T-\tau) = \frac{V_k(T-2t)}{V_T}(dist)_x + P(\tau)_x = 7.7$$

$$Py(\tau) = \frac{1}{2V_t} at^2 (dist)_y + P_{0y} = 0 \qquad Py(T-\tau) = \frac{V_k(T-2t)}{V_T}(dist)_y + P(\tau)_y = 0$$

$$Pz(\tau) = \frac{1}{2V_t} at^2 (dist)_z + P_{0z} = 20 \qquad Pz(T-\tau) = \frac{V_k(T-2t)}{V_T}(dist)_z + P(\tau)_z = 20$$

The first segment of the path (from P0 to $P(\tau)$) is calculated using Eq. (17):

$$fx(t)=\sum_{i=1}^{k/2}\frac{V_k}{2t\left\|P(t)-P_0\right\|_x}t_i^2\left(P(t)_x-P_{0x}\right)+P_{0x}$$

$$fy(t)=\sum_{i=1}^{k/2}\frac{V_k}{2t\left\|P(t)-P_0\right\|_y}t_i^2\left(P(t)_y-P_{0y}\right)+P_{0y}\quad 0\leq t\leq\tau$$

$$fz(t)=\sum_{i=1}^{k/2}\frac{V_k}{2t\left\|P(t)-P_0\right\|_v}t_i^2\left(P(t)_z-P_{0z}\right)+P_{0z}$$

The time vector t is obtained as: $t=\left[0:\frac{1}{V_k}\frac{2\left\|P(\tau)-P_0\right\|}{k}:\frac{1}{V_k}2\left\|P(\tau)-P_0\right\|\right]$ which is:

t = [0 0.5250 1.0500 1.5750 2.1000 2.6250 3.1500 3.6750 4.2000].

For the second segment (at constant velocity V_k), the time vector is calculated as:

$t=\left[0:\frac{1}{n-2k}:1\right]$ = [0 0.0833 0.1667 0.25 0.3333 0.4167 0.5 0.5833 0.6667 0.75 0.83 0.9167]

$$fx(t)=\sum_{i-1}^{kc}t_i\left(P(T-\tau)_x-P(\tau)_x\right)+P(\tau)_x$$

$$fy(t)=\sum_{i-1}^{kc}t_i\left(P(T-\tau)_y-P(\tau)_y\right)+P(\tau)_y\quad \tau<t\leq T-\tau$$

$$fz(t)=\sum_{i-1}^{kc}t_i\left(P(T-\tau)_z-P(\tau)_z\right)+P(\tau)_z$$

Finally the last segment drives the vehicle from Vk to zero end velocity. The time vector is:

$$t=\left[0:\frac{1}{V_k}\frac{2\left\|P(T-\tau)-P(\tau)\right\|}{k}:\frac{1}{V_k}2\left\|P(T-\tau)-P(\tau)_0\right\|\right]=\begin{bmatrix}0 & 0.287 & 0.575 & 0.862 & 1.15 & 1.43\\ 1.72 & 2.01 & 2.3\end{bmatrix}$$

yielding the following function for each coordinate of motion:

$$f(t)_x=\sum_{i=1}^{k/2}\frac{-V_kt_i^2+V_kt_i}{2\left\|P(T)-P(T-t)\right\|_x}\left(P(T)_x-P(T-t)_x\right)+P(T-t)_x$$

$$f(t)_y=\sum_{i=1}^{k/2}\frac{-V_kt_i^2+V_kt_i}{2\left\|P(T)-P(T-t)\right\|_y}\left(P(T)_y-P(T-t)_y\right)+P(T-t)_y\quad T-t<t\leq T$$

$$f(t)_z=\sum_{i=1}^{k/2}\frac{-V_kt_i^2+V_kt_i}{2\left\|P(T)-P(T-t)\right\|_z}\left(P(T)_z-P(T-t)_z\right)+P(T-t)_z$$

The results are presented in Fig. 9:

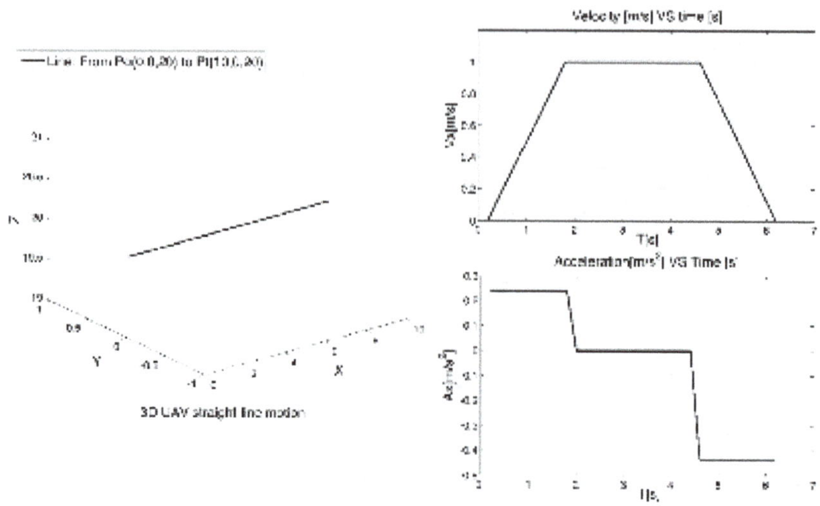

Figure 9. UAV straight-line motion with trapezoidal velocity profile

The end of Example 3: UAV straight-line trajectory with trapezoidal velocity profile

3.3 UAV Guidance

The geometrical representation of a UAV trajectory has been presented in the previous sub-section. Complex trajectories may be described using third-fourth order degree splines, or simple maneuvers may be generated using common lines and circumferences functions with some parametric features defined by the end-user. But in order to complete the generation of trajectories for a single UAV a guidance module is required.

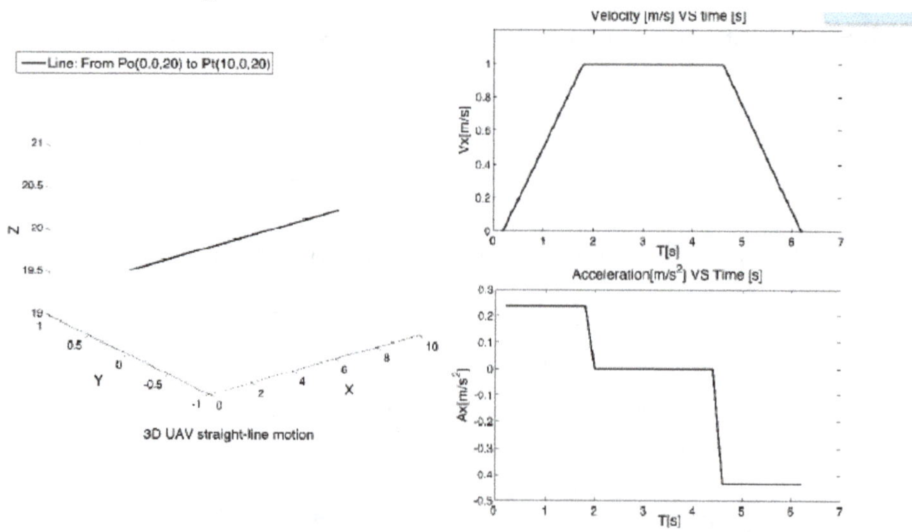

Figure 10. TG2M guidance scheme

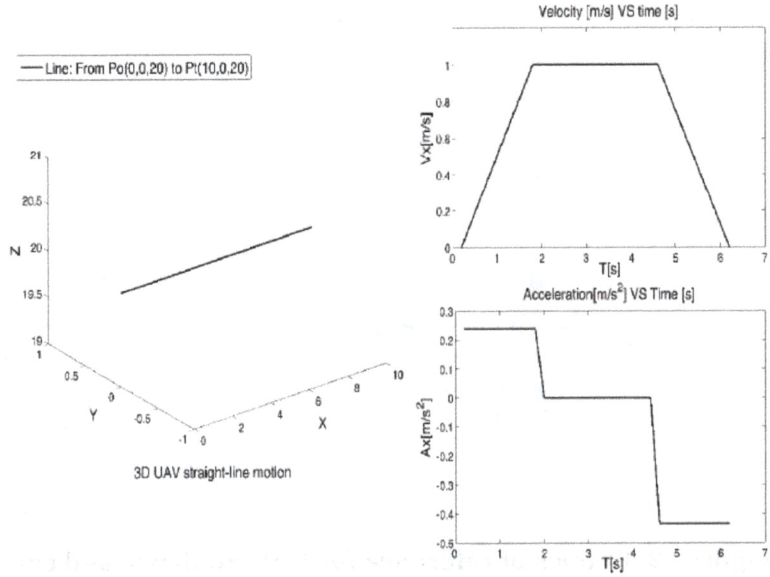

Figure 11. Velocity error command

The AVCL simulation module contains a dynamic model of a mini-helicopter (Valero, 2005) and its associated controllers: attitude, velocity and position. The velocity controller is based

on a Proportional-Integrative "PI" design; it receives the *velocity error command* generated by the guidance module shown in Fig. 10. These velocity error references are named *Vxref* and *Vyref*, with respect to either the world or vehicle's frame, and are calculated with the UAV's error position. In Fig. 11 the theoretical (or ideal) UAV velocity vector is represented by the *Vt* term. Due to wind and other perturbations during flight, a velocity error vector *Ve* must be considered in order to set the final velocity references to send to the vehicle's controller. This vector is derived from the UAV position error between the current and desired positions. Finally, the velocity references *Vxref* and *Vyref* are the components of the vector *Vt* +*Ve*.

The guidance module must take into account that the helicopter's orientation (yaw) is *independent* from the direction of travel. The high-level modules must generate a set of orientations across the vehicle's trajectory. The built-in AVCL control module (see Fig. 10) is capable of receiving *velocity* and *yaw* angle orientation commands from the TG2M module and generating the needed commands for the attitude controller that governs the vehicle's roll and pitch. The TG2M's guidance module focuses on the generation of *yaw* references, and use a simple Proportional "P" controller for smooth *yaw* angle transition during the flight. Two methods are use to generate the *yaw* angle references: for simple maneuvers the *yaw* angle is calculated using simple trigonometric relations due to the path displacement. For complex trajectories using splines we introduce a feasible method to calculate a smooth *yaw* angle evolution. This method also shows how to calculated roll and pitch references due to the vehicle trajectory and its velocity. For *roll, pitch* and *yaw* angles calculation, the following frame of reference is used:

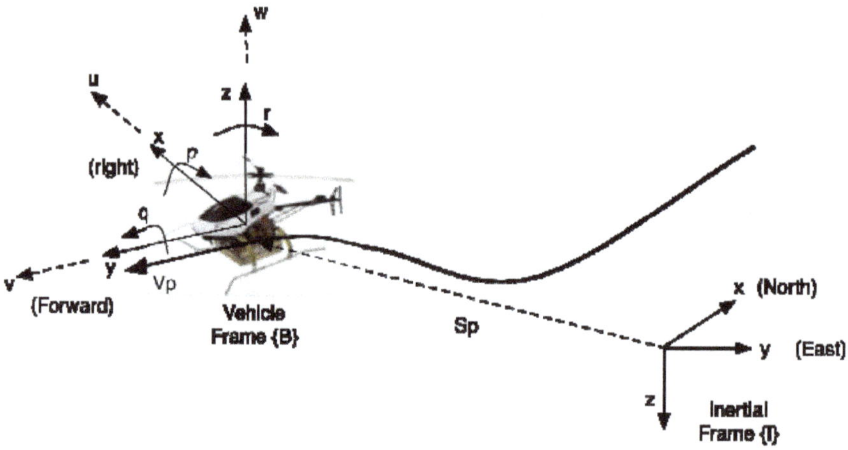

Figure 12. Frames of reference for UAV guidance and control

From Fig. 12, the following vectors can describe the motion of the UAV vehicle in 6 DOF:

$$h = [h_1, h_2]^T = [x, y, z, f, q, y]^T$$

$$v = [v_1, v_2]^T = [u, v, w, p, q, r]^T \tag{25}$$

In Eq. (25), h_1 denotes the position of the center of mass CM of the vehicle and *h* 2 its orientation described by the *Euler angles* with respect to the inertial frame {I}. The vector *v*1 refers to the linear velocity and *v*2 to the angular velocity of vehicle frame {B} with respect to inertial frame {I}. In order to express the velocity quantities between both frames of references (from {B} to {I} and vice versa), the following transformation matrix is used3:

$$\eta_1' = R_B^I v_1$$

$$\eta_1' = \begin{bmatrix} c\psi c\theta & -s\psi c\phi + c\psi s\theta s\phi & s\psi s\phi + c\psi c\theta s\phi \\ s\psi c\theta & c\psi c\phi + s\phi s\theta s\psi & -c\psi s\phi + s\theta s\psi c\phi \\ -s\theta & c\theta s\phi & c\theta c\phi \end{bmatrix} v_1 \tag{26}$$

The body-fixed angular velocities and the rate of the Euler angles are related through:

$$\eta_2' = \begin{bmatrix} 1 & s\phi t\theta & c\phi t\theta \\ 0 & c\phi & -s\phi \\ 0 & s\phi / c\theta & c\phi / c\theta \end{bmatrix} v_2 \tag{27}$$

The position and the magnitude of the velocity vector at a point *P* on the trajectory are given by 4:

$$S_p = [x, y, z]^T$$
$$V_p = \|S_p'\| = \sqrt{x'^2 + y'^2 + z'^2} \tag{28}$$

The method to define the Euler angles is based on the Frenet-Serret theory (Angeles, 1997). To every point of the curve we can associate an orthonormal triad of vectors (a set of unit vectors that are mutually orthogonal) namely the tangent *et*, the normal *en* and the bio- normal *eb* (see Fig. 13). The Frenet-Serret theory says that by properly arranging these vectors in a matrix ∈ℜ3*x* 3 , we obtain a description of the curve orientation due to the position, velocity and acceleration of the UAV while tracing out the path. The unit vectors are then defined as:

$$e_t = \frac{S_p'}{V_p}, \quad e_b = \frac{(S_p' \times S_p'')}{\|S_p' \times S_p''\|}, \quad e_n = e_b \times e_t \tag{29}$$

In the definition of a frame associated with the curve the original definition of the Frenet frame for counterclockwise rotating curves is used; in the case of a clockwise rotating curve, the zDaxis of the Frenet frame points in the opposite direction upwards than the inertial {I} frame. So in order to define small relative rotation angles for the orientation of a vehicle rotating clockwise and having its *zb* axis pointing downwards, we define a reference frame associated with the curve as previously, but rotated with respect to the Frenet by an angle of 180 degrees about the *x*-axis of the Frenet frame (see Fig. 13).

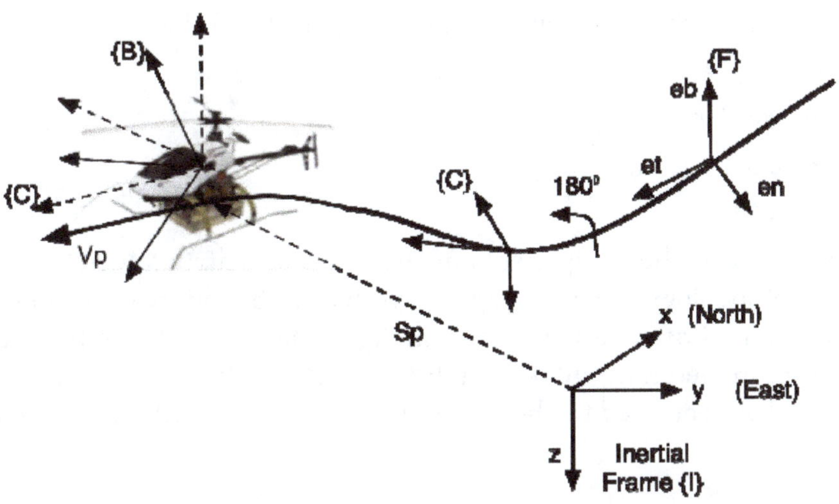

Figure 13. The inertial, the Frenet, the vehicle and the curve Frames

Collectively we denote the Frenet and the rotated frame as the "curve" frame {C}. According to the notation of rotational transformations used in robotics literature, we can express the coordinates of a vector given in the curve frame {C} to the {I} frame with the matrix:

$$R_C^I = \begin{bmatrix} e_t & e_n & e_b \end{bmatrix}$$

$$R_I^C = R_C^{I\ T} \tag{30}$$

For a counterclockwise rotation: $RC = Rx\ (180°)\ [e_t\ e_n\ e_b]$. Likewise, the rotation of the {B} frame from the {C} frame to the reference {R} frame can be expressed using customary aeronautical notation by considering the sideslip angle β and angle of attack α, (Siouris, 2004):

$$\beta = \sin^{-1}\left(\frac{v_R}{V_p}\right)$$

$$\alpha = \tan^{-1}\left(\frac{w_R}{u_R}\right) \tag{31}$$

The vector v_R refers to the *y*-axis velocity component in the reference frame and w_R, u_R to the *z* and *x* - axis respectively. The overall rotation is composed by a rotation about body z_B axis through the angle β , followed by a rotation about the body y_B through the angle β , which is expressed as:

$$RC = R_y^T(\tau)R_z^T(-\beta) \tag{32}$$

Finally, the *roll, pitch* and *yaw* angles can be deduced as follows:

$$R_I^R = R_C^R R_I^C$$

$$\phi = atan2\ (r_{23}\ , r_{33})$$

$$\theta = atan2\ (-r_{13}\ ,\ \sqrt{r^2_{23} + r^2_{33}})$$

$$\psi = atan2\ (r_{12},\ r_{33}) \tag{33}$$

Where *r* represent the components of the rotation matrix $R_I^R \in \Re^{3x3}$.Computing the previous methodology, we use 3D splines with fixed boundary conditions in order to generate a complicated path as shown in Fig. 11. Seven knot-points have been used (distance are in meters): P=[0 0 0; 5 1 2; 10 5 5; 15 10 10; 10 15 15; 5 10 20; 0 8 20; 5 0 20]. Eq. (33) has been used to obtain the UAV orientation with respect to the Inertial Frame as:

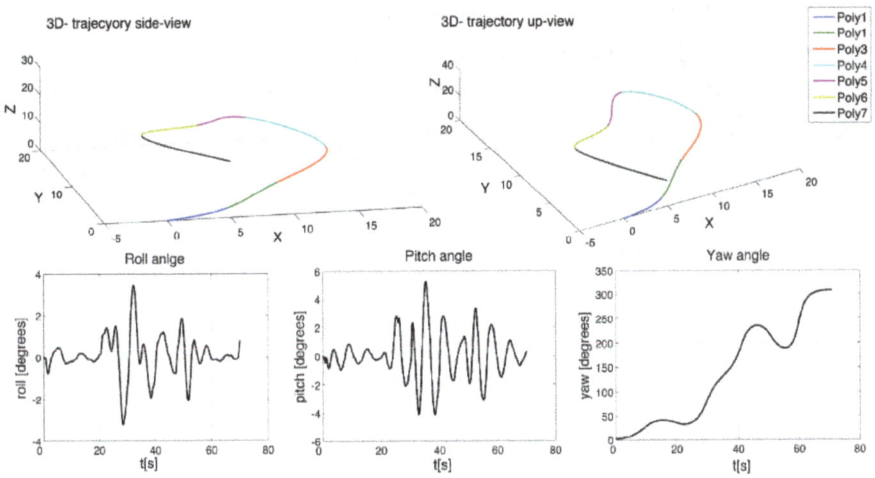

Figure 14. Roll, Pitch and Yaw angle references for UAV guidance

For complex maneuvers the Frenet theory allowed to define smooth *yaw* references as well as *roll* and *pitch* angles if it is required. Nevertheless, the computational cost of calculating those equations could decrease the system performance if the number of knot-points of the path is large. For this reason, normal maneuvers such as straight-lines use simple trigonometric theory to obtain the UAV orientation. On the other hand, the end-user is able to define the kind of orientation of the vehicle, this means that the vehicle is not constrained to be oriented just by the trajectory direction, hence, it will be able to trace out the trajectory oriented towards to any fixed point defined. The *yaw* angle defined by the term *y* is given by:

$$\psi = \tan^{-1}\left(\frac{y_{diff}}{x_{diff}}\right)$$

Where x_{diff}, y_{diff} correspond to the difference between the target fixed point and the current position of the UAV. In addition, depending of the motion quadrant, the term *y* in Eq. (34) must be fixed, this means that for the {x-,y+} and the {x-,y-} quadrant, the *yaw* angle is

$y = y^{+}$, otherwise, for the $\{x^{+},y^{-}\}$ quadrant, $y = y + 2\pi$.

Figure 15 shows a circumference-arc generated using Eq. (24) as well as the *yaw* evolution of the UAV oriented towards the center of the arc located at [0,0] coordinate in the *x-y* plane using the previosly theory described in Eq. (34).

This section has successfully introduced the mathematical treatment and methods for the generation of complex trajectories and simple maneuvers using the available theory reported in specialized literature (LaValle S.M, 2006), (Jaramillo-Botero et al., 2004). Geometrical trajectory generation and some techniques for its parameterization based on polynomial splines and function with trapezoidal velocity profile are an interest solution for this problem, actually, some of these methodologies are used for complex UAV trajectory definition nowadays. The novel solution presented in this book is the integration of these methods into a powerful environment that allows high-level user control of complex missions. The AVCL definitively brings those features and the next section will introduce some tests using the AVCL interpreter and the simulation environment.

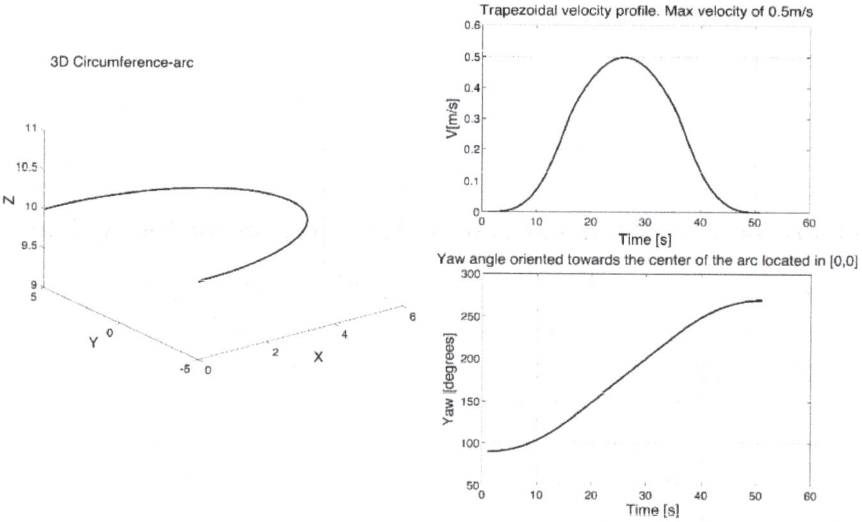

Figure 15. Circumference 3D motion, trapezoidal velocity profile and yaw angle evolution

4. TG2M SIMULATION RESULTS

As shown in Fig. 1 the Mission Planner (MP) has two similar loops for mission planning and simulation/supervision. The difference is that in the Planning Loop the interpreter sends the projected waypoints back to the MP's Enhanced Reality, while in the Simulation Loop the interpreter commands the simulated vehicle, which in turn sends the simulated positions to

the MP. Our research group has developed a Simulink-based model of a UAV helicopter named *Vampira*, which includes a position controller and is capable of real-time simulation. This simulator has been used with the Mission Planning and Simulation tool to test the TG2M. For Mission Supervision the AVCL commands would be sent to the real vehicle, and its position plotted alongside the projected and/or simulated paths. The *Vampira* helicopter was built within the framework of the project: "Guidance and Control of an Unmanned Aerial Vehicle" DPI 2003 01767 (VAMPIRA), granted by the Spain Ministry of Science and Technology, and it will be used for the real-world tests of the built-in TG^2M framework. Figure 16 shows the *Vampira* prototype, which includes: a GPS, Wi-Fi link, IMU, and a PC104 computer for the low-level control (main rotor and tail servos). The Vampira's dynamics model has been obtained, identified and validated in previous works (Valero, 2005), (del Cerro et al., 2004). This work takes advantage from the AVCL simulation capabilities in order to validate the TG²M framework theory for trajectory planning.

Figure 16. The Vampira's Helicopter prototype

Three test scenarios showcase the TG²M validation process. These tests involve the whole methodology previously presented in the other sections of this chapter, as well as the numerical simulation results using the AVCL environment and the embedded dynamics and control algorithms for the Vampira's helicopter. Two complex maneuvers are presented using 3D and 4D splines respectively and a simple last test using analytical function to generate a parameterized circumference motion.

1). *Semi-spiral using 3D splines for the velocity profile generation and the Frenet theory for UAV orientation:* In this first test, we used a 3D spline to joint three knot control points: (P_0(0, 0, 0), P_1(3, 5, 10), P_2(6, -7, 20)) at the desire time (given in seconds) for each point: (t(0, 10, 20)) and the desire initial and final speed (given in m/s): (V0(0, 0, 0), V_f(0, -0.2, 0.4)):

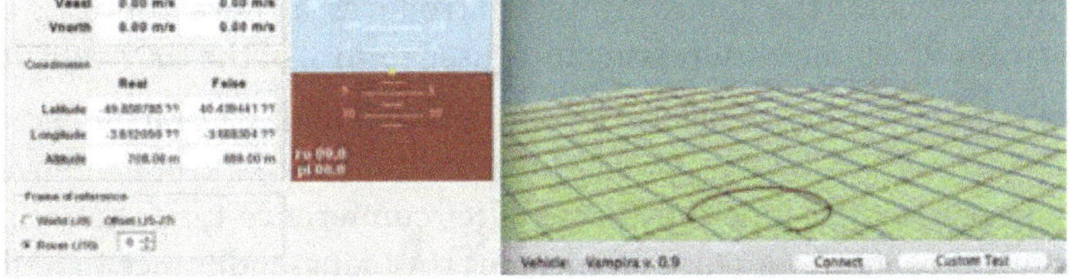

Figure 17. The AVCL simulation environment: Vampira's helicopter executing a semi-spiral motion using 3D splines

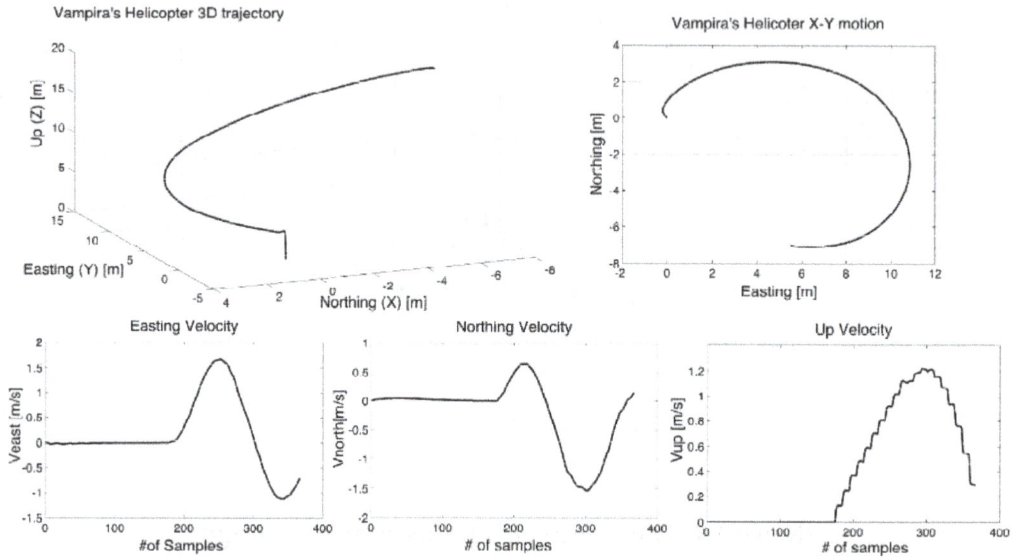

Figure 18. Test1: Cartesian UAV position and velocities with respect to the Inertial Frame

The UAV started from initial point located at (0, 0, 0) coordinate and finished its trajectory at (6, -7, 20). Visual simulation depicted in Fig 17, showed smooth motion across the trajectory due to the 3D spline approach. Nonetheless, 3D splines just allow the user to define the initial and final velocities of the motion, lacking of velocity control for the rest of the knot- points.

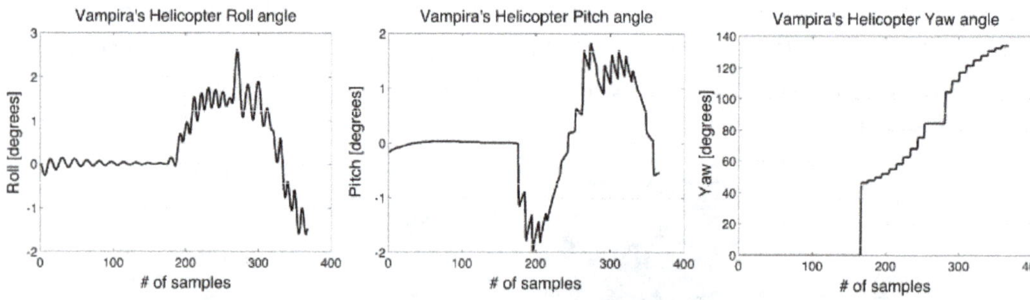

Figure 19. Test1: UAV orientation (Euler angles evolution)

To solve this problem, the following test introduces a more complex trajectory generation using 4D splines, addressing total user control of the UAV velocity profile.

2). Complex trajectory using 4D splines for the velocity profile generation and the Frenet theory for UAV orientation. This trajectory includes different kind of maneuvers joined into a single polynomial function (take-off, circumference-type motion and slow down in spiral-type motion). This test includes UAV long-endurance to high altitude (150 meters above ground) and a maximum easting displacement about of 60 meters. The following knot-control points (given in meters) have been defined: (P_0(0, 0, 0),

$P_1(0, 0, 20)$, $P_2(0, 0, 40)$, $P_3(0, 0, 60)$, $P_4(10, 2, 80)$, $P_5(20, 4, 110)$, $P_6(25, -7, 130)$, P7(30, -10, 150), P8(35, -5, 140) , P9(30, 16, 125) , P10(20, 5, 130) , $P_{11}(33, -10, 145)$, $P_{12}(40, -5, 135)$, $P_{13}(55, -6, 125)$):

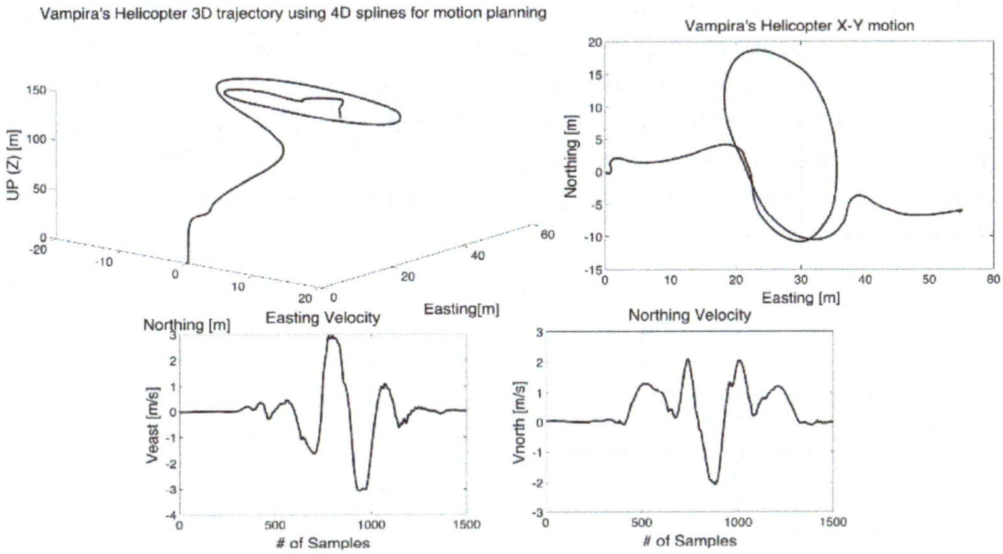

Figure 20. Test2: smooth 4D spline for complex manoeuvre

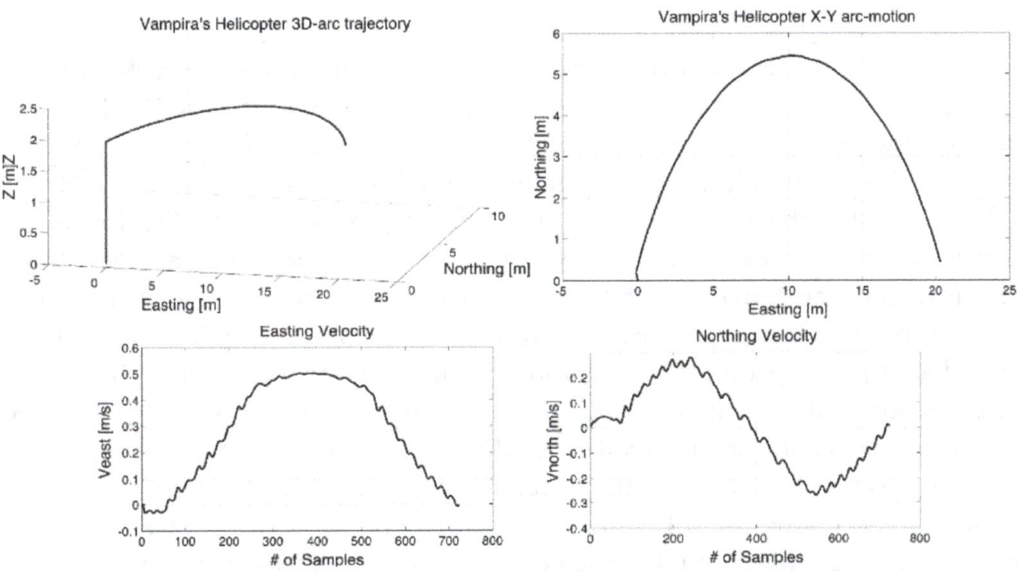

Figure 21. Test3: Arc-type motion using analytical functions

The advantage about using 4D splines relies in the possibility of defining feasible paths that matches with the knot-control points defined (with less match error percentage than the 3D polynomial splines). In addition, the user is able to define the set of velocities for each of the knot-points during the motion. The set of velocities (given in m/s) are: (V_0(0, 0, 0), V_1(0, 0, 0.8), V_2(0, 0, 1), V_3(0, 0, 1.2), V_4(0.5, 1, 1.4), V_5(0, 0.5, 1.7) , V_6(2, 1.5, 2.5), V_7(3, 2.2, 3), V_8(2, 1.2, 2) , V_9(1, 0.5, 1.5) , V_{10}(-0.5, -1, 0.8) , V_{11}(-3, -2, 0.4) , V_{12}(-1, -0.5, 0.8), V_{13}(0,_0, 0)).

3). *Simple arc-type maneuver with trapezoidal velocity profile parameterization:* for the analytical AVCL feature of trajectory planning, the TG²M module supports straight-lines and circumferences motions. An arc defined by: P_0(0, 0, 2), P_1(10, 5.5, 2), P_2(20, 0, 2)) with a maximum velocity of 0.5m/s is tested using the AVCL interpreter that allows the user to define the trapezoidal velocity profile configuration. For this case, the acceleration slopes of the curves (see Fig. 21) have been set to the 30% of the total motion.

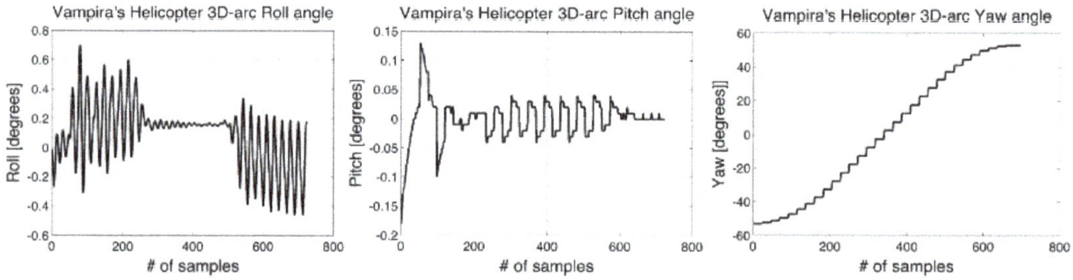

Figure 22. Test3: UAV orientation (Euler angles evolution)

5. OBSERVATIONS

Numerous analytical functions and polynomial interpolation techniques are accessible for modeling continuous Cartesian trajectories in the AVCL, all of which can be utilized in various combinations. The TG2M module manages the specification of trajectories through knot control points and the integration of path constraints. It also facilitates the description of intricate jobs necessitating the formulation of trajectories composed of various primitive routes in any amalgamation of analytical and interpolated functions. User-created spatial trajectories can be seen in three dimensions on the display window or plotted against time utilizing the included plotting library. Simulation findings indicate that the TG2M module functions optimally for the development and testing of a diverse range of smooth trajectories, providing the user with elevated control over the mission through the AVCL interpreter. The three distinct scenarios employed for testing confirmed the efficacy of the mathematical framework utilized for trajectory development and steering during simulated flight. The percentage errors during maneuver execution were negligible, keeping the UAV inside the specified velocity limits and designated trajectory. We also implemented corrections for velocity errors during flight. In high altitude experiments, wind velocity serves as a crucial external disruption force. The TG2M module has wind perturbation adjustment. The Guidance module rectifies velocity commands during real-time flight maneuvers, thereby reducing positional tracking errors. In the three scenario tests, the AVCL simulation environment incorporates normal wind conditions during the simulated flight, bringing minor disturbances into the UAV equations of motion. The results indicate that the disturbances were mitigated, enabling the UAV to adhere to the target trajectory with minimal inaccuracy. The Frenet-Serret formulas utilized for UAV orientation provided an effective method for achieving a smooth rotation rate during flight. Employing basic trigonometric principles to ascertain and delineate the UAV's orientation profile (Yaw angle) is impractical for intricate operations. Splines often necessitate numerous control points for effective trajectory guiding; hence, employing these polynomial equations, the Frenet approach facilitated smooth angle transitions between control points, which could not be achieved through basic trigonometric angle calculations.

REFERENCES

1. JAA & Eurocontrol,. A concept for the European Regulations for Civil Unmanned Aerial Vehicles. *UAV Task-Force Final Report*. 2004
2. Coifman, B., McCord, M., Mishalani, M., Redmill, K., Surface Transportation Surveillance from Unmanned Aerial Vehicles. *Proc. of the 83rd Annual Meeting of the Transportation Research Board*, 2004.
3. Held, Jason M; Brown, Shaun and Sukkarieh, Salah. Systems for Planetary Exploration. *15th NSSA Australian Space Science Conference*, pp. 212-222, ISBN: 0864593740. RMIT University, Melbourne, Australia, from 14 to 16 September 2005.
4. Gutiérrez, P., Barrientos, A., del Cerro, J., San Martin., R. Mission Planning and Simulation of Unmanned Aerial Vehicles with a GIS-based Framework; *AIAA Guidance*, Navigation and Control Conference and Exhibit. Denver, EEUU, 2006.
5. Rysdyk, R. UAV path following for constant line-of-sight. In 2th AIAA Unmanned Unlimited. *Conf. and Workshop and Exhibit*, San Diego, CA, 2003.
6. Price, I.C, Evolving self organizing behavior for homogeneous and heterogeneous swarms of UAVs and UCAVS. *PhD Thesis*, Graduate School of Engineering and Management, Air Force Institute of Technology (AU), Wright-Patterson AFB, OH, March 2006.
7. Herwitz, S. Developing Military COEs UAV applications. *UAV Applications Center – NASA Ames Research Center*, Moffett Field, CA. 2007.
8. Alison A. P., Bryan G., Suresh K. K., Adrian A. K., Henrik B. C. and Eric N. J. Ongoing Development of an Autonomous Aerial Reconnaissance System at Georgia Tech . *UAV Laboratory*, School of Aerospace Engineering. Georgia Institute of Technology, 2003

9. Jaramillo-Botero A., Correa J.F., and Osorio I.J., Trajectory planning in ROBOMOSP, *Robotics and Automation Group* GAR, Univ. Javeriana, Cali, Colombia, Tech. Rep. GAR-TR-10-2004, 2004.

10. LaValle, S.M.: *Planning Algorithms*. Cambridge University Press (2006)

11. Moitra, A., Mattheyses, R.M., Hoebel, L.J., Szczerba, R.J., Yamrom, B.: Multivehicle 5reconnaissance route and sensor planning. *IEEE Transactions on Aerospace and Electronic Systems*, 37 (2003).

12. Zheng, C., Li, L., Xu, F., Sun, F., Ding, M.: Evolutionary Route Planner for Unmanned Air Vehicles. *IEEE Transactions on Robotics* 21 (2005) 609–620

13. Frazzoli, E. Maneuver-based motion planning and coordination for single and multiple UAVs. *AIAA's 1st technical conference and workshop on unmanned aerospace vehicles*. University of Illinois at Urbana-Champaign, il 61801. S. Portsmouth, Virginia. May 2002.

14. Angeles J., Fundamentals of Robotic Mechanical Systems. *Theory, Methods, and Algorithms*. Springer, New York, 1997.

15. Siouris, G. M. Missile Guidance and Control Systems. Springer, New York, 2004.

16. Valero, Julio. Modelo dinámico y sistema de supervisión de vuelo de un helicóptero autónomo. *Proyecto de fin de carrera*. ETSIIM-UPM. 2005.

5 Chapter PROTOTYPING OF UNMANNED HELICOPTER MODELING AND CONTROL

1. INTRODUCTION: AN OVERVIEW

The utilization of UAVs (Unmanned Aerial Vehicles) in civilian applications has generated novel potential for enterprises focused on inspections, surveillance, and aerial photography, among others. Nonetheless, the primary disadvantage of employing such vehicles in civilian contexts is the exorbitant cost, insufficient safety, and evolving legislation. The decline in the expense of sensors, including Global Positioning System receivers (GPS) and non-strategic Inertial Measurement Units (IMU), the affordability of computer systems, and the availability of economical radio-controlled helicopters have fostered the development of a market for small aerial vehicles suitable for diverse applications. The deficiency in safety mostly stems from two principal factors: mechanical robustness and control robustness. The initial factor pertains to the platform utilized in constructing the UAV to minimize system costs, typically a radio-controlled helicopter necessitating cautious maintenance by specialists. The second factor arises from the intricacies of helicopter dynamics, which are influenced by fluctuations in altitude, meteorological conditions, and alterations in the vehicle's configuration, such as weight, payload, or fuel quantity. These scenarios impede the modeling process and, consequently, influence the systematic creation of control systems, leading to laborious and crucial heuristic adjustment procedures. Global researchers provide many modeling methodologies or tactics for the dynamic simulation of helicopters. Numerous studies on helicopter modeling, such those by Godbole et al. or Mahony et al. (2000), Gavrilets et al. (2001), Metler et al. and La Civita et al. (2002), and Castillo et al. (2007), demonstrate extensive methodologies employed in this engineering domain. The absence of an identifying technique in certain instances, coupled with a limited scope of use in others, occasionally complicates their utilization. This chapter describes both the modeling and the identification process that has been successfully tested. The suggested model is defined utilizing a hybrid analytical and heuristic algorithm, leveraging flight dynamics knowledge and addressing essential elements through heuristic equations, enabling real-time simulations without convergence issues. Evolutionary algorithms have been employed for parameter identification. The suggested model has been validated throughout several phases of aircraft flight: hovering, lateral, longitudinal, and vertical, utilizing a Vario Benzin Trainer with a primary rotor diameter of 1.5 meters and a payload capacity of around five kilos.

Figure 1. Benzin trainer by Vario with GNC (Guidance Navigation & Control) system onboard.

A comprehensive control structure has been devised and evaluated utilizing a proposed rapid design methodology based on Matlab-Simulink for simulation and calibration to illustrate the efficacy of the provided tool. The suggested architecture delineates low-level control (servo actuator level), attitude control, position, velocity, and maneuvers. Consequently, there are directions for straight flight with a fixed speed or maneuvers such as circular flight. The findings have validated that the hybrid model utilizing evolutionary algorithms for identification yields exceptional outcomes in helicopter modeling. Real-time simulation facilitates rapid development by generating direct code for onboard controllers, hence minimizing the possibility of programming errors.

2. HELICOPTERS FLIGHT PRINCIPLES

The main rotor only generates the necessary force for helicopter flight (both elevation and forward motion). To acquire a thorough understanding of the forces acting on the main rotor, a detailed description is necessary. The forces acting on the main rotor induce torsion and flexure in the blades, together with lift forces and various forms of aerodynamic drag.

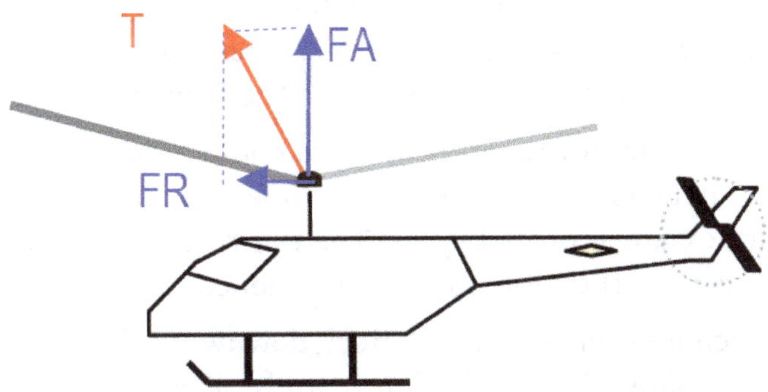

Figure 2. Basic Helicopter Thrust analysis

Initially, just a support force and a resistance force, independent of any external influences, could be considered. The thrust (T) produced by the air pressure on the blade is oriented relative to the plane of rotation. This force can be categorized into vertical force (FA) and resistance (FR), which is exerted in the horizontal plane opposing the direction of rotation (Figure 2). Helicopters utilize devices to adjust the pitch angle of the primary rotor blades. It facilitates the regulation of the fuselage's movement via the tilt of the rotation plane. The attack angle of the blade is not temporally nor spatially consistent. It undergoes continual alteration as the blade rotates, as indicated by the azimuth angle. The assault angle can be seen as the sum of two components: The initial term refers to the mean attack angle over a full rotation of the blade, known as the collective angle. The second component depends on the azimuth angle. When the azimuth angle is 0º or 180º, the blade has the roll cyclic angle. When 90 º or 270 º, the blade has the pitch cyclic angle (Figure 3).

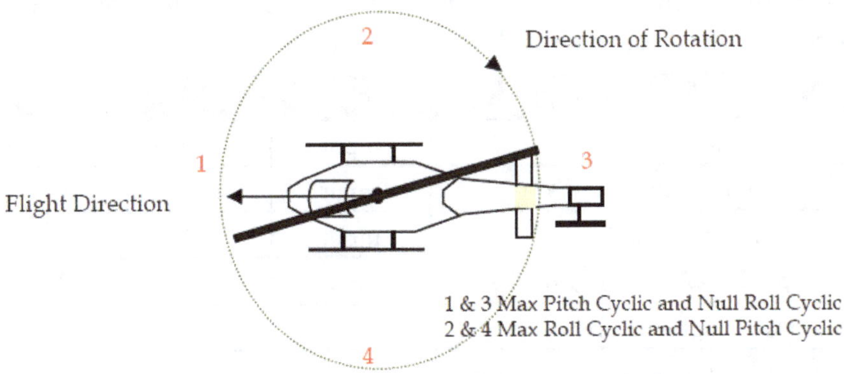

Figure 3. Attack angle during a blade revolution

Using these three signals (collective, roll and pitch cyclic), a pilot is able to control the main rotor. In addition to these signals, the pilot also controls the attack angle of tail rotor blades and the engine throttle.

Typically, radio-controlled helicopters rely upon a commercial control system to maintain the speed of the main rotor constant. The vertical control of the helicopter is done by changing the collective attack angle in the main rotor.

On other hand, the mission of the tail rotor is to compensate the torque that main rotor creates on the helicopter fuselage. The compensation torque can be adjusted by changing the attack angle of its blades. The tail rotor typically requires values between 5 to 30 per cent of the total power of the helicopter.

A lot of physical principles such as ground effect, downwash, flapping or atmospheric effects have to be considered in an in-depth study of helicopter flight dynamics. Taking into account the application scope for the proposed model, which is a small helicopter with small capabilities, no change of air density is considered. Moreover, the blades of the proposed model are also considered as solid bodies.

3. SPECIFICATION OF THE MODEL

Helicopter dynamics can be modeled using either analytical or empirical equations. The former pertains to the physical principles of aerodynamics, whereas the later seeks to characterize the observed phenomena using more straightforward mathematical formulations. The suggested model is a hybrid analytic-empirical framework that integrates the benefits of both approaches: the rapidity and simplicity of the empirical method, alongside the accuracy and physical significance of the analytic equations.

3.1 Inputs and outputs

The proposed model tries to replicate the behavior of a small radio controlled helicopter, therefore the inputs and outputs have been selected as the controls that the pilot relies when using a commercial emitter. Cyclic (Roll and Pitch) and collective controls have been considered as inputs.

Denomination	I/O	Symbol	Units
Collective	Input	θcol	Degrees
Roll Cyclic	Input	θRoll	Degrees
Pitch Cyclic	Input	θPitch	Degrees
Rotational speed over a main rotor shaft.	Input	ω_{hz}	Degrees/s
Acceleration (Helicopter reference frame)	Output	a	m/s2
Velocity (Helicopter reference frame)	Output	v (ua,va,wa)	m/s

Radio-controlled helicopters typically utilize electronic systems that depend on gyroscopes for tail stabilization. The yaw angle is regulated by issuing directives for rotating speed. Consequently, the yaw rate has been regarded as one of the inputs to the model. Helicopters often depend on a main rotor speed control system. Thus, the rotor sustains the speed, and vertical control is thereafter executed by adjusting the collective angle. The assumption of constant speed simplifies the model while preserving its integrity. Moreover, the utilization of this hardware controller reduces the amount of inputs as no throttle command is necessary. The model produces accelerations and rotational speeds as outputs. Table 1 delineates the model›s inputs and outputs.

3.2 Model block diagram

A block diagram of the proposed model is described in Figure 5. A brief definition of every part is also shown in the following sections.

3.3 Main rotor

It is modeled with analytic equations, derived form research on full-scale helicopters [Heffley 2000], which can be used to model small helicopters without fly-bars. The iterative algorithm computes thrust and induced wind speed while taking into account geometrical parameters, speed of the blades, commanded collective, roll and pitch cyclic.

One of the most important concepts to be known is the relationship among the Torque, Power and Speed. Such response arises from the induced velocity of the air passing through the disk of the rotor.

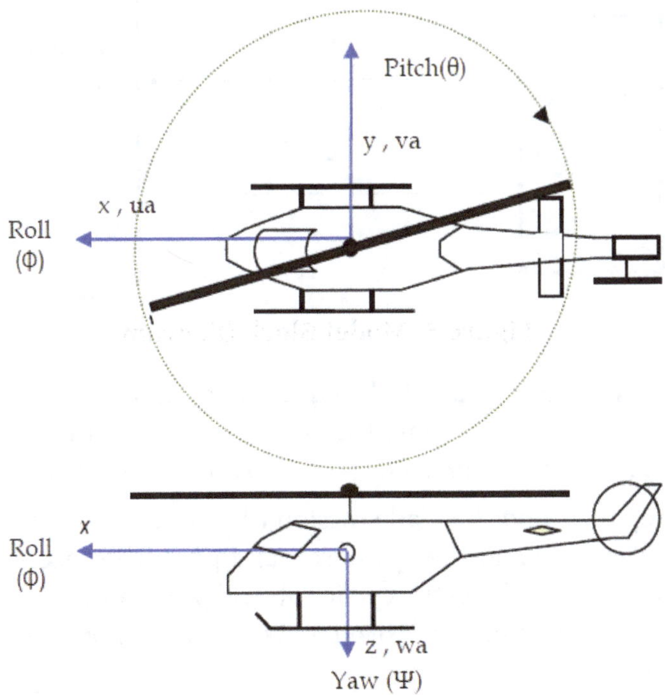

Figure 4. Outputs of the model

The downward airflow provides a vertical force, in accordance with the action-reaction principle, that sustains the helicopter in flight. The engine generates the torque necessary for the blades to revolve and produce airflow. Despite numerous factors complicating the precise measurement of the relative speed between the helicopter and the airflow generated by the revolving rotor disk, a first-order technique is feasible. Thus, it is feasible to describe utilizing classical momentum theory and to estimate the force and induced velocity by a feedback aerodynamic block. Nonetheless, calculus proves to be challenging due to the very non-linear nature of the feedback. The induced speed also affects surrounding surfaces, influenced by the ailerons and other fuselage components, and varies with the speed and direction of flight. This model assumes a uniform distribution of air moving through the rotor disk for torque and induced speed calculations.

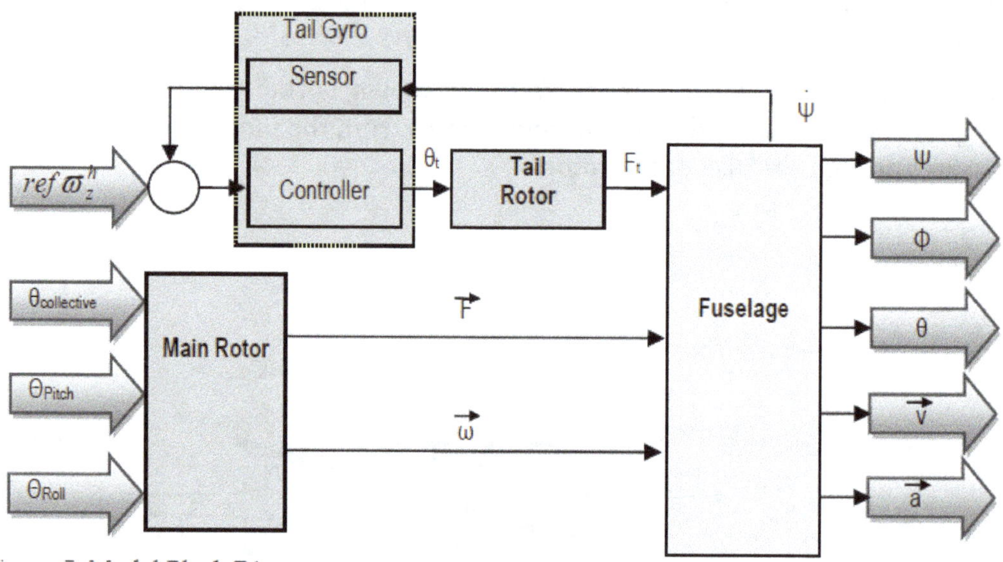

Figure 5. Model Block Diagram

The computation of the torque and induced speed is based on the classic momentum theory but using a recursive scheme that allows to reach a fast convergence. The equations used to model the main rotor have two groups of inputs, as Figure 6 shown.

The collective step (θ_{Col}) and the blade torsion (θ_{Twist}), compose the first branch. The rotor axis inclination (Is) and the cyclic roll and pitch the second one. The attack angles a_1 and b_1 are derived from the cyclic roll and pitch references. The wind speed relative to the helicopter is present through its three components: ua, va, and wa.

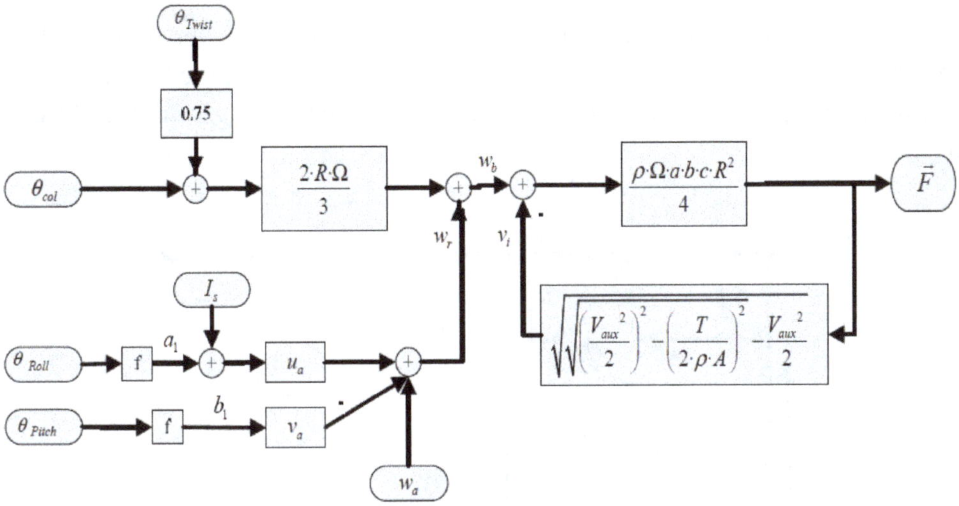

Figure 6. Main rotor block diagram

The equations corresponding to this part are:

$$w_r = w_a + (a_1 + I_s)\, u_a - b_1 \cdot v_a \tag{1}$$

$$w_b = w_r + \frac{2 \cdot \Omega \cdot R}{3}\left(\theta_{col} + \frac{3}{4}\theta_{Twist}\right) \tag{2}$$

The output of the block is the rotor's thrust (F). R and Ω represent the radius of the rotor and its angular speed respectively; ρ is the air density, and a, b and c are geometrical blade factors. The relationship between thrust and angular rates has been derived from observation; therefore the model is also empirical.

$$Vaux^2 = ua^2 + va^2 + w_r\,(w_r - 2v_i) \tag{3}$$

$$T = (w_b - v_i)\,\frac{\rho \cdot \Omega \cdot R \cdot a \cdot b \cdot c \cdot R}{4} \tag{4}$$

$$v_1^2 = \sqrt{\left(\frac{V_{aux}^2}{2}\right)^2 - \left(\frac{T}{2 \cdot \rho \cdot \prod \cdot R^2}\right)^2} - \frac{V_{aux}^2}{2} \tag{5}$$

When the helicopter flies close to the ground (distances less than 1.25 times the diameter of the rotor) the ground effect turns to be very important. This effect has been modeled using the parameter η defined in (6) where h is the distance from the helicopter to the ground.

$$\eta = \frac{h}{2R} \tag{6}$$

In these cases, the thrust is modified using (7) where Th' is the resulting thrust after correcting Th. Values of To, T1 and T2 have been calculated for no creating a discontinuity when h is 1.25 times the diameter of the rotor.

$$Th' = Th\,(T_0 + T_1\eta + T_2\eta^2) \qquad (7)$$

By other hand, the commanded Roll and Pitch cyclic have been considered as reference for rotations in x and y axes considering the helicopter frame. In addition to this, a coupling effect has been considered for simulating real coupling in the helicopter.

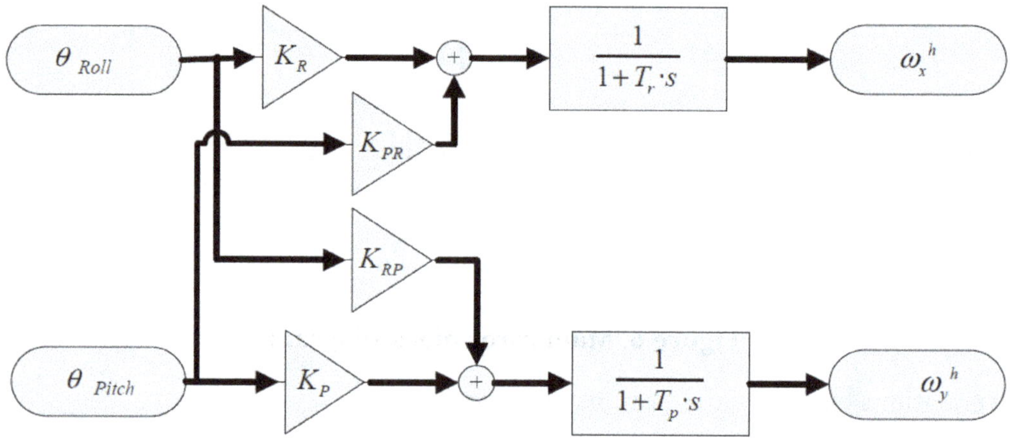

Figure 7. Roll and Pitch dynamic

As it was mentioned in the last section, RC helicopters usually rely upon commercial speed controllers in the main rotor. These devices have been modeled using an ideal dynamic response. On the other hand, engine has been modeled as a first order system. Therefore, variations in the speed of the rotor have been considered only due to changes in the collective angle as Figure 8 shows.

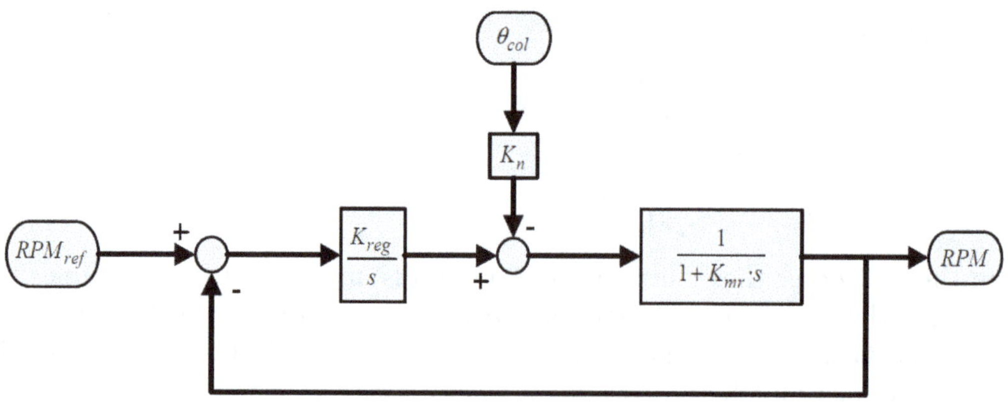

Figure 8. Engine model

3.2.2 Tail rotor

The algorithm for estimating the thrust provided by tail rotor is similar to the main one but only the pitch angle of the blades has been considerated as input. This signal is provided by the hardware controller that is in charge of stabilization of the tail. A PI classical controller has been used to model the controller and the sensor has been considered as no dynamic as Figure 9 shows.

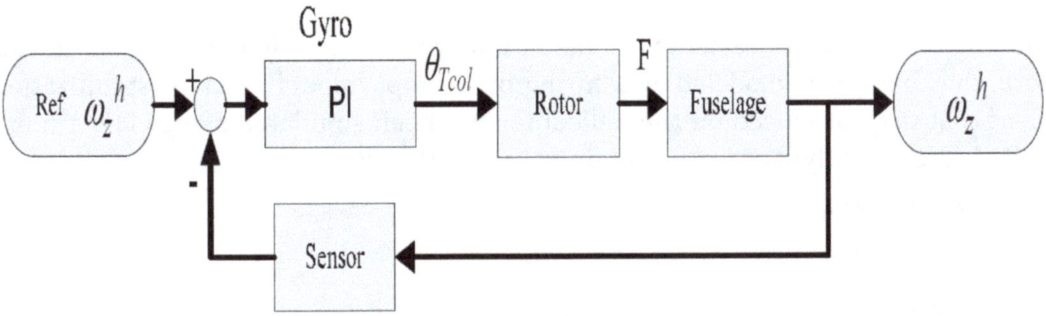

Figure 9. Tail rotor model

3.2.3 Fuselage

In a first step, all the forces (main and tail rotor, gravity, friction and aerodynamic resistances) have to be taking into account for computing the movement of the fuselage of the helicopter. After that, accelerations and velocities can be estimated.

Forces due to the aerodynamic frictions are estimated using the relative velocity of the helicopter with respect to the wind applying (8), where Ar is the equivalent area and vw is the wind velocity.

The resulting equations are summarized in (9).

$$F = \frac{1}{2} A_r \rho v_w^{\ 2} \tag{8}$$

$$\begin{aligned} Fx^h &= Gx^h + Fus_x^{\ h} \\ Fy^h &= Gy^h + Fus_y^{\ h} \\ Fz^h &= Gz^h + Fus_z^{\ h} + T^h \\ M_z^{\ h} &= M_d^{\ h} + M_t^{\ h} + M_{vt}^{\ h} \end{aligned} \tag{9}$$

Where the prefix Fus means aerodynamic forces on fuselage and T the main rotor thrust. The prefix M means Torque where suffix d denotes main rotor, the t denotes tail rotor and the v is the effect of the air in the tail of the helicopter. The prefix G means the gravity components.

Once forces and torques have been calculated, accelerations can be obtained and therefore velocities. Once the velocity referred to helicopter frame (v^h) is calculated, a transformation is required in order to obtain an absolute value by using (10).

$$\begin{bmatrix} v_x \\ v_y \\ v_z \end{bmatrix} = \begin{bmatrix} \cos\theta\cos\psi & -\cos\phi\sin\psi+\sin\theta\sin\phi\cos\psi & \sin\phi\sin\psi+\cos\phi\sin\theta\cos\psi \\ \cos\theta\sin\psi & \cos\phi\cos\psi+\sin\theta\sin\phi\sin\psi & -\sin\phi\cos\psi+\cos\phi\sin\theta\sin\psi \\ -\sin\theta & \sin\phi\cos\theta & \cos\phi\cos\theta \end{bmatrix} \cdot \begin{bmatrix} v_x^h \\ v_y^h \\ v_z^h \end{bmatrix}$$

Flybars are important elements to take into account in the dynamic model. In this work, they have been modeled by using empiric simple models, and the stabilization effect that they produced on the helicopter has been simulated using (11). A more realistic model of flybars can be obtained in [Mettler-2003].

$$\varphi = \varphi - k1\ \varphi$$
$$\theta.' = \theta. - k_2\theta \qquad (11)$$

3.3 Identification of the parameters

Once the mathematic equations have been obtained, a procedure to give values to the parameters that appear into the model is required. Some of the parameters can be easily obtained by simple measurements or weighs but some of them turn to be very difficult to obtain or estimate.

Table 2 describes the parameter list to be identified.

Some methods have been studied for performing the parameters identification, such as multi-variable systems procedures (VARMAX), but they are difficult to apply because there is an iterative process into the model. Due to this, the selected method was evolutionary algorithms, after trying unsuccessfully stochastic methods.

Genetic algorithms may be considered as the search for a sub-optimal solution of a specific cost- based problem. The parameters are codified as chromosomes that are ranked with a fitness function, and the best fits are reproduced through genetic operators: crossover and mutation. This process continues until the fitness of the best chromosome reaches a preset threshold.

Table 2. Parameter to identify list

#	Name	Meaning
1	TwstMr	Main rotor blade twist.
2	TwstTr	Tail rotor blade twist.
3	KCol	Collective step gain.
4	IZ	Moment of inertia around the z axis.
5	HTr	Vertical distance from tail rotor to centre of mass of the helicopter.
6	WLVt	Vertical position of the aerodynamic centre of the tail.
7	XuuFus	Frontal effective area of the helicopter.

#	Name	Meaning
8	YvvFus	Lateral effective area of the helicopter.
9	ZwwFus	Effective area of the helicopter
10	YuuVt	Frontal area of the tail.
11	YuvVt	Lateral area of the tail.
12	Corr1	Correction parameter for Roll.
13	Corr2	Correction parameter for Pitch.
14	Tproll	Time constant for Roll response.
15	Tppitch	Time constant for Pitch response.
16	Kroll	Gain for Roll input.
17	Kpitch	Gain for Pitch input.
18	Kyaw	Gain for Yaw input.
19	DTr	Horizontal distance from centre of the tail rotor to mass centre of the helicopter.
20	DVt	Horizontal distance from the aerodynamic centre of the tail and mass centre of the helicopter.
21	YMaxVt	Saturation parameter (no physical meaning).
22	KGyro	Parameter of commercial gyro controller. (gain)
23	KdGyro	Parameter of commercial gyro controller. (derivative)
24	Krp	Cross gain for Roll and Pitch coupling.
25	Kpr	Cross gain for Pitch and Roll coupling.
26	OffsetRoll	Offset of Roll input (trimmer in radio transmitter).
27	OffsetPitch	Offset of Pitch input (trimmer in radio transmitter).
28	OffsetCol	Offset of Collective input (trimmer in radio transmitter).

The main steps of the process for identification using GA's have been:

3.3.1 Parameters codification

The parameters are codified with real numbers, as it is more intuitive format than large binary chains. Different crossover operations have been studied:

- The new chromosome is a random combination of two chains.

Asc1	0	1	2	3	4	5	6	7	8	9
Asc2	10	11	12	13	14	15	16	17	18	19
Desc	0	11	12	3	14	5	6	17	18	9

- The new chromosome is a random combination of random genes with values in the range defined by the ascendant genes.

Asc1	0	1	2	3	4	5	6	7	8	9
Asc2	10	11	12	13	14	15	16	17	18	19
Desc	6.5	7.6	2	13	4	15	9.2	8.5	8	19

The first operator transmits the positive characteristics of its ascendants while the second one generates diversity in the population as new values appear. In addition to the crossover operator there is a mutation algorithm. The probability of mutation for the chromosomes is 0.01, and the mutation is defined as the multiplication of random genes by a factor between 0.5 and 1.5.

When the genetic algorithms falls into a local minimum, (it is detected because there is no a substantial improvement of the fitness in the best chromosome during a large number of iterations), the probability of mutation have to be increased to 0.1. This improves mutated populations with increased probability of escaping from the local minimum.

3.3.2 Initial population

The initial population is created randomly with an initial set of parameters of a stable model multiplied by random numbers selected by a Monte-Carlo algorithm with a normal distribution with zero mean and standard deviation of 1.

The genetic algorithm has been tested with different population sizes, between thirty and sixty elements. Test results showed that the bigger population did not lead to significantly better results, but increased the computation time. Using 20 seconds of flight data and a population of 30 chromosomes, it took one hour to process 250 iterations of the genetic algorithm on a Pentium IV processor. Empirical data suggests that 100 iterations are enough to find a sub-optimal set of parameters.

3.3.3 Fitness function

The fitness function takes into consideration the following state variables: roll, pitch, yaw and speed. Each group of parameters (chromosome) represents a model, which can be used to compute simulated flight data from the recorded input commands. The difference between simulated and real data is determined with a weighted least-squares method and used as the fitness function of the genetic algorithm.

In order to reduce the effect of the error propagation to the velocity due to the estimated parameters that have influence in attitude, the global process has been decomposed in two steps: Attitude and velocity parameters identification.

The first only identifies the dynamic response of the helicopter attitude, and the genetic algorithm modifies only the parameters related to the attitude. Once the attitude-related parameters have been set, the second process is executed, and only the velocity-related parameters are changed. This algorithm uses the real attitude data instead of the model data so as not to accumulate simulation errors. Using two separate processes for attitude and velocity yields significantly better results than using one process to identify all parameters at the same time.

The parameters related to the attitude process are: TwstTr, IZ, HTr, YuuVt, YuvVt, Corr1, Corr2, Tproll, Tppitch, Kroll, Kpitch, Kyaw, DTr, DVt, YMaxVt, KGyro, KdGyro, Krp, Kpr, Offset Roll, OffsetPitch and Offset Yaw. The parameters related to the vehicle's speed are TwstMr, KCol, XuuFus, YvvFus, ZwwFus and Offset Col.

The fitness functions are based on a weighted mean square error equation, calculated by comparing the real and simulated responses for the different variables (position, velocity, Euler angles and angular rates) applying a weighting factor.

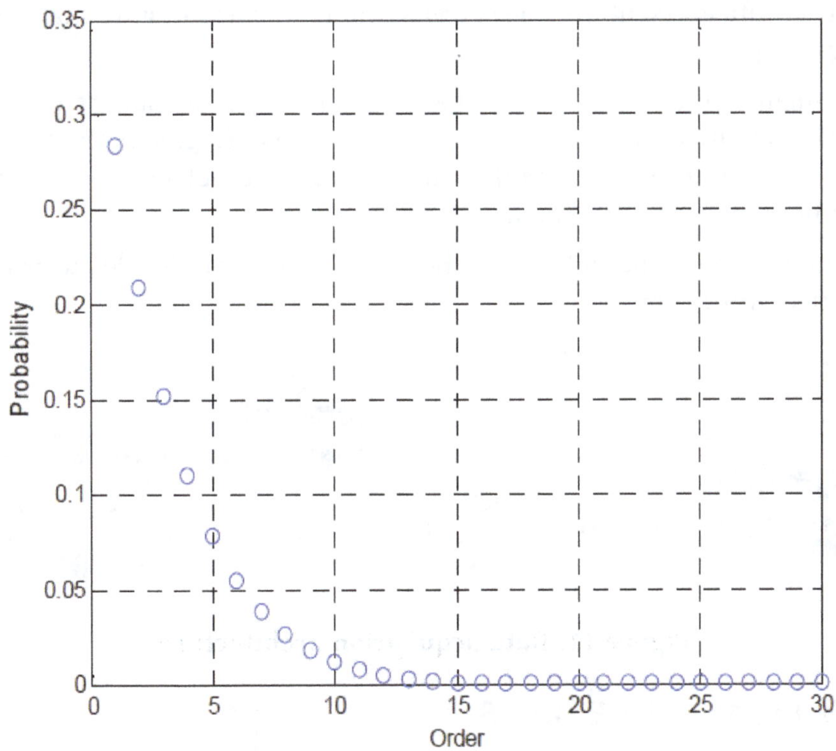

Figure 10. Probability function for selection of elements

The general process is described bellow:

- To create an initial population of 30 elements.
- To perform simulations for every chromosome;
- Computation of the fitness function.
- Classification the population using the fitness function as the index. The ten best elements are preserved. A Monte-Carlo algorithm with the density function shown in Figure 10. Probability function for selection of elements

is used to determine which pairs will be combined to generate 20 more chromosomes. The 10 'better' elements are more likely (97%) to be combined with the crossover operators.

- To repeat from step 2 for a preset number of iterations, or until a preset value for the fitness function of the best element is reached.

3.3.4 Data acquisition

The data acquisition procedure is shown in Figure 11. Helicopter is manually piloted using the conventional RC emitter. The pilot is in charge to make the helicopter performs different maneuvers trying to excite all the parameters of the model. For identification data is essential to have translational flights (longitudinal and lateral) and vertical displacements.

All the commands provided by the pilot are gathered by a computer trough to a USB port by using a hardware signal converter while onboard computer is performing data fusion from sensors and sending the attitude and velocity estimation to the ground computer using a WIFI link.

In this manner inputs and outputs of the model are stored in files to perform the parameters identification.

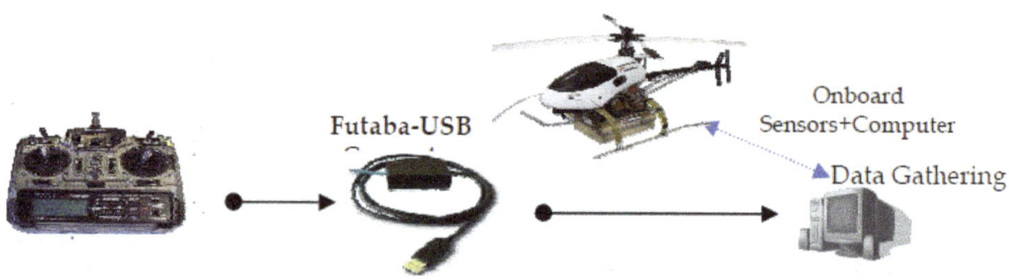

Figure 11. Data acquisition architecture

4. IDENTIFICATION RESULTS

The identification algorithm was executed several times with different initial populations and, as expected, the sub-optimal parameters differ. Some of these differences can be explained by the effect of local minimum. Although the mutation algorithm reduces the probability of staying inside a local minimum, sometimes the mutation factor is not big enough to escape from it.

The evolution of the error index obtained with a least-squares method in different cases is shown in Figure 12. The left graph shows the quick convergence of the algorithm in 50 iterations. On the other hand the right graph shows an algorithm that fell in a local minimum and had an overall lower convergence speed. Around the 50th step a mutation was able to escape the local minimum, and the same behavior is observed in the 160th step.

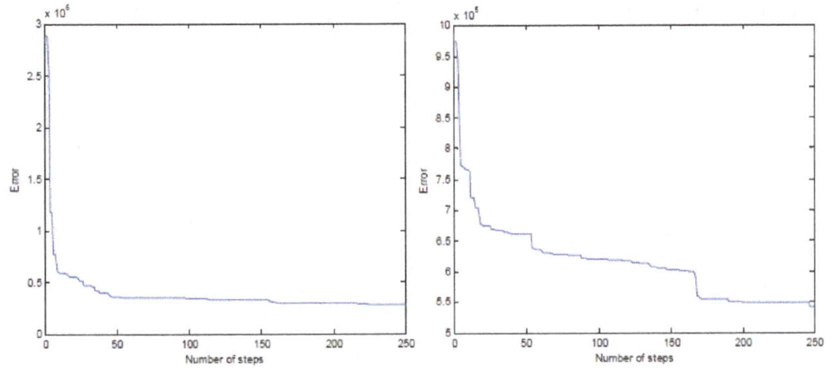

Figure 12. Error evolution for two different cases

The result may be used as the initial population for a second execution of both processes. In fact, this has been done three times to obtain the best solution.

The analysis of cases where mutation was not able to make the algorithm escaped from local minimum, led to the change of the mutation probability from 0.1 to 1 when detected.

On the other hand, not all the parameters were identified at the same time, actually two iterative processes were used to identify all parameters. The first process used 100 steps to identify the parameters related to the modeling of the helicopter's attitude, beginning with a random variation of a valid solution. The second process preserved the best element of the previous process' population, and randomly mutated the rest. After 100 steps the parameters related to the helicopter's speed were identified.

The simulated attitude data plotted as a blue line against real flight data in red, is shown for roll, pitch, yaw angles in Figure 13. They give you an idea about how simulations follow real tendencies even for small attitude variations.

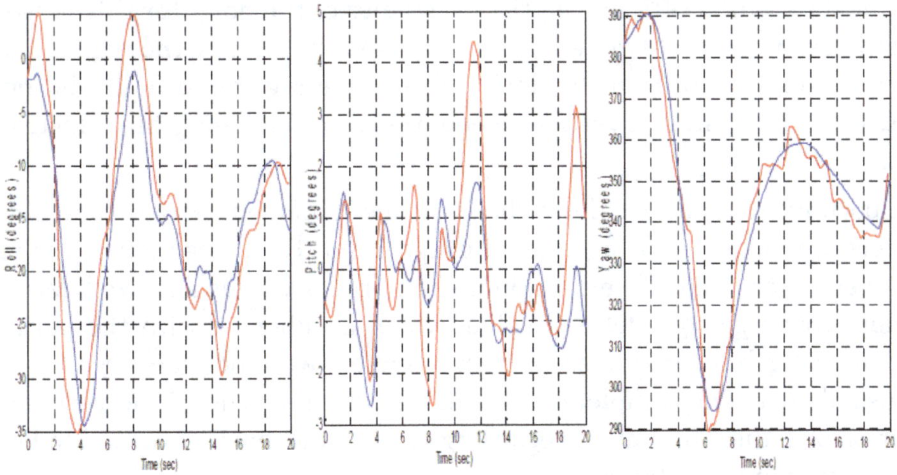

Figure 13. Roll, Pitch and Yaw real vs simulated

On the other hand, the results obtained for the velocity analysis are shown in Figure 14.

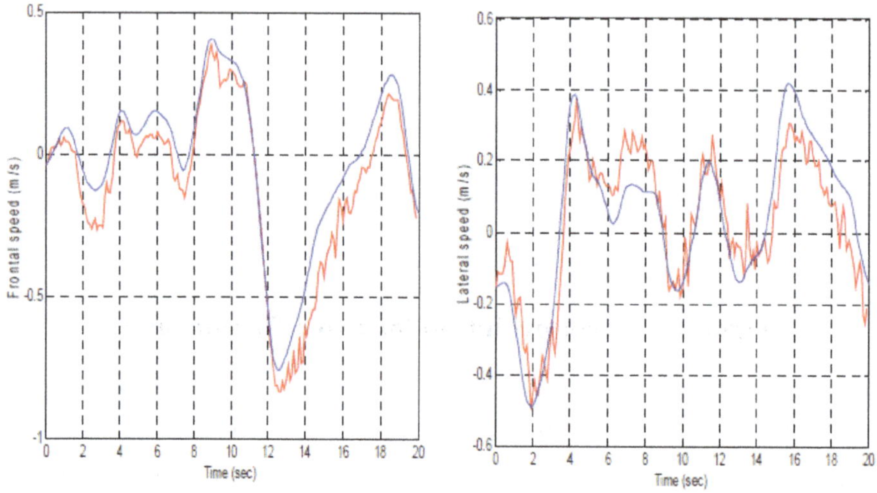

Figure 14. Velocity Simulation Analysis

The quality of the simulation results is the proof of a successful identification process both for attitude and speed.

The simulated roll and yaw fit accurately the registered helicopter response to input commands. The simulated pitch presents some problems due to the fact that the flight data did not have a big dynamic range for this signal. Nevertheless the error is always below three degrees.

For the simulation of the vehicle's velocity good performance was obtained for both frontal and lateral movement. The unsuccessful modeling of vertical velocity can be linked to the sensors available onboard the helicopter. Vertical velocity is registered only by the differential GPS, and statistical studies of the noise for this sensor show a standard deviation of 0.18 meters per second. In other words, the registered signal cannot be distinguished from the noise because the registered flight did not have a maneuver with significant vertical speed; therefore modeling this variable is not possible.

It is important to analyze the values of the obtained parameters since the parameters are identified using a genetic algorithm. The dispersion of the value of the parameters for five different optimization processes was analyzed. Most of the parameters converge to a defined value, which is coherent with the parameter's physical meaning, but sometimes, some parameters were not converged to specific values, usually when no complete set of flights was used. Thus, if no vertical flights were performed, the parameters regarding vertical movements turned to be with a great dispersion in the obtained values.

This conclusion can be extended to different optimization processes for different groups of flight data. In other words, several flights should be recorded, each with an emphasis on the behaviors associated to a group of parameters. With enough flight data it is trivial to identify all the parameters of the helicopter model.

5. CONTROL PROTOTYPING

To facilitate the development and testing of control algorithms, the suggested model has been enclosed within a Matlab-Simulink S-Function, as seen in Figure 15, serving as a prototyping template. Additional modules have been utilized for conducting realistic simulations, hence a sensor model of GNC systems has been developed. Auxiliary modules for automatic-manual switching have been necessitated for testing reasons.

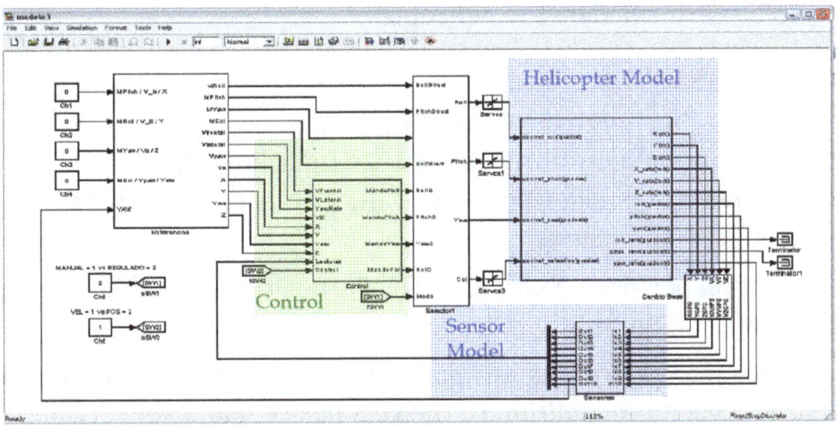

Figure 15. Control Developing template

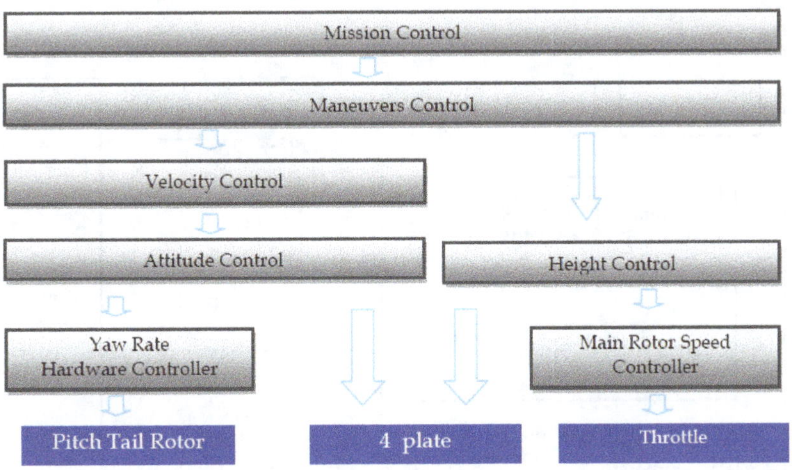

Figure 16. Control Architecture

The suggested control architecture is founded on a hierarchical structure comprising five control levels, as illustrated in Figure 16. The bottom level consists of hardware controls for rotor speed and yaw rate. The top level pertains to attitude control. This is the fundamental level for achieving helicopter stability during flying. Various approaches, including classical PI and fuzzy controllers, have been evaluated on it. The subsequent factor is the velocity and altitude level. The fourth level is the maneuver level, tasked with executing a series of predefined simple movements, while the highest level is mission control.

5.1 Mission control

In this level, the operator designs the mission of the helicopter using GIS (Geographic Information System) software. These tools allow visualizating the 3D map of the area by using Digital Terrain Model files and describing a mission using a specific language. (Gutiérrez et al -06). The output of this level is a list of maneouvers (parametric commands) to be performed by the helicopter.

5.2 Maneuver Control

This control level is in charge of performing parametric manoeuvres such as flight maintaining a specific velocity during a period of time, hovering with a fix of changing yaw, forward, backward or sideward flights and circles among others. The output of this level are velocity commands.

Internally, this level performs a prediction of the position of the helicopter, applying acceleration profiles (Figure 17).

Velocity and heading references are computed by using this theoretical position and the desired velocity in addition to the real position and velocity obtained from sensors.

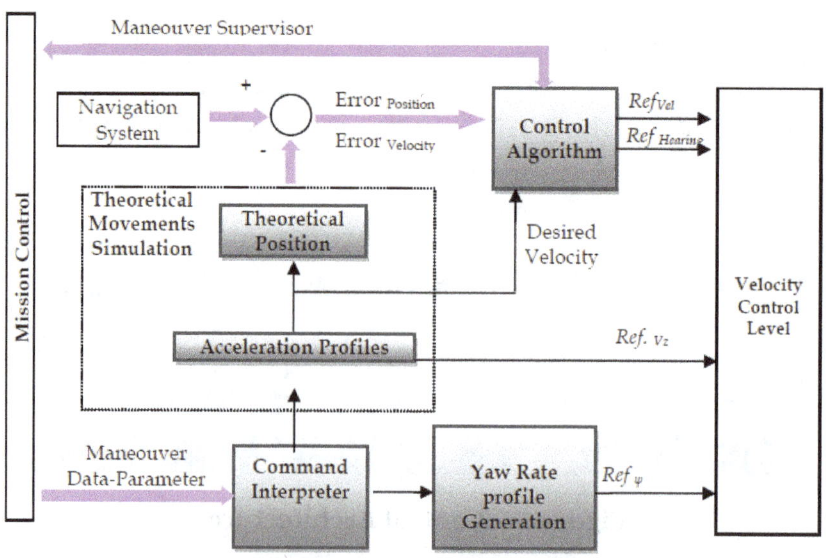

Figure 17. Maneuvers Control Scheme

A vertical control generates references for altitude control and Yaw control.

In manoeuvres that no high precision of the position is required, i.e. forwared flights, the position error is not taken into account, because the real objective of the control is maintain the speed. This system allow manage a weighting criteria for defining if the main objetive is the trajectory or the speed profile. The manoeuvres that helicopter is able to perform are an upgradeable list.

5.3 Velocity Control

The velocity control is in charge of making the helicopter maintains the velocity that maneouvers control level computes.

The velocity can be referred to a ENU (East-North-Up) frame or defined as a module and a direction. Therefore, a coordinate transformation is required to obtain lateral and frontal velocities (vl, vf) that are used to provide Roll and Pitch references. This transformation is very sensitive to the yaw angle estimation.

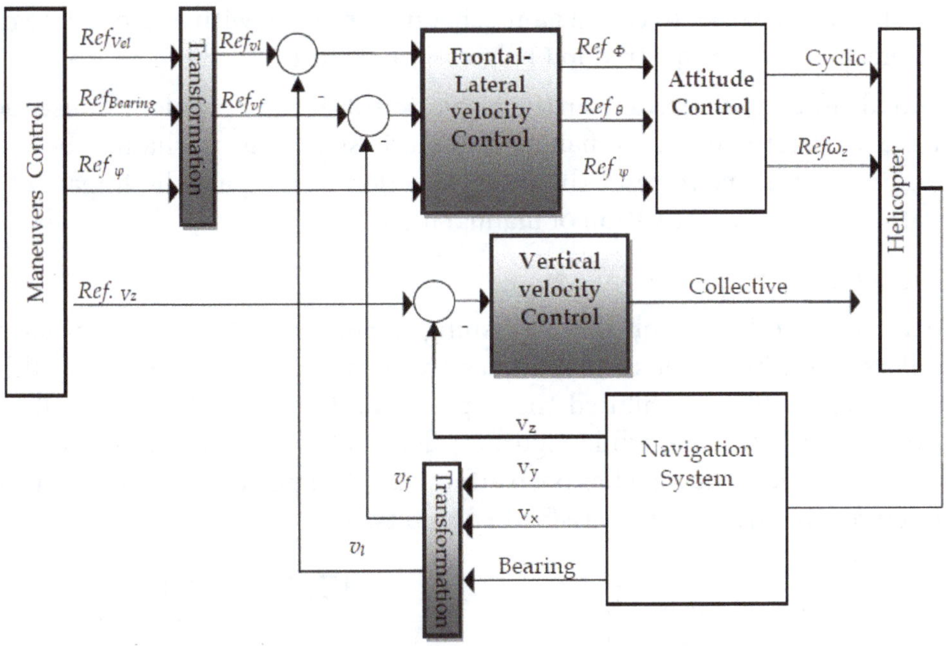

Figure 18. Velocity control structure

Control maneuver has also to provide with the yaw reference due to the capability of helicopters for flying with derive (different bearing and heading).

Vertical velocity is isolated from the horizontal because it is controlled by using the collective command.

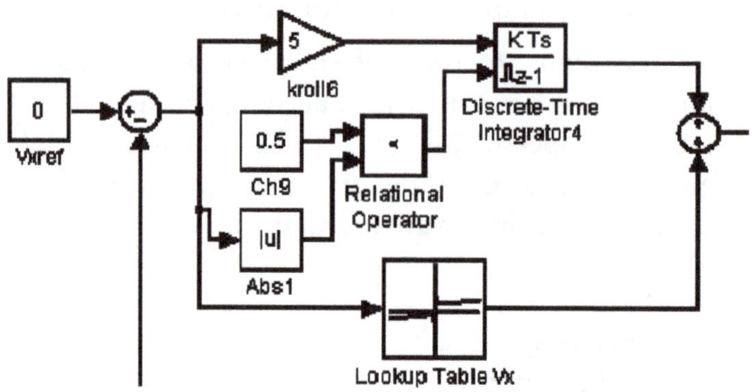

Figure 19. Velocity control example

Concerning to the control algorithms used to test the control architecture, two main problems have been detected and solved:

- The first one is based on the fact that the speed is very sensitive to the wind and payload. Due to this, a gain-scheduller scheme with a strong integration efect has been requiered in PI algorithm. (Figure 19)

The second difficulty emerges from the resolution of the first problem. Consequently, when the operator executes a manual-automatic switch using channel nine of the emitter, it is necessary to reset the integral action to mitigate the impact of error integration during the activation of manual mode.

5.4 Attitude Control

The attitude control is responsible for issuing commands to the servos (or hardware controllers). Multiple control techniques were evaluated at this level, although fuzzy controllers likely exhibited the superior performance. Figure 20 illustrates the proposed architecture, while Figure 21 depicts the control surfaces utilized by the fuzzy controller. The Yaw has been effectively decoupled from the multivariable Roll-Pitch controller, yielding satisfactory results.

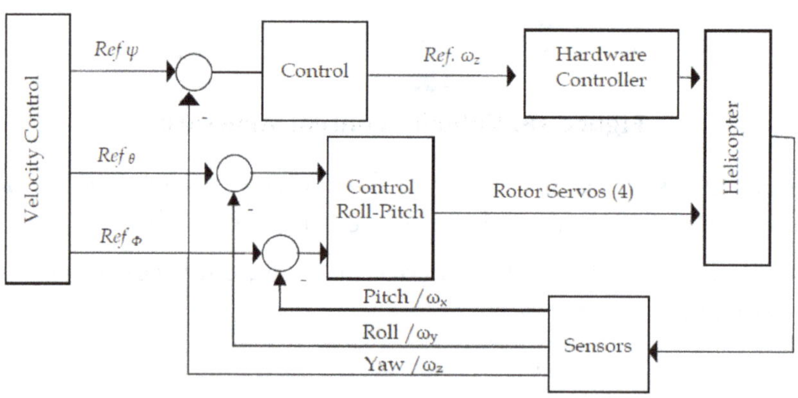

Figure 20. Attitude control structure

Once the control structure has been designed, the first step is to perform simulations in order to analyze if the control fulfills all the established requirements.

Considering the special characteristics of the system to control, they main requirements to be taken into account are:

- No (small) oscillations for attitude control. In order to transmit a confidence in the control system to the operator during the tests.
- No permanent errors in velocity control. This is an important concept to consider for avoiding delays in long distance missions.
- Adjustable fitness functions for maneuvers control. This allows characterizing the mission where position or velocity is the most important reference to keep.

The operator relies upon all the tools that Matlab-Simulink provides and an additional 3D visualization of the helicopter as Figure 22 shows for test performing.

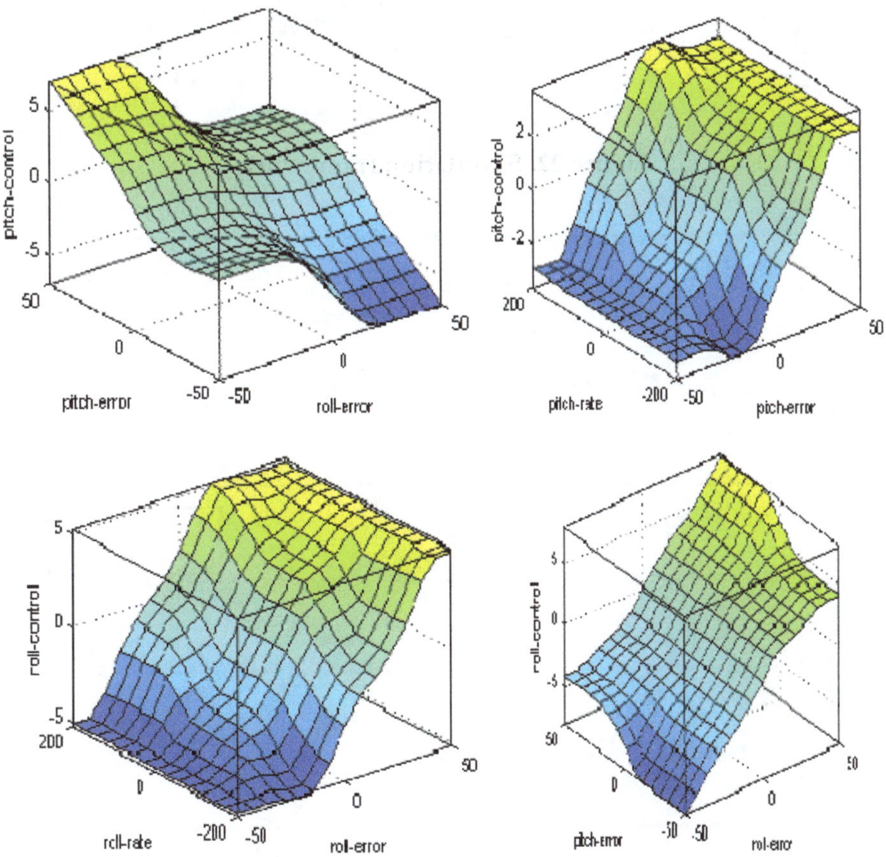

Figure 21. Control surfaces for a Attitude level

The controller block has to be isolated for automatic C codification by using the embedded systems coder of Matlab-Simulink when simulation results are satisfactory. Then, the obtained code can be loaded into onboard computer for real testing (Del-Cerro 2006).

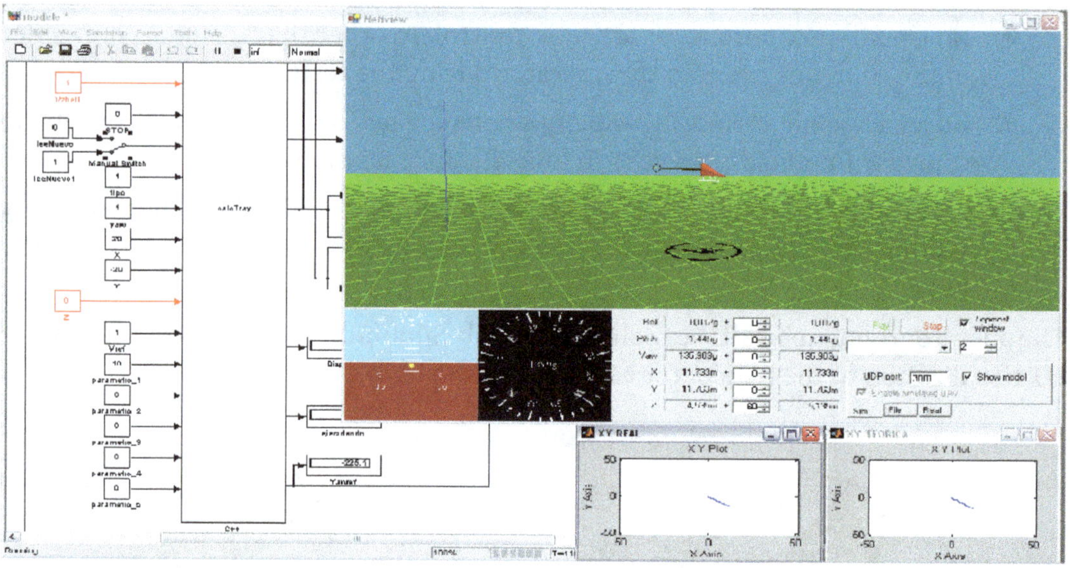

Figure 22. Simulation framework

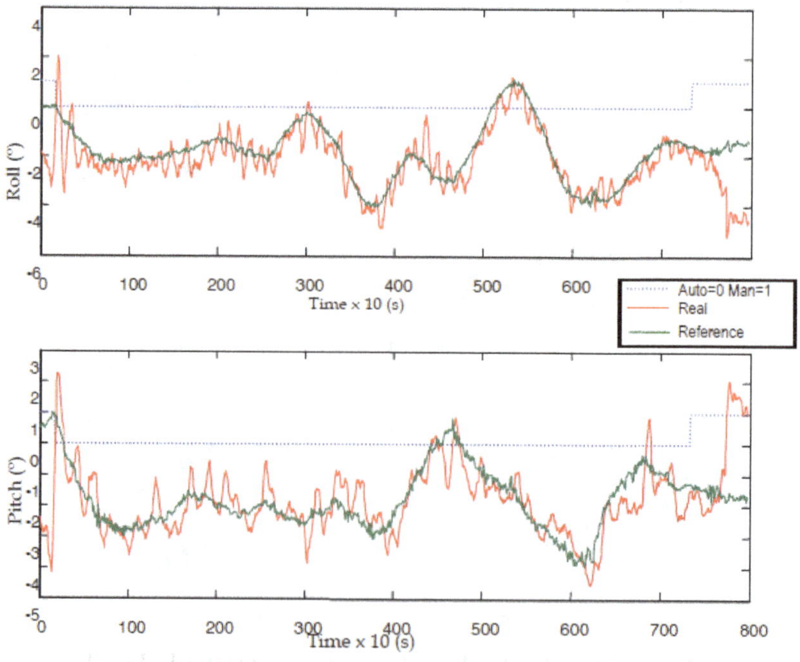

Figure 23. Roll and pitch Control results during real flight

6. CONTROL RESULTS

Figure 23 shows the results obtained for attitude control during an eighty-seconds real flight with the Vario helicopter.

Lines in red show the real attitude of the helicopter while green ones are the references provided by the velocity control level.

Dotted blue line indicates when the helicopter is in manual (value of 1) or automatic (value of 0). It can be observed that, when helicopter is in manual mode, the red line does not follow the references of the automatic control, because the pilot is controlling the helicopter. Figure 24 shows the obtained results for velocity control. The same criteria that in last graphic has been used, thus red line indicates real helicopter response and green means references from maneuvers control level. Dotted blue line is used to show when the helicopter flies autonomously or remote piloted. Small delays can be observed.

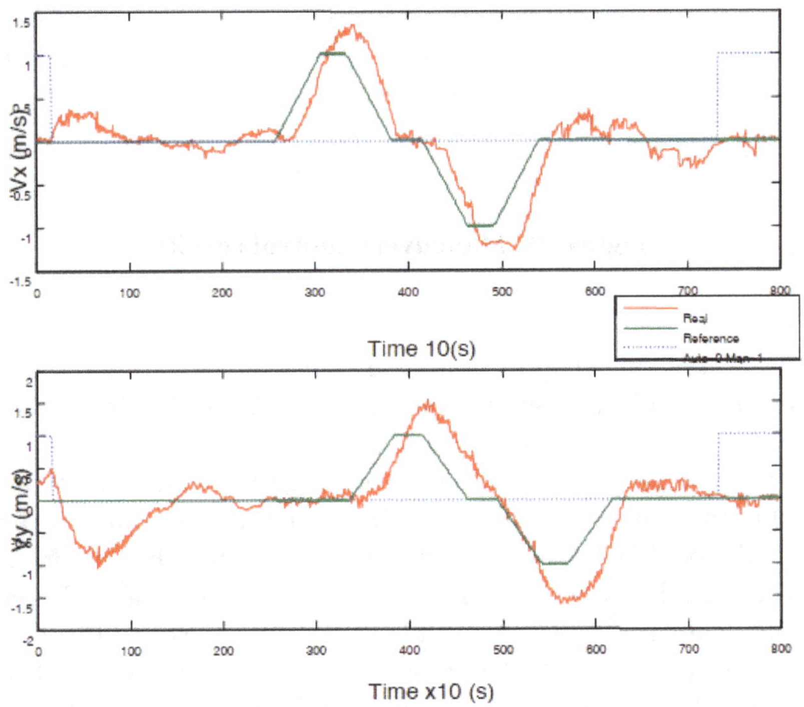

Figure 24. Velocity control results

Figure 25 shows two examples of the maneuvers control performance.

The right side graph shows the position when helicopter is moving in a straight line. It can be observed that maximum error in transversal coordinates is less than one meter. This result can be considered as very good due to is not a scalable factor. The precision of the position control reduces the error when trajectory is close to the end by weighting the error position against the velocity response.

On the other hand, the left side shows a square trajectory. The speed reference performs smooth movementes and therefore the position control is less strong that in the first case.

Even in this case, position errors are smaller than two meters in the worst case.

In general, a high precision flight control has been developed, because sub-metric precision has been reached during the tests with winds slower than twenty kilometers per hour.

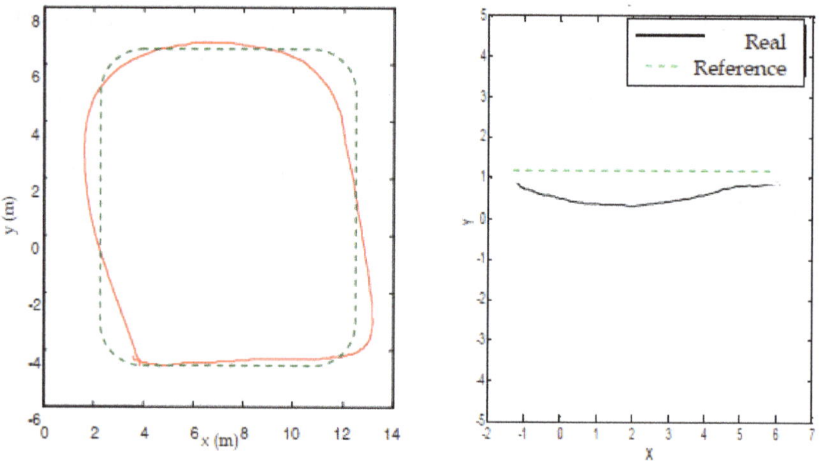

Figure 25. Maneuvers control results

7. CONCISE SUMMARY

The initial section of this chapter delineates a methodology for the modeling and identification of a small helicopter. The model has been developed using existing literature, although modifications made to the original model have enhanced its stability without compromising precision. The suggested identification methodology utilizing evolutionary algorithms is a versatile procedure with extensive applications. The established model facilitates realistic simulations, and the outcomes have been validated. The model has been employed for the design of actual controllers on a real aircraft through a rapid prototyping technique described in section 4. The model has demonstrated resilient performance in response to alterations in flight conditions. It is effective for hovering and both frontal and lateral non-aggressive flight. The precision and utility of a parametric model are significantly influenced by the quality of its parameters, and the identification process frequently necessitates extensive knowledge of both the model and the phenomena being modeled. The suggested identification procedure eliminates the intricacies of model tuning; it just necessitates high-quality flight data within the designated simulation range. The genetic algorithm has successfully identified suitable values for the model's 28 parameters, values that align with their physical significance and provide an accurate model. This chapter does not aim to determine the optimal control technique. The objective is to illustrate that the suggested model is sufficiently robust to conduct simulations suitable for control design and implementation utilizing an automatic coding tool.

REFERENCES

Castillo P, Lozano R, Dzul A. (2005) Modelling and Control of Mini-Flying Machines. Springer Verlag London Limited 2005. ISBN:1852339578.

Del-Cerro J, Barrientos A, Artieda J, Lillo E, Gutiérrez P, San- Martín R, (2006) Embedded Control System Architecture applied to an Unmanned Aerial Vehicle International Conference on Mechatronics. Budapest-Hungary

Gavrilets, V., E. Frazzoli, B. Mettler. (2001) Aggresive Maneuvering of Small Autonomous Helicopters: A. Human Centerred Approach. *The International Journal of Robotics Research.* Vol 20 No 10 , October 2001. pp 795-807

Godbole et al (2000) Active Multi-Model Control for Dynamic maneuver Optimization of Unmanned Air Vehicles. *Proceedings of the 2000 IEEE International conference on Robotics and Automation.* San Francisco, CA. April 2000.

Gutiérrez, P., Barrientos, A., del Cerro, J., San Martin., R. (2006) Mission Planning and Simulation of Unmanned Aerial Vehicles with a GIS-based Framework; *AIAA Guidance, Navigation and Control Conference and Exhibit.* Denver, EEUU, 2006.

Heffley R (1988). Minimum complexity helicopter simulation math model. *Nasa center for AerosSpace Information* . 88N29819.

La Civita M, Messner W, Kanade T. Modeling of Small-Scale Helicopters with Integrated First-Principles and System-Identification Techniques. *American Helicopter Society 587h Annual Forum*, Montreal, Canada, June 11-13,2002.

Mahony-R Lozano-R, Exact Path Tracking (2000) Control for an Autonomous helicopter in Hover Manoeuvres. *Proceedings of the 2000 IEEE International Conference on Robotics & Automation.* San Francisco, CA. APRIL 2000. P1245-1250

Mettler B.F., Tischler M.B. and Kanade T.(2002) System Identification Modeling of a Model- Scale Helicopter. *T. Journal of the American Helicopter Society*, 2002, 47/1: p. 50-63. (2002).

Mettler B, Kanade T, Tischler M. (2003) System Identification Modeling of a Model-Scale Helicopter. *Internal Report* CMU-RI-TR-00-03.

6 Chapter

STABILIZATION OF SCALE MODELS OF VERTICAL TAKE-OFF AND LANDING VEHICLES WITHOUT MEASURING THEIR SPEEDS

1. INTRODUCTION: AN OVERVIEW

Miniature rotorcraft-based Unmanned Aerial Vehicles (UAVs) have garnered increasing attention in industrial and academic research. Due to their hover and vertical take-off and landing (VTOL) capabilities, they are exceptionally suited for various civil missions, including video monitoring of road traffic, surveillance of urban areas, victim localization following natural disasters, fire detection, and building maintenance inspections. The development of guidance, navigation, and control algorithms for the autonomous operation of small rotorcraft-based UAVs presents significant challenges because to their nonlinear dynamics and heightened sensitivity to aerodynamic disturbances. Diverse control strategies, including backstepping (Bouabdallah & Siegwart, 2005; Frazzoli et al., 2000; Mahony & Hamel, 2004), nonlinear model predictive control (Kim et al., 2002; Bertrand et al., 2007a), and sliding modes (Bouabdallah & Siegwart, 2005), have been effectively utilized for the stabilization and trajectory tracking of UAV models. Nonetheless, the majority necessitate comprehensive state information for feedback control design.For robotic systems it may be useful, for cost or payload reasons, to limit the number of embedded sensors. For a miniature UAV, the nature of the mission itself may also directly impact the choice of the sensors that will be used, and therefore the type of measurements that will be available for the vehicle control.

In constrained environments, for example, the use of a vision based sensor may be preferred to a GPS to estimate the relative position of the vehicle with respect to its environment. In that case, linear velocity measurements may not be available.

Another example is the case of a test bench design, where a "ready-to-use" radio controlled vehicle is used along with external sensors that do not require structural modifications of the vehicle. Such external sensors are for example motion capture systems (Kondak et al., 2007), (Kundak & Mettler, 2007), (Valenti et al., 2006), or magnetic field based sensors (Castillo et al., 2004). With such equipments, only the position and the attitude angles of the vehicle can be directly measured. Nevertheless, knowledge of the vehicle state components (positions, linear velocities, attitude angles and angular velocities) is required for control.

A practical approach may consist in computing the velocities from the position measurements by finite differentiations. This method is used in (Kondak et al., 2007) to compute the linear velocity of rotorcraft-based miniature UAVs, and in (Castillo et al., 2004) to compute both linear and angular velocities to control a four-rotors vehicle. However, no theoretical stability guarantee is provided.

One way to theoretically deal with partial state measurement is to define an observer. In (Do et al., 2003) the problem of trajectory tracking for a planar Vertical Take-Off and Landing (VTOL) aircraft with only position and attitude angle measurements is solved by designing a full-order observer. Changes of coordinates are then used to put the system in a triangular form so that a backstepping technique can be used to develop a velocity-independent stabilizing controller.

However, the use of an observer may introduce additional computational burden in the control loop. It is also necessary to prove firstly its own convergence. In addition, compatibility between the frequency of the observer and the frequency of the controller must be checked to ensure the closed loop stability of the complete observer-based controlled system.

Another approach that can be used to avoid computational burden or complexity due to the introduction of an observer is partial state feedback: the controller is designed directly from the available measurements.

Early work on partial state feedback has been done in the context of rigid-link robot manipulators when no velocity measurement is available. In (Burg et al., 1996) and (Burg et al., 1997) the velocity measurement is replaced by a velocity-related signal generated by a linear filter based only on link position measurement. An extension of this work, using a nonlinear filter, can be found in (Dixon et al., 2000). The same method has been applied to solve the problem of attitude tracking of rigid bodies. A velocity-related signal generated by a linear filter is indeed employed in (Wong et al., 2000), where a kinematic representation using modified Rodrigues parameters has been chosen. In (Costic et al., 2000), a unit- quaternion-based representation is adopted and a nonlinear filter generates a signal replacing the angular velocity measurement in the feedback controller.

First-order dynamic attitude feedback controllers were proposed in (Arkella, 2001) and (Astolfi & Lovera, 2002) to address the attitude tracking problem for rigid bodies and spacecraft utilizing magnetic actuators. The kinematic representations employed in these two papers are based on modified Rodrigues parameters and quaternions, respectively. A unit-quaternion representation is utilized in (Tayebi, 2007), where a feedback controller based on an estimation error quaternion is developed to address the issue of inflexible spacecraft attitude monitoring. The attitude control of rigid bodies without angular velocity measurement is discussed in (Lizarralde & Wen, 1996) and (Tsiotras, 1998), where a passivity-like characteristic of the system is employed to design feedback controllers for kinematic representations based on unit-quaternions and Rodrigues parameters, respectively. Both employ a filtering approach to circumvent the necessity of velocity measurement. This chapter addresses the issue of position and attitude stabilization for a six degrees of freedom VTOL UAV model in the absence of measurements for linear and angular velocities. In contrast to prior studies, our kinematic representation utilizes the SO(3) group and associated manifold. The proposed solution relies on the incorporation of virtual states into the system dynamics, eliminating the need for observer design. The remainder of the chapter is structured as follows. Section 2 presents the notations and mathematical identities that will be employed in the chapter. Section 3 delineates the dynamics of the VTOL UAV model and the hierarchical configuration of the controller. Section 4 elaborates on the design of the position controller, whereas Section 5 delineates the attitude controller. Section 6 analyzes the closed-loop stability of the system, whereas Section 7 presents the simulation results. The concluding remarks are presented in the final section of this chapter.

2. NOTATIONS AND MATHEMATICAL BACKGROUND

Let $SO(3)$ denote the special orthogonal group of $\mathbb{R}^{3x3}$ and $so(3)$ the group of antisymmetric matrices of R . We define by $(.)_\times$ the operator from $\mathbb{R}^{3\,x3} \rightarrow so(3)$ such that

$$\forall b \in \mathbb{R}^3,\ b_\times = \begin{bmatrix} 0 & -b_3 & b_2 \\ b_3 & 0 & -b_1 \\ -b_2 & b_1 & 0 \end{bmatrix} \tag{1}$$

where bi denotes the ith component of the vector b .

Let $V(.)$ be the inverse operator of $(.)_\times$, defined from $so(3) \rightarrow \mathbb{R}^3$, such that

$$\forall b \in \mathbb{R}^3, V(b_\times) = b \qquad \forall B \in so(3), V(B)_\times = B \tag{2}$$

For a given vector $b \in \mathbb{R}^3$and a given matrix $M \in \mathbb{R}^{3x3}$, let us consider the following notations and identities:

$$P_a(M)=\frac{M-M^T}{2} \qquad P_s(M)=\frac{M+M^T}{2} \tag{3}$$

$$\mathrm{tr}(P_a(M)P_s(M)) = 0 \tag{4}$$

$$\frac{1}{2}\,\mathrm{tr}(b_\times M)=-b^T V(P_a(M)) \tag{5}$$

The following identity will also be used:

$$\forall (A_a, B_a) \in \mathfrak{so}(3)^2,\ \frac{1}{2}\,\mathrm{tr}(A_a^T B_a) = V(A_a)^T V(B_a) \tag{6}$$

Denote by (γR ,nR) the angular-axis coordinates of a given matrix $R \in SO(3)$, and by Id the identity matrix of $\mathbb{R}$ 3x3. One has:

$$\forall R \in SO(3),\ \mathrm{tr}(Id - R) = 2(1 - \cos(\gamma R)) \tag{7}$$

3. UAV MODEL AND CONTROL STRATEGY

3.1 VTOL UAV Model

The VTOL UAV is represented by a rigid body of mass m and of tensor of inertia $I = diag\ (I_1,I_2,I_3)$ with I_1, I_2 and I_3 strictly positive. To describe the motion of the UAV, two reference frames are introduced: an inertial reference frame (I) associated with the vector basis (e_1, e_2, e_3) and a body frame (I) attached to the UAV and associated with the vector basis (e_1^b , e_1^b , e_1^b) (see Figure 1).

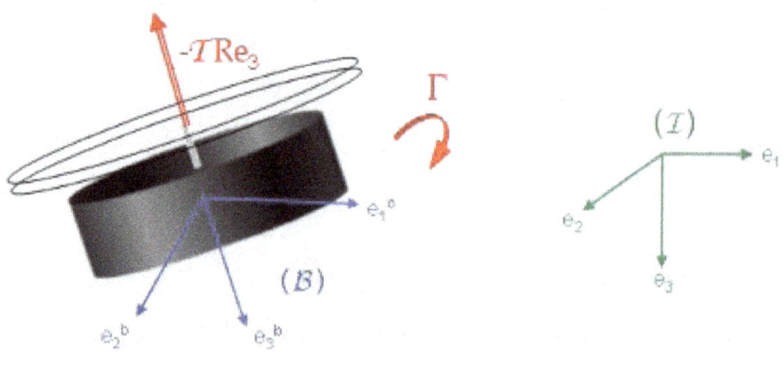

Figure 1. Reference frames

The position and the linear velocity of the UAV in (I) are respectively denoted $\chi=[\chi_x \chi_y \chi_z]^T$ and $v=[v_x\ v_y\ v_z]^T$.

The orientation of the UAV is given by the orientation matrix $R \in SO(3)$ from $(\mathcal{I})$to (β) , usually parameterized by Euler's pseudo angles y, θ, f (yaw, pitch, roll):

$$R=\begin{bmatrix} c_\theta\, c_\psi & s_\phi\, s_\theta\, c_\psi - c_\phi\, s_\psi & c_\phi\, s_\theta\, c_\psi + s_\phi\, s_\psi \\ c_\theta\, s_\psi & s_\phi\, s_\theta\, s_\psi + c_\phi\, c_\psi & c_\phi\, s_\theta\, s_\psi - s_\phi\, c_\psi \\ -s_\theta & s_\phi\, c_\theta & c_\phi\, c_\theta \end{bmatrix} \tag{8}$$

with the trigonometric shorthand notations $ca = \cos(a)$ and $sa = \sin(a)$, $\forall a \in \mathbb{R}$. Let $\Omega=[\omega p\ \omega q\ \omega r\ ]^T$ be the angular velocity of the UAV defined in (β) .

The dynamics of a rigid body can be described as:

$$\begin{cases} \dot{\chi} = v \\ m\dot{v} = F \\ \dot{R} = R\Omega\times \\ I\dot{\Omega} = -\Omega\times I\Omega + \Gamma \end{cases} \tag{9}$$

where the inputs are a translational force $F \in \mathbb{R}^3$ and a control torque $\Gamma \in \mathbb{R}^3$.

For the VTOL UAV, the translational force F combines thrust, lift, drag and gravity components. In quasi-stationary flight we can reasonably assume that the aerodynamic forces are always in direction $e_3^b\,(=Re_3)$, since the lift force predominates the other components (Hamel & Mahony, 2004). By separating the gravity component mge_3 from the combined aerodynamic forces, the dynamics of the VTOL UAV are rewritten as:

$$\begin{cases} \dot{\chi}=v \\ m\dot{v}=-T\,Re_3+mge_3 \\ \dot{R}=R\Omega_\times \\ I\dot{\Omega}=-\Omega\times I\Omega+\Gamma \end{cases} \tag{10}$$

where the inputs are the scalar $I \in \mathbb{R}$ representing the magnitude of the external forces applied in direction e_3^b, and the control torque $\Gamma = [\Gamma_1\ \Gamma_2\ \Gamma_3]^T$ defined in (β) .

3.2 Control Strategy

In this chapter, we consider the problem of the vehicle stabilization around a desired position $c\,d$ assumed to be constant ($c\,d = 0$).

For control design, let us define the position error $c = (c - c\,d)$. The system (10) becomes:

$$\begin{cases} \dot{\xi}=v \\ m\dot{v}=-T\,Re_3+mge_3 \\ \dot{R}=R\Omega_\times \\ I\dot{\Omega}=-\Omega\times I\Omega+\Gamma \end{cases} \tag{11}$$

Designing a controller for the model (11) can be realized by a classical backstepping approach applied to the whole dynamical system. In that case, the input vector - (Tlm) $Re3$ must be dynamically extended (Frazzoli et al., 2000), (Mahony et al., 1999). To avoid such a dynamical extension, the singular perturbation theory can be used to split the system dynamics into two reduced order subsystems (Khalil, 2002), (Calise, 1976). This approach leads to a time-scale separation between the translational dynamics (slow time-scale) and the orientation dynamics (fast time-scale). Reduced order controllers can therefore be designed to stabilize the system dynamics (Njaka et al., 1994).

We introduce the scaling parameter $\varepsilon \in (0,1]$ such that:

$$\begin{cases} \dot{\xi}=v \\ \dot{v}=-\frac{T}{m}Re_3+ge_3 \end{cases} \tag{12}$$

$$\begin{cases} \varepsilon\dot{R}=R\Omega_\times \\ \varepsilon I\dot{\Omega}=-\Omega\times I\Omega+\Gamma \end{cases} \tag{13}$$

For $\omega=1$, we obtain the full order system. Setting $\omega=0$ leads to the slow time-scale reduced- order system, where the orientation dynamics satisfy a quasi steady state condition $\Omega=0$.

For the translational dynamics (12), we will define the full vectorial term position control input. We will assign its desired value

$$\left(TRe_3\right)^d = f(\xi, v) \tag{14}$$

Assuming that the actuator dynamics can be neglected with respect to the rigid body dynamics of the UAV, the value T^d is considered to be instantaneously reached by T. Therefore, we have $(TRe_3)^d = TR^d e_3$, where Rd is the desired orientation of the vehicle. The vector $TR^d e^3$ will then be split into its magnitude

$$T=\|f(\xi, v)\| \tag{15}$$

Representing the first control input, and its direction.

$$R^{d} e_3 = \frac{1}{T} f(\xi, v) \tag{16}$$

Representing the desired orientation.

Remark 1: The desired orientation R^d *can then be deduced from the given direction* $R^d e3 = \frac{1}{T} f(\xi, v)$, *solving for* (ψ, θ, φ) *for a given specified yaw value* $\psi\ d$ *and using (8) (Hamel, 2002).*

Since $\varepsilon \ll 1$ for the considered system, the design of the position controller can be done in the slow time-scale, i.e. for $R \equiv R^d$.

For the orientation dynamics (13), we will assign the control torque Γ such that the orientation R of the UAV converges asymptotically to the desired orientation Rd, and such that the angular velocity Ω converges to Ω^d defined by:

$$R^d = R^d \, \Omega^d_{\times} \tag{17}$$

Therefore, the assigned control law will be of the form

$$\Gamma = h(R, R^d, \Omega, \Omega^d) \tag{18}$$

The design of the attitude controller can be done in the fast time-scale, assuming $\Omega^d = 0$. Indeed, defining $\Lambda = R - Rd$ and using the singular perturbation theory, we get

$$\varepsilon\Lambda = R\Omega\times - \varepsilon R^d \, \Omega^d_{\times} \tag{19}$$

Since $\varepsilon \ll 1$ for the considered system (the translational dynamics are characterized by a slow time-scale with respect to the orientation dynamics), the term $\varepsilon\, Rd\, \Omega\, d$ can therefore be ignored.

The structure of the controller we will develop is summarized in Figure 2. It is defined as a cascaded combination of the position controller and the attitude controller.

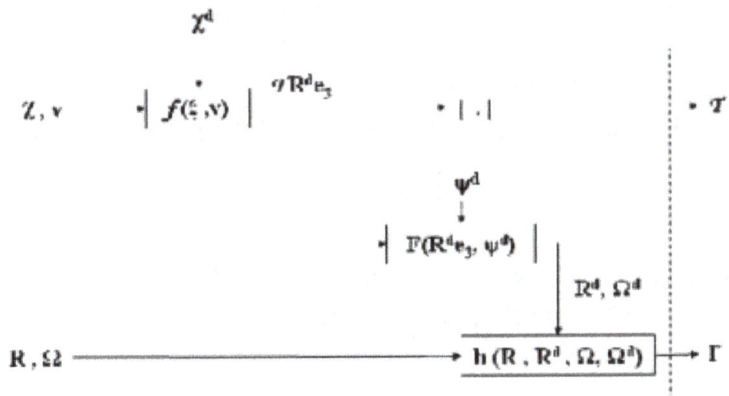

Figure 2. Cascaded structure of the controller

Remark 2: *Note that in the considered case where no velocity measurements are available, the functions f and h defining the control laws will not depend on v nor on* Ω.

4. POSITION CONTROLLER

Consider the translational dynamics. According to the above discussion of Section 3.2, we assume, for control design, that $Tre_3 \equiv TR^de_3$ is the control input of the translational dynamics.

Let $q, w \in \setminus \mathbb{R}^3$ be two virtual states and let $\delta \in \mathbb{R}^3$ be a virtual control such that:

$$\begin{cases} \dot{\xi} = v \\ \dot{v} = \dfrac{T}{m} Rde_3 + ge_3 \\ \dot{q} = -w \\ \dot{w} = \delta \end{cases} \tag{20}$$

Lemma 1: Consider the system dynamics (20). Let us define the control vector

$$TR^de_3 = \frac{m}{k_v}\{k_x\ \xi + k_1(\xi - q) + k_2(\xi - q + w)\} + mge_3 \tag{21}$$

and the virtual control

$$\delta = -\frac{1}{k_2}\{k_2 w + k_1(\xi - q) + k_2(\xi - q + w)\} \tag{22}$$

where k_x, k_v, k_1 and k_2 are strictly positive gains.

Consider the Lyapunov function candidate

$$S = \frac{1}{2}k_x\|\xi\|^2 + \frac{1}{2}k_v\|v\|^2 + \frac{1}{2}k_1\|\xi - q\|^2 + \frac{1}{2}k_2\|\xi - q + w\|^2 \tag{23}$$

Define k_{min} (k_x, k_v, k_1, k_2) and kmax (k_x, k_1, k_2).

Then, for any initial conditions $\xi(0)$, $v(0)$, and $q(0)=\xi(0)$ and $w(0)=0$ verifying

$$S(0) < \frac{1}{18}\frac{g^2 k_v^2 k_{min}}{k_{max}^2} \tag{24}$$

the control vector (21) along with the virtual control (22) exponentially stabilizes the translational dynamics (20), and the input T is strictly positive and bounded.

Proof:

The time derivative of the Lyapunov function candidate S is

$$\dot{S} = k_x \xi^T v + k_v v^T\left(-\frac{T}{m}R^d e_3 + ge_3\right) + k_1(\xi - q)^T(v + w) + k_2(\xi - q + w)^T(v + w + \delta) \tag{25}$$

which can be expressed as:

$$\dot{S}=v^T\left\{k_x\xi-k_v\frac{T}{m}R^d e_3+k_v g e_3+k_1(\xi-q)+k_2(\xi-q+w)\right\}+k_1 w^T(\xi-q)+k_2(\xi-q+w)^T(w+\delta) \quad (26)$$

Taking the control vector $TR^d e_3$ as defined in (21), we get:

$$\dot{S}=k_1 w^T(\xi-q)+k_2(\xi-q+w)^T(w+\delta) \quad (27)$$

We introduce (x -q) in the first term to obtain:

$$\dot{S}=k_1(w+(\xi-q)-(\xi-q))^T(\xi-q)+k_2(\xi-q+w)^T(w+\delta) \quad (28)$$

which leads to:

$$\dot{S}=-k_1\|\xi-q\|^2+(\xi-q+w)^T(k_1(\xi-q)+k_2(w+\delta)) \quad (29)$$

Choosing the virtual control δ according to (22) makes $\dot{S}$ become:

$$S=-k_1\,|x-q|^2-k_2\,|x-q+w|^2 \quad (30)$$

The application of La Salle's principle leads to $x \to q$ and $w \to -(w-q)$, i.e. $w \to 0$. By continuity, we get $\xi \to \dot{q}$ &, that is $v \to -w$. Since $w \to 0$, it yields $v \to 0$ and $\xi \to 0$, and by continuity $\dot{v} \to 0$. Combining the second equation of system (20) with (21) and the fact that $\dot{v} \to 0$ leads to:

$$\left(k_x\xi+k_1(\xi-q)+k_2(\xi-q+w)\right) \to 0 \quad (31)$$

Using ($w \to q$) $\to 0$ and $w \to 0$, we finally get $w \to$ and $q \to 0$. Therefore, the closed-loop system (20) is asymptotically stable, and since it is linear, we can conclude that it is exponentially stable.

It remains to show that the input T is strictly positive and bounded.

From (21), we have:

$$T=\left\|\frac{m}{k_v}\{k_x\,\xi+k_1(\xi-q)+k_2(\xi-q+w)\}+mge_3\right\| \quad (32)$$

By triangular inequality, we get:

$$T \ge mg-\frac{m}{k_v}k_{max}\left\{\|\xi\|+\|\xi-q\|+\|\xi-q+w\|\right\} \quad (33)$$

$$T \ge mg-3\frac{m}{k_v}k_{max}\left\{\|\xi\|^2+\|v\|^2+\|\xi-q\|^2+\|\xi-q+w\|^2\right\}^{\frac{1}{2}} \quad (34)$$

Using the definition of S and k_{min}, we have:

$$\frac{1}{2}k_{min}\left\{\|\xi\|^2+\|v\|^2+\|\xi-q\|^2+\|\xi-q+w\|^2\right\}\leq S \tag{35}$$

Since S is decreasing, we have:

$$S\leq S(0) \tag{36}$$

where, taking $q(0)=x\rightarrow(0)$ and $w(0)=0$, the initial value S (0) is defined by:

$$S(0)=\frac{1}{2}k_x\|\xi(0)\|^2+\frac{1}{2}k_v\|v(0)\|^2 \tag{37}$$

Using (34) along with (35) and (36), we get:

$$T\geq mg-3\frac{m}{k_v}k_{max}\sqrt{\frac{2S(0)}{k_{min}}} \tag{38}$$

Using condition (24), we obtain T >0. Let us finally show that T is bounded. From equation (32) and by triangular inequality, we also get:

$$T\leq mg+\frac{m}{k_v}k_{max}\{\|\xi\|+\|\xi-q\|+\|\xi-q+w\|\} \tag{39}$$

$$T\leq mg+3\frac{m}{k_v}k_{max}\left\{\|\xi\|^2+\|v\|^2+\|\xi-q\|^2+\|\xi-q+w\|^2\right\}^{\frac{1}{2}} \tag{40}$$

Using (35) and (36) leads to:

$$T\leq mg+3\frac{m}{k_v}k_{max}\sqrt{\frac{2S(0)}{k_{min}}} \tag{41}$$

and, therefore, the input T is bounded.

Lemma 1 ensures that the control vector (21) along with the virtual control (22) exponentially stabilizes the translational dynamics (20) without using any measurement of the linear velocity.

Remark 3: *From our previous discussion, Section 3.2, equation (32) can be directly used to provide the control input* T. *Furthermore, since* T *is strictly positive, the direction given by*

$$R^d e_3=\frac{1}{T}\left(mge_3+\frac{m}{k_v}\{k_x\,\xi+k_1(\xi-q)+k_2(\xi-q+w)\}\right) \tag{42}$$

is well defined and can be used to compute the desired orientation R^d.

Due to the position controller we developed, the closed-loop translational dynamics are exponentially stable for $R=R^d$. However, since the orientation R will not converge instantaneously to the desired value R^d, an orientation error term is introduced in the translational dynamics:

$$m\dot{v}=-TR^{d}e_3+mge_3-T(R-R^{d})e_3 \quad (43)$$

Therefore an attitude controller must be designed to allow, at least, asymptotic convergence of R to R^d.

5. ATTITUDE CONTROLLER

Consider the orientation dynamics (13) and assume that measurements on the angular velocity Ω are not available. Let us introduce derivative of a given function g will be denoted by $\tau = t/$. In the fast time-scale, the time

$$\overset{\circ}{g} = \frac{d}{d\tau}g = \varepsilon\frac{d}{dt}g \quad (44)$$

Similarly to the translational dynamics, we introduce two virtual states $Q\in SO(3)$, $W\in\mathbb{R}^3$ and a virtual control $\Delta \in \mathbb{R}^3$ for the orientation dynamics, such that:

$$\begin{cases} \overset{\circ}{R}=R\Omega_\times \\ I\overset{\circ}{\Omega}=-\Omega\times I\Omega+\Gamma \\ \overset{\circ}{Q}=-QW_\times \\ \overset{\circ}{W}=\Delta \end{cases} \quad (45)$$

For a given desired orientation R^d we define

$$\tilde{R}=(R^d)^T R \quad (46)$$

$$\tilde{Q}=Q^T\tilde{R} \quad (47)$$

According to the previous discussion of Section 3.2, we assume $\Omega^d = 0$ for control design, i.e. $\overset{\circ}{R^d}=0$.

Using (46) and (47), we rewrite (45) as:

$$\begin{cases} \overset{\circ}{\tilde{R}}=\tilde{R}\Omega_\times \\ I\overset{\circ}{\Omega}=-\Omega\times I\Omega+\Gamma \\ \overset{\circ}{\tilde{Q}}=W_\times\tilde{Q}+\tilde{Q}\Omega_\times \\ \overset{\circ}{W}=\Delta \end{cases} \quad (48)$$

Lemma 2: *Consider the orientation dynamics (48).*

Define the control torque

$$\Gamma=\frac{1}{k_\omega}\left\{-k_r V(P_a(\tilde{R}))-k_3 V(P_a(\tilde{Q}))+k_4 V(P_a(M))+k_4 V(P_a(N))\right\} \tag{49}$$

and the virtual control

$$\Delta=-\frac{1}{k_4}V\left(\frac{1}{2}k_3 P_a(\tilde{Q})+\frac{1}{2}k_4(W_\times\tilde{Q}+\tilde{Q}^T W_\times)+\frac{1}{2}k_5(W_\times+P_a(\tilde{Q}))\right) \tag{50}$$

where

$$M=(W_\times+P_a(\tilde{Q}))^T\tilde{Q} \tag{51}$$

$$N=\tilde{Q}^T(W_\times+P_a(\tilde{Q}))^T \tag{52}$$

and k_r, $k\omega$, k_3, k_4 and k_5 are strictly positive gains with

$$k_r<k_3 \tag{53}$$

Consider the control Lyapunov function candidate

$$\mathcal{L}=\frac{1}{2}k_r\ \mathrm{tr}(I_d-\tilde{R})+\frac{1}{2}k_\omega\Omega^T I\Omega+\frac{1}{2}k_3\ \mathrm{tr}(I_d-\tilde{Q})+\frac{1}{2}k_4\ \mathrm{tr}\left\{(W_\times+P_a(\tilde{Q}))^T(W_\times+P_a(\tilde{Q}))\right\} \tag{54}$$

Then, for any initial condition $\tilde{R}(0)$, $\Omega(0)$, with $Q(0)=\tilde{R}(0)$ and $W(0)=0$, such that

$$\mathcal{L}(0)<2k_r \tag{55}$$

the control torque (49) along with the virtual control (50) asymptotically stabilizes the orientation dynamics (48).

Proof : Consider the Lyapunov function candidate $\mathcal{L}$ defined by (54). It›s time derivative along the trajectories of (48) is given by:

$$\mathring{\mathcal{L}}=-\frac{1}{2}k_r\ \mathrm{tr}(\mathring{\tilde{R}})+k_\omega\Omega^T\{-\Omega\times I\Omega+\Gamma\}-\frac{1}{2}k_3\ \mathrm{tr}(\mathring{\tilde{Q}})+k_4\ \mathrm{tr}\left\{(W_\times+P_a(\tilde{Q}))^T\left(\Delta_\times+\overset{\circ}{\overbrace{P_a(\tilde{Q})}}\right)\right\} \tag{56}$$

where

$$\overset{\circ}{\overbrace{P_a(\tilde{Q})}}=\frac{1}{2}\left\{W_\times\tilde{Q}+\tilde{Q}\Omega_\times+\tilde{Q}^T W_\times+\Omega_\times\tilde{Q}^T\right\} \tag{57}$$

Using (57) and the fact that $\Omega^T(\Omega\times I\Omega)=0$, it yields:

$$\begin{aligned}\mathring{\mathcal{L}}=&-\frac{1}{2}k_r\ \mathrm{tr}(\tilde{R}\Omega_\times)+k_\omega\Omega^T\Gamma-\frac{1}{2}k_3\ \mathrm{tr}(W_\times\tilde{Q}+\tilde{Q}\Omega_\times)\\&+k_4\ \mathrm{tr}\left\{(W_\times+P_a(\tilde{Q}))^T\left(\Delta_\times+\frac{1}{2}\left\{W_\times\tilde{Q}+\tilde{Q}\Omega_\times+\tilde{Q}^T W_\times+\Omega_\times\tilde{Q}^T\right\}\right)\right\}\end{aligned} \tag{58}$$

Using identity (4) we get:

$$\mathring{\mathcal{L}}=-\frac{1}{2}k_r \operatorname{tr}(\Omega_\times P_a(\tilde{R}))+k_\omega \Omega^T\Gamma-\frac{1}{2}k_3 \operatorname{tr}(W_\times P_a(\tilde{Q}))-\frac{1}{2}k_3 \operatorname{tr}(\Omega_\times P_a(\tilde{Q})) \\ +k_4 \operatorname{tr}\left\{(W_\times+P_a(\tilde{Q}))^T\left(\Delta_\times+\frac{1}{2}\left\{W_\times\tilde{Q}+\tilde{Q}\Omega_\times+\tilde{Q}^T W_\times+\Omega_\times\tilde{Q}^T\right\}\right)\right\} \tag{59}$$

Also using (6) we obtain:

$$\mathring{\mathcal{L}}=\Omega^T\left\{k_r V(P_a(\tilde{R}))+k_\omega\Gamma+k_3V(P_a(\tilde{Q}))\right\}-\frac{1}{2}k_3 \operatorname{tr}(W_\times P_a(\tilde{Q}))+\frac{1}{2}k_4 \operatorname{tr}\left\{(W_\times+P_a(\tilde{Q}))^T\tilde{Q}\Omega_\times\right\} \\ +\frac{1}{2}k_4 \operatorname{tr}\left\{(W_\times+P_a(\tilde{Q}))^T\Omega_\times\tilde{Q}^T\right\}+k_4 \operatorname{tr}\left\{(W_\times+P_a(\tilde{Q}))^T(\Delta_\times+\frac{1}{2}(W_\times\tilde{Q}+\tilde{Q}^T W_\times))\right\} \tag{60}$$

Using (51) we have:

$$\frac{1}{2}k_4 \operatorname{tr}\left\{(W_\times+P_a(\tilde{Q}))^T\tilde{Q}\Omega_\times\right\}=\frac{1}{2}k_4 \operatorname{tr}\{M\Omega_\times\}=-k_4\Omega^T V(P_a(M)) \tag{61}$$

In the same way, we use (52) to get:

$$\frac{1}{2}k_4 \operatorname{tr}\left\{(W_\times+P_a(\tilde{Q}))^T\Omega_\times\tilde{Q}^T\right\}=-k_4\Omega^T V(P_a(N)) \tag{62}$$

Therefore, the time derivative of $\mathcal{L}$ can be simplified:

$$\mathring{\mathcal{L}}=\Omega^T\left\{k_r V(P_a(\tilde{R}))+k_\omega\Gamma+k_3V(P_a(\tilde{Q}))-k_4V(P_a(M))-k_4V(P_a(N))\right\}-\frac{1}{2}k_3 \operatorname{tr}(W_\times P_a(\tilde{Q})) \\ +k_4 \operatorname{tr}\left\{(W_\times+P_a(\tilde{Q}))^T(\Delta_\times+\frac{1}{2}(W_\times\tilde{Q}+\tilde{Q}^T W_\times))\right\} \tag{63}$$

Choosing Γ according to (49) leads to:

$$\mathring{\mathcal{L}}=-\frac{1}{2}k_3 \operatorname{tr}(W_\times P_a(\tilde{Q}))+k_4 \operatorname{tr}\left\{(W_\times+P_a(\tilde{Q}))^T(\Delta_\times+\frac{1}{2}(W_\times\tilde{Q}+\tilde{Q}^T W_\times))\right\} \tag{64}$$

As $W_\times^T=-W_\times$, and introducing $P_a(\tilde{Q})$ in the first term:

$$\mathring{\mathcal{L}}=\frac{1}{2}k_3 \operatorname{tr}\left\{\left(W_\times-P_a(\tilde{Q})+P_a(\tilde{Q})\right)^T P_a(\tilde{Q})\right\}+k_4 \operatorname{tr}\left\{(W_\times+P_a(\tilde{Q}))^T(\Delta_\times+\frac{1}{2}(W_\times\tilde{Q}+\tilde{Q}^T W_\times))\right\} \tag{65}$$

$$\mathring{\mathcal{L}}=-\frac{1}{2}k_3 \operatorname{tr}(P_a(\tilde{Q})^T P_a(\tilde{Q}))+\operatorname{tr}\left\{(W_\times+P_a(\tilde{Q}))^T\left(\frac{1}{2}k_3P_a(\tilde{Q})+k_4\Delta_\times+\frac{1}{2}k_4(W_\times\tilde{Q}+\tilde{Q}^T W_\times)\right)\right\} \tag{66}$$

Taking Δ as defined in (50), one has:

$$\mathring{\mathcal{L}}=-\frac{1}{2}k_3 \operatorname{tr}\left\{P_a(\tilde{Q})^T P_a(\tilde{Q})\right\}-\frac{1}{2}k_5 \operatorname{tr}\left\{(W_\times+P_a(\tilde{Q}))^T(W_\times+P_a(\tilde{Q}))\right\} \tag{67}$$

Using again identity (6), we finally have:

$$\mathring{\mathcal{L}}=-k_3\left\|V(P_a(\tilde{Q}))\right\|^2-k_5\left\|V(W_\times+P_a(\tilde{Q}))\right\|^2 \tag{68}$$

ensuring that $\mathcal{L}$ is strictly decreasing until $P_a(\tilde{Q})\rightarrow 0$ and $W_\times\rightarrow -P_a(\tilde{Q})$, i.e. $W_\times\rightarrow 0$. Denote by $(\gamma_{\tilde{Q}}, n_{\tilde{Q}})$ the angle-axis coordinates of $\tilde{Q}$. Using (7), one has:

$$k_3(1-\cos(\gamma_{\tilde{Q}}))=\frac{1}{2}k_3\ \mathrm{tr}(I_d-\tilde{Q})\leq\mathcal{L} \qquad (69)$$

Since $\mathcal{L}$ is decreasing, we have $\mathcal{L}\leq\mathcal{L}(0)$. Using (55) it yields:

$$k_3(1-\cos(\gamma_{\tilde{Q}}))\leq\mathcal{L}\leq\mathcal{L}(0)<2k_r \qquad (70)$$

Using (53), we get:

$$1-\cos(\gamma_{\tilde{Q}})<2\frac{k_r}{k_3}<2 \qquad (71)$$

From $P_a(\tilde{Q})\to 0$, we have $\gamma_{\tilde{Q}}=0$ or $\gamma_{\tilde{Q}}=\pm\pi$. The second possibility is excluded by (71). Therefore we have $\tilde{Q}\to I_d$. By (47), it yields $\tilde{R}\to Q$. By continuity and using La Salle's principle, we get $\mathring{\tilde{R}}\to\mathring{Q}$. Using the first equation of (48) and the third equation of (45), one has $\tilde{R}\Omega_\times\to -QW_\times$. Since $\tilde{R}$ is orthogonal, we get $\Omega_\times\to-\tilde{R}^TQW_\times$. Using $W_\times\to 0$ it yields $\Omega_\times\to 0$ and then $\Omega\to 0$. Therefore, using the first equation of (48), we get $\mathring{\tilde{R}}\to 0$. By continuity $\mathring{\Omega}\to 0$ and then, by the second equation of system (48), $\Gamma\to 0$. Knowing that $P_a(\tilde{Q})$ and $W_\times$ converge to zero, one can ensure that, respectively from (51) and (52), *M* and *N* converge to zero. Combining the above discussion with the fact that $\Gamma\to 0$, equation (49) ensures that $P_a(\tilde{R})\to 0$.

Similarly to the previous analysis on $\tilde{Q}$, let us denote $(\gamma_{\tilde{R}}, n_{\tilde{R}})$. $\tilde{R}$ the angle-axis coordinates of One has:

$$k_r(1-\cos(\gamma_{\tilde{R}}))=\frac{1}{2}k_r\ \mathrm{tr}(I_d-\tilde{R})\leq\mathcal{L}\leq\mathcal{L}(0)<2k_r. \qquad (72)$$

It yields

$$1-\cos(\gamma_{\tilde{R}})<2 \qquad (73)$$

From $P_a(\tilde{R})\to 0$, we have $\gamma_{\tilde{R}}=0$ or $\gamma_{\tilde{R}}=\pm\pi$. The second possibility is excluded by (73). Therefore, we finally have $\tilde{R}\to I_d$ and $\tilde{R}\to R^d$.

Lemma 2 ensures that the control (49) along with the virtual control (50) asymptotically stabilizes the orientation dynamics (48) without using any measurement of the angular velocity.

Remark 4: *The time-scale parameter ε > 0 is chosen such that the deviation $\tilde{R}$ converges to I_d faster than the translational dynamics (ε « 1).*

Remark 5: *The condition (55) is not very conservative. Choosing condition (55) can be simplified: W (0)=0 and Q(0)= $\tilde{R}(0)$, the condition (55) can be simplified:*

$$\frac{k_r}{k_\omega}>\frac{E_\omega(0)}{(2-E_r(0))} \qquad (74)$$

with $E_r(0)=\frac{1}{2}\ \mathrm{tr}(I_d-\tilde{R}(0))$ *and* $E_\omega(0)=\frac{1}{2}\Omega(0)^T I\Omega(0)$

6. STABILITY ANALYSIS

We consider the full dynamics of the system along with virtual states and with the orientation error term in the translational dynamics:

$$\begin{cases} \dot{\xi}=v \\ \dot{v}=-\frac{T}{m}R^{d}e_3+ge_3-\frac{T}{m}(R-R^{d})e_3 \\ \dot{q}=-w \\ \dot{w}=\delta \\ \overset{\circ}{\tilde{R}}=\tilde{R}\Omega_{\times} \\ I\overset{\circ}{\Omega}=-\Omega\times I\Omega+\Gamma \\ \overset{\circ}{\tilde{Q}}=W_{\times}\tilde{Q}+\tilde{Q}\Omega_{\times} \\ \overset{\circ}{W}=\Delta \end{cases} \quad (75)$$

with $\tilde{R}$ and $\tilde{Q}$ respectively defined by (46) and (47).

Proposition 1: *Consider the system dynamics (75). Under the conditions (24), (53) and (55), the control laws (21) and (49), along with the virtual controls (22) and (50), asymptotically stabilize the system (75).*

Sketch of the proof:

By Lemma 2, under the conditions (53) and (55), the closed loop orientation dynamics are asymptotically stable when (49) and (50) are respectively used as control and virtual control. By Lemma 1, under the condition (24), the input T is bounded. Therefore, the orientation error term $-\frac{T}{m}(R-R^{d})e_3$ asymptotically converges to zero.

Since, from Lemma 1, the control of the translational dynamics is exponentially stabilizing for $R\equiv R^{d}$, we can use (Khalil, 2002) to conclude that the control of the translational dynamics is asymptotically stabilizing in presence of the orientation error term.

Therefore, the system (75) is asymptotically stable when the control laws (21) and (49) are used along with the virtual control laws (22) and (50).

By introducing virtual states, we have been able to design stabilizing controllers for the position and attitude of the VTOL UAV model using no measurement of the linear velocity v nor of the angular velocity Ω .

Remark 6: *In the case where only the linear velocity v of the vehicle is not measured, a detailed proof using the singular perturbation theory can be found in (Bertrand et al., 2008).*

Remark 7: *Control laws for trajectory tracking, in the case where the linear and angular velocities are not measured, have been proposed in (Bertrand et al., 2007b).*

7. SIMULATION RESULTS

The VTOL UAV is described by the following parameters: m = 2.5 kg , I_1=I_2 = 0.13 kg.m^2 and I_3= 0.16 kg.m^2. The gravitational acceleration is g = 9.81 m.s^{-2}.

Simulation results are provided for stabilization at hover around the desired position $\chi^d = [3\ -1\ \ 1]^T$ (m), starting from the initial condition $\chi_0 = [5\ -3\ \ 4]^T$ (m) $[\psi_0\ \theta_0\ \phi_0] = [0\ -8\ -10]^T$ (deg), $v_0 = 0$ and $\Omega_0 = 0$. The desired yaw ψ^d equal to zero.

The values of the gains are: k_x = 0.2, k_v = 3.0, k_1 = 0.8, k_2 = 0.8, k_r = 0.74, k_ω = 3.3, $k3$ = 12, k_4 = 0.25, k_5 = 6.1.

Figure 3 presents the coordinates of the position error ξ = [$x\,y\,z$] and attitude angles. Stabilization of the UAV model is achieved from the given initial condition with satisfying behaviour performances. The input T and the components of the control torque Γ are plotted in Figure 4.

The evolution of the angular deviation terms $\tilde{\phi}=\phi-\phi^d$, $\tilde{\theta}=\theta-\theta^d$ and $\tilde{\psi}=\psi-\psi^d$ are presented in Figure 5. It can be verified that these terms converge faster than the closed loop translation, hence validating the time scale separation approach used for the design of the controllers.

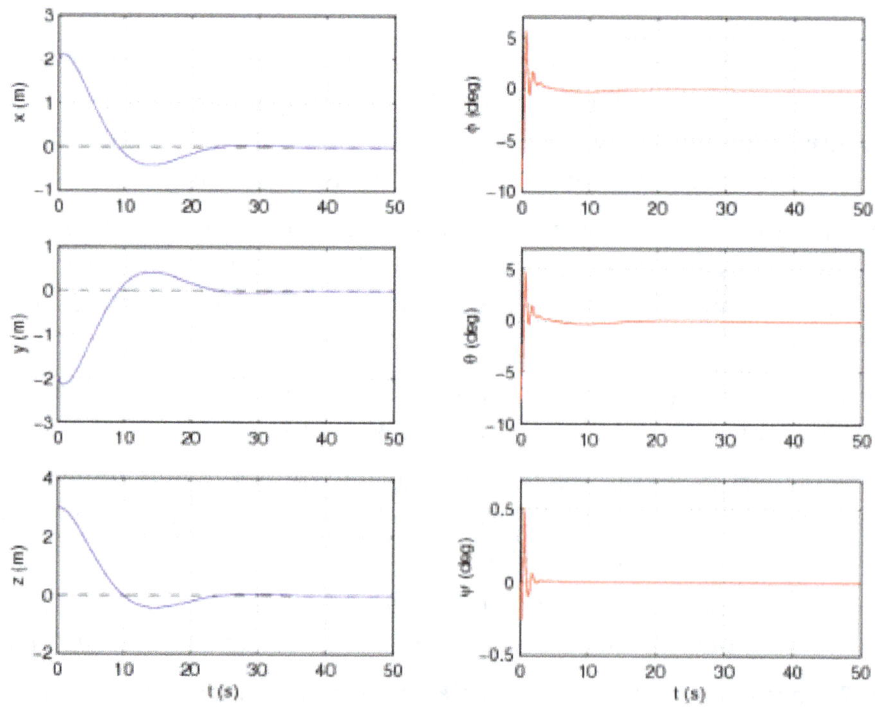

Figure 3. Position error and attitude angles

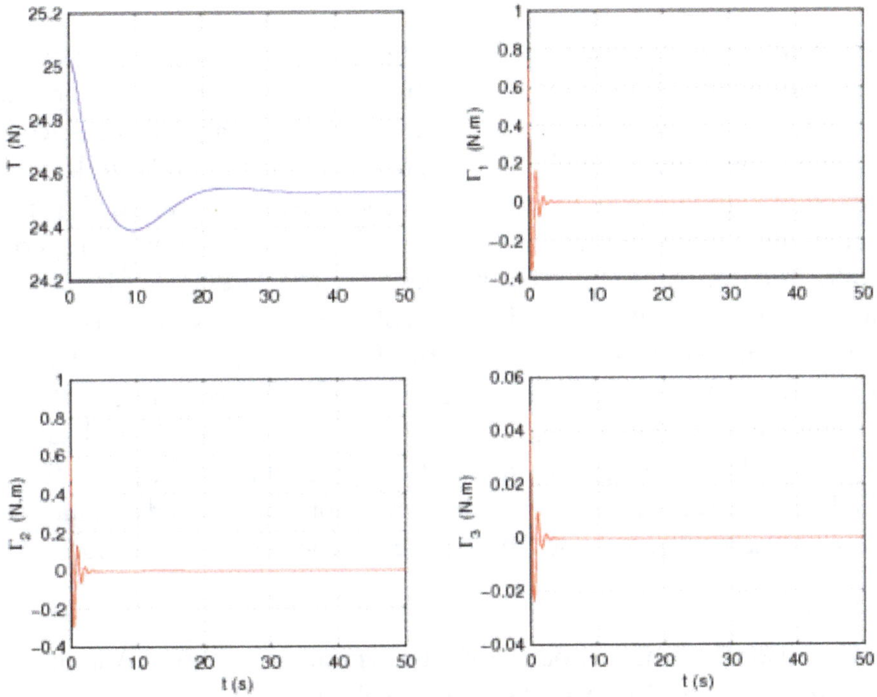

Figure 4. Control inputs

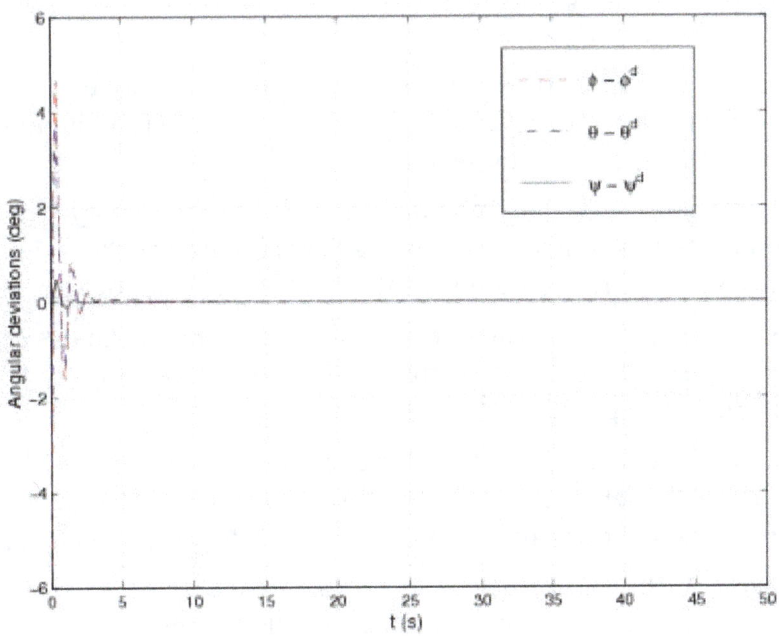

Figure 5. Angular deviation terms

8. CONCISE SUMMARY

This chapter introduces a method for designing guidance and control rules to stabilize a scale-model VTOL UAV in the absence of measurements for linear and angular velocity. The controller is constructed in two phases, influenced by the model's cascade structure and the single perturbation method, while accounting for a time-scale separation between translational and orientation dynamics. The position controller calculates the magnitude of external pressures, regarded as a control input for the translational dynamics, and the intended orientation of the UAV. The attitude controller provides control torque that guarantees asymptotic convergence of the actual orientation to the intended orientation. The suggested approach involves the creation of these two feedback controllers through the incorporation of virtual states in the system dynamics, without the utilization of any observer. This study is founded on a kinematic representation of the SO(3) group and its manifold. Stability analysis elements have been presented, and simulation results have been supplied to demonstrate the efficacy of the proposed method.

REFERENCES

Arkela, M. R. (2001). Rigid Body Attitude Tracking without Angular Velocity Feedback. Systems & Control Letters, Vol. 42, pp. 321-326, 2001

Astolfi, A. & Lovera, M. (2002). Global Spacecraft Attitude Control using Magnetic Actuators, Proceedings of the American Control Conference, Vol. 2, pp. 1331-1335, Anchorage, USA, 2002

Bertrand, S.; Piet-Lahanier, H. & Hamel, T. (2007a). Contractive Model Predictive Control of an Unmanned Aerial Vehicle, 17th IFAC Symposium on Automatic Control in Aerospace, Toulouse, France, 2007

Bertrand, S.; Hamel, T. & Piet-Lahanier, H. (2007b). Trajectory Tracking of an Unmanned Aerial Vehicle Model using Partial State Feedback, Proceedings of the European Control Conference 2007, pp. 307-314, Kos, Greece, 2007

Bertrand, S.; Hamel, T.; Piet-Lahanier, H. (2008). Stability Analysis of an UAV Controller Using Singular Perturbation Theory, Proceedings of the 17th IFAC World Congress, pp. 5706-5711, Seoul, Korea, 2008

Bouabdallah, S. & Siegwart, R. (2005). Backstepping and Sliding-Mode Techniques Applied to an Indoor Micro Quadrotor, Proceedings of the 2005 IEEE International Conference on Robotics and Automation, pp 2259-2264, Barcelona, Spain, 2005

Burg, T.; Dawson, D.; Hu, J. & de Queiroz, M. (1996). An Adaptive Partial State-Feedback Controller for RLED Robot Manipulators. IEEE Transactions on Automatic Control, Vol. 41, No. 7, pp. 1024-1030, 1996

Burg, T.; Dawson, D. & Vedagarbha, P. (1997). A Redesigned DCAL Controller without Velocity Measurements: Theory and Demonstration. Robotica, Vol. 15, pp. 337-346, 1997

Calise, A. J. (1976). Singular Perturbation Methods for Variational Problems in Aircraft Flight. IEEE Transactions on Automatic Control, Vol. 41, No. 3, pp. 345-353, 1976

Castillo, P.; Dzul, A. & Lozano, R. (2004). Real-Time Stabilization and Tracking of a Four- Rotor Mini Rotorcraft. IEEE Transactions on Control Systems Technology, Vol. 12, No. 4, pp. 510-516, 2004

Costic, B. T.; Dawson, D. M.; de Queiroz, M. S. & Kapila, V. (2000). A Quaternion-Based Adaptive Attitude Tracking Controller Without Velocity Measurements, Proceedings of the 39th IEEE Conference on Decision and Control, Vol. 3, pp. 2424-2429, Sydney, Australia, 2000

Dixon, W. E.; Zergeroglu, E.; Dawson, D. M. & Hannan, M. W. (2000). Global Adaptive Partial State Feedback Tracking Control of Rigid-Link Flexible-Joints Robots. Robotica, Vol. 18, pp. 325-336, 2000

Do, K. D.; Jiang, Z. P. & Pan, J. (2003). On Global Tracking Control of a VTOL Aircraft without Velocity Measurements. IEEE Transactions on Automatic Control, Vol. 48, No. 12, pp. 2212-2217, 2003

Frazzoli, E.; Dahleh, M.A. & Feron, E. (2000). Trajectory Tracking Control Design for Autonomous Helicopters using a Backstepping Algorithm, Proceedings of the American Control Conference, Vol. 6, pp. 4102-4107, Chicago, USA, 2000

Hamel, T. & Mahony, R. (2004). Pure 2D Visual Control for a Class of Under-Actuated Dynamic Systems, Proceedings of the 2004 IEEE International Conference on Robotics and Automation, Vol. 3, pp. 2229-2235, New Orleans, USA, 2004

Hamel, T.; Mahony, R; Lozano, R. & Ostrowski, J. (2002). Dynamic Modelling and Configuration Stabilization for an X4-Flyer, Proceedings of the 15th IFAC World Congress, Vol. 15, Barcelona, Spain, 2002

Khalil, H. K. (2002). Nonlinear Systems, Prentice Hall, ISBN: 0130673897, New Jersey, USA Kim, H. J.; Shim, D. H. & Sastry, S. (2002). Nonlinear Model Predictive Tracking Control for Rotorcraft-Based Unmanned Aerial Vehicles, Proceedings of the American Control Conference, Vol. 5, pp. 3576-3581, Anchorage, USA, 2002

Kondak, K.; Bernard, M.; Meyer, N. & Hommel, G. (2007). Autonomously Flying VTOL-Robots: Modeling and Control, Proceedings of the 2007 IEEE International Conference on Robotics and Automation, pp. 736-741, Roma, Italy, 2007

Kundak, N. & Mettler, B. (2007). Experimental Framework for Evaluating Autonomous Guidance and Control Algorithms for Agile Aerial Vehicles, Proceedings of the European Control Conference 2007, pp. 293-300, Kos, Greece, 2007

Lizarralde, F. & Wen, J. T. (1996). Attitude Control without Angular Velocity Measurement: A Passivity Approach. IEEE Transactions on Automatic Control, Vol. 41, No. 3, pp. 468-472, 1996

Mahony, R. & Hamel, T. (2004). Robust Trajectory Tracking for a Scale Model Autonomous Helicopter. International Journal of Robust and Nonlinear Control, Vol. 14, pp. 1035- 1059, 2004

Mahony, R.; Hamel, T. & Dzul, A. (1999). Hover Control via Approximate Lyapunov Control for a Model Helicopter, Proceedings of the 38th IEEE Conference on Decision and Control, Vol. 4, pp. 3490-3495, Phoenix, USA, 1999

Njaka, C. E.; Menon, P. K. & Cheng, V. H. L. (1994). Towards an Advanced Nonlinear Rotorcraft Flight Control System Design, 13th AIAA/IEEE Digital Avionics Systems Conference, Phoenix, USA, 1994

Tayebi, A. (2007). A Velocity-free Attitude Tracking Controller for Rigid Spacecraft, Proceedings of the 46th IEEE Conference on Decision and Control, pp. 6430-6434, New Orleans, USA, 2007

Tsiotras, P. (1998). Further Passivity Results for the Attitude Control Problem. IEEE Transactions on Automatic Control, Vol. 43, No. 11, pp. 1597-1600, 1998

Valenti, M.; Bethke, B.; Fiore, G.; How, J. P. & Feron, E. (2006). Indoor Multi-Vehicle Flight Testbed for Fault Detection, Isolation, and Recovery, AIAA Guidance, Navigation, and Control Conference and Exhibit, Keystone, USA, 2006

Wong, H.; de Queiroz, M. S. & Kapila, V. (2000). Adaptive Tracking Control Using Synthesized Velocity from Attitude Measurements, Proceedings of the American Control Conference, Vol. 3, pp. 1572-1576, Chicago, USA, 2000

Chapter 7
OPTIMIZATION OF FLIGHT CONTROL SYSTEM DESIGN USING GENETIC PROGRAMMING

1. INTRODUCTION: AN OVERVIEW

This chapter outlines a methodology used to construct a controller that meets the requirements of shipboard recovery for a fixed-wing UAV. The methodology is thorough and should be easily applicable to diverse UAV models and a range of task objectives. The methodology can be used to a wide array of control problems with suitable adjustments to the control law representation. The creation of the recovery controller for the UAV Ariel serves as a design illustration to endorse the process. This chapter examines the adaption of Evolutionary Algorithms for aircraft control issues. The investigation of standard control laws and control design methodologies initiates the process. The UAV controller's architecture and the formulation of control laws appropriate for evolutionary design are subsequently established. This is succeeded by the formulation of the generic evolutionary design algorithm, subsequently applied to the UAV recovery issue. The results given exhibit the strong characteristics of the designed control system.

2. AIRCRAFT FLIGHT CONTROL

2.1 Types of feedback control

Not unlike the generic control approach, aircraft flight control is built around a feedback concept. Its basic scheme is shown in Fig. 1. The controller is fed by the difference between the commanded reference signal r and the system output y. It generates the system control inputs u according to one or another algorithm.

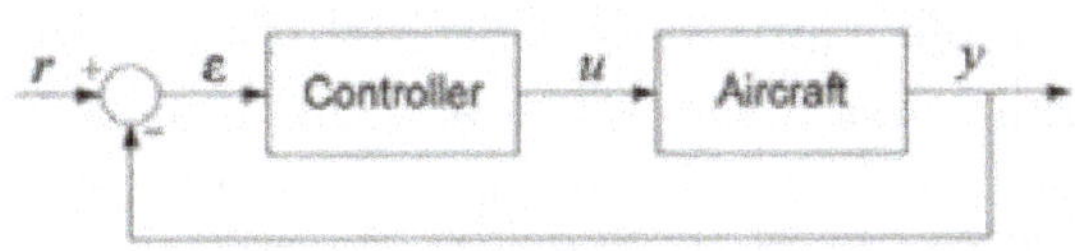

Figure 1. Feedback system (one degree of freedom)

One of the main tasks of flight control as an engineering discipline is design of the controllers which enable a given aircraft to complete a defined mission in the most optimal manner, where optimality is based on mission objective. A number of techniques of producing a required controller have been developed over the last hundred years since the first altitude hold autopilot was introduced by Sperry Gyroscope Company in 1912. Most of the techniques make certain assumptions about the controlled system (i.e. the aircraft), most notably linearity of its dynamics and rigidity of the airframe, to simplify synthesis of the controller. A brief overview of types of feedback control is presented below.

2.1.1 On-Off control

The simplest feedback control is the *on-off control*, also referred to among engineers as *bang- bang control*. This control law can be expressed as follows:

$$u = u_0 + \begin{cases} k, z < a \\ 0, z > a \end{cases} \tag{1}$$

where u_0 and k are arbitrary offset and gain suitable for the given system, and a is the threshold (set point). It can be extended to a multiple choice situation.

This law is not particularly suitable for direct flight control in normal situations. Indeed, with only two (or several) discrete control input levels, the system output will tend to oscillate about the set point, no matter how well damped the system is, because the control signal will not switch until the set point is already passed. Moreover, if the dynamics of the system is fast or a significant amount of noise in the system output is present, the controller will be switching also fast ('hunting'), possibly causing extensive wear to the control actuators.

To prevent such behaviour, a special 'dead zone' (or 'deadband') is established around the set point where no switching occurs. Obviously, this reduces the accuracy of control. However, the on-off control comes into play when event handling is required. A classic example of application of such control for aircraft is stall recovery. When angle of attack exceeds a specified value, a nose-down elevator command is issued. A similar logic can be implemented for overload protection or ground collision avoidance.

Another important area in the aerospace field where on-off rules can be successfully applied is internal switching between the controllers (or controller parameters). This approach is known as Gain scheduling (Rugh, 2000). The technique takes advantage of a set of relatively simple controllers optimised for different points of the flight envelope (or other conditions). However, it was found that rapid switching may cause stability and robustness problems (Shamma & Athans, 1992). One of the popular simple solutions is to 'blend' (interpolate) the output of two or more controllers, which effectively turns simple on-off switching into a complicated control method.

2.1.2 PID control

The proportional-integral-derivative (PID) control is probably the most widely used type of control, thanks to the simplicity of its formulation and in most cases, predictable characteristics. In a closed-loop system like that in Fig. 1, its control law is

$$u=K_p\varepsilon+K_I\int\varepsilon dt+K_D\dot{\varepsilon} \tag{2}$$

where the parameters K_P, K_I and K_D are coefficients for proportional, integral and derivative components of the input error signal respectively. By adjusting these three parameters, the desired closed-loop dynamics can be obtained.

Probably the most problematic issue with the PID control is due to the differential term. While being important for good response time and high-speed dynamics, the differential component keenly suffers from both instrumental noise and calculation errors. Indeed, even a small amount of noise can greatly affect the slope of the input signal. At the same time, numerical calculation of the derivative (for a digital controller in particular) must be done with a fairly small time step to obtain correct slope at a given point. Although lowpass filtering applied to the input signal smoothens the signal, it severely compromises the usefulness of the derivative term because the low-pass filter and derivative control effectively cancel each other out.

In contrast, the integral term averages its input, which tends to eliminate noise. However, a common problem associated with the integral control owes exactly to its 'memory'. When a large input value persists over a significant amount of time, the integral term becomes also large and remains large even after the input diminishes. This causes a significant overshoot to the opposite values and the process continues.

In general, integral control has a negative impact on stability and care must be taken when adjusting the integral coefficient. Limiting the integrator state is a common aid for this problem. A more elegant approach involves 'washing out' the integrator state by incorporating a simple gain feedback, effectively implementing a low-pass filter to the input signal. PID control found wide application in the aerospace field, especially where near- linear behaviour takes place, for example, in various hold and tracking autopilots such as attitude hold and flight path following. The techniques of selecting the optimal PID coefficients are well established and widely accepted. Typically, they use the frequency domain as a design region and utilise phase margins and gain margins to illustrate robustness. However, design of a PID controller for nonlinear or multi-input multi-output (MIMO) systems where significant coupling between the different system inputs and outputs exists is complicated. As the aircraft control objectives evolved, new design and control techniques were developing. Although many of them essentially represent an elaborated version of PID control, they are outlined in the following sections.

2.1.3 ear optimal control

The concept of optimality in mathematics refers to minimisation of a certain problem dependent cost functional:

$$J=\int_0^T g(t, x,u)dt+h(x(T)) \tag{3}$$

where T is the final time, u is the control inputs and x is the state of the controlled system. The last term represents the final cost that depends on the state in which the system ends up. Optimal control is, therefore, finding the control law $u(t)$ that minimises J. In general, for an arbitrary system and cost functional, only numerical search can find the optimal (or near optimal) solution.

There is a number of types of linear optimal control, such as the *Linear Quadratic Regulator (LQR),* the *Linear Quadratic Gaussian with Loop Transfer Recovery (LQG/LTR)* and the extended Kalman filter (EKF). The theory and application of these control techniques and Kalman filtering is detailed in many common control and signal processing textbooks, such as for example (Brown & Hwang, 1992; Bryson & Ho, 1975; Kalman, 1960; Maciejowski, 1989).

The Kalman filter relies on the model of the system (in its predicting part), and the guaranteed performance of the controller can easily be lost when unmodelled dynamics, disturbances and measurement noise are introduced. Indeed, robustness is a challenging issue of modern control designs. Also, the well known in classic linear design phase and gain margin concepts cannot be easily applied to the multivariable systems that modern control is so suited for. These problems led to the introduction of robust modern control theory.

2.1.4 Robust modern control

Unlike traditional optimal control, robust optimal control minimises the influence of various types of uncertainties in addition to (or even instead of) performance and control energy optimisation. Generally, this implies design of a possibly low gain controller with reduced sensitivity to input changes. As a consequence, robust controllers often tend to be conservative and slow. On the other hand, they may be thought as the stabilising controllers for the whole set of plants (which include a range of uncertainties) and not only for the modelled system, which is a more demanding task.

The majority of modern robust control techniques have origins in the classical frequency domain methods. The key modification of the classic methods is shifting from eigenvalues to singular values (of the transfer function that describes the system), the singular value Bode plot being the major indicator of multivariable feedback system performance (Doyle & Stein, 1981).

Probably the most popular modern robust control design techniques (particularly in the aerospace field) are H_2 and H_∞ control, also known as the frequency-weighted

LQG synthesis and the small gain problem respectively. These techniques and the underlying theories are thoroughly described in several works, notably (Doyle at al., 1989; Kwakernaak, 1993; Zhou & Doyle, 1998).

The H_∞ control design found extensive use for aircraft flight control. One of the first such applications was the development of controllers for the longitudinal control of a Harrier jump jet (Hyde, 1991). This work has been subsequently extended in (Postlethwaite & Bates, 1999) to fully integrated longitudinal, lateral and propulsive control. Other works include (Kaminer et al., 1990), where a lateral autopilot for a large civil aircraft is designed, and (Khammash & Zou, 1999), where a robust longitudinal controller subject to aircraft weight and c.g. uncertainty is demonstrated.

A mixed H_2 / H_∞ approach is applied in (Shue & Agarwal, 1999) to design an autoland controller for a large commercial aircraft. The method employed here utilises the H_2 controller for slow trajectory tracking and the $H\infty$ controller for fast dynamic robustness and disturbance rejection.

Several H_∞ controllers have been tried to accomplish the UAV shipboard launch task (Crump, 2002). It has been found that these controllers perform quite well in nominal situations. However, in the presence of large disturbances which place the aircraft well beyond its linear design operating point, the controllers performed poorly (sometimes extremely). At the same time, inability to include even simple static nonlinearities such as time delays and saturations made it difficult to synthesise a practical controller for this task within linear approach. Another deficiency found is common to all frequency domain techniques: the frequency domain performance specifications cannot be rigidly translated into time and spatial domain specifications. Meanwhile, time and especially spatial (trajectory) constrains are crucial for both launch and recovery tasks.

The time domain performance can be accounted for directly in the time domain $l1$ design (Blanchini & Sznaier, 1994; Dahleh & Pearson, 1987).

It is often extended to include the frequency domain objectives, resulting in a mixed norm approach. However, $l1$ design is plagued by the excessive order of the generated controllers. This is usually solved by reducing the problem to suboptimal control, imposing several restrictions on the system and performance specifications (Sznaier & Holmes, 1996). The application of $l1$ approach to flight control is discussed in (Skogestad & Postlewaite, 1997), with the conclusion that controllers with excessive order will generally be produced when using practical constraints.

There have been attempts to solve the $H\infty$ optimal control problems for nonlinear systems. However, these methods usually rely on very limiting assumptions about the model, uncertainties and disturbance structure. The mathematical development of nonlinear $H\infty$ control can be found in (Dalsamo & Egeland, 1995). An example of nonlinear control of an agile missile is given in (Wise, 1996).

Despite a very complicated solution algorithm, this work is limited by a linear assumption on the vehicle aerodynamics, reducing the benefits gained from the use of nonlinear control.

2.1.5 Nonlinear control

Linear control design techniques have been used for flight control problems for many years. One of the reasons why aircraft can be controlled quite well by linear controllers is that they behave almost linearly through most of their flight envelope. However, when the aircraft is required to pass through a highly nonlinear dynamic region or when other complicated control objectives are set, it has been found by several researchers that it is difficult to obtain practical controllers based on linear design techniques. The UAV shipboard launch and recovery tasks are substantially nonlinear problems. The sources of nonlinearities are the aerodynamic forces generated at low airspeeds and high angles of attack (especially when wind disturbances are present); trajectory constraints imposed due to proximity of ground (water) and ship installations; kinematic nonlinearities when active manoeuvring is required; actuator saturations and some more.

In contrast to the linear systems, the characteristics of the nonlinear systems are not simply classified and there are no general methods comparable in power to those of linear analysis. Nonlinear techniques are quite often designed for individual cases, regularly with no mathematical justification and no clear idea of re-applicability of the methods.

Some of the more popular nonlinear control techniques are covered in textbooks (Atherton, 1975; Graham & McRuler, 1971).

The most basic nonlinear control laws and the On-off control and Gain scheduling, noting that these controllers often lack robustness when the controllers are scheduled rapidly.

Another modern control technique remotely related to the on-off control is *variable structure* (also known as *sliding mode*) control (DeCarlo et al., 1988; Utkin, 1978). In this approach, a hypersurface (in state space) called *sliding surface* or *switching surface* is selected so that the system trajectory exhibits desirable behaviour when confined to this hypersurface. Depending on whether the current state is above or below the sliding surface, a different control gain is applied. Unlike gain scheduling, the method involves high speed switching to keep the system on the sliding surface.

A completely different approach is to enable applicability of the well known linear control methods to control nonlinear systems. This can be achieved using *nonlinear dynamic inversion*. This process, also known as *feedback linearisation*, involves online approximate linearisation of a nonlinear plant via feedback. Dynamic inversion gained particular attention in aviation industry in the late 1980s and 90s, aiming to control high performance fighters during high angle of attack manoeuvres (known as *supermanoeuvres*). One of the early applications is NASA High Angle of Attack Vehicle (HARV) (Bugajski & Enns, 1992). In this work, quite good simulation performance results are obtained; however, with the inversion based on the same simulation model, any possible discrepancies are transferred into the controller, leading to questionable results in physical implementation with respect to incorrectly modelled dynamics.

2.1.6 Intelligent control

Intelligent control is a general and somewhat bold term that describes a diverse collection of relatively novel and non-traditional control techniques based on the so called *soft computing* approach. These include neural networks, fuzzy logic, adaptive control, genetic algorithms and several others. Often they are combined with each other as well as with more traditional methods; for example, fuzzy logic controller parameters being optimised using genetic algorithms or a neural network driving a traditional linear controller.

Neural network (NN), very basically, is a network of simple nonlinear processing elements (neurons) which can exhibit complex global behaviour determined by element parameters and the connections between the processing elements. The use of artificial neural networks for control problems receives an increased attention over the last two decades. It has been shown that a certain class of NN can approximate any continuous nonlinear function with any desired accuracy (Spooner, 2002). This property allows to employ NN for *system identification* purposes, which can be performed both offline and online. This approach is used in (Coley, 1998; Kim & Calise, 1997) for flight control of a tilt-rotor aircraft and the F-18 fighter.

Another area of intensive application of NN is fault tolerant control, made possible due to online adaptation capability of NN. In the recent work (Pashilkar et al., 2006), an 'add-on' neural controller is developed to increase auto-landing capabilities of a damaged aircraft with one of two stuck control surfaces. A significant increase in successful landings rate is shown, especially when the control surfaces are stuck at large deflections.

Fuzzy logic control also gained some popularity among flight control engineers. This type of control relies on approximate reasoning based on a set of rules where intermediate positions between 'false' and 'truth' are possible. Fuzzy logic control may be especially useful when a decision must be made between several controversial conditions. For example, in (Fernandez-Montesinos, 1999) fuzzy logic is used for windshear recovery in order to decide whether energy should be given to altitude or airspeed, based upon the current situation.

Adaptive control term covers a set of various control techniques that are capable of online adaptation. A good survey of adaptive control methods is given in (Astrom &

Wittenmark, 1995). The applications of adaptive control is generally biased towards control for large time scales so that the controller has sufficient time to learn how to behave. This makes the relatively short-time recovery process unsuitable for online adaptation.

Evolutionary and *genetic algorithms* (EAs, GAs) are global optimisation techniques applicable to a broad area of engineering problems. They can be used to optimise the parameters of various control systems, from simple PID controllers (Zein-Sabatto & Zheng, 1997) to fuzzy logic and neural network driven controllers (Bourmistrova, 2001; Kaise & Fujimoto, 1999). Another common design approach is evolutionary optimisation of trajectories, accompanied by a suitable tracking controller (e.g. (Wang & Zalzala, 1996)). An elaborated study of applications of EAs to control and system identification problems can be found in (Uzrem, 2003).

Unlike the majority of other techniques, Genetic Algorithms (in the form of *Genetic Programming*) are able to evolve not only the parameters, but also the *structure* of the controller.

In general, EAs require substantial computational power and thus are more suitable for offline optimisation. However, online evolutionary-based controllers have also been successfully designed and used. The *model predictive control* is typically employed for this purpose, where the controller constantly evolves (or refines) control laws using an integrated simulation model of the controlled system. A comprehensive description of this approach is given in (Onnen et al., 1997).

2.2 Flight control for the UAV recovery task

Aircraft control at recovery stage of flight can be conventionally separated into two closely related, but distinctive tasks: guidance and flight control. Guidance is the high-level ('outer loop') control intended to accomplish a defined mission. This may be path following, target tracking or various navigation tasks. Flight control is aimed at providing the most suitable conditions for guidance by maintaining a range of flight parameters at their optimal levels and delivering the best possible handling characteristics.

2.2.1 Traditional landing

In a typical landing procedure, the aircraft follows a defined glide path. The current position of the aircraft with respect to the glidepath is measured in a variety of ways, ranging from pilot's eyesight to automatic radio equipment such as the Instrument Landing System (ILS). Basically, the objective of the pilot (or autopilot) is to keep the aircraft on the glidepath, removing any positioning error caused by disturbances and aircraft's dynamics. This stage of landing is known as *approach*. Approach may be divided into *initial approach*, in which the aircraft makes contact with ('fixes') the approach navigation system (or just makes visual contact with the runway) and aligns with the runway; and *final approach*, when the aircraft descends along a (usually) straight line. In conventional landing, final approach is followed by a flare manoeuvre or nose-up input to soften the touchdown; however, flare is typically not performed for shipboard landing due to random ship motion and severe constraints on the landing space.

The guidance task during final approach involves trajectory tracking with both horizontal and vertical errors (and their rates) readily available. It is important to note that the errors being physically measured are *angular* deviations of the aircraft position (Fig.2) as seen from the designated touchdown point (or more precisely, from where the radars or antennas or other guidance systems are located). They can be converted to linear errors Δh if the *distance* L to the aircraft is known.

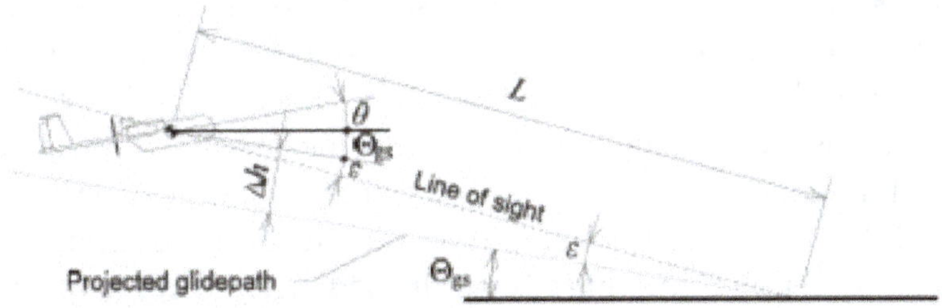

Figure 2. Angular and linear glidepath errors

However, in many applications precise distance measurement is unavailable. Nevertheless, successful landing can be carried out even without continuous distance information. This comes from the fact that exactly the angular errors are relevant to precise landing. Indeed, the goal is to bring the aircraft to a specified point on the runway (or on the deck) in a certain state. The choice of the landing trajectory only serves this purpose, considering also possible limitations and secondary objectives such as avoiding terrain obstacles, minimising noise level and fuel consumption and so on. The angular errors provide an adequate measurement of the current aircraft position with respect to the glidepath.

However, if the landing guidance system takes no account of distance and is built around the angular error only, it may cause stability problems at close distances, because the increasing sensitivity of the angular errors to any linear displacement effectively amplifies the system gain.

2.2.2 UAV control for shipboard recovery

As it was seen from the discussion in the previous section, landing of an aircraft is a well established procedure which involves following a predefined flight path. More often than not, this is a rectilinear trajectory on which the aircraft can be stabilised, and the control interventions are needed only to compensate disturbances and other sources of errors.

The position errors with respect to the ideal glidepath can be measured relatively easily. Shipboard landing on air carriers is principally similar; the main differences are much tighter error tolerances and absence of flare manoeuvres before touchdown. The periodic ship motion does have an effect on touchdown; however, it does not affect significantly the glidepath, which is projected assuming the average deck position. The choice of the landing deck size, glideslope, aircraft sink rate and other parameters is made to account for any actual deck position at the moment of touchdown. For example, a steeper glideslope (typically 4°) is used to provide a safe altitude clearance at the deck ramp for its worst possible position (i.e. ship pitched nose down and heaved up). This makes unnecessary to correct the ideal reference trajectory on the fly.

For the UAV shipboard recovery, ship oscillations in high sea cause periodic displacement of the recovery window (the area where capture can be done) several times greater than the size of the window itself. This fact (and also the assumption that the final recovery window position cannot be predicted for a sufficient time ahead) makes it impossible to project an optimal flight path when the final approach starts. Instead, the UAV must constantly track the actual position of the window and approach so that the final miss is minimised. Therefore, it turns out that the UAV recovery problem resembles that of homing guidance rather than typical landing. While stabilisation on a known steady flight path can be done relatively easy with a PID controller, homing guidance to a moving target often requires a more sophisticated control.

Not surprisingly, homing guidance found particularly wide application in ballistic missiles development, hence the accepted terminology owes to this engineering area. In the context of UAV recovery, the UAV moves almost straight towards the 'target' from the beginning (compared to typical missile intercept scenarios) and thus the velocity vector and the line of sight almost coincide.

However, there are two major difficulties that can compromise the effectiveness of Proportional Navigation (PN) for UAV recovery. First, PN laws are known as generating excessive acceleration demands near the target. For a UAV with limited manoeuvrability, such demands may be prohibitive, especially at low approach airspeeds. On the other hand, the PN guidance strategy does not change during the flight. Several alternative guidance strategies with more favourable acceleration demands exist, e.g. augmented proportional navigation. However, it is unlikely they can sufficiently improve the guidance to an oscillating target such as ship's deck.

2.3 UAV controller structure

The objective is therefore to synthesise such guidance strategy that enables reliable UAV recovery, and to produce a controller that implements the target tracking guidance strategy. The evolutionary design (ED) method applied for this task allows to evolve automatically both the structure and the parameters of the control laws, thus potentially enabling to generate a 'full' controller, which links available measurements directly with the aircraft control inputs (throttle, ailerons, rudder and elevator) and implements both the guidance strategy and flight control (Fig. 3):

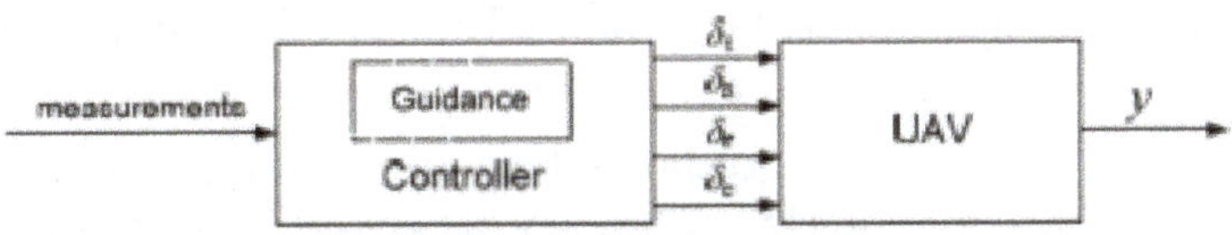

Figure 3. Full controller with embedded guidance strategy

However, this approach, even though appealing at first and requiring minimum initial knowledge, proves to be impractical as the computational demands of the evolutionary algorithms (EAs) soar exponentially with the dimensionality of the problem. It is therefore desirable to reduce complexity of the problem by reducing the number of inputs/outputs and limiting, if appropriate, possible structures of the controllers.

Another difficulty is the evaluation of the controller's performance. In ED, performance or fitness is multiobjective. Therefore, it is highly desirable to decompose this complex task into several simpler problems and to solve them separately.

A natural way of such decomposition is separating the trajectory control (guidance) and flight control. The guidance controller issues commands *ug* to the flight controller, which executes these commands by manipulating the control surfaces of the UAV (Fig. 4). These two controllers can be synthesised separately using appropriate fitness evaluation for each case.

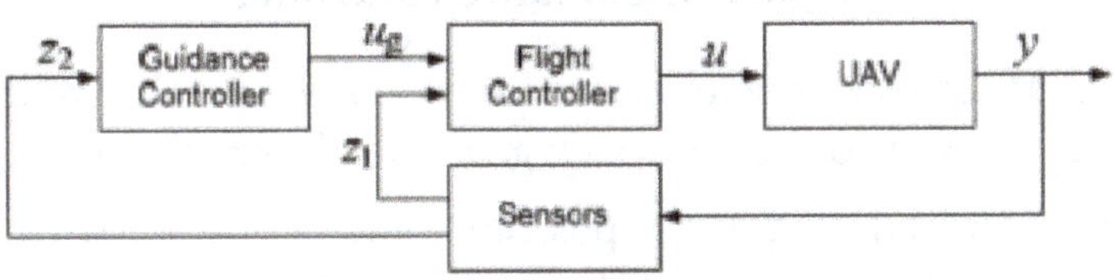

Figure 4. UAV recovery control diagram

2.3.1 Guidance controller

The internal structure of the controller is defined by the automatic evolutionary design based on predefined set of inputs and outputs. It is desirable to keep the number of inputs and outputs to minimum, but without considerably compromising potential performance.

An additional requirement to the outputs u_g is that these signals should be straightforwardly executable by the flight controller. This means that u_g should represent a group of measurable flight parameters such as body accelerations, velocities and Euler angles, which the flight controller can easily track.

The structure of the output part of the guidance controller is as shown in Fig. 5.

With this scheme, the guidance laws produce general requests to change trajectory in horizontal and vertical planes. The kinematic converter then recalculates these requests to the form convenient for the flight controller. Both the bank angle and normal body load factor n_y can be relatively easily tracked, with the sensors providing direct measurements of their actual values. At the same time, this approach allows to evolve the horizontal and vertical guidance laws separately, which may be desirable due to different dynamics of the UAV's longitudinal and lateral motion and also due to computational limitations.

Input measurements to the guidance controller should be those relevant to trajectory. First of all, this is all available positioning information, pitch and yaw angles and airspeed. They do not account for steady wind, but still provide substantial information regarding the current 'shape' of trajectory.

The yaw angle fed into the controllers is corrected by the 'reference' yaw ψ_0, which is perpendicular to the arresting wire in the direction of anticipated approach. A zero yaw indicates that the UAV is pointed perpendicularly to the arresting wire (ignoring ship oscillations), which is the ideal condition in the absence of side wind and when the UAV moves along the ideal glidepath. This is similar to rotating the ground reference frame $O_g x_g y_g z_g$ by the correction yaw angle ψ_0. The rotated frame is referred as *approach ground reference frame.*

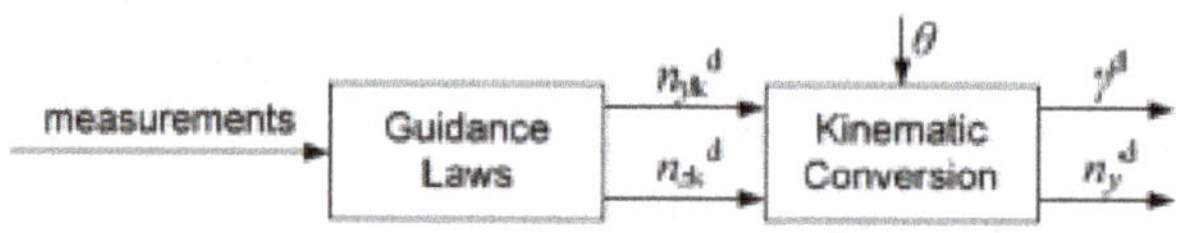

Figure 5. Guidance controller

The derived quantities from raw measurements are the vertical and lateral velocity components with respect to the approach ground reference frame.

Determination of the current UAV position and velocities with respect to the arresting wire is crucial for successful recovery. While previously discussed flight parameters may only help to improve the guidance quality, positioning carries direct responsibility for recovery.

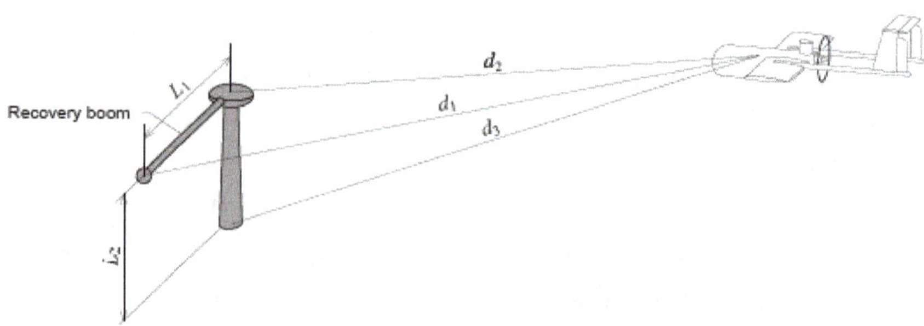

Figure 6. Positioning scheme

The system is based on radio distance metering and provides ten independent raw measurements (Fig. 6): three distances d_1, d_2 and d_3 from the UAV to the radio transmitters located at both ends of the recovery boom which supports the arresting wire and at the base of recovery mast; three rates of change of these distances; distance differences ($d_1 - d_2$) and ($d_3 - d_2$); and rates of change of the differences.

The guidance laws evolution process is potentially capable to produce the laws directly from raw measurements, automatically finding necessary relationships between the provided data and the required output.

The target spot elevation is chosen to be a constant, and the value is determined in view of the expected cable sag obtained from simulation of the cable model. For normal approach speed, the cable sag varies between 3.5 and 4.5 m. Accordingly, the target spot elevation is chosen to be hT = 2 m above the arresting wire.

The recovery procedure, if successful, lasts until the cable hook captures the arresting wire. This happens when the UAV have moved about the full length of the cable *past* the recovery boom. However, position measurements may be unavailable beyond the boom threshold, and even shortly before the crossing the readings may become unreliable. For these reasons, the terminal phase of approach, from the distance about 6–10 m until the capture (or detection of a miss), should be handled separately.

It is possible to disconnect the guidance controller several metres before the recovery boom without affecting the quality of guidance. The allowed error (approximately 2 m in all directions, determined by the lengths of the arresting wire and the cable) should absorb the absence of controlled guidance in the last 0.3 to 1 second (depending on the headwind) in most situations.

2.3.2 Flight controller

Flight controller receives two inputs from the guidance controller: bank angle demand γ^d and normal body load factor demand n_y^d . It should track these inputs as precisely as possible by manipulating four aircraft controls: throttle, ailerons, rudder and elevator.

The available measurements from the onboard sensors are body angular rates ω_x, ω_y, ω_z from rate gyros, Euler angles γ, ψ, θ from strapdown INS, body accelerations *nx, ny, nz* from the respective accelerometers, airspeed V_a, aerial angles α and β, actual deflection of the control surfaces δa, δr, δe, and engine rotation speed N_{rpm}.

For simplicity of design, the controller is subdivided into independent longitudinal and lateral components. In longitudinal branch, elevator tracks the input signal n_y^d, while throttle is responsible for maintaining a required airspeed. In lateral control, naturally, ailerons track γ^d, while rudder minimises sideforce by keeping n_z near zero.

3. EVOLUTIONARY DESIGN(ED)

The Evolutionary Design outlined in this part makes no assumptions about the system, allowing its application to a diverse range of issues, including nonlinear systems with indeterminate structures. This technology is fresh in many regions, and the application of ED to UAV navigation and control issues illustrates the promise of this design methodology. The essence of evolutionary design is a specifically customized evolutionary algorithm (EA) that changes both the architecture and parameters of the control rules. The algorithm's performance is of secondary concern since it is utilized solely for creative tasks during the design phase, provided that the computations are completed within a reasonable timeframe. The primary needs for automatic design approaches are the quality of the outcomes and the capacity for investigation. The fundamental structure of an EA is rather uncomplicated; nonetheless, three essential components must be established prior to the algorithm's functionality. They are as follows:

- representation of phenotype (control laws in our case) suitable for genetic operations (genome encoding);
- simulation environment, which enables to implement the control laws within the closed loop system;
- fitness evaluation function, which assesses the performance of given control laws. These elements, as well as the whole algorithm outline, are addressed below.

Parallel evolution of both the structure and the parameters of a controller can be implemented in a variety of ways. One of the few successfully employed variants is the *block structure controller evolution* (Koza et al., 2000).

In this work the ED algorithm enables to evolve suitable control laws within a reasonable time by utilising gradual evolution with the principle of strong casualty. This means that structure alterations are performed so that the information gained so far in the structure of the control law is preserved. Addition of a new block, though being random, does not cause disruption to the structure. Instead, it adds a new dimension and new potential which may evolve later during numerical optimisation. The principle of strong casualty is often regarded as an important property for the success of continuous evolution (Sendhoff et al., 1997).

The addition of new points or blocks is carried out as a separate dedicated operation (unlike sporadic structure alterations in the sub-tree crossover), and is termed *structure mutation*. Furthermore, in this work structure mutation is performed in a way known as *neutral structure mutation*. That's when the new block should be placed initially with zero coefficient. This will not produce any immediate improvement and may even deteriorate the result slightly because more blocks are used for the same approximation. However, further numerical optimisation should fairly quickly arrive at a better solution. The usefulness of neutral mutations has been demonstrated for the evolution of digital circuits (Van Laarhoven & Aarts, 1987) and aerodynamic shapes (Olhofer et al., 2001).

As a result, the ED algorithm basically represents a numerical EA with the inclusion of structure mutations mechanism.

3.1 Representation of the control laws

Control laws are represented as a combination of static functions and input signals, which are organised as a dynamic structure of *state equations* and *output equations* in form of continuous representation.

The controller being evolved has *m* inputs, *r* outputs and *n* states. The number of inputs and outputs is fixed. The algorithm allows varying number of states; however, in this work, the number of states is also fixed during the evolution. As a result, the controller comprises of *n* state equations and *r* output equations:

$$\begin{matrix} \dot{x}_1 = g_1(x,u) \\ \dot{x}_2 = g_2(x,u) \\ \vdots \\ \dot{x}_n = g_n(x,u) \end{matrix} \quad \text{and} \quad \begin{matrix} y_1 = f_1(x,u) \\ y_2 = f_2(x,u) \\ \vdots \\ y_n = f_n(x,u) \end{matrix} \tag{4}$$

where u is size *m* vector of input signals, x = $[x_1,x_2,...x_n]$ is size *n* vector of state variables, $y_{1...r}$ are controller outputs. Initial value of all state variables is zero. All *n+r* equations are built on the same principle and are evolved simultaneously. For structure mutations, a random equation is selected from this pool and mutated.

3.1.1 Representation of input signals

Input signals delivered to each particular controller are directly measured signals as well as the quantities derived from them.

Within each group, inputs are organised in the subgroups of 'compatible' parameters. Compatible parameters are those which have close relationship with each other, have the same dimensions and similarly scaled. The examples of compatible parameters are the pairs (n_x, n_{xg}), (ω_y, ψ), (V_a, V_{CL}). As a rule, only one of the compatible parameters is needed in a given control law. For this reason, the probability of selection of such parameters for the structure mutation is reduced by grouping them in the subgroups. Each subgroup receives equal chances to be selected. If the selected subgroup consists of more than one input, a single input is then selected with uniform probability.

Therefore, every controller input may be represented by a unique *code* consisting of three indices: the number of group *a*, the number of subgroup *b* and the number of item in the subgroup *c*. The code is designated as u(a,b,c).

3.1.2 Representation of control equations and the structure mutation

Each of the control equations (4) is encoded as described above. To this end, only one single output equation of the form $y = f(u)$ will be considered in this section. State variables x are considered as special inputs and have no effect on the encoding.

This is done to speed up the search, which could otherwise be hampered by the multitude of possible operations. It proved to be more effective to include all meaningful quantities derived from source measurements as independent inputs than to implement all the functions and operations which potentially allow to emerge all necessary quantities automatically in the course of evolution.

The encoding should allow a simple way to insert a new parameter in any place of the equation without disrupting its validity and in a way that this insertion initially does not affect the result, thus allowing neutral structure mutations.

Conceptually, the equation is a sum of input signals, in which:

- every input is multiplied by a numeric coefficient or another similarly constructed expression;
- the product of the input and its coefficient (whether numeric or expression) is raised to the power assigned to the input;
- a free (absolute) term is present.

The simplest possible expression is a constant:

$$y = k_0 \tag{5}$$

A linear combination of inputs plus a free term is also a valid expression:

$$y = k2u_2 + k_1u_1 + k_0 \tag{6}$$

Any numeric constant can be replaced with another expression. An example of a full featured equation is

$$y = ((k_4u_4 + k_3)u_3)-0.5 + k_2u_2 + (k_1u_1)^2 + k_0 \tag{7}$$

This algorithm can be illustrated by encoding the example (7). The respective internal representation of this expression is:

- Equation: $y = ((k_4u_4 + k_3)u_3)-0.5 + k_2u_2 + (k_1u_1)^2 + k_0$
- Expression: { u(3,-0.5) u(4) 1 2 *u*(2) 3 u(1,2) 4 5 }
- Object parameters: [$k_4\ k_3\ k_2\ k_1\ k_0$]
- Strategy parameters: [$s_4\ s_3\ s_2\ s_1\ s_0$]

This syntax somewhat resembles Polish notation with implicit '+' and '*' operators before each variable. The representation ensures presence of a free term in any sub-expression, such as *k*3 and *k*0 in the example above.

The algorithm of structure mutation is presented below.

- Select an input (or a state variable) at random: u(a,b,c).
- Obtain the initial values of the numeric coefficient and the strategy parameter (initial step size). The initial coefficient *k* is selected as described above, it can be either 0 or 10^6.
- Append the object parameters vector with the initial coefficient, and the strategy parameters vector with the initial step size.
- Form a sub-expression consisting of the selected variable code and the obtained index: { u(a,b,c) *n* }.
- With 40% probability, set the insertion point (locus) to 1; otherwise, select a numeric value in the expression at random with equal probability among all numeric values present and set the locus to the index of this value.
- Insert the sub-expression into the original expression at the chosen locus (*before* the item pointed).

This procedure may produce redundant expressions when the selected variable already exists at the same level. Thus an algebraic simplification procedure is implemented. It parses given expression, recursively collects the factors of each variable encountered and then rebuilds the expression.

3.2 Simulation environment

Fitness evaluation of the controllers is based on the outcome of one or more simulation runs of the models involved with the controller being assessed in the loop.

Simulation environment for this study is constructed in the MATLAB/Simulink software. All the models are implemented as Simulink library blocks (Khantsis, 2006). The models participating in the evolution include:

- The *Ariel* UAV model;
- The atmospheric model;
- The ship model;
- The static cable model.

The sample rate should be kept at minimum, ensuring, however, numerical stability. The fastest and thus the most demanding dynamics in the model is contained in

- sensors (both the lags and noise);
- control actuators;
- turbulence model;
- and potentially, controllers.

These components should be especially carefully examined when adjusting the sample rate. The experimentally determined minimum sample rate for the model is 100 Hz. The model is integrated using the 4th order Runge-Kutta method.

3.3 Fitness evaluation

Fitness evaluation is determined by objectives that are unique to each controller, hence calculated on an individual basis. The utilization of multiple specialized controllers facilitates the delineation of their distinct purposes and diminishes the quantity of these objectives in contrast to a singular, multifunctional autopilot. The assessment of a controller's performance can be categorized into two primary phases. The initial step involves preparing the sample task and simulating the model with the controller included into the loop. The second stage involves analyzing the data derived from the simulation and assessing the fitness accordingly. The controller frequently repeats these stages, doing diverse tasks to achieve a comprehensive assessment of its performance through multi-task fitness evaluation.

Fitness value may reflect different levels of performance achieved by the controller, including 'hard bounds.' For example, if the controller crashes the UAV, a high penalty score may be assigned.

Other parameters of flight taken into account is control usage (or control effort): it is desirable to keep control usage at minimum. Tracking abilities, if applicable, can be estimated by the weighted sum of the tracking error.

The total fitness value is calculated as the weighted sum of all estimates:

$$F = W_c C_c + W_f C_f + W_e C_e + \dots \tag{8}$$

The exact value of the weighting coefficients W, as well as the number of estimates taken into account, is individual for each controller. As a rule, the weighting coefficients are chosen empirically. It should be noted that the signals are dimensionalised and thus the coefficients may vary significantly. For example, aerodynamic surfaces deflection is measured in degrees and may range from −16 to 16 (for rudder and ailerons); considering 10-fold saturation limits, the peak values may reach ±160. At the same time, throttle range is 0.01 to 1, thus the cost for its usage will be about 30 times as low (for similar control dynamics).

3.4 Evolutionary design algorithm outline

The Evolutionary Design algorithm is implemented as a function with two input arguments: population size and number of generations to run. Other, more specific parameters are initialised within the program. The algorithm framework is as follows.

- Initialise all parameters. If the evolution is continued, go to step 3.
- Create initial population.

- Evaluate fitness of each individual.
- Save the full state to a temporary file for emergency recovery.
- If the required number of generations is reached, return.
- Sort the population according to the fitness of each individual.
- If elitism is enabled, copy the best scored individual into the new population.
- Until the new population is filled up:
- Select the next individual from the sorted list (starting from the elite member).
- Every (*ks*)th generation, with probability *Ps*, perform structure mutation of the selected individual.
- Reproduce *n* offspring individuals from the selected (and possibly mutated) individual.
- Put these *n* new individuals to the new population.
- Continue the loop from step 8.
- Increase the generation counter.
- Continue from step 3.

Several steps of this algorithm need further clarification.

Initial population is created according to the specified task. By default, it is initialised with the control laws of the form $y = const$ with randomly chosen constants. Most of the controllers, however, are initialised with more meaningful control laws. For example, tracking controllers may be initialised with the laws $y = k_{1\varepsilon} + k_0$, where ε is the respective error signal and the coefficients k are sampled at random (taking into account default step sizes for the respective signal).

It can be seen that *selection* is performed deterministically. This is the most commonly used way in ES. The populations used in this study were of moderate size, usually 24 to 49 members. Selection pressure is determined by the number of the offspring *n* of each individual. The smaller *n*, the lower the selection pressure. For nearly all runs in this work $n = 2$, which means that half of the population is selected. This is a rather mild level of selection pressure.

The parameters determining *structure mutation* occurrence, k_s and P_s, both change during the evolution. As discussed previously, the number of structure mutations should decrease as the complexity of the controllers grows and more time to optimise the coefficients is required. The probability of structure mutation P_s is normally high in the beginning (0.7 to 1.0) and then decrease exponentially to moderate levels (0.4 to 0.6) with the exponent 0.97 to the generation number. Default value of k_s is set according to overall complexity of the controller being evolved. For simpler single-output controllers, as a rule, $k_s = 10$; for more complex controllers $k_s = 20$. However, in the beginning of evolution, k_s is reduced by half until the generation 20 and 100 respectively.

Reproduction is performed simultaneously with mutation, as it is typically done in ES, with the exception that this operation is performed separately for each selected member.

4. CONTROLLER SYNTHESIS AND TESTING

The UAV control system is synthesised in several steps. First, the flight controller is produced. This requires several stages, since the flight controller is designed separately for longitudinal and lateral channels. When the flight controller is obtained, the guidance control laws are evolved.

Application of the ED algorithm to control laws evolution is fairly straightforward: 1) preparation of the sample task for the controller, 2) execution of the simulation model for the given sample task and 3) analysis of the obtained performance and evaluation of the fitness value. When both the model and fitness evaluation are prepared, the final evolution may be started. Typically, the algorithm is run for 100–200 generations (depending on complexity of the controller being evolved). The convergence and the resulting design is then analysed and the evolution, if necessary, is continued.

4.1 Step 1: PID auto throttle

Initially a simple PID variant of auto throttle is evolved - to ensure a more or less accurate airspeed hold. At the next stage, its evolution is continued in a full form together with the elevator control law. The PID structure of the controller may be ensured by appropriate initialisation of the initial population and by disabling structure mutations. Therefore, the algorithm works as a numerical optimisation procedure. The structure of the auto throttle control law is following:

$$\begin{aligned} \dot{x}_1 &= k_1 x_1 + k_2 \Delta V_a \\ \delta_t &= k_3 x_1 + k_4 \Delta V_a + k_5 \dot{V}_a + k_6 \end{aligned} \tag{9}$$

where $\Delta V_a = V^d - V^a$ is the airspeed error signal, δ_t is the throttle position command and $k_{1...6}$ are the coefficients to be optimised.

A 30-second flight is allocated for performance measurement. Such a long flight is needed because throttle response is quite slow. Since robustness of the PID controller to discrepancies in the UAV model is not topical at this stage (it will be addressed in further evolution), and because structure of the controller is fixed so that irrelevant measurements cannot be attracted, only a single simulation run for each fitness evaluation is performed.

Elevator inputs provide the main source of disturbances for training the autothrottle. A 2.5- degree nose-up step input is executed at time t = 5 s. It produces nearly steady climb without reaching stall angles of attack and saturation on the throttle (except for, possibly, dynamic transition moments).

At t = 14 s, a similar nose-down elevator step is commanded, entering the UAV into glide with small descent angle. In addition, a 3 m/s tailwind gust is imposed at t = 23 s. Generally, manoeuvres and wind disturbances are the most prominent sources of airspeed variations, therefore they are included for auto throttle assessment.

Initial airspeed is set to 23 m/s, i.e. 1 m/s above the required airspeed. This provides an additional small scale disturbance. Initial altitude is 100 m, providing enough elevation to avoid crash for any sensible throttle control.

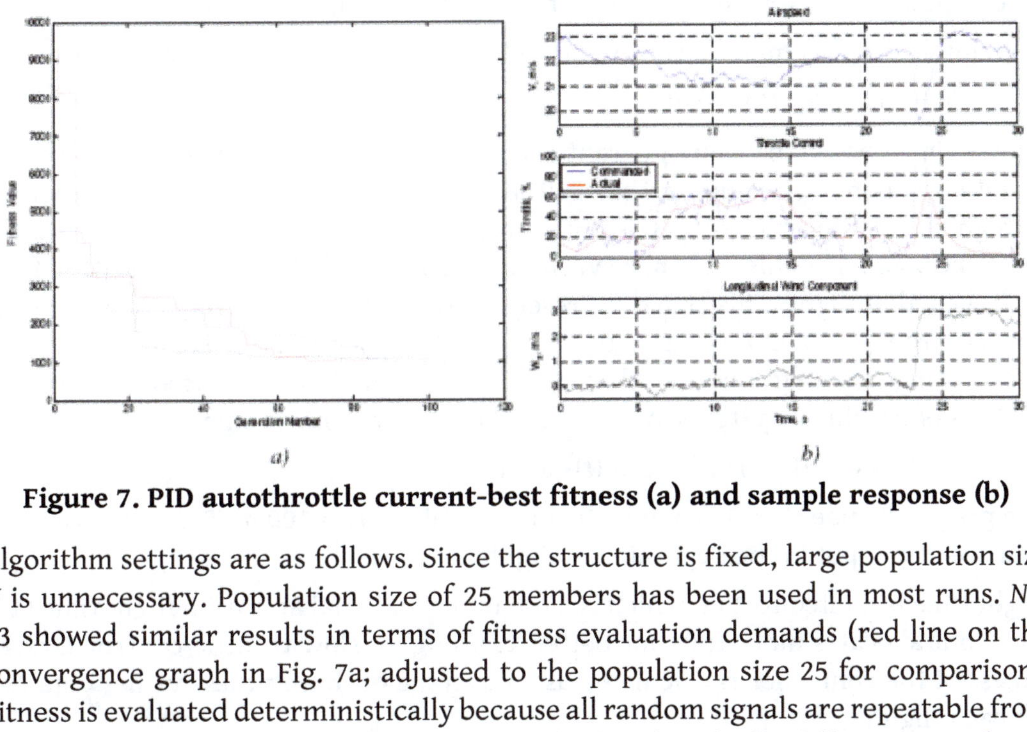

Figure 7. PID autothrottle current-best fitness (a) and sample response (b)

Algorithm settings are as follows. Since the structure is fixed, large population size N is unnecessary. Population size of 25 members has been used in most runs. N = 13 showed similar results in terms of fitness evaluation demands (red line on the convergence graph in Fig. 7a; adjusted to the population size 25 for comparison). Fitness is evaluated deterministically because all random signals are repeatable from run to run. Elitism is enabled. Fitness is calculated with the following weighting coefficients:

$$F_t = 2000C_e(V_e) + 1000C_c(\delta_t) + 2000C_f(\delta_t) \tag{10}$$

It should be noted that the *commanded* signal evaluation. δ_t (which enters the actuator) is used for Convergence graphs for three independent runs of the ED algorithm are presented in Fig. 7a. They show the best fitness values for each generation. On average, 2000 to 2500 simulation runs is needed to achieve satisfactory performance. Because this is only an intermediate controller, full optimisation is not necessary at this stage. The best controller obtained in the experiments is the following:

$$x_1 = -2.9966x_1 + 85.57\Delta V_a$$
$$\delta_t = 0.012616x_1 + 0.089324\Delta V_a - 0.044004V_a + 0.23827 \qquad (11)$$

Sample response of this controller is illustrated in Fig. 7b.

4.2 Step 2: Longitudinal control

With a simple autothrottle available, elevator control can be developed to provide tracking of the normal body load factor demand n_y^d.

It is desirable that the sample training task closely reflect the conditions the controller will be subject to in a real guidance task. Unfortunately, this is not possible at the early stages of multistage design process as guidance laws can be evolved only after a capable flight controller has been produced. In the meantime, a replacement pseudoguidance task must be used for the flight controller training.

As a suitable replacement of the guidance law, a simple static altitude hold is used:

$$n_y^d = 0.03(H^d - H) + \cos\theta \qquad (12)$$

A low gain is chosen to ensure guaranteed stability. The $\cos\theta$ term takes into account gravity component, assuming that roll angles are small. However, real guidance in high sea will most likely require a more active control, especially closer to the recovery boom. For this reason, a synthetic 'chirp' signal $c(t)$ is added starting from 10th second of flight. This signal represents a sine wave which frequency increases linearly from f_0 = 0.01 Hz at t_0 = 10 s to f_1 = 0.5 Hz at t_1 = 20 s. For t < 10 s, $c(t)$ = 0, providing 'training' for smooth control. Chirp signal is often used for response analysis of nonlinear systems. Altogether, the reference input signal is

$$n_y^d(t) = 0.03(H^d - H(t)) + \cos\theta(t) + 0.2c(t) \qquad (13)$$

The preset altitude H^d is 5 m above the initial altitude H_0 = 100 m. This causes a small positive step input in the beginning of flight.

Flight time is limited to 20 s, which is close to normal final approach flight time (10–15 s), but allocates more time for better 'training' of slow throttle control. Overall fitness of the controller is calculated as an average of three fitness values obtained for each flight.

Elevator control law is initialised as follows:

$$x_2 = 0$$
$$\delta_e = k_1x_2 + k_2\Delta n_y + k_3 \qquad (14)$$

where $\Delta n_y = n_y^d - ny$ is the load factor error and the coefficients $k_{1\ldots3}$ are sampled at random for the initial population. For simplicity, first-order dynamic control law is used. However, when combined with the throttle control, both control laws may include both state variables $x1$ and $x2$, forming a tightly coupled longitudinal control.

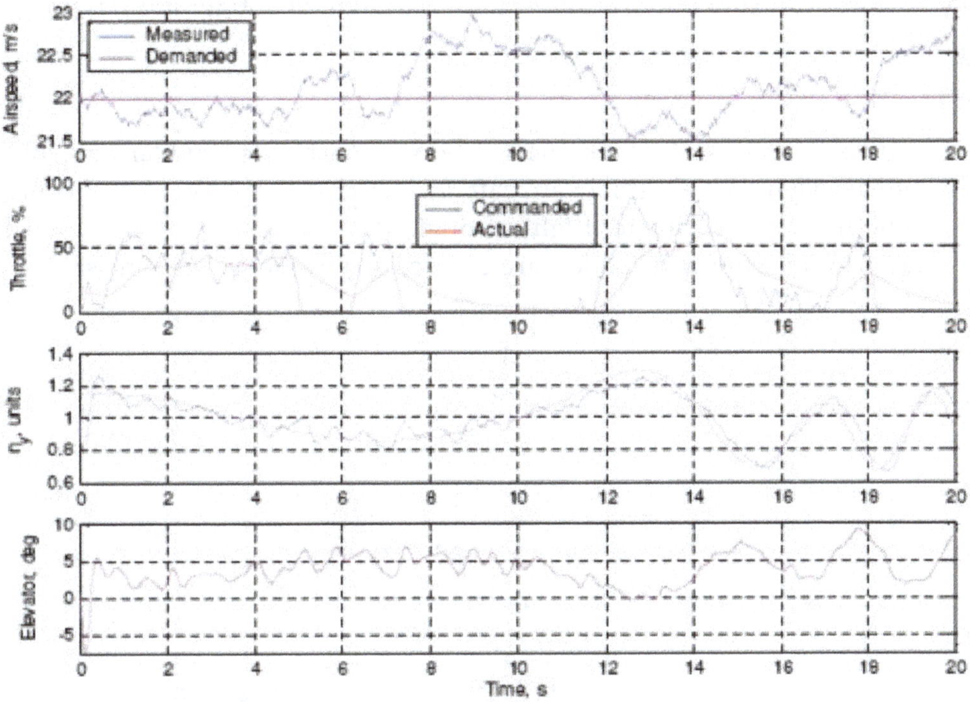

Figure 8. Longitudinal control sample response

Fitness estimation combines both throttle control fitness and elevator control fitness. They are constructed in a similar manner with the following weighting coefficients:

$$F_{te} = F_t + F_e = F_t + 500000Ce\,(n_y) + 50C_c\,(\delta_e) + 5000C_f\,(\delta_e) \tag{15}$$

The weights put slightly more emphasis on elevator control than on throttle control, because the former is more important for guidance. Autothrottle only supplements it to provide safe flight conditions.

Relatively large population size of 49 members has been used in all runs. Initial structure mutation probability P_s was set to 1.0; final value $P_s \to 0.6$. Structure mutations happened every (k_s = 5)th generation for the first 20 generations, then k_s = 10 until 100th generation, then k_s = 20. Satisfactory controllers were obtained after about 100 to 120 generations, which requires approximately 15000 simulation runs. The best performing controller has the following form:

$$\begin{aligned}
x_1 &= ((-252.611\omega_z - 172.0765)a - 4.411)x_1 + 75.0613\Delta V_a \\
x_2 &= -0.77101x_2 - 228.8894\Delta n_y \\
\delta_t &= 0.0011588x_1 + 1.0565\Delta V_a - 0.6989V_a + 0.24651 \\
\delta_e &= 41.5594\theta + (-140.8115a)^{0.5} + 0.89016x_2 - 26.1045\Delta n_y + 50.1835\omega_z + 3.871
\end{aligned} \tag{16}$$

Response of this controller to the sample input (in default configuration) is presented in Fig. 8.

4.3 Step 3: Lateral control

Lateral control consists of two channels: ailerons control and rudder control. As a rule, for an aerodynamically stable aircraft such as the *Ariel* UAV, lateral control is fairly simple and is not as vital for flight as longitudinal control. For this reason, both control laws, for ailerons and rudder, are evolved simultaneously in one step.

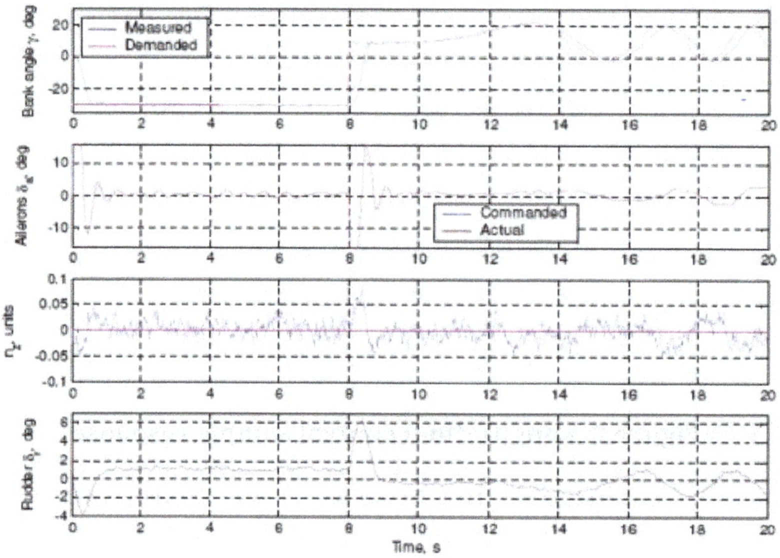

Figure 9. Lateral control sample response

The set-up is largely similar to that used in the previous step. The just evolved longitudinal control laws (16) are connected to throttle and elevator. Both ailerons and rudder control laws are initialised in a similar manner to (14):

$$\begin{aligned} \dot{x}_3 &= 0 \\ \dot{x}_4 &= 0 \\ \delta_a &= k_{11}x_3 + k_{12}\Delta\gamma + k_{13} \\ \delta_r &= k_{21}x_4 + k_{22}n_z + k_{23} \end{aligned} \qquad (17)$$

Fitness is calculated as follows:

$$F_{ar} = F_a + F_r = 50000C_e(\gamma) + C_c(\delta_a) + 2000(\delta_a) + \ldots 200000C_e(n_z) + 10C_c(\delta_r) + 2000C_f(\delta_r) \qquad (18)$$

Algorithm settings are the same as for longitudinal control design. Fairly quick convergence (within 70–80 generations) was observed, with slow further progress. Considering that the initial state is zero, state variable x_3 is always zero. The controller is obtained as follows:

$$\dot{x}_4 = -39.2775x_4 - 44.57\omega_x + \left((4.6845x_4)^{0.5} + 69.3478\right)n_z$$
$$\delta_a = -26.7812\Delta\gamma + 8.1188\omega_x + 0.57032 \quad (19)$$
$$\delta_r = -1.6916x_4 - 1.33n_z + 4.1764\omega_y + 21.2955\beta - 0.65184$$

Sample response of this controller is presented in Fig. 9.

4.4 Step 4: Stall prevention mechanism

Active guidance control may produce high acceleration demands n_y^d , which may lead to exceeding the maximum allowed angle of attack a_{max} and result in stall.

However, fitness evaluation of the control law with stall prevention would be more complex and time consuming. Separate estimation is required because the objectives are contradictory: adequate limitation of a invariably results in tracking degradation and vice versa.

Though, a separate control law which limits angle of attack is developed. The limiter should work only when dangerous angle of attack is reached, being deactivated in normal conditions.

The control law is evolved in a similar manner to other laws described above. The objective of the control law is to maintain $a = a_{max}$ by correcting n_y^d as long as this demand is greater than required for safe flight. Also, free term is disregarded in both state and output equations.

Fitness is calculated similarly to all tracking controllers with respect to the command $a = a_{max}$. Two exceptions are made, however. First, overshoot is penalised twice as heavily as undershoot, because exceeding the maximum angle of attack is considered to be hazardous. Second, since the limiter has no direct control on elevator, the amount of longitudinal oscillations is taken into account as the variance of pitch rate. Altogether, the fitness value is calculated as

$$F_a = 10\,C_a + 0.1C_c\,(\delta e) + 1000C_c\,(\omega_z) + 1000C_f\,(\delta_e) \quad (20)$$

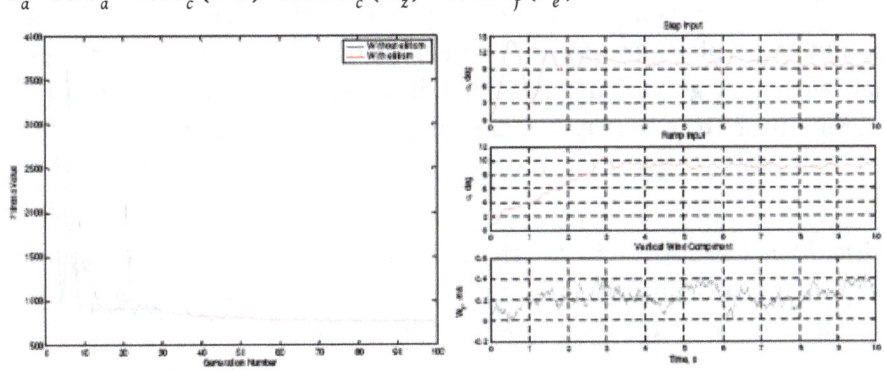

a) b)

Figure 10. Angle of attack limiter current-best fitness (a) and sample responses (b)

Algorithm settings and initialisations are similar to the case of longitudinal control evolution, except that smaller population of 24 or 25 members is used because only one control law is evolved. Both elitist and non-elitist strategies have been tried with largely similar results. Quick convergence, within 40–50 generations, was observed in all cases, with slow further progress. Examples of current-best convergence are presented in Fig. 10a. A simpler solution obtained in elitist strategy has been selected:

$$\begin{aligned} &\dot{x}_5 = (93.01\omega_z - 15.0798)(\alpha - \alpha_{max}) \\ &n_y^{d.corr} = n_y^{d0} - \Delta n_y^d = n_y^{d0} - 0.63208x_5 - 25.246(\alpha - \alpha_{max}) \\ &K_{max} = 60 \end{aligned} \quad (21)$$

Where K_{max} is a damping coefficient and $n_y{}^{d}0$ is a frozen value for n^d. Sample responses of this controller are presented in Fig. 10.

4.5 Step 5: Guidance

At this point, flight controller synthesis is completed and guidance laws can be evolved. Guidance controller comprises two control laws, for the vertical and horizontal load factor demands, n^d_{yk} and n^d_{zk} respectively. This is equivalent to acceleration demands

$$a_{yk}^d = gn_{yk}^d \text{ and } a_{zk}^d = gn_{zk}^d \quad (22)$$

These demands are passed through the kinematic converter to form the flight controller inputs n^d_y and γ^d.

It is important to note that when the controller is evolved in a stochastic environment with many random factors, the evolution progress becomes highly stochastic as well. It is desirable to obtain the fitness value on the basis of multiple simulation runs, reducing thereby sampling errors and making fitness value more deterministic and reliable. Excessive number of simulation runs, however, may slow down the evolution considerably, thus an optimal number should be found.

Second, for the same reason, stagnation of progress of the current best fitness is not a reliable indicator of stagnation of the evolution. Every generation may produce individuals with very good fitness which will not sustain further selections. In some cases, current best fitness may appear stagnated from the very first generation. Average population fitness may be considered as a better measure of evolution progress.

Third, elitism has little sense in a sufficiently stochastic environment and is usually not employed. If it is desirable, for some reason, to use elitist strategy, fitness of the elite solution must be recalculated every generation.

In the guidance task, the random factors include initial state of the UAV; Sea State, atmospheric parameters, and initial state of the ship.

Fitness is calculated as follows:

$$F_g = 40\Delta h_1^2 + 20\Delta z_1^2 + 50|\psi_1| + 25|\gamma_1| + 500C_c(n_{yk}^d) + 500C_c(n_{zk}^d) + 100C_f(n_{zk}^d) \tag{23}$$

where Δh_1 and Δz_1 are vertical and horizontal miss distances, and ψ1 and γ1 are final yaw and bank angles. Greater weight for vertical miss than that for horizontal miss is used because vertical miss allowance is smaller (approximately 3–4 m vs. 5 m) and also because vertical miss may result in a crash into the boom if the approach is too low.

Three simulation runs are performed for each fitness evaluation. To provide a more comprehensive estimation of guidance abilities, initial positions of the UAV in these N runs are maximally distributed in space, maintaining, at the same time, a sufficient random component.

Algorithm initialisation is the same as for longitudinal flight control laws evolution, except that elitism is not used and the population size is 48 members. The initial population is sampled with the control laws of the form

$$\begin{aligned}
\dot{x}_6 &= 0 \\
\dot{x}_7 &= 0 \\
a_{yk}^d &= k_{11}x_6 + k_{12} \\
a_{zk}^d &= k_{21}x_7 + k_{22}
\end{aligned} \tag{24}$$

where all coefficients k are chosen at random. For convenience, the control laws are expressed in terms of accelerations, which are then converted to load factors according to (22). Since these control laws effectively represent 'empty' laws $y = const$ (plus a low-pass output filter with randomly chosen bandwidth), structure mutation is applied to each member of the initial population. Further structure mutations are performed as specified above. With such initialisation, no assumption about possible guidance law is taken.

From the final populations, the best solution is identified by calculating fitness of each member using $N = 25$ simulation runs, taking into account also success rate. The best performing controller is the following (unused state variables are removed) with 100% success rate in the 25 test runs with average fitness 69.52:

$$\begin{aligned}
\dot{x}_6 &= -3.548\omega_v \\
a_{yk}^d &= 0.916x_6 - 54.3\omega_v - 0.1261 \\
a_{zk}^d &= -107.68\omega_h + 0.0534V_{zT} + 0.4756
\end{aligned} \tag{25}$$

The other attempted approach is the pure proportional navigation (PPN) law for recovery which was compared with the obtained solutions (Duflos et al., 1999; Siouris & Leros, 1988). The best PN controller have been selected using the fitness values of each member of the final population averaged over 100 simulation runs. It is the following:

$$\begin{aligned} n_{yk}^{d} &= -3.27 V_{CLa} \omega_V \\ n_{zk}^{d} &= -3.18 V_{CLa} \omega_h \end{aligned} \tag{26}$$

This controller received the fitness 88.85 and showed success rate 95%.

5. CONTROLLER TESTING

Testing is a crucial stage of any controller development process as the designer and end user must be satisfied that the controller meets its performance requirements. In addition, an allowable operational envelope must be obtained. Testing of a developed controller can be performed either within the physical system or in a comprehensive simulation environment. In physical implementation, the controller is working on a true model and hence there are no modelling errors to introduce uncertain effects into the system. In addition, physical testing accurately reflects the operational use of the controller. On the other hand, physical testing involves large time and cost demands for multiple tests. Moreover, the failure of the controller at any stage may lead to a dangerous situation and even to loss of the aircraft.

Testing within a simulation environment has the benefit that many different tests may be applied rapidly with no severe consequences in the event of controller failure. The whole range of operating conditions may be tested. On the downside is that potentially large modelling errors may be introduced, which may bias the results and cause an otherwise excellent controller to fail when implemented physically. In addition, testing situations which cannot occur in practice are possible in the simulation environment.

In this work, two main types of simulation tests are conducted - robustness and performance. Results are presented for robustness test, which is aimed at ensuring the controller has good robustness to modelling uncertainties and to test whether the controller is sensitive to specific perturbations.

5.1 Robustness tests

Testing the controller for robustness to model uncertainty involves perturbing the model and determining the effect upon controller performance. The perturbations can be performed in several ways. Physical quantities such as mass and wing area can be changed directly. Dynamics of the system can be varied by introducing additional dynamic elements and by changing internal variables such as aerodynamic coefficients. For a realistic test, all perturbations should be applied simultaneously to identify the worst case scenario. However, single perturbation tests (the *sensitivity analysis*) allow to analyse the degree of influence of each parameter and help to plan the robustness test more systematically. After single perturbation tests multiple perturbation tests are carried out, where all parameters of the model are perturbed randomly within the identified limits.

5.2 Single perturbation tests

In this type of test, a single model variable is perturbed by a set amount and the effect upon the performance of the controller is determined. Performance evaluation can be measured in a manner similar to fitness evaluation of the guidance controller (equation (23)).

The additional parameters taken into account in performance measurement are impact speed V_{imp} and minimum altitude H_{min} attained during the approach.

Altogether, the performance cost (PC) is calculated as follows:

$$PC = 40\Delta h_1^2 + 20\Delta z_1^2 + 50|\psi_1| + 25|\gamma_1| + 50 f_V + 20 f_H + 10C_c(\delta_a) + 10C_c(\delta_r) + C_c(\delta_e) + \ldots \\ + 200C_f(\delta_a) + 200C_f(\delta_r) + 200C_f(\delta_e) \quad (27)$$

Where

$$f_V = \begin{cases} V_{imp} - 30, & V_{imp} > 30 \\ 0, & V_{imp} \le 30 \end{cases}, \quad f_H = \begin{cases} 10 - H_{min}, & H_{min} > 10 \\ 0, & H_{min} \le 10 \end{cases}. \quad (28)$$

Impact speed V_{imp} and minimum altitude H_{min} are measured in m/s and metres respectively. Other designations are as in (23). Unlike fitness evaluation in the flight controller evolution, the commanded control deflections δ_a, δ_r and δ_e are saturated as required for control actuators.

The absolute value of the PC obtained using (27) is not very illustrative for comparison between the results. Smaller values indicate better performance, but the 'ideal' zero value is unreachable because a minimum level of control activity is always present even in very calm environment. For this reason, a *Normalised Performance Cost* (NPC) will be used:

$$NPC = \frac{PC}{PC_{ref}} \quad (29)$$

where PC_{ref} is the reference Performance Cost obtained for the reference (unperturbed) model with the guidance controller being considered. NPC > 1 indicates deterioration of performance. However, the performance with the perturbed model may be better than the reference performance, thus NPC < 1 is also possible.

The environment delivers a great deal of uncertainty. To obtain a reliable estimate of performance, several tens (up to a hundred) of simulation runs in various conditions should be performed at every point. However, this would imply a prohibitively high computational cost. Meanwhile, there is no single 'typical' scenario that could encompass most of the real flight conditions. For these reasons, in this work few different scenarios were used for evaluation of a single Performance Cost to cover most of the possible real world conditions. The range of disturbances and the initial ship phase are chosen to provide a moderately conservative estimation. All random parameters (which include the turbulence time history and the ship initial state) are reproducible between the PC estimations, therefore PC is calculated deterministically. Results presented in this work are for calm environment:

Calm environment. No wind, no turbulence; initial position of the UAV is at the ideal reference point: distance 300 m, elevation 14 m, zero sideways displacement. A small amount of ship motion corresponding to Sea State 2 (SWH)2 is, however, included. This scenario is useful to analyse the performance in benign environment and also to soften the conservative bias to the difficult conditions.

The first test determines the robustness to time delays in the measurement signals.

The tests are carried out for the two controllers selected in Section 4.5. The controller #1 is

(26) (pure Proportional Navigation guidance) and the controller #2 is (25). Fig. 11a shows the NPC evaluated for these controllers for varying time delay.

Time delays occur in a physical system due to several factors, which include delays associated with the measurement devices, delays in encoding, decoding and transmitting the signals to the controllers, delays in controller computation and delays in the actuator systems.

The miss distance (averaged over the four scenarios) is shown in Fig. 11b. It can be seen that in terms of guidance accuracy, delays up to 0.07 s for the second controller and up to 0.09 s for the first controller cause little effect. However, considering the NPC value, which also takes into account control activity, sufficient degradation is observable from 0.035 second delay. This is expectable, because delays usually cause oscillations in the closed-loop system.

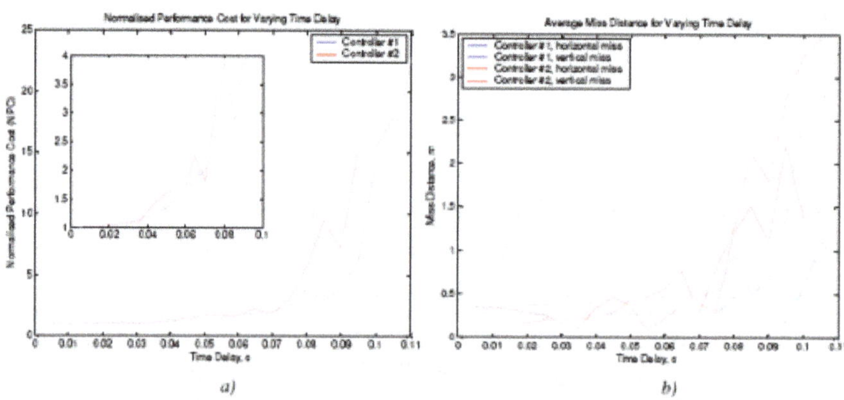

Figure 11.10 NPC (a) and miss distance (b) for controllers with time delays

For the selected NPC threshold, the maximum time delay can be defined as 0.065 s for the controller #1 and 0.06 s for the controller #2.

It should be noted, however, that perturbation of any one of the parameters does not provide a comprehensive picture of the effect on performance. For this reason, the effect of perturbation of empty mass m_{empt} will be closely examined as well. Unlike time delay, this perturbation should primarily affect the trajectory rather than the control activity.

m_{empt} is varied from 50% of the reference value (20.45 kg) to 200% linearly with 2% steps. This is done separately for increasing and decreasing the variable until the threshold NPC =1.8 is crossed, or a failure occurs, or the limit (200% or 50% respectively) is reached.

The parameters corresponding to aircraft geometry and configuration are tested in a similar manner. The range is increased to the scale factors between 0 and 10 (0 to 1000%) with step 0.05. NPC is linearly interpolated at the intermediate points. The allowable perturbations (as factors to the original values) are summarised in Table 1, where * denotes the extreme value tested.

Table 1. Allowable perturbations of UAV inertial properties and geometry

Parameter	Lower limit factor		Upper limit factor	
	Controller 1	Controller 2	Controller 1	Controller 2
Empty mass (m_{empt})	0.50*	0.50*	1.54	1.52
Rolling moment of inertia (I_x)	0.20	0.20	10*	2.43
Yawing moment of inertia (I_y)	0*	0*	10*	10*
Pitching moment of inertia (I_z)	0*	0*	2.17	1.91
xy cross product of inertia (I_{xy})	0*	0*	10*	10*
Wing area (S)	0.74	0.87	1.55	1.60

Fig. 12 demonstraites results for calm environment. Control signals for the worst two cases (with NPC > 3.8) are not shown, they exhibit extremely aggressive oscillations. Dashed cyan and green lines on the trajectory graphs represent the 'unrolled' along the flight path traces of the tips of the recovery boom (lateral position on the top view and vertical position on the side view). The bar on the right hand side illustrates the size of recovery window. The height and position of the window may be slightly different for different trajectories because it is determined by the shape of the arresting cable-hook. However, in most cases this is barely noticeable on the graphs since the difference in flight time and terminal part of trajectory is small. Note that the horizontal and vertical scale on the trajectory graphs is different.

The other parameters evaluated in this research were the perturbations of the power unit parameters, the aircraft aerodynamics parameters and sensor noise (Khantsis, 2006).

Overall, these tests have not exposed any significant robustness problems within the controllers. Large variations in single aircraft parameters caused very few control problems. However, such variations cannot realistically judge the performance of the controllers under simultaneous perturbations of multiple parameters. In order to compare the practical robustness of the aircraft, the following tests are performed.

5.3 Simultaneous multiple perturbation tests

These tests involve simultaneous perturbation of all the aircraft variables by a random amount. This style of testing requires many simulations to ensure adequate coverage of the testing envelope. Normally distributed random scale factors are used for perturbations.

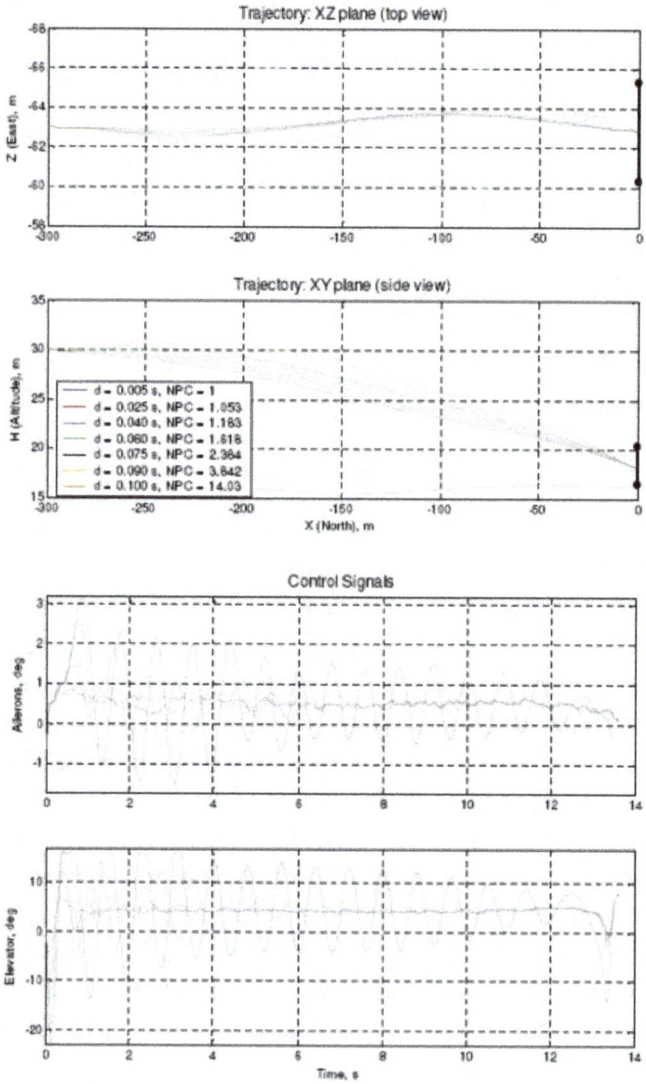

Figure 12. Flight path and control signals with time delays

The chosen standard deviations are approximately one tenth of the maximum allowed perturbations identified in the previous robustness tests.

For each of the two controllers, 1000 tests (4000 simulations considering four scenarios) have been carried out. Note that the perturbations were the same for both controllers in each case. This allows to compare the performance between the controllers in similar conditions. The second controller showed much poorer robustness. 90.4% of test cases showed NPC < 1.8. At the same time, only 69.8% of the tests reported successful recovery. Such a large proportion of cases which acceptable NPC value but without capture (about 20%) indicates that the controller delivers sufficiently inferior guidance strategy when subject to model uncertainty. This could be expected because this controller uses more positioning measurements-. The majority of these occurrences indicate a failure in the tailwind scenario, underscoring a possible guidance issue. The incidence of incidents involving a crash in one or more scenarios is 2.8%, representing a substantial increase compared to the first controller. The Proportional Navigation controller is evidently superior based on these testing. Nonetheless, it is imperative that it undergo testing across a diverse array of scenarios.

6 CONCISE SUMMARY

This chapter demonstrates an application of Evolutionary Design (ED). The objective of the design was to create a controller that facilitates the recovery of a fixed-wing UAV onto a vessel amongst the complete spectrum of disturbances and uncertainties inherent in real-world environments. The synthesis of the controller is a multistage procedure. Nonetheless, the methodology utilized for the synthesis of each block is quite analogous. The evolutionary algorithm serves as a mechanism to develop and enhance control laws. This methodology's primary advantage is its minimal or nonexistent reliance on prior understanding of control systems, initiating synthesis from fundamental proportional control or even from 'null' control laws. Throughout evolution, increasingly sophisticated and effective laws arise autonomously. The ensuing control laws indicate that evolution does not typically yield parsimonious solutions. The approach has significant robustness in convergence, suggesting that a near-optimal solution can be attained. In certain instances, the evolutionary process may need excessive time to identify the essence of a potentially ideal solution, resulting in a lack of convergence. This frequently indicates an inadequate selection of algorithm parameters. The paramount and challenging issue in Evolutionary Design is the formulation of the fitness evaluation technique with predetermined specialized intermediate challenges. Computational factors are also critically significant. The robustness of evolutionary algorithms incurs a significant computing expense, necessitating numerous fitness evaluations. The simulation testing encompasses the complete operational environment and identifies several scenarios when recovery poses significant risks. All environmental factors—ocean waves, wind velocity, and turbulence—significantly influence the likelihood of success. The interplay of multiple elements can create highly adverse conditions, even if each factor independently does not cause failure. Winds reaching 12 m/s do not impede recovery in a calm sea, and significant ship motion associated with Sea

State 5 likewise poses minimal risk in low wind conditions. Simultaneously, vigorous winds in elevated Sea States may pose risks to the aircraft. The fundamental aspect in this study environment is the legitimacy of the Evolutionary Design technique. Although it is clear that ED developed an effective controller that addresses the initial problem statement, numerous significant points can be noted. Initially, in the lack of comparable solutions, the proximity of the outcome to the optimum remains ambiguous. Given the exceptional robustness of evolutionary algorithms in achieving the global optimum, one can conclude that the controller is near the optimal solution. Nonetheless, an evaluation of the performance of the two created guidance controllers across various configurations indicates that there remains potential for enhancement. The primary cause for this is thought to be the restricted parameters of the testing circumstances employed in the evolutionary process, primarily attributable to a lack of computer resources. Overall, Evolutionary Design is an effective and potent instrument for intricate nonlinear control design. In contrast to most other design techniques, it seeks to autonomously resolve the existing problem rather than just optimizing a predetermined form. While ED does not negate the need for comprehensive testing, it can yield a nearly perfect answer if the complete spectrum of situations is considered in the fitness assessment. In theory, no particular understanding of the system is necessary, and the controllers may be regarded as 'black boxes' whose internal mechanisms are irrelevant. The successful design of the controller for the complex task of shipboard recovery illustrates the significant potential of this innovative technology.

6.1 Constraints of Evolutionary Design

It is important to recognize that evolutionary design (ED), like any evolutionary approach, does not yield fundamentally robust solutions for performance and unmodeled uncertainties. The behavior of the solutions under unproven evolutionary settings may be unpredictable. Comprehensive coverage of all situations in the evolutionary process necessitates numerous simulations, resulting in significant computational expenses. When the level of uncertainties in the system and environment escalates significantly, the computational cost may become excessively expensive. The simultaneous segmentation of the problem into smaller components and the reduction of environmental variables to mitigate uncertainties necessitate a comprehensive understanding of the system. This significantly undermines one of the primary benefits of ED, which asserts that minimal, if any, a priori knowledge of the system is required. Another constraint is that evolutionary approaches possess minimal analytical foundation. Their use is mostly grounded in empirical experience. Specifically, the majority of algorithm parameters, including population size, selection pressure, and the frequency of genetic operators, are calibrated using a trial and error approach. Although EAs are often resilient to these configurations, it is anticipated that the parameters employed for controller synthesis in this study may be suboptimal for alternative applications, and their optimization would require some time. The primary constraints of the controller developed in the ED approach originate from the restricted validity of the given system models. Higher quality models will be required for a comprehensive control system design.

6.2 Research opportunities on the Evolutionary Design

The ED approach demonstrated ease of adoption and exceptional suitability for contemporary use. Nevertheless, the primary focus has been on the actual implementation and use of this design process, with minimal exploration of its overall functioning and efficiency. An examination of optimal algorithm configurations, including population sizes and genetic operator probabilities, will also be advantageous. The enhancement and optimization of control rules representation could broaden the range of potential applications and increase the algorithm's efficiency. Significant potential for enhancement exists in the computational execution of the method. Although considerable effort has been devoted to enhancing the internal algorithmic efficiency in this work, substantial optimization of the code is achievable. This thesis outlines a methodology for creating a highly efficient controller that enables autonomous recovery of the UAV in challenging settings. The Evolutionary Design technique, central to this methodology, possibly facilitates the creation of controllers for a diverse array of control issues. It enables the automatic resolution of issues even with minimal or no knowledge of the controlled system, rendering it a powerful instrument for addressing complex control challenges.

REFERENCES

Astrom, K. J. & Wittenmark B. (1995). A survey of adaptive control applications, *Proceedings of IEEE Conference on Decision and Control* , pp. 649-654, ISBN: 0-7803-2685-7, New Orleans, LA, USA, December 1995

Atherton, D. (1975). *Nonlinear Control Engineering,* Van Nostrand Reinhold Company, ISBN: 0442300174

Blanchini, F. & Sznaier, M. (1994). Rational L1 suboptimal compensators for continuous time systems,. *IEEE Transactions on Automatic Control*, Vol. 39, No 7, page numbers (1487-1492), (July 1994), ISSN: 0018-9286

Bourmistrova, A. (2001). Knowledge based control system design for autonomous flight vehicle, *PhD thesis*, RMIT University: Melbourne, September 2001

Brown, R. G. & Hwang, P. Y. C. (1992). *Introduction to Random Signals and Applied Kalman Filtering.* 2nd ed. John Wiley & Sons, ISBN-10: 0471128392, ISBN-13: 978-0471128397

Bryson, A. J. & Ho Y.-C. (1975). *Applied Optimal Control.* Hemisphere, Washington, DC, 1975

Bu, T., Sznaier, M. & Holmes, M. (1996). A linear matrix inequality approach to synthesizing low order L1 controllers, *Proceedings of IEEE Conference on Decision and Control* proceedings, pp. 1875-1880 vol.2, ISBN: 0-7803-3590-2, Kobe, Japan, December 1996

Bugajski, D. J. & Enns, D. F. (1992). Nonlinear control law with application to high angle- ofattack flight, *Journal of Guidance, Control and Dynamics*, Vol. 15, No 3, (May/June 1992), page numbers (761-767), ISSN 0731-5090

Coley, D. A. (1998). *An Introduction to Genetic Algorithms for Scientists and Engineers.* University of Exeter, ISBN 9810236026, 9789810236021, Published by World Scientific, 1999

Crump, M. R. (2002). The dynamics and control of catapult launching Unmanned Air Vehicles from moving platforms, *PhD thesis*, RMIT University: Melbourne, 1992

Dahleh, M. A. & Pearson, J. B. (1987). L1 optimal feedback compensators for continuous- time systems, *Proceedings of IEEE Transactions Automatic Control,* pp. 889-895, Vol AC-32, ISSN: 0018-9286, October 1987

Dalsamo, M. & Egeland, O. (1995). H∞ control of nonlinear passive systems by output feedback, *Proceedings of IEEE Conference on Decision and Control* proceedings, pp. 351 - 352 vol. 1, ISBN: 0-7803-2685-7, New Orleans, LA, December 1995

DeCarlo, R., Zak, S. & Mathews, G. (1988). Variable structure control of nonlinear multivariable systems: a tutorial. *IEEE*, Vol. 76, No 3, (March 1988), page numbers (212-232), ISSN: 0018-9219

Doyle, J. C., Glover, K., Khargonekar P. P. & Francis B. A. (1989). State-space solutions to standard H2 and H∞ control problems, *IEEE Transactions on Automatic Control*, Vol. 34, No. 8, (August 1989), page numbers (831-847), ISSN: 0018-9286

Doyle, J. C. & Stein, G. (1981). Multivariable feedback design: concepts for a classical/ modern synthesis,. *IEEE Transactions on Automatic Control*, Vol. AC-26, No 1, (February 1981), page numbers (4-16), ISSN: 0018-9286

Duflos, E., Penel, P. & Vanheeghe, P. (1999). 3D guidance law modelling, *IEEE Transactions on Aerospace and Electronic Systems*, Vol. 35, No 1, page numbers (72-83), (January 1999), ISSN: 0018-9251

Graham, D. & McRuler, D. (1971). *Analysis of Nonlinear Control Systems,* Dover, ISBN-10: 0486610144, ISBN-13: 9780486610146

Hyde, R.(1991). The Aplication of Robust Control to VSTOL Aircraft, *PhD thesis*, Cambridge

Kaise, N. & Fujimoto, Y. (1999). Applying the evolutionary neural networks with genetic algorithms to control a rolling inverted pendulum, *Lecture Notes in Computer Science : Simulated Evolution and Learning*, Volume 1585/1999, (January

1999), page numbers (223-230), Springer Berlin / Heidelberg, ISBN 978-3-540-65907-5

Kalman, R. E. (1960). A new approach to linear filtering and prediction problems, *Transaction of the ASME—Journal of Basic Engineering*: p. 35-45, March 1960

Kaminer, I., Khargonekar, P. P. & Robel, G. (1990). Design of a localizer capture and track modes for a lateral autopilot using H∞ synthesis, *IEEE Control Systems Magazine*, Vol. 10, No 4, (June 1990), page numbers (13-21), ISSN: 0272-1708

Khammash, M. & Zou, L., Almquist, J.A.; Van Der Linden, C (1999). Robust aircraft pitch- axis control under weight and center of gravity uncertainty, *Proceedings of the 38th IEEE Conference on Decision and Control*, pp. 1970 - 1975 vol.2, ISBN: 0-7803-5250- 5, Phoenix, AZ, December 1999

Khantsis, S., (2006). Control System Design Using Evolutionary Algorithms for Autonomous Shipboard Recovery of Unmanned Aerial Vehicles, *PhD thesis*, RMIT University: Melbourne, 1992

Kim, B. S. & Calise, A. J. (1997). Nonlinear flight control using neural networks,. *Journal of Guidance, Control and Dynamics*, Vol. 20, No 1, page numbers (26-33), (January 1997), ISSN 0731-5090

Koza, J. R., Keane, M. A., Yu, J., Bennett III, F. H. & Mydlowec, W. (2000). Automatic creation of human-competitive programs and controllers by means of genetic programming, *Genetic Programming and Evolvable Machines*, Vol. 1, No 1, page numbers (121-164), (January 2000), ISSN: 1389-2576

Kwakernaak, H. (1993). Robust control and H∞ optimization - tutorial paper, *Automatica*, Vol. 29, No 2, page numbers (255-273), (February 1993), ISSN: 0005-1098

Maciejowski, J. (1989). *Multivariable Feedback Design.* Addison Wesley, ISBN-10: 0201182432, ISBN-13: 978-0201182439, 1989

MathWorks Inc. *Robust Control Toolbox User's Guide.* 2001. Available online at http://www.mathworks.com/access/helpdesk/help/pdf_doc/robust/robust.p df

Olhofer, M., Jin, Y. & Sendhoff, B. (2001). Adaptive encoding for aerodynamic shape optimization using Evolution Strategies, *Proceedings of Congress on Evolutionary Computation*, pp. 576-583 vol. 1, Seoul, South Korea, 2001, ISBN:1-59593-010-8

Onnen, C., Babuska, R., Kaymak, U., Sousa, J. M., Verbruggen, H. B. & Isermann, R. (1997). Genetic algorithms for optimization in predictive control, *Control Engineering Practice*, Vol. 5, No 10, page numbers (1363-1372), (October 1997), ISSN: 0967-0661

Pashilkar, A. A., Sundararajan, N. & Saratchandran, P. (1999). A fault-tolerant neural aided controller for aircraft auto-landing, *Aerospace Science and Technology*, Vol. 10, No 1, page numbers (49- 61), (January 2006), ISSN: 1270-9638

Postlethwaite, I. & Bates, D. (1999). Robust integrated flight and propulsion controller for the Harrier aircraft, *Journal of Guidance, Control and Dynamics*, Vol. 22, No 2, page numbers (286-290), (February 1999), ISSN 0731-5090

Rugh, W. J. & Shamma, J. S. (2000). Research on gain scheduling,. *Automatica*, Vol. 36, No 10, page numbers (1401- 1425), (October 2000), ISSN: 0005-1098

Sendhoff, B., Kreuz, M. & von Seelen, W. (1997). A condition for the genotype-phenotype mapping: Casualty, *Proceedings of 7th International Conference on Genetic Algorithms*, pp. 73-80, Michigan USA, July 1997, Morgan Kaufmann

Shamma, J. S. & Athans, M. (1992). Gain scheduling: potential hazards and possible remedies, *IEEE Control Systems Magazine*, Vol. 12, No 3, (June 1992), page numbers (101-107), ISSN: 0272-1708

Shue, S. & Agarwal, R. (1999). Design of automatic landing system using mixed H2 / H control, *Journal of Guidance, Control and Dynamics*, Vol. 22, No 1, page numbers (103-114), (January 1999), ISSN 0731-5090

Siouris, G. M., Leros, P. (1988). Minimum-time intercept guidance for tactical missiles, *Control Theory and Advanced Technology*, Vol. 4, No 2, page numbers (251-263), (February, 1988), ISSN 0911-0704

Skogestad, S. & Postlethwaite I. (1997). *Multivariable Feedback Control.* John Wiley and Sons, ISBN-10: 0471943304, ISBN-13: 978-0471943303

Spooner, J. T., Maggiore, M., Ordonez ,R. & Passino, K. M. (2002). *Stable Adaptive Control and Estimation for Nonlinear Systems: Neural and Fuzzy Approximator Techniques.* John Wiley and Sons, ISBN-10: 0471415464, ISBN-13: 978-0471415466

Ursem, R. K. (2003). Models for evolutionary algorithms and their applications in system identification and control optimization, *PhD dissertation*, Department of Computer Science, University of Aarhus, Denmark, 2003

Van Laarhoven, P. J. M. & Aarts, E. H. L. (1987). *Simulated annealing: theory and applications.* Dordrecht, The Netherlands: D. Reidel, ISBN-10: 9027725136

Wang, Q. & Zalzala, A. M. S. (1996). Genetic control of near time-optimal motion for an industrial robot arm, *Proceedings of IEEE International Conference on Robotics and Automation*, pp. 2592-2597 vol. 3, ISBN: 0-7803-2988-0, Minneapolis, MN, April 1996

Wise, K. & *Nonlinear*, S. J. (1996). $H\infty$ optimal control for agile missiles, *Journal of Guidance, Control and Dynamics*, Vol. 19, No 1, page numbers (157-165), (January 1996), ISSN 0731-5090

Zhou, K. & Doyle, J. C. (1998). *Essentials of Robust Control.* Prentice Hall, ISBN-10: 0135258332, ISBN-13: 978-0135258330

Chapter 8

FLY-THE-CAMERA PERSPECTIVE: MANAGEMENT OF A REMOTELY OPERATED QUADROTOR UAV AND CAMERA SYSTEM

1. INTRODUCTION: AN OVERVIEW

This chapter outlines a mission-focused methodology for regulating the optical axis of a video camera affixed to a camera positioner and secured to a remotely driven quadrotor vehicle. The methodology posits that for video collecting duties, a sole operator should be capable of managing the system by "flying-the-camera," meaning they gather video data from the viewpoint of the operator, who acts as the camera's pilot. This will enable the integration of quadrotor and camera manipulator controls into a unified robotic manipulation issue, whereby the camera is positioned by the four-degree-of-freedom (DOF) quadrotor and the two DOF camera positioner, facilitating comprehensive six DOF actuation of the camera view. The design of a closed-loop controller to execute this strategy is illustrated using a Lyapunov-type study. Results from computer simulations are presented to illustrate the proposed controller. Historically, the principal impetus for UAV reconnaissance capabilities has been military uses; now, we seem to be reaching a point where the cost and capabilities of these systems have become appealing for civilian use. The success of recent UAV systems has heightened expectations for accelerated technological advancement in essential areas, including cost-effective, dependable sensors, airframe design, stronger and lighter materials, higher energy-density battery technologies, and user-friendly interfaces that necessitate minimal operator training. The camera positioner, comprising the camera, camera base, and multi-axis servo platform, is a fundamental technology. The efficacy of UAVs equipped with camera positioners has been firmly demonstrated across various applications, including firefighting, emergency response, military and civilian surveillance, agricultural monitoring, and geographical registration. Numerous scientific and business entities have shown compelling evidence of the efficacy of UAVs in these applications. Most commercial systems are standardly equipped with camera positioners; nevertheless, the camera's operation is not integrated with the UAV's control. The standard configurations of a camera positioner comprise pan-tilt, tilt-roll, or pan/tilt/roll revolute joints or multi-axis gimbals. In the context of the camera gimbal actuator, rate gyroscopes or encoders are employed to assess orientations. If the system is compact and lightweight, the actuator dynamics may be disregarded in the control of the UAV. Heavier systems, in relation to the UAV, may need the incorporation of the camera positioner's interaction with the airframe into the control system. This chapter presents a method that integrates the location of a camera system with the control of the UAV. This technique is anticipated to establish the foundation for integrating dynamic control of the two systems.

Figure 1. Clemson UAV with Pan-Tilt Camera

1.1 Motivation

A standard camera system is equipped with a 2-axis gimbal or pan-tilt mechanism affixed to the UAV, as illustrated in Fig. 1. This camera positioner, equipped with actuators, is often affixed to the UAV's rigid body and commonly mounted at the front of the aerial vehicle for purposes such as surveillance, monitoring, or object targeting. The system is typically governed by an open-loop control mechanism utilizing a controller or joystick at the ground station. In this scenario, as the UAV tilts vertically to achieve position control objectives, the camera will similarly orient upwards and downwards, as illustrated in the upper plots of Fig. 2. If uncorrected, the camera loses the target and does not fulfill the surveillance aim. Adjusting the camera's field of view, as illustrated in the lower section of Fig. 2, indicates that the camera positioner is repositioned in response to the platform's movement. In the most basic approach, the UAV operator manually does this correction.

When navigation or surveillance activities get complex, two individuals may be necessary to accomplish the camera targeting objective, which is managed independently from the vehicle: a pilot to operate the UAV and a camera operator. In automated systems, feature tracking software can identify areas of interest inside the field of vision and subsequently create camera position commands to maintain these features within the field of view. Examining the activities of the two participants in the pilot/camera operator scenario is enlightening for hypothesizing a novel operating mode.

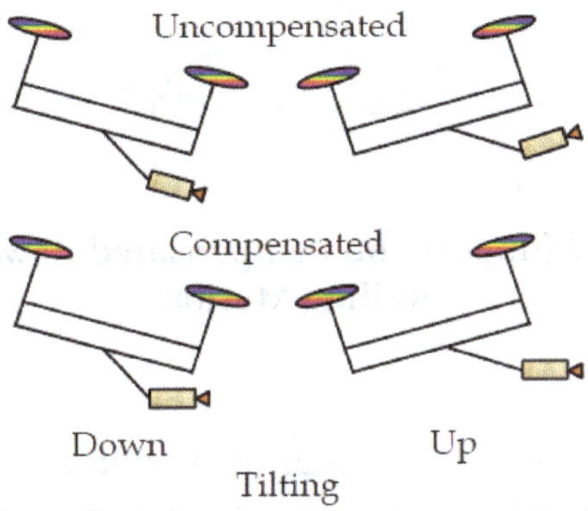

Figure 2. Uncompensated (upper) Comensated (lower) Camera Platform Position for UAV Airframe Tilting Motion

The pilot will maneuver the aircraft to circumvent obstacles and align the aerial vehicle, enabling the camera operator to monitor the camera feed and adjust the camera positioner to follow a target or survey a designated region. A crucial step by the camera operator that renders this scenario viable is the need to adjust for the UAV's movements that steady the camera's focus. Fig. 3 further describes the effect of uncompensated camera platform motion on the camera axis in the upper figures and the bottom figures show a scheme where the camera positioner is used to compensate for the UAV body motion and maintain the camera view. Additionally, there must be communication between the pilot and the camera operator so that the camera platform is correctly positioned or moved to meet the video acquisition objective. More specifically, the camera positioning problem is split between the pilot and the camera operator. Since the operator is not in full control of positioning the camera, she must rely on commands to the pilot to provide movement of the camera platform for motions not included in the camera positioner. For example, if the camera positioner is a simple pan-tilt and the camera operator requires translation of the camera, then a request must be made to the pilot to move the camera platform. The potential shortcomings of this typical operational scenario can be summarized as:

- multiple skilled technicians are typically required,
- the camera operator must compensate for the actions of the pilot, and
- it is not intuitive for a camera operator to split the camera targeting tasks between actions of the camera positioner controlled by the operator and commands to the pilot.

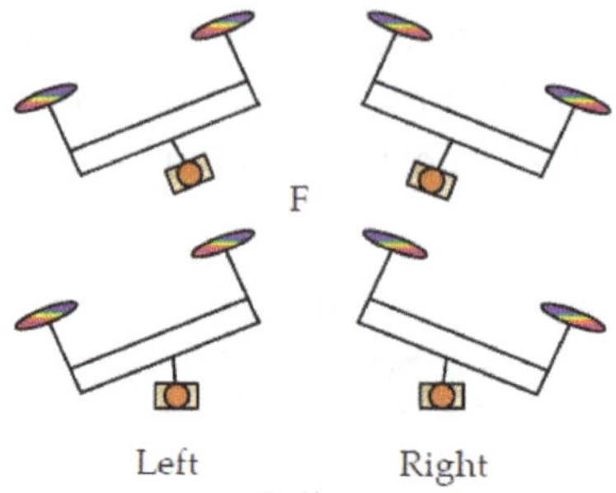

Figure 3. Uncompensated (upper) and Compensated (lower) Camera Platform for Rolling Motion

1.2 Previous Research

The current study examines research that contextualizes the UAV-camera control issue. The user's report of challenges in object localization suggests the necessity for an actuated roll axis as outlined in Adelstein & Ellis (2000). A 3 DOF camera platform was developed to enhance telepresence by integrating roll motion with yaw and pitch movements in an experiment. Mahony et al. (2002) investigated hardware and software designed to autonomously maintain the target within the camera's field of view and stabilize the image. This study proposed a «eye-in-hand» visual servo methodology for a robotic manipulator equipped with a camera on the end effector and incorporated nonlinear gain into the orientation feedback kinematics to guarantee that the target picture remains within the visual field. Jakobsen & Johnson (2005) demonstrated the management of a 3-axis Pan-Tilt-Roll camera system with customized servos and optical encoders affixed to a UAV, featuring three distinct operational modes. Lee et al (2007) presented the modelling of a complete 3 link robotic positioner and suggested a single robotic body system by combining it with a quadrotor model.

The authors, Yoon & Lundberg (2001), presented equations of motion for a two-axis, pan- tilt, gimbal system to simplify the gimbal control and to further illustrate the properties of the configuration. In Sharp et al. (2001), the authors implemented a realtime vision system for a rotorcraft UAV during landing by estimating the vehicle state. A pan-tilt camera vision system is designed to facilitate the landing procedure based on geometric calculations from the camera image. The work in Stolle & Rysdyk (2003) proposed a solution to the problem of the limited range of the camera mechanical positioner. The system attempts to keep a target in the camera field-of-view by applying a circle-based flight path guidance algorithm with a nose-mounted pan-tilt camera. Pieniazek (2003) presented a software-based camera control and stabilization for two degree-of-freedom onboard camera so as to stabilize the image when the aircraft attitude was disturbed by turbulence or attitude changes. Quigley et al. (2005) presented a field-tested mini-UAV gimbal mechanism, flightpath generation algorithm, and a human-UAV interface to enhance target acquisition, localization, and surveillance.

Procerus Technologies manufactures the OnPointTM targeting system for vision-based target engagement. The system employs feature tracking and video stabilization software, a stationary camera, and lacks a GPS sensor to accomplish these objectives. The Micropilot firm produces a stabilized pan-tilt-zoom camera system that maintains stability in yaw, roll, and elevation, ensuring a clear image at high zoom levels. This stabilized gimbal system is affixed to a mechanical interface that may be tailored to align the UAV's objectives with the ground control software HORIZONmp.

1.3 Camera Stabilization and Targeting

Figure 4 illustrates a 2-axis camera positioner suitable for surveillance or trajectory tracking. The challenge of delivering an intuitive interface for an operator to maneuver a camera positioner, enabling a video camera to track a target image, is prevalent in several contexts. The difficulty of moving a system that follows a subject with a video camera was recently addressed in Cooke et al. (2003) where operating a multilink, redundant-joint camera boom for the movie and television industry is described. The interesting result from this work is that an integrated control strategy, using a vision servoing approach to reduce the number of links controlled by the operator, can improve the use of the system. The final outcome demonstrates that an inexperienced operator attained the same tracking results as a seasoned operator; thus, the control method has made the system more user-friendly for the operator. The key aspect of the control method is the independent macro- and micro-positioning of the camera, where the operator manages coarse positioning and the vision system oversees precise positioning. The authors propose that the identical methodology may be applicable to alternative camera platforms; nonetheless, it is essential for the system to include redundant placement axes. An automated vision servoing system may be unsuitable for broad reconnaissance when the target is unidentified.

A different perspective to this basic camera targeting problem was presented in Chitrakaran (2006) and Neff et al. (2007) where the camera platform, a quadrotor UAV, and the camera positioning unit are considered to be controlled concurrently. In this work a controller was developed which simultaneously controls both the quadrotor and the camera positioning unit in a complimentary fashion. Both works show combining the four degrees-of-freedom provided by motion of the quadrotor helicopter with two degrees-of-freedom provided by a camera positioner to provide arbitrary six degree-of-freedom positioning of the on-board video camera. The work in Chitrakaran (2006) is actually directed towards providing an automated means of landing the quadrotor through the vision system but provides an important mathematical framework for analyzing the combined quadrotor/camera system. The work in Neff et al. (2007) builds on Chitrakaran (2006) using a velocity controller and first introduced the "fly-by-camera"concept by combining UAV and 2-axis tilt-roll camera for the combined quadotor/camera system that works from operator commands generated in the camera field-of-view to move both elements, presenting a new interface to the pilot.

Figure 4 Two-axis Camera Positioner

The research in Lee et al (2007) is exploited to provide an integrated system combining quadrotor UAV with 3-DOF camera system to present a single robotic unit. The paper designed a complete Pan-Tilt-Roll model which can be used for two camera optical axis; looking forward which can be used for surveillance, tracking, or targeting and looking downward axis for vision-based landing and monitoring the ground situation. In addition, the work suggests a position-based controller to show the upper bounds of position and velocity tracking error which yields Globally Uniformly Ultimately Bounded (SGUUB). The stabilization of the camera comes from this proposed perspective, which will be referred to as the fly-the-camera perspective, where the pilot commands motion from the perspective of the on-board camera - it is as though the pilot is riding on the tip of the camera and commanding movement of the camera ala a six-DOF flying camera. This is subtly different from the traditional remote control approach wherein the pilot processes the camera view and then commands an aircraft motion to create a desired motion of the camera view. The work proposed here exploits this new perspective for fusing vehicle and camera control. In this exposition, a quadrotor UAV model will be combined with a two-DOF camera kinematic model to create a fully actuated camera frame and a positioner controller will be designed.

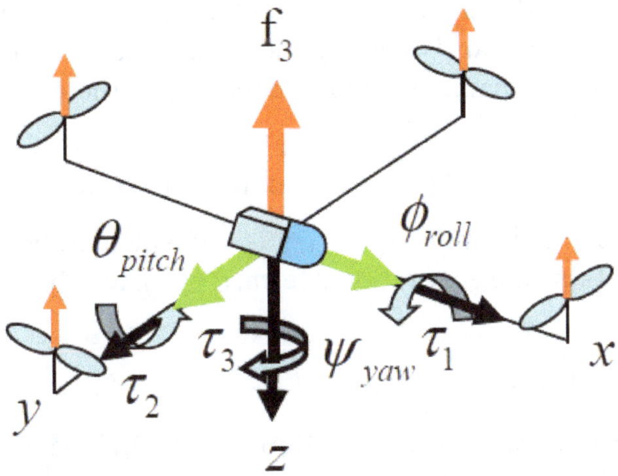

Figure 5. Underactuated Quadrotor Helicopter

2. SYSTEM MODELING

2.1 Quadrotor UAV

The elements of the quad-rotor unmanned aerial vehicle model are shown in Fig. 5. This system can only directly produce a thrust, f_3, along the z-axis (vertical) and torques about the roll, pitch, and yaw angle axes (τ_1, τ_2, and τ_3); hence, the system is underactuated and the translational motion in the x- and y-directions is indirectly achieved. The quadrotor is assumed to be a rigid body on which thrust and torque act uniformly through the body and that the quadrotor body fixed frame, F, is chosen to coincide with the center of gravity, which implies that it has a diagonal inertia matrix. The kinematic model of a quadrotor expressed in the inertial reference frame, denoted as North-East-Downward (NED), is given by

$$\begin{bmatrix} \dot{p}_F \\ \dot{\Theta}_F \end{bmatrix} = \begin{bmatrix} R_F^I & O_3 \\ O_3 & T_F^I \end{bmatrix} \begin{bmatrix} v^F \\ \omega^F \end{bmatrix} \in R^6 \tag{1}$$

where $v^F(t)$ and $\omega^F = [\omega_x, \omega_y, \omega_y]^T \in R^3$ denote the linear velocity and angular velocity of the quadrotor body-fixed frame, F , with respect to the earth-fixed inertial frame, I , expressed in the body-fixed frame, F . The position and angle, p_F (t) and Θ_F(t), and velocities $v^F(t)$ and ω^F(t) are assumed to be measurable. The $\dot{p}_F(t)$ in (1) is the velocity v^F (t) of the quadrotor transformed by the spatial rotation matrix $R_F^I(\Theta) \in SO(3)$ where $\Theta_F = [\phi, \theta, \psi]^T \in R^3$ are roll, pitch, and yaw angles about x-, y-, and z-axes (Fossen, 2002)

$$\dot{R}_F^I = R_F^I S\left(\omega^F\right). \tag{2}$$

$\dot{\Theta}_F(t)$ in (1) represents the angular velocity $\omega^F(t)$ transformed into the inertial frame by the angular orientation matrix $T_F^I(\Theta) \in R^{3\times3}$ (Fossen, 2002) which represents the modeling assumption that angular velocity of the quadrotor is calculated directly in lieu of modeling the angular dynamics; that is, $\omega^F(t)$ is considered as the system input. The translational dynamic modeling equation of the quadrotor is based on Chitrakaran (2006) and given by

$$m\dot{v}^F = F^f - mS(\omega^F)v^F + N_1(v^F) + G(R_I^F) \tag{3}$$

where the rotational angular velocity, $\omega^F(t)$, is implemented by control command (see (57)) and (3) contains the gravitational term, $G(R_I^F)$, which is represented in the body- fixed frame as

$$G(R_I^F) = mg(R_I^F)E_3 \in R^3 \tag{4}$$

where $g \in R^1$ denotes gravitational acceleration, $E_3 = [0, 0,1]^T$ denotes the unit vector in the coordinates of the inertial frame, $m \in R^1$ is the known mass of the quad-rotor, $N_1(v^F) \in R^3$ represents a bounded, unknown, nonlinear function (i.e, aerodynamic damping) and $S(.) \in R^{3x3}$ is a general form of the skew-symmetric matrix as follows

$$S(\xi) = \begin{bmatrix} 0 & -\xi_3 & \xi_2 \\ \xi_3 & 0 & -\xi_1 \\ -\xi_2 & \xi_1 & 0 \end{bmatrix}, \xi = [\xi_1, \xi_2, \xi_3]^T \in R^3. \tag{5}$$

As mentioned at the start of this modeling section, the quadrotor shown in Fig. 5 is underactuated in the sense that it has one translational force along the z-axis. The vector $F^f(t) \in R^3$ refers to the quadrotor translational forces but in reality represents the single translational force which is created by summing the forces generated by the four rotors and is expressed as

$$F^f = B_1 u_1 = \begin{bmatrix} 0 & 0 & u_1 \end{bmatrix}^T \tag{6}$$

where $B_1 = I_3$ is a configuration matrix (actuator dynamics are beyond the scope of this design) and $u1(t) = f_3(t) \in R^1$. The modeling nomenclature is summarized in 7.3.

2.2 Model Assumptions

The following assumptions are made regarding the system model:

- A1: $T_F^I(\Theta)$ are full rank, i.e., $s(t) \neq \pm\frac{\pi}{2}$. This will ensure that the orientation angle, θa, (defined in (38)) remains within the range $-\pi < \theta a < \pi$ about the rotation axis $\mu(t)$ and will ensure that det ($L\omega$) exists (see (43)).
- A2: Aerodynamic damping term $N_1(v^F)$ in (3) can be replaced by the parameter linearizable form; i.e., $Y_1(v^F)\theta_1 = N_1(v^F)$ where $\|Y_1(v^F)\| \leq \zeta_0\|v^F\|$, ζ_0 is a positive constant, $Y_1(v^F) \in R^{3xn}$ is a known regression matrix, and $\theta_1 \in R^n$ is a known parameter vector. Additionally, $|Y_1(v^F)\theta_1| \leq \zeta_1(|v^F|) \leq \xi_1|v^F|$ where $\zeta 1$ is a positive function and non-decreasing in $|v^F|$ and $\xi_1 \in R^1$ is a positive constant.

2.3 Camera Positioner

Here we suggest a camera system with two revolute joints to cover forward and downward optical axes as shown in Fig. 6. Considering the integrated system, UAV and camera positioner, the yaw action of the UAV provides the third orientation angle for the camera. To create a general analytic framework for modeling the configurations of a two-link revolute positioner, a 2-axis robot positioner model is proposed. This general two-axis camera positioner will provide the forward kinematics for a tilting-rolling motion and panning-tilting motion where the angles $\theta_{tilt}(t)$ and $\theta_{roll}(t)/\theta_{pan}(t)$ are assumed to be measurable. The Tilt-Roll configuration can be used to compensate for the quadrotor body roll and pitch when the camera is facing forward for tasks such as general navigation or surveillance. On the other hand, the Pan-Tilt camera positioner configuration can be used to compensate for the quadrotor pan and pitch motion when the camera is looking downward for tasks like landing, monitoring, or surveillance.

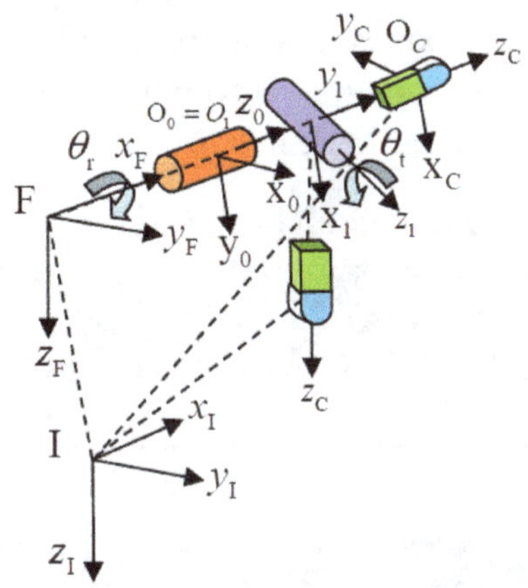

Figure 6. Quadrotor with a Camera Positioner Configurable for Tilt-Roll and Pan-Tilt Operating Modes

The dynamics of the camera unit will be considered negligible. The standard coordinate system definitions are made, specifically, O_0 is used to represent the origin of the coordinate system at the base (B) of the camera, O_1 is the origin of the second link, and O_c is used to represent the camera frame and can be denoted as O_2. Note that the link lengths are assumed to be zero so O_0, O_1, and O_2 are coincident. The camera positioned kinematics will be calculated using Denavit-Hartenberg (D-H) Convention (Spong, 2006). From the convention the rotation matrix R_i^{i-1} from the ith coordinate system to the (i -1)th frame is given

$$R_i^{i-1} = \begin{bmatrix} \cos\theta & -\sin\theta\cos\alpha & \sin\theta\sin\alpha \\ \sin\theta & \cos\theta\cos\alpha & -\cos\theta\sin\alpha \\ 0 & \sin\alpha & \cos\alpha \end{bmatrix} \qquad (7)$$

where θ is the rotated angle, pan/tilt/roll, and α is the twisted angle between the links.

2.3.1 Case 1: Tilt-Roll Camera Configuration (camera looking forward)

A feasible 2-axis Tilt-Roll camera positioner is pictured in Fig. 7. Note that the matrix O_0 represents a static mounting on the quadrotor as a camera base frame denoted B and is given by

$$R^F_{0=B} = R_{\alpha,\frac{\pi}{2}} R_{\theta,\frac{\pi}{2}} = \begin{bmatrix} 0 & 0 & 1 \\ 1 & 0 & 0 \\ 0 & 1 & 0 \end{bmatrix} \tag{8}$$

where θ and α angles are each +90o in order calculated by (7) and $z_0 = [1\ 0\ 0]^T$ where z_{i-1} (i is the link origin) is the last column of the rotation matrix which will be used in the Jacobian matrix. The Denavit-Hartenberg (D-H) table for the Tilt- Roll configuration when the optical axis is looking forward, as shown in Fig. 7, is in Table I. The camera positioner unit is considered to have coincident rotational links, thus the link lengths are zero (a=0 in Table I).

Table 1. D-H for Tilt-Roll Camera Positioner while looking forward optical axis

Link (i)	Offset	Length	Twist	Angle
	D	a	α	θ
1	0	0	+900	$\theta_r + 900$
2 (forward axis)	0	0	–900	θ_t

Figure 7. Quadrotor with Tilt-Roll Camera Positioner

The rotation matrices for first and second links are obtained as

$$R_1^B = \begin{bmatrix} -s\theta_r & 0 & c\theta_r \\ c\theta_r & 0 & s\theta_r \\ 0 & 1 & 0 \end{bmatrix} \quad \text{and} \quad R_2^1 = \begin{bmatrix} c\theta_t & 0 & -s\theta_t \\ s\theta_t & 0 & c\theta_t \\ 0 & -1 & 0 \end{bmatrix} \tag{9}$$

where c· = cos(·), s· = sin(·) were used, and the rotation matrix from the base frame to first link frame is given

$$R_1^F = R_B^F R_1^B = \begin{bmatrix} 0 & 1 & 0 \\ -s\theta_r & 0 & c\theta_r \\ c\theta_r & 0 & s\theta_r \end{bmatrix} \tag{10}$$

where $z_1 = [0 \ c\theta_r \ T \ s\theta_r]^T$. The total rotation matrix from the quadrotor through the second link can be obtained as

$$R_{2=C}^F = R_B^F R_1^B R_2^1 = \begin{bmatrix} \sin\theta_t & 0 & \cos\theta_t \\ -\sin\theta_r\cos\theta_t & -\cos\theta_r & \sin\theta_r\sin\theta_t \\ \cos\theta_r\cos\theta_t & -\sin\theta_r & -\cos\theta_r\sin\theta_t \end{bmatrix}. \tag{11}$$

The angular velocity of the camera frame can be written as follows

$$\omega_{BC}^F = R_B^F \omega_{BC}^B = J_C \dot{\theta}_C \tag{12}$$

where the positioner joint angles, $\dot{\theta}_C(t) \in R^3$, are given by

$$\dot{\theta}_C = \begin{bmatrix} \dot{\theta}_r \\ \dot{\theta}_t \end{bmatrix} \text{ and } \theta_C = \begin{bmatrix} \theta_r \\ \theta_t \end{bmatrix}. \tag{13}$$

The Jacobian matrix $J_C(\theta_C) \in R^{3\times2}$ result given by can be built from the rotation matrices with the final

$$J_C = \begin{bmatrix} z_0 & z_1 \end{bmatrix} = \begin{bmatrix} 1 & 0 \\ 0 & c\theta_r \\ 0 & s\theta_r \end{bmatrix} \tag{14}$$

where the reader is referred to Murray et al. (1994) for the Jacobian matrix of a manipulator.

2.3.2 Case 2: Pan-Tilt Camera Configuration (camera looking downward)

A similar approach to that for case 1, 2.3.1, can be followed to develop the kinematics for the Pan-Tilt configuration when the optical axis is looking downward as shown in Fig. 8.

Figure 8. Quadrotor with Pan-Tilt Camera Positioner

The systematic Denavit-Hartenberg convention is used to obtain the rotation matrices of the Pan-Tilt camera positioner configuration and the detail procedure is shown in Appendix 7.1.

2.4 Position and Orientation of the Camera Positioner

The objective is to control the motion of the camera optical axis. Towards this end, the kinematic relationships in (1) will be extended to include the action of the camera positioning unit and to obtain the position and orientation of the camera. The derivative of the position of the camera in the camera frame, C, with respect to the inertial frame, I, and expressed in the inertial frame, $\dot{p}_C(t) \in R^3$, is defined as

$$\dot{p}_C = R_C^I v^C \tag{15}$$

where $v^C(t) \in R^3$ is the linear velocity in the camera frame, C, referred to inertial frame, I, expressed in the camera frame, C. The rotation matrix R_C^I from the camera frame to the inertial frame can be obtained using the previous equations as

$$R_C^I = R_F^I R_C^F . \tag{16}$$

The velocity $v^C(t)$ can be divided into two components as follows

$$v^C = v^{IF} + v^{FC} = R_F^C v^F \tag{17}$$

where $v^{IF}(t)$ is the UAV velocity expressed in camera frame (C) and $v^{FC}(t) = 0 = 0$ since the camera positioning unit only has rotational axes and does not translate from the quadrotor body. Thus, substituting (17) for $v^C(t)$ in (15) yields

$$\dot{p}_C = R_C^I R_F^C v^F = R_F^I v^F . \tag{18}$$

The desired camera position trajectory, $\dot{p}_D(t) \in R^3$, is generated via

$$\dot{p}_D = R_D^I v^D \tag{19}$$

where $v^D(t) \in R^3$ is a desired input velocity vector and the desired rotation matrix expressed in inertial frame, $R_D^I(\Theta, \theta_C, \theta_d) \in R^{3x3}$, can be defined in the following manner

$$R_D^I = R_F^I R_C^F R_D^C \tag{20}$$

where $R_D^C(\theta_d) \in R^{3x3}$ represents a rotation matrix from D to desired frame C as shown in Fig. 9 for the subsequent control development which is the tracking of the camera frame towards the desired camera frame using angle-axis representation (see Appendix 7.2). Utilizing (20), (19) yields

$$\dot{p}_D = R_F^I R_C^F R_D^C v^D . \tag{21}$$

Hence, to quantify the mismatch between the camera (actual) and desired attitudes, we define the rotation matrix $\tilde{R}(\Theta, \theta_C, \theta_d) \in R^{3\times3}$ which can be obtained using measurable rotations and using the angle-axis representation (Spong, 2006) as

$$\tilde{R} = R_C^I (R_D^I)^T . \tag{22}$$

A similar result for the camera angles, equation of (1) is now shown as $\Theta_C(t)$, in the camera frame from the second

$$\dot{\Theta}_C = T_F^I(\Theta)\,\omega_{IC}^F = T_F^I(\Theta)\,R_C^F \omega^C \tag{23}$$

where an angular transformation $T_F^I(\Theta)$ is used. Decomposing the angular velocity,

$$\omega^C(t) = \omega^{IF} + \omega^{FB} + \omega^{BC} = R_F^C \omega^F + R_F^C \omega_{FB}^F + R_B^C \omega_{BC}^B , \tag{24}$$

yields

$$\dot{\Theta}_C = T_F^I(\Theta)\,R_C^F (R_F^C \omega^F + R_F^C \omega_{FB}^F + R_B^C \omega_{BC}^B). \tag{25}$$

Then, (25) yields

$$\dot{\Theta}_C = T_F^I(\Theta)\,\omega^F + T_F^I(\Theta)\,J_C \dot{\theta}_c \tag{26}$$

where $R_C^F R_F^C = 1$ is used and $\omega_{FB}^F = 0$ since the camera base is rigidly mounted on the quadrotor frame. Finally, following the same approach the desired camera angle, $\Theta_D(t)$, is obtained from the desired angular velocity of the camera in the quadrotor frame, $\omega^D(t)$, as

$$\dot{\Theta}_D = T_F^I(\Theta)\,R_D^F \omega^D . \tag{27}$$

The changing rate of $R_C^I(\Theta)$ is obtained as follows

$$\dot{R}_C^I = R_C^I S(\omega^C) \quad \text{and} \quad \dot{R}_I^C = S^T(\omega^C) R_I^C = -S(\omega^C) R_I^C . \tag{28}$$

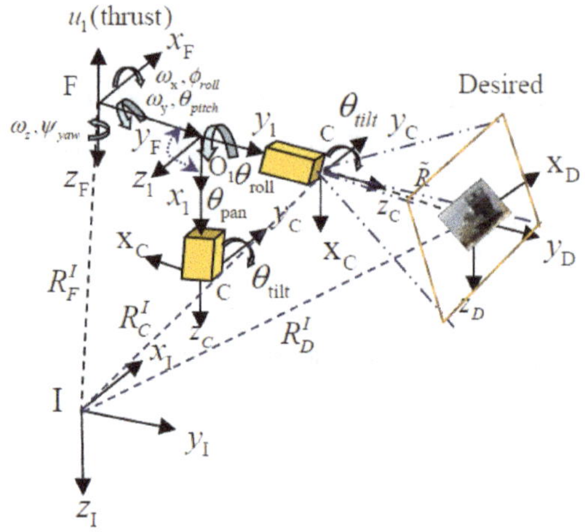

Figure 9. Tracking of the Camera Frame Towards the Desired Camera Frame

3. CONTROL METHOD

Fig. 10 demonstrates the culmination of the modeling effort to combine the quadrotor and camera positioner to create a means of fully actuating the camera optical axis. In this diagram, it can be seen that the camera is positioned and oriented by using the two camera positioner angles (selected from $\theta_{tilt}(t)$, $\theta_{roll}(t)$, and $\theta_{pan}(t)$, $\theta_{tilt}(t)$, according to the configuration of the camera optical axis), the quadrotor linear force, F (t), and the quadrotor angles $\varphi_{roll}(t)$, $\theta_{pitch}(t)$, and $\psi_{yaw}(t)$, through $\omega_x(t)$, $\omega_y(t)$, and $\omega_z(t)$. In keeping with the fly-the-camera objective, a controller will be designed based on these inputs to move the camera optical axis along a desired trajectory. A control strategy will be proposed to control the camera translational position error, ep(t) ∈ R3, and the camera orientation error, eθ (t) ∈ R3.

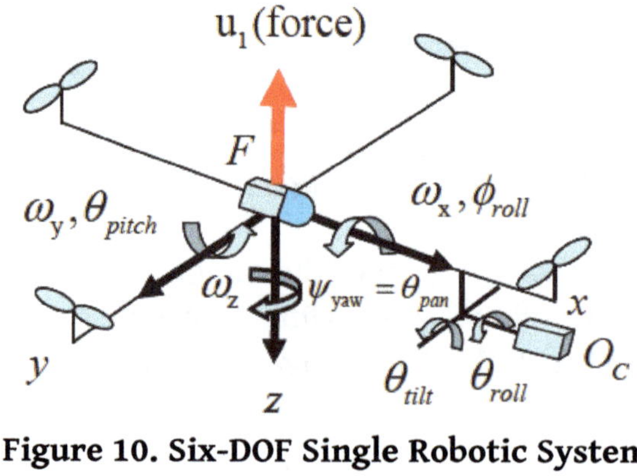

Figure 10. Six-DOF Single Robotic System

3.1 Tracking Error Formulation

The camera translational position error, $e_p(t) \in R^3$, is defined in the camera frame (C) as the transformed difference between the inertial frame based camera position, pC(t), and the inertial frame based desired camera position, denoted as $p_D(t) \in R^3$, as follows

$$e_p = R_I^C (p_C - p_D). \tag{29}$$

The camera translational position error rate, $\dot{e}_p(t) \in R^3$, is obtained by taking the time derivative of (1) to yield

$$\dot{e}_p = \dot{R}_I^C (p_C - p_D) + R_I^C (\dot{p}_C - \dot{p}_D) \tag{30}$$

where $\dot{p}_C(t)$ and $\dot{p}_D(t)$ were introduced in (18) and (19), respectively. Substituting (28) into the first term in (30) yields

$$\dot{R}_I^C (p_C - p_D) = -S(\omega^C) e_p, \tag{31}$$

and the term as $R_I^C\, \dot{p}_C(t)$ in (30) can be rewritten in terms of the quadrotor velocity, $v^F(t)$,

$$R_I^C\, \dot{p}_C = R_I^C (R_C^I v^C) = R_F^C v^F \tag{32}$$

where all translation of the camera is the result of the quadrotor translation. The final term in (30) is rewritten as

$$R_I^C\, \dot{p}_D = R_I^C (R_D^I v^D) = R_D^C v^D. \tag{33}$$

Substituting (31), (32), and (33) into (30) yields

$$\dot{e}_p = -S(\omega^C) e_p + R_F^C v^F - R_D^C v^D. \tag{34}$$

To further the controller development, a filtered error, $r(t) \in R^3$, is introduced as

$$r = [r_p^T \quad e_\theta^T]^T \tag{35}$$

where the filtered position error, $r_p(t) \in R^3$, is defined as

$$r_p = k_p e_p + R_F^C v^F + R_F^C \delta, \tag{36}$$

in which $\delta = [0 \quad 0 \quad \delta_3]^T \in R^3$ is a constant design vector. The orientation tracking signal, $e_\theta(t) \in R^3$, is defined in terms of the angle-axis representation of the rotation matrix between the desired and actual camera orientations, $\tilde{R}(\Theta, \theta_c, \theta_d)$, as

$$e_\theta = \theta_a \mu \tag{37}$$

where the scalar angle θa is obtained from

$$\theta_a = \cos^{-1}(\frac{1}{2}(Tr(\tilde{R}) - 1)) \in R^1, \tag{38}$$

in which $Tr(\tilde{R})$ rotation μ ∈ R defines the trace of the matrix $\tilde{R}(\Theta, \theta_c, \theta_d)$ and the unit length axis of rotation $\mu \in R^3$ defined as

$$\mu = \frac{1}{2\sin\theta_a}\left[(r_{32}-r_{23}),(r_{13}-r_{31}),(r_{21}-r_{12})\right]^{T}, \tag{39}$$

for which $\|\mu\|^2 = 1$. In this representation, the rotation angle θ_a is assumed to stay within the range $-\pi < \theta_a < \pi$. Note that the terms on the right-hand side of the definition in (39) come from $\tilde{R}(\Theta,\theta_C,\theta_d)$ as

$$\tilde{R}(\Theta,\theta_C,\theta_d) = \begin{bmatrix} r_{11} & r_{12} & r_{13} \\ r_{21} & r_{22} & r_{23} \\ r_{31} & r_{32} & r_{33} \end{bmatrix}. \tag{40}$$

Substituting the axis-angle representation (38) and (39) into (37) yields

$$e_\theta = \frac{1}{2sinc\{\cos^{-1}(\frac{1}{2}(Tr(\tilde{R})-1))\}}\begin{bmatrix} r_{32}-r_{23} \\ r_{13}-r_{31} \\ r_{21}-r_{12} \end{bmatrix} \tag{41}$$

where $sinc(\theta_a) = \frac{\sin(\theta_a)}{\theta_a}$. The angular error rate can be obtained by taking the time derivative of (37) as

$$\dot{e}_\theta = L_\omega \omega^{CD} \in R^3, \tag{42}$$

and $L_\omega \in R^{3x3}$ can be defined as

$$L_\omega = I_3 - \frac{\theta_a}{2}S(\mu) + \left(1 - \frac{sinc(\theta_a)}{sinc^2(\frac{\theta_a}{2})}\right)S^2(\mu), \tag{43}$$

in which $S(\mu)$ is a skew symmetric matrix and details for obtaining

$$S(\mu) = \frac{(\tilde{R}-(\tilde{R})^T)}{2\sin\theta_a}, \tag{44}$$

for obtaining (42) can be found in Malis (1998). The term $\omega^{CD}(t) \in R^3$ introduced in (42) represents the angular velocity of the desired camera frame relative to the actual camera frame as

$$\omega^{CD} = \omega^{CI} + \omega^{ID} = \omega^D - \omega^C = \omega^D - (\omega^C_{IF} + \omega^C_{FC}) \tag{45}$$

where $\omega^{ID}(t) = \omega^D(t)$ and $\omega^{CI}(t) = -\omega^{IC}(t) = -\omega^C(t)$. Substituting (45) along with (12) - (14) into (42) produces

$$\dot{e}_\theta = L_\omega \omega^D - L_\omega R^C_F \omega^F - L_\omega R^C_F J_c \dot{\theta}_c. \tag{46}$$

Taking the time derivative of $r_p(t)$ in (36) yields

$$\dot{r}_p = R_F^C \dot{v}^F + \dot{R}_F^C v^F + k_p \dot{e}_p + \dot{R}_F^C \delta. \quad (47)$$

Substituting from (3) and (34) into (47), utilizing the fact that

$$\dot{R}_F^C(\Theta) = R_F^C S(\omega_{CF}^F) = -R_F^C S(\omega_{FC}^F),$$

and subtracting and adding $R_F^C S(\omega^F)\delta$ yields

$$\dot{r}_p = R_F^C[\frac{1}{m} N_1(v^F) - S(\omega^F)v^F + gR_I^F E_3 + \frac{1}{m} F^f] + R_F^C S(\omega^F)\delta$$
$$+k_p[R_F^C v^F - S(\omega^C)e_p - R_D^C v^D] - R_F^C S(\omega_{FC}^F)v^F - R_F^C\left(S(\omega_{FC}^F) + S(\omega^F)\right)\delta. \quad (48)$$

Combining the angular velocities represented by the second term in the first bracket and the mid-term in the last row of (48) yields

$$-R_F^C S(\omega^F + \omega_{FC}^F)v^F = -R_F^C S(\omega_{IC}^F)v^F. \quad (49)$$

The right-hand side of (49) can be further clarified using $\omega_{IC}^F = R_C^F \omega^C(t)$ and

$$S(\omega_{IC}^F) = R_C^F S(\omega^C) R_F^C, \quad (50)$$

to yield

$$-R_F^C R_C^F S(\omega^C) R_F^C v^F = -S(\omega^C) R_F^C v^F. \quad (51)$$

Combining the angular velocity terms $\omega_{FC}^F(t)$ and $\omega^F(t)$ in the last terms of the last row in (48) with $\omega_{IC}^F(t)$ yields

$$-R_F^C S(\omega^F)\delta - R_F^C S(\omega^{FC})\delta = -R_F^C S(R_C^F \omega^C)\delta = -S(\omega^C) R_F^C \delta. \quad (52)$$

Multiplying both sides of (36) by $S(\omega^C)$ yields

$$-S(\omega^C) r_p = -S(\omega^C)(k_p e_p + R_F^C v^C + R_F^C \delta). \quad (53)$$

Substituting (51) through (53) into (48) yields

$$\dot{r}_p = \frac{R_F^C N_1(v^F)}{m} - S(\omega^C) r_p + gR_I^C E_3 + \frac{R_F^C F^f}{m} + k_p R_F^C v^F - k_p R_D^C v^D - R_F^C S(\delta)\omega^F. \quad (54)$$

By taking the time derivative of r(t) in (35) and substituting (46) and (54) it can be obtained that

$$\dot{r} = \begin{bmatrix} \frac{R_F^C F^f}{m} - R_F^C S(\delta)\omega^F \\ -L_\omega R_F^C(\omega^F + J_c \dot{\theta}_c) \end{bmatrix} - \begin{bmatrix} S(\omega^C) r_p \\ O_{3x1} \end{bmatrix} + \begin{bmatrix} \frac{R_F^C N_1(v^F)}{m} + k_p(R_F^C v^F - R_D^C v^D) + gR_I^C E_3 \\ -L_\omega \omega^D \end{bmatrix}. \quad (55)$$

Arranging the first term in (55) yields

$$\begin{bmatrix} \frac{1}{m} R_F^C F^f - R_F^C S(\delta)\omega^F \\ -L_\omega R_F^C \omega^F - L_\omega R_F^C J_c \dot{\theta}_c \end{bmatrix} = \overline{L}_\omega B\overline{U} = \overline{B}\overline{U} \tag{56}$$

where

$$\overline{L}_\omega = \begin{bmatrix} R_F^C & O_{3x3} \\ O_{3x3} & -L_\omega R_F^C \end{bmatrix}, B = \begin{bmatrix} \frac{B_1}{m}, -S(\delta), O_{3x2} \\ O_{3x1}, I_{3x3}, J_c \end{bmatrix} \in R^{6x6}, \overline{U} = \begin{bmatrix} u_1 \\ \omega^F \\ \dot{\theta}_C \end{bmatrix} \in R^6, \tag{57}$$

along with the new term U (t) where $U_1(t) \in R^3$ and $U_2(t) \in R^3$, defined also further clarify the control design procedure.

$$U = \overline{B}\overline{U} = [U_1, U_2]^T \in R^6. \tag{58}$$

Substitution of (57) to (58) into (56) and then substitution of the resulting form of (56) using (55) produces the open-loop filtered error dynamics as follows

$$\dot{r} = \begin{bmatrix} -S(\omega^C) r_p \\ O_{3x1} \end{bmatrix} + \begin{bmatrix} U_1 \\ U_2 \end{bmatrix} - \begin{bmatrix} k_p R_D^C v^D \\ L_\omega \omega^D \end{bmatrix} + \begin{bmatrix} \frac{R_F^C N_1(v^F)}{m} + k_p R_F^C v^F + g R_I^C E_3 \\ O_{3x1} \end{bmatrix} \tag{59}$$

where it is clear that the control inputs $U_1(t)$ and $U_2(t)$ will be designed to meet the control objective. Note that implementation of the control will require $\overline{U}(t)$ which is obtained from

$$\overline{U} = \overline{B}^{-1} U \in R^6, \tag{60}$$

which requires $\overline{B}^{-1} = B^{-1}(\overline{L}_\omega)^{-1}$ where

$$(\overline{L}_\omega)^{-1} = \frac{1}{\Delta(\overline{L}_\omega)} \begin{bmatrix} -L_\omega R_F^C & O_3 \\ O_3 & R_F^C \end{bmatrix}, \tag{61}$$

in which $\Delta(\overline{L}_\omega) = \left| -R_F^C L_\omega R_F^C \right|$ is the determinant of the matrix $\overline{L}_\omega(t)$ and $\Delta(\overline{L}_\omega) \neq 0$ due to Assumption 1 (A1).

3.2 Lyapunov-based Control Design

The non-negative scalar function V(t) is chosen as

$$V = \frac{1}{2} e_p^T e_p + \frac{1}{2} r^T r.$$

Differentiating yields

$$\dot{V} = e_p^T \dot{e}_p + r_p^T \dot{r}, \tag{62}$$

by substituting (34) and (59) into (62) it can be obtained that

$$\dot{V} = [r_p^T, e_\theta^T]^T [\begin{bmatrix} -S(\omega^C) r_p \\ O_{3x1} \end{bmatrix} + \begin{bmatrix} U_1 \\ U_2 \end{bmatrix} + \begin{bmatrix} \frac{R_F^C N_1(v^F)}{m} + k_p R_F^C v^F + g R_I^C E_3 - k_p R_D^C v^D \\ L_\omega \omega^D \end{bmatrix}] \quad (63)$$
$$+ e_p^T [-S(\omega^C) e_p + (r_p - k_p e_p - R_F^C \delta) - v^D]$$

where (36) was utilized. The terms in (63) can be collected to yield

$$\dot{V} = [r_p^T U_1 + \frac{1}{m} r_p^T R_F^C N_1(v^F) + e_p^T r_p + r_p^T k_p R_F^C v^F + r_p^T g R_I^C E_3$$
$$- r_p^T k_p R_D^C v^D - k_p e_p^T e_p - e_p^T R_F^C \delta - e_p^T R_D^C v^D + e_\theta^T U_2 - e_\theta^T L_\omega \omega^D]. \quad (64)$$

Equation (64) will be utilized to design the control inputs $U_1(t)$ and $U_2(t)$. From the upper equation in (64) we can design $U_1(t)$ to subtract out four terms, add stabilizing feedback, and add robust compensatation for the unknown nonlinear term as follows

$$U_1 = -k_r r_p - \frac{r_p \zeta_1^2(\|v^F\|)}{\varepsilon_0} - k_p R_F^C v^F - g R_I^C E_3 - e_p \quad (65)$$

where ε0 defined by is a positve constant, A2 is used for $\zeta_1(\|v^F\|)$, and the nonlinear term

$$\bar{N}_1 = \frac{1}{m} R_F^C N_1(v^F). \quad (66)$$

From the lower equation in (59), $U_2(t)$ can be designed as

$$U_2 = L_\omega \omega^D - k_\theta e_\theta. \quad (67)$$

Substituting (65) and (67) into (64) yields

$$\dot{V} = r_p^T (\bar{N}_1 - \frac{r_p \zeta_1^2(\|v^F\|)}{\varepsilon_0}) - k_r r_p^T r_p - r_p^T k_p R_D^C v^D - k_\theta e_\theta^T e_\theta - k_p e_p^T e_p - e_p^T (R_F^C \delta + R_D^C v^D). \quad (68)$$

According to sufficient condition, finally it yields

$$\dot{V} \le -\lambda_2 \|\eta\|^2 + \varepsilon_4. \quad (69)$$

4. STABILITY ANALYSIS

The closed-loop control law of (65) and (67) ensure that the tracking error is Globally Uniformly Ultimately Bounded (GUUB) in the manner

$$\|\eta\| \le \sqrt{\|\eta(0)\|^2 e^{-2\lambda_2 t} + \frac{\varepsilon_4}{\lambda_2}} \quad (70)$$

where

$$\eta = [r_p^T, e_\theta^T, e_p^T]^T, \tag{71}$$

ε_4 is a positive constant, and λ2 is a positive constant given by the following form

$$\lambda_2 = \min\left\{ (k_r - \frac{\lambda_1}{2}), (k_p - \frac{\lambda_0}{2}), k_\theta \right\} \tag{72}$$

where λ0, λ1 are positive constants under the conditions that

$$k_r > \frac{\lambda_1}{2} \quad \text{and} \quad k_{p>} \frac{\lambda_0}{2}. \tag{73}$$

A proof of the theorem can be proven using Lyapunov-type stability analysis (This is beyond the scope).

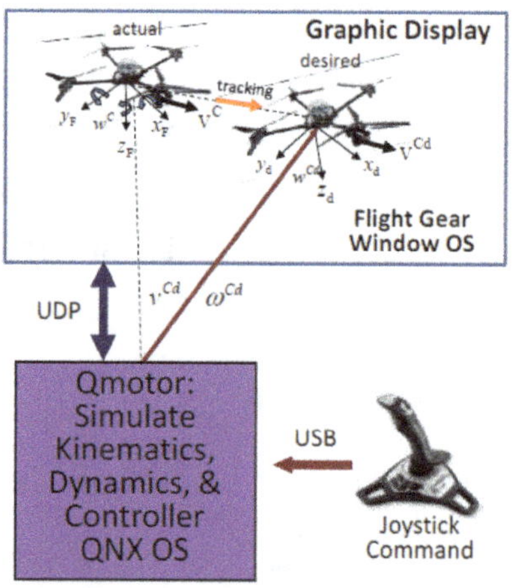

Figure 11. System Overview

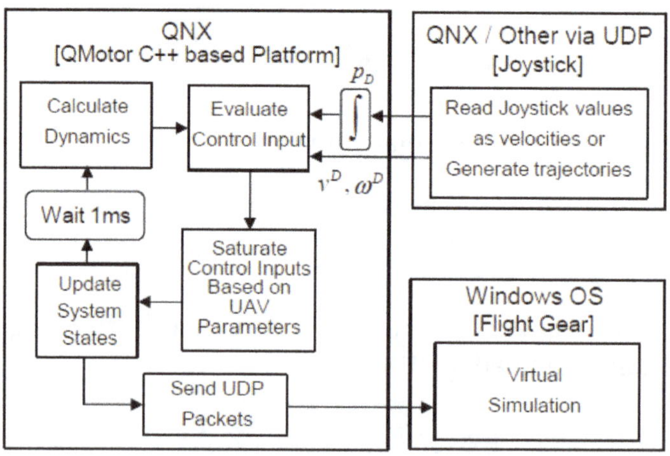

Figure 12. System Interal Outline in Simulation System

5. SIMULATION SYSTEM

The simulation system overview is given Fig. 11. A two computer system, QNX and Windows, was built to simulate the proposed controller. The QNX software systems are configured to run QNX Real-Time Operating System (RTOS) and host the QMotor (2000) control and simulation package while the Windows XP computer is configured to run and host the FlightGear (v.0.9.10) (2006) open-source flight simulator package. A QMotor program was written to simulate the rigid body kinematics, dynamics, and the proposed control. The output of the dynamics simulation is sent via UDP to set the aircraft/camera position and orientation in the FlightGear virtual world (note that FlightGear is used only as a graphics processor). The desired trajectories are input by the operator using a 6 DOF joystick (Logitech Extreme 3D Pro, 2006). Specifically, the operator indirectly supplies the desired position trajectory used by the controller through monitoring the simulated camera view and using the joystick to command the velocities that move the camera view in the virtual world. A detailed block diagram showing the three inputs on the joystick, labeled as x, y, and twist, are used to generate and control either three translational velocities, $v^D(t)$, or the three angular velocities, $\omega^D(t)$, depending on trigger position. That is, the magnitude and direction of these quantities is derived from the joystick position. The velocity commands are then integrated to produce the desired position trajectory used by the controller. A typical scene from FlightGear (2006) is shown in Fig. 13 where the quadrotor is tilted but the camera view seen by the operator remains level. The quadrotor simulation was developed to approximate the parameters of the DraganFlyer X-Pro (2005).

Parameters such as mass (m) and saturation limits for control inputs, and the control gains are chosen to be

$$k_r = k_\theta = diag(1,1,1),\ k_p = diag(1,1,5),\ m = 2.72[kg] \text{ and } g = 9.81[m^3 / kgs^2],$$
$$u_{1\_\max} = 35.586\ [N],\ \theta_{tilt\_\max} = \theta_{roll\_\max} = 4.067\ [Nm].$$

A short timespan of the simulation was captured to demonstrate the operation of the system. The simulation results are shown in Fig. 14 through Fig. 19. The easiest to appreciate operational point is found at t=50 [sec] in Fig. 17 and Fig. 18 where the orientations of the camera and quadrotor rotate in opposite directions measured from the x-, y-, and z-axes to achieve the fly-the-camera perspective. This also can be shown by the torque inputs of UAV roll, the second plot in Fig. 16, and camera roll, the last plot in Fig. 16, which acts in opposite directions as the UAV torque produces the required lateral motion and the camera compensates for the resulting roll. This is also seen in the pitch input in the third plot of Fig. 16 and the camera tilt input in the fourth plot of Fig. 16. Fig. 14 shows the position tracking of the quad-rotor to the desired trajectory and Fig. 15 shows the position errors about the coordinates. The actual quad-rotor trajectory represented by the dotted line follows the desired trajectory represented by the solid line which is commanded to go up at the first time and then, move to forward and again go forward near the end.

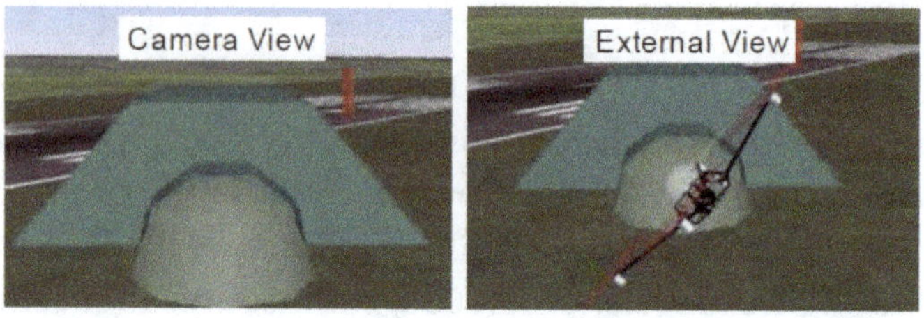

Figure 13. The "fly-the-camera" view used by the operator and an outside view of the quadrotor position

5.1 Simulation Results

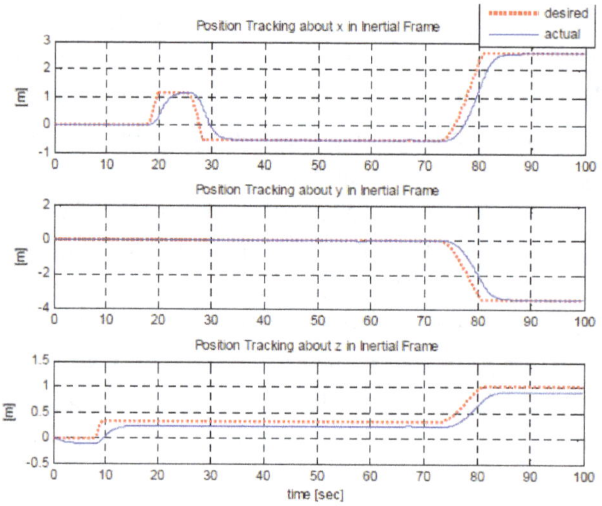

4

Figure 14. Position Tracking

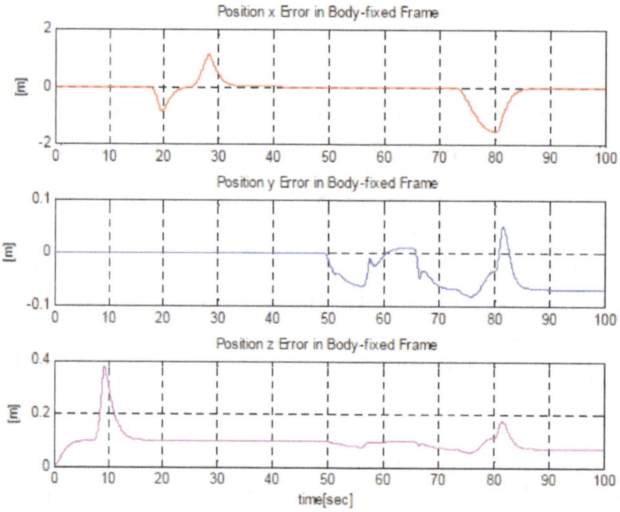

4

Figure 15. Tracking Errors

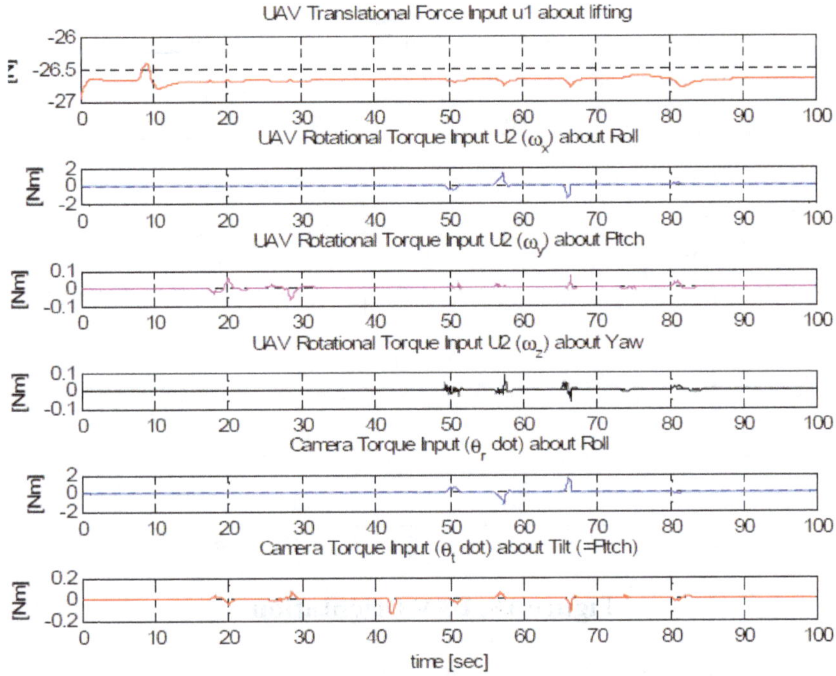

Figure 16. Control Inputs

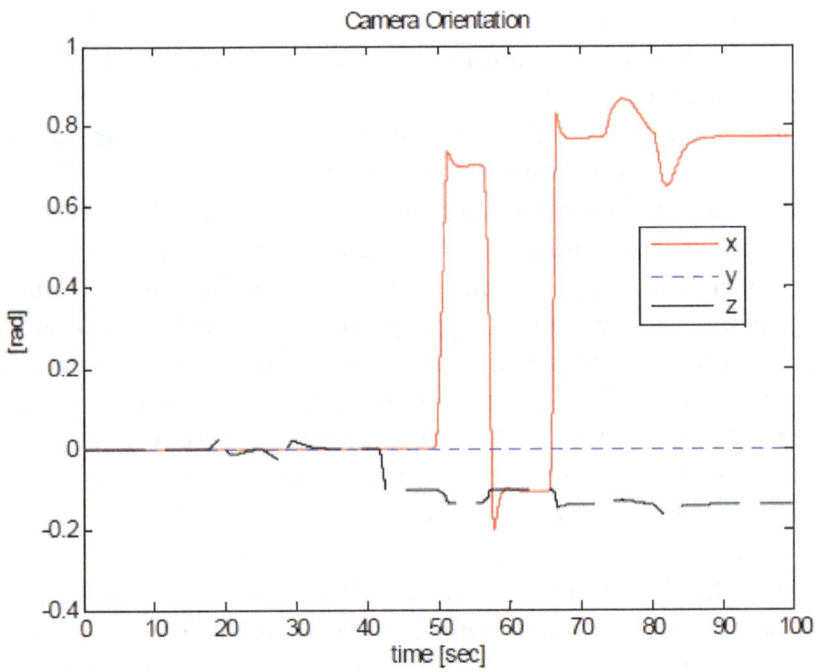

Figure 17. Camera Orientation

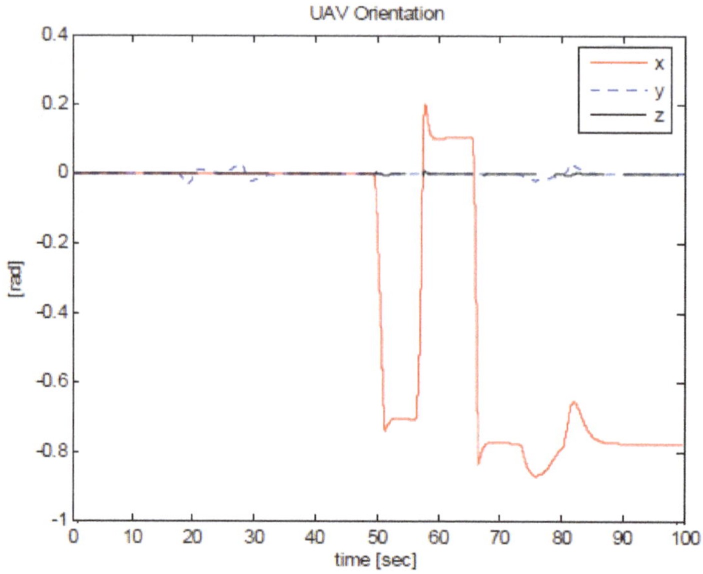

Figure 18. UAV Orientation

6. CONCISE SUMMARY

We have articulated the fly-the-camera methodology for the simultaneous control of a camera positioner and an unmanned aerial vehicle in a user-friendly manner. We elaborated on the design of a nonlinear controller as one potential manifestation of this concept. The fly-the-camera methodology examines a singular robotic dynamic entity including an underactuated aerial aircraft and two complementing camera axes to create a fully actuated camera targeting system. The fly-the-camera method is expected to offer a more natural viewpoint for a solitary remote operator to manage the quadrotor vehicle and camera for surveillance, navigation, and landing/monitoring functions. The method integrates the often distinct functions of vehicle navigation and camera targeting into a unified task, allowing the pilot to perceive and operate the system as if traversing along the camera›s optical axis. The controller described herein demonstrated the capability of position and angle-based tracking, yielding a Globally Uniformly Ultimately Bounded (GUUB) outcome. The simulation results illustrated the efficacy of the proposed technology. The advancement of the fly-the-camera perspective method can offer a foundation for various sectors, including commercial, industrial, and research endeavors. The exhibition of the «fly-the-camera» concept and a control mechanism establishes the basis for the advancement of this approach. Two salient features would enhance the applicability of this work. Initially, only the kinematic model of the camera has been incorporated. This estimate is valid only when the mass of the camera and the acceleration forces are minimal in comparison to the airframe. Secondly, cognitive load tests ought to be conducted to assess the genuine potential of this methodology.

7. APPENDIX

7.1 Case 2: Pan-Tilt Camera Configuration (camera looking downward)

The Denavit-Hartenberg (D-H) table for the Pan-Roll configuration when the optical axis is looking downward as shown in Fig. 7 is given in Table II.

Table 2. D-H for Pan-Tilt Camera Positioner while the optical axis is looking forward

Link (i)	Offset	Length	Twist	Angle
	D	a	α	θ
1	0	0	+90o	θp + 90o
2 (downward axis)	0	0	-90o	θ t - 9o

The rotation matrices of the Pan-Tilt camera positioning unit for first and second links as shown in Fig. 8 are obtained as

$$R_1^B = \begin{bmatrix} -s\theta_p & 0 & c\theta_p \\ c\theta_p & 0 & s\theta_p \\ 0 & 1 & 0 \end{bmatrix} \quad \text{and} \quad R_2^1 = \begin{bmatrix} s\theta_t & 0 & c\theta_t \\ -c\theta_t & 0 & s\theta_t \\ 0 & -1 & 0 \end{bmatrix} \tag{74}$$

where the first link rotation matrix for panning when the camera looking downward is the same as that of matrix for rolling motion when looking forward as shown in Fig. 6 and 7. Next, we can obtain the matrix from the camera base to first link (also same) but the total rotation matrix by combining all those matrices from quadrotor through the second link to yield

$$R_1^F = \begin{bmatrix} 0 & 1 & 0 \\ -s\theta_p & 0 & c\theta_p \\ c\theta_p & 0 & s\theta_p \end{bmatrix} \quad \text{and} \quad R_{2=C}^F = \begin{bmatrix} -c\theta_t & 0 & c\theta_t \\ -s\theta_t s\theta_p & -c\theta_p & -c\theta_t s\theta_p \\ s\theta_t c\theta_p & -s\theta_p & c\theta_t c\theta_p \end{bmatrix} \tag{75}$$

where $R_1^F = R_B^F R_1^{B=O_0}$ and $R_{2=C}^F = R_1^F R_{2=C}^1$ were used to calculate and the third column in R_1^F is the vector $z_1(t)$ in the Jacobian matrix $J_C(t)$. Thus, the positioner joint angles, $\dot{\theta}_C(t) \in R^3$, are given by

$$J_C = [z_0 \quad z_1] = \begin{bmatrix} 1 & 0 \\ 0 & c\theta_p \\ 0 & s\theta_p \end{bmatrix}, \; \dot{\theta}_C = \begin{bmatrix} \dot{\theta}_{p=r} \\ \dot{\theta}_t \end{bmatrix}, \text{and } \theta_C = \begin{bmatrix} \theta_{p=r} \\ \theta_t \end{bmatrix}. \tag{76}$$

Note that Jacobian matrix of the Pan-Tilt camera manipulator in (76) is the same one used in the Tilt-Roll configuration due to the configuration of the same camera base and the first link. In addition, the pan angle in Pan-Tilt configuration is actually the same of roll used in Tilt-Roll configuration due to the actuation of the second link.

7.2 Rotation Matrix

A rotation matrix, $R_D^C(\theta_d)$ denoted in (20), from D to desired frame C for the tracking of the camera frame to desired frame can be obtained by considering the camera frame with regard to the desired frame as another end link as follows:

Table 3. D-H Table to find the rotation matrices between the camera and desired frames which yields the following rotation matrix, respectively (for simulation, $\theta_d = 0°$)

Link (i)	Offset	Length	Twist	Angle
	D	a	α	θ
C (looking forward)	0	0	+90o	θd + 90o
C (looking downward)	0	0	0o	θd + 90o

$$R_D^C = \begin{bmatrix} -s\theta_d & 0 & c\theta_d \\ c\theta_d & 0 & s\theta_d \\ 0 & 1 & 0 \end{bmatrix}_{(forward)} \quad \text{and} \quad R_D^C = \begin{bmatrix} -s\theta_d & -c\theta_d & 0 \\ c\theta_d & -s\theta_d & 0 \\ 0 & 0 & 1 \end{bmatrix}_{(downward)} . \tag{77}$$

7.3 Notation and Nomenclature

R_i^{i-1} Rotation matrix from the origin i to the origin i -1 coordinate frame

T_F^I Orientation matrix from the origin of F to the origin of frame I

p_i, $\dot{p}_i$, Camera (C) or UAV (F) position and position rate represented in Inertial frame I - ground-based - quantities denoted using subscript

Θ_i, $\dot{\Theta}_i$ Camera or UAV angle ($[\varphi, \theta, \psi]^T$) and angular rates about roll, pitch, and yaw represented in I frame -ground-based-quantities using subscript

Θ_C Camera angles about pan, tilt, and roll ($[\theta_p, \theta_t, \theta_r]^T$) expressed in the Inertial frame

v^i, ω^i Camera (C), Camera base (B) or UAV linear and angular velocities about x, y, and z-axis which is airborne quantities denoted using superscript

v^{ij}, ω^{ij} Velocities expressed in Camera frame (C) between the origin i and another origin j coordinate frame

v^i_{jk}, ω^i_{jk} UAV or Camera base frame velocities (i) denoting specific quantities between j frame and k frame

REFERENCES

Adelstein, B. D. ; Ellis, S. R. (2000). Rotation and Direction Judgment from Visual Images Head- Slaved in Two and Three Degrees-of-Freedom, IEEE Transactions on Systems, Man and Cybernetics – Part A: Systems and Humans, Vol. 30, No. 2, March 2000, pp. 165 – 173.

Chitrakaran, V.; Dawson, D. M.; Kannan, H. (2006). Vision Assisted Autonomous Path Following for Unmanned Aerial Vehicles, in Proc of 45th IEEE Conference on Decision and Control, pp. 63 - 68, Dec. 2006.

Cooke, N; Pringle, H. ; Pedersen, H.; Connor, O. (2006). Human Factors of Remotely Operated Vehicles, ELSVIER JAI, 2006

DraganFlyer X-Pro (2005): http://www.rctoys.com/ (cited 10/2008). FlightGear (2006). http://www.flightgear.org/ (cited 10/2008).

Fossen, T. I. (2003). Marine Control Systems : Guidance, Navigation, and Control of Ships, Rigs, and Underwater Vehicles, Marine Cybernetics, 2003.

Jakobsen, O. C.; Johnson, E. N. (2005). Control Architecture for a UAV - Mounted Pan/ Tilt/Roll Camera Gimbal, AIAA 2005-7145, Infotech@Aerospace, Arlington, Virginia, Sep. 2005, pp. 01 – 10.

Lee, D.; Chitrakaran, V.; Burg, T.C.; Dawson, D. M.; Xian, B. (2007). Control of a Remotely Operated Quadrotor Aerial Vehicle and Camera Unit Using a Fly-The-Camera Perspective, in Proc of 46th IEEE Conference on Decision and Control, New Orleans, LA, Dec. 2007, pp. 6412-6417. Logitech Extreme 3D Pro Joystick (2006); http://www.logitech.com

Mahony, R.; Hamel, T.; Chaumette, F. (2002). A Decoupled Image Space Approach to Visual Servo Control of a Robotic Manipulator, in IEEE Intl Conference on Robotics and Automation, Washington D. C., May 2002, pp. 3781 – 3786.

Malis, E. (1998) Controbutions a la modelisation et a la commande en asservissement visuel, Ph.D. Thesis, University of Rennes I, IRISA, France, 1998. Micropilot.; http://www.micropilot.com/ (9/2008 cited)

Murray, R. M.; Li, Z.; Sastry S. S. (1994). A Mathematical Introduction to Robotic Manipulator, CRC Press, 1994.

Neff, A.; Lee, D.; Chitrakaran, V.; Dawson, D. M.; Burg, T. C. (2007). Velocity Control of a Quadrotor UAV Fly-By-Camera Interface, in Proceedings of IEEE Southeast Conference, pp. 273-278, Richmond, VA, Mar. 2007.

Pieniazek, J. (2003). Software-based Camera Stabilization on Unmanned Aircraft,

Aircraft Engineering and Aerospace Technology: An International Journal, Vol. 75, No. 6, 2003, pp. 575 – 580. Procerus Technologies.; http://www.procerusuav.com/ (9/2008 cited)

Qmotor Simulator (2000). A QNX-based C++ Real-time Control Environment, http://www.ece.clemson.edu/crb/research/realtimesoftware/qmotor/index.htm (cited 10/2008).

QNX 6, QNX Neutrino 6.2.1. (2006). QNX Software Systems, Real-Time Control Software. http://www.qnx.com (cited 10/2008)

Quigley, M. ; Goodrich, M. A. ; Griffiths, S. ; Eldredge A. ; Beard, R. W. (2005). Target Acquisition, Localization, and Surveillance Using a Fixed-Wing Mini-UAV and Gimbaled Camera, in IEEE Intl Conference on Robotics and Automation, Barcelona, Spain, April 2005, pp. 2611 – 2616.

Sharp, C. S.; Shakernia, O. & Sastry S. S. (2001). A Vision System for Landing an Unmanned Aerial Vehicle, in IEEE Intl Conference on Robotics and Automation, Seoul, Korea, May 2001, pp. 1720-1727.

Spong, M. W.; Hutchinson, S.; Vidyasagar M. (2006). Robot Modeling and Control, John Wiley and Sons, Inc:, 2006.

Stanciu, R. & Oh. P. (2007). Human-in-the-Loop Camera Control for a Mechatronic Broadcast Boom, IEEE/ASME Transactions on Mechatronics, Vol. 12, No. 1, February 2007.

Stolle, S. & Rysdyk, R. (2003). Flight Path Following Guidance for Unmanned Air Vehicles with Pan-Tilt Camera for Target Observation, in Proceedings of the AIAA Digital Avionics Systems Conference, Oct. 2003, pp 01 – 12.

Yoon, S.; Lundberg, J. B. (2001). Equations of Motion for a Two-Axes Gimbal System, IEEE Transactions on Aerospace and Electronic Systems, Vol. 37, No. 3, July 2001, pp. 1083 - 1091.

9 Chapter

DRAGONFLIES CAN TEACH US ABOUT HOW TO FLY INTELLIGENT AERIAL VEHICLES

1. INTRODUCTION: AN OVERVIEW

The dragonfly is among the most agile insects and one of the oldest flying species on the planet. Studying the flight techniques of certain insects is crucial for developing insect-like Micro Aerial Vehicles, as their flight performance much surpasses that of other insects. They can hover, cruise up to 54km/h, turn 180° in three wing beats, fly sideways, glide, and even fly backwards (Alexander, 1984; Appleton, 1974; Whitehouse, 1941). They intercept prey in the air with amazing speed and accuracy. Their thorax are equipped with wing muscles which accounts for 24% (Aeshna) of its body weight, compared to 13% of those of the honey bees (Appleton, 1974). Most dragonflies change their wing motion kinematics for different flight modes such as hovering, cruising and turning. Among these kinematic parameters, the most interesting one is the phase difference (γ) between forewing and hindwing. It is defined as the phase angle by which the hindwing leads the forewing. When hovering, dragonflies employ 180° phase difference (anti-phase) (Alexander, 1984; Norberg, 1975; Rüppell, 1989), while 54~100° are used for forward flight (Azuma and Watanabe, 1988; Wang et al., 2003). When accelerating or performing aggressive maneuvers, they use 0° (in- phase) phase difference (Alexander, 1984; Rüppell, 1989; Thomas et al., 2004). Of various phase differences, 270° is rarely observed in dragonfly flight.

Figure 1. Phase difference in dragonfly

The fact that flapping in-phase (0°) appears in situations requiring large acceleration suggests that in-phase might produce higher forces (Alexander, 1984; Rüppell, 1989).

The film sequences made by Alexander (Alexander, 1984) showed that in-phase is employed during take-off and sharp turning. It was also found that in a rising flight of a dragonfly, lift was increased during downstroke and drag was increased during upstroke when flying in-phase (Azuma et al., 1985). The conclusion was derived by using the momentum theory and the blade element theory, combined with a numerical method modified from the local circulation method.

However, it was argued that as two wings in tandem are brought closer together, the lift force produced by each wing is reduced (Alexander, 1984). Therefore, forewing and hindwing flapping in-phase would produce less lift because they are closer together than when beating anti-phase. Alexander believed that the reason of dragonflies using in-phase flight may be due to physiological reason as well as the preference of peak forces enhancement at the cost of the mean forces reduction (Alexander, 1984).

Counterstroking (180° or anti-phase) produces uniform flight, whereas flight produced by parallel stroking (0°) is irregular (Rüppell, 1989). This is because inequalities in the aerodynamic effects of the upstroke and downstroke can be compensated to some extent in countertroking. As one pair of wing's upstroke with a steep angle of attack generates strong thrust, the other pair's downstroke with a small angle of attack mainly generates lift. Therefore the net thrust and lift production remains relatively constant during flight due to the alternating force generation on two wings (Rüppell, 1989).

In a recent computational study, (Wang and Russell, 2007) calculated dragonfly's aerodynamic force and power as a function of forewing-hindwing phase difference. They found that anti-phase flapping consumes nearly minimal power while generating sufficient force to balance body weight, and that in-phase motion provides an additional force to accelerate (Wang and Russell, 2007). Furthermore, they proposed an analogy to explain the results by analyzing a model of two cylinders moving in parallel next to each other.

Other computational studies include (Wang and Sun, 2005) and (Huang and Sun, 2007), where they calculated the aerodynamic effects of forewing-hindwing interactions of a specific dragonfly (Aeshna juncea) in hover and slow forward flight. They showed that the interaction is detrimental to force generation in almost all cases. At hovering with $\gamma = 180°$, the reduction is 8~15%, compared with the force without interaction. The force on hindwing is greatly influenced by the forewing at $\gamma = 180\sim360°$, with the lift coefficient decreased by 20~60%. Furthermore, they proposed a mechanism to explain the effect of forewing on hindwing force reduction: the forewing in each of its downstroke produces a downward "jet" behind it; when the hindwing lags the forewing, it moves into the jet and its effective angle of attack is reduced, resulting in a decrease in its aerodynamic force.

Previous computational studies include (Lan, 1979), where the unsteady quasi-vortex-lattice method was applied to the study of dragonfly aerodynamics, and the results showed that dragonfly can produce high thrust with high efficiency if hindwing leads the forewing by 90°, and that hindwing was able to extract wake energy from the forewing under this condition.

Direct force measurements on tethered dragonflies showed that peak lift increases from approximately 2 to 6.3 times body weight when the animal decreases the phase difference between both flapping wings (Reavis and Luttges, 1988). But maybe this enhancement on peak lift may due to overlap of peak lifts on forewing and hindwing, not necessarily due to wing-wing interaction.

Experimental investigations of the aerodynamic effect of wing-wing interaction was previously performed in (Maybury and Lehmann, 2004), where a pair of robotic wings were vertically stacked to simulate dragonfly hovering flight with horizontal stroke plane. They found that the lift production of the forewing remains approximately constant, while hind wing lift production is reduced to some extent and its maximum value occurs at a phase difference $\gamma = 90°$. They attributed the wing-wing interaction to two reasons: LEV destruction and local flow condition (Maybury and Lehmann, 2004). Their results explained the hovering behavior of dragonflies using horizontal stroke plane (Sympetrum Sanguineum), while many other dragonfly species employ a 20~70° inclined stroke plane (Alexander, 1984; Norberg, 1975; Rüppell, 1989) and employ an aerodynamic mechanism of “drag based lift generation” (Wang, 2004), which is quite different from that for horizontal stroke plane flight (Wang and Russell, 2007). This study examines the wing-wing interaction and the mechanisms involved in species with an angled stroke plane. The influence of forewing-hindwing interactions in dragonflies has been examined through several computational and experimental research; nonetheless, the conclusions remain limited and inconsistent. This project involved the construction of a pair of robotic dragonfly wings to examine the aerodynamic impact of wing-wing interactions during both hovering and forward flight. This equipment facilitates the examination of inclined stroke plane species, such as Aeshna juncea, at a 60° inclined stroke plane and with various forward velocities. The wing bases are in close proximity to replicate dragonfly wings. We systematically manipulate the phase relationship between the forewing and hindwing to ascertain the rationale behind dragonflies’ selection of specific phase discrepancies in various flying modes. This research’s result would be essential in the creation of a dragonfly-inspired Micro Air Vehicle (M.A.V).

2. A ROBOTIC DRAGONFLY MODEL

The Micro Robotics Lab at the University of Delaware developed two dynamically-scaled robotic wing models (flappers) to emulate dragonfly wing motion and quantify the instantaneous aerodynamic forces during flapping in a tank containing mineral oil (Fig. 2).

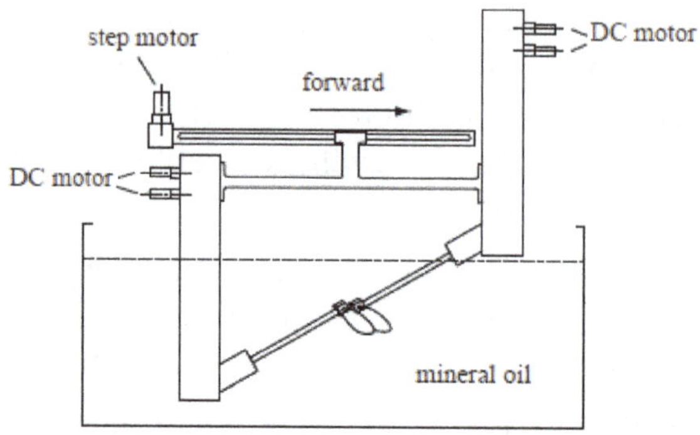

Figure 2 & 3. Sketch and image of the experimental setup

A bevel-geared robotic wrist was constructed for each flapper to produce movements with two degrees of freedom (translation and rotation). A collection of bevel gears conveys motion from coaxially driven shafts to the wing holder, facilitating wing translation and rotation (Fig. 3). The two wings are positioned in close proximity, resembling those of an authentic dragonfly. Furthermore, the two flappers are affixed to a linear stage propelled by a stepper motor to provide simultaneous forward movement.The drive shafts were powered by 16 mm, 0.3 Nm torque DC brush motors (Maxon, Sachseln, Switzerland) equipped with gear heads to reduce speed and magnetic encoders to provide kinematic feedback to ensure motion fidelity. The motors were driven along kinematic patterns provided by a custom MATLAB (Mathworks, Natick, MA) Simulink program with WinCon software (Quanser Consulting, Ontario, Canada). This software provided commands to the real-time control and data acquisition board (Quanser Consulting, Ontario, Canada) communicating with the hardware. We used Proportional- Integral-Derivative (PID) controllers to run the motors with precision of 0.1°. Motion commands from the computer were amplified by analog amplifier units (Advanced Motion Control) which directly controlled the input current received by the motor.

Wings were made from Mylar plastic film with a thickness of 0.25 cm which behaves as a rigid wing in our experiments. A carbon fiber rod was glued on the plastic film to serve as the leading edge. The end of the carbon fiber rod was affixed on to the force sensor. The wings have identical geometry as the dragonfly wings but are four times larger with a length of 19 cm for forewing and 18.5 cm for hindwing. The wing length is calculated as the distance from wing tip to the flapping axis.

The wings along with part of the device were immersed into a tank (46cm × 41cm × 152cm) filled with mineral oil (Kinematic viscosity= 3.4 cSt at 20 oC, density=830 kg/ m^3). This configuration allowed us to operate the wings according to a specified dragonfly kinematics while concurrently monitoring the stresses on both the forewing and hindwing (Fig. 2).

3. FLIGHT PATTERN OF A TYPICAL DRAGONFLY

High-speed photos of the dragonfly (Aeshna juncea) in hovering flight were taken by Norberg (Norberg, 1975). The body is held almost horizontal, and the wing stroke plane is tilted 60° relative to the horizontal line. For both forewing and hindwing, the chord is almost horizontal during the downstroke and is close to being vertical during the upstroke (Fig. 4); the stroke frequency is 36 Hz, the stroke amplitude is 60°; the translational angle is from 35° above the horizontal to 25° below for forewing, and is from 45° above to15° below for hindwing; the hindwing leads the forewing in phase by 180°. The mass of the insect is 754mg; forewing length is 4.74 cm; hindwing length is 4.60 cm; the mean chord lengths of the forewing and hindwing are 0.81 cm and 1.12 cm, respectively; the moment of inertial of wing-mass with respect to the fulcrum is 4.54 g cm-2 for the forewing and 3.77 g cm-2 for the hindwing (Norberg, 1972).

Due to the low frame rate of the camera used at that time (80 Hz), Norberg's data does not consist of a detailed continuous trajectory of the wing kinematics. Azuma instead, (Azuma et al., 1985) successfully filmed a slow climbing flight of a dragonfly (Sympetrum Frequens) with a high speed camera (873 frames per second). He showed that the flapping trajectory can be well represented by a sinusoidal function and the rotation trajectory can be represented by a third harmonic function. Since the two species (Aeshna juncea and Sympetrum frequens) share similar values for many kinematic parameters such as translational amplitude, rotational amplitude and stroke plane angle (Azuma et al., 1985; Norberg, 1975), here we assume they also employ similar wing motion trajectories. This assumption of kinematic trajectories can be reasonably applied to other species with highly inclined stroke planes without affecting the main results of this study. We developed a pair of wing kinematic trajectories by matching those for Sympetrum Frequens in (Azuma et al., 1985), as shown in Fig. 5.

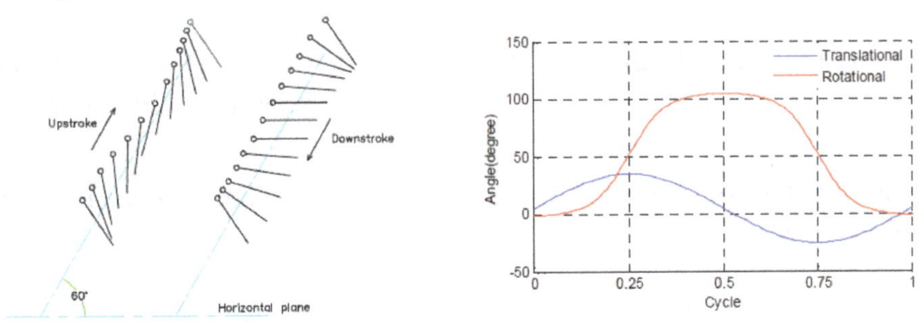

Figure 4 & 5. (left) Sketches of the dragonfly wing kineamtics during upstroke and downstroke based on Aeshna juncea data (Norberg, 1975). Side views with circles denote the leading edge. Both forewing and hindwing apply; (right) Wing kinematics developed by matching those for Sympetrum Frequens from (Azuma et al., 1985) and (Norberg, 1975). Translational angle (blue) and rotation angle (red). This is for forewing only; the kinematics for hindwing is almost the same except a 10° offset for translational angle

4. FORCE MEASUREMENTS AND COMPUTATIONS

For each wing, we measured the instantaneous lift and thrust forces and calculated the average forces. The forces as well as torques on the moving wings can be measured by a six- component force sensor (ATI NANO-17, Apex, NC), with a range of ±12 N for force and ±0.5 Nm for torque along three orthogonal axes. Through suitable trigonometric transformations, the force data were subsequently transformed to lift and thrust forces in the terrestrial coordinate system. In this study, lift force is defined as the vertical component of the aerodynamic force acting on the wing, whereas thrust force is the horizontal component in the global coordinate frame. The magnitude of aerodynamic forces acting on an actual dragonfly, F_{fly}, is related to those measured in the robotic model, 2005): Frobot , according to the following scaling rule (Fry et al., 2005)

$$F_{fly} = F_{robot} \cdot \frac{\rho_{air}}{\rho_{oil}} \cdot \left(\frac{n_{fly}}{n_{robot}}\right)^2 \cdot \left(\frac{r_{fly}}{r_{robot}}\right)^2 \cdot \frac{S_{fly}}{S_{robot}} \cdot \frac{\hat{r}^2_{2fly}}{\hat{r}^2_{2robot}} \quad (1)$$

where ρ is fluid density, n is stroke frequency, r is wing length, S is wing area, and $\hat{r}^2_2$ is the normalized second moment of wing area. $\hat{r}^2_{2fly}$ and $\hat{r}^2_{2robot}$ differ slightly, due to a small null length at the base of the robotic wing required to accommodate the force sensor.

In this study, lift force is defined as the vertical component of the aerodynamics force on the wing; thrust force is the horizontal component. Lift coefficient and thrust coefficient are respectively defined in the following way, similar to (Wang and Sun, 2005):

$$C_l = L / [0.5 \rho U^2 (S_f + S_h)] \quad (2)$$

$$C_t = T / [0.5 \rho U^2 (S_f + S_h)] \quad (3)$$

L and T denote the total lift and the total thrust. The reference velocity is calculated to be $U = 2\Phi n r_2 = 2.1 \text{ms}^{-1}$; S_f and S_h are the areas of forewing and hindwing, respectively.

Accordingly, we define lift coefficient and thrust coefficient for a single wing. For example, for forewing:

$$C_{l,f} = L_f / [0.5 \rho U^2 (S_f + S_h)] \quad (4)$$

$$C_{t,f} = T_f / [0.5 \rho U^2 (S_f + S_h)] \quad (5)$$

The total lift coefficient and total thrust coefficient are as follows:

$$C_l = C_{l,f} + C_{l,h} \quad (6)$$

$$C_t = C_{t,f} + C_{t,h} \quad (7)$$

The lift coefficient on a dragonfly should be 2 times Cl, since it has two forewings and two hindwings.

5. AERODYNAMIC FORCE FOR ANTI-PHASE HOVERING FLIGHT

We first replay the hovering kinematics of dragonfly on the robotic flappers and measured lift and thrust forces on the both wings. Here the forewing and hindwing are anti-phase (180°), which is most commonly observed in dragonfly hovering flight. The results show an average lift force of 0.0711 N on the forewing and 0.082 N on the hindwing during one beat cycle. The average thrust forces are 0.001 N on the forewing and 0.003 N on the hindwing.

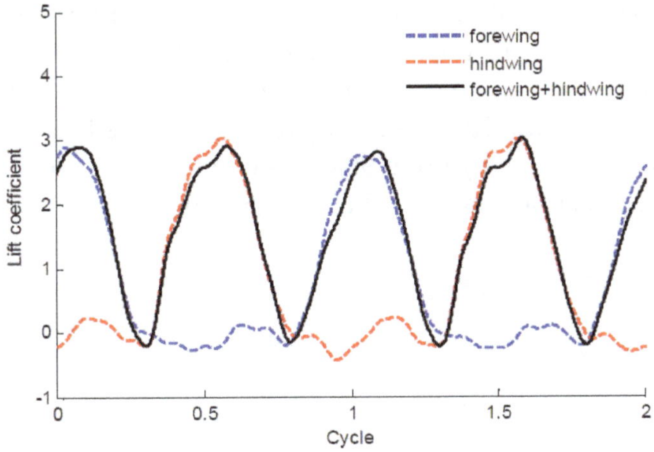

Figure 6. Time trace of lift force coefficients generated when hovering with γ = 180°

We then apply (1) to scale the lift force back to those of a true dragonfly. The resulted average lift force is 187 mg on a forewing and 216 mg on a hindwing. The total average lift on a four-wing dragonfly is therefore (187+216)*2=806 (mg). This result is comparable to the 754 mg body mass measured in (Norberg, 1972). On the other hand, the total average thrust force is almost zero when compared with the average lift force, which follows the hovering condition that the thrust should be zero.

The time traces of instantaneous lift forces are shown in Fig.6. Both forewing and hindwing produce lift peaks during downstrokes. Because the flapping of forewing and hindwing are anti-phase, they generate lift force alternatively during one wingstroke. The average lift force on hindwing is 1.15 times that on forewing, while the area of hindwing is 1.32 times that of forewing. This inconsistency is probably caused by the fact that, compared to forewing, hindwing has more area distributed at positions close to the body. It was proposed that for dragonfly flight, the hindwing acts as power wing which provide more lift force while the forewing is the steering wing (Azuma and Watanabe, 1988; Wang et al., 2003). As we can see, the total lift force is generated alternatively by the forewing and hindwing during one wingbeat.

By doing so the insect is able to hover with regular forces which reduce the vibration of the body.

6. IMPACT OF PHASE DIFFERENCE ON HOVERING FLIGHT

We measured the instantaneous forces with the same kinematics of hovering but systematically vary the phase differences from 0° to 360° in steps of 30°. The average lift coefficients of both wings according to varying phase difference are plotted in Fig. 7.

Fig. 7 shows the averaged force coefficients in hover. As the phase difference tends to 0°, forces tend to be higher, and reach their maximum on 0°. The forces get lower values when γ is around 180°. An interesting point is that not all values are below the single wing force, that is, the wing-wing interaction is not always detrimental to force generation. We can conclude here, in range 330° to 30°, the wing-wing interaction enhances the total lift force by up to 6%; in range 150° to 180°, interaction decreases the total lift force by up to 9%. The thrust forces are much smaller than the lift forces, which is reasonable for hovering flight.

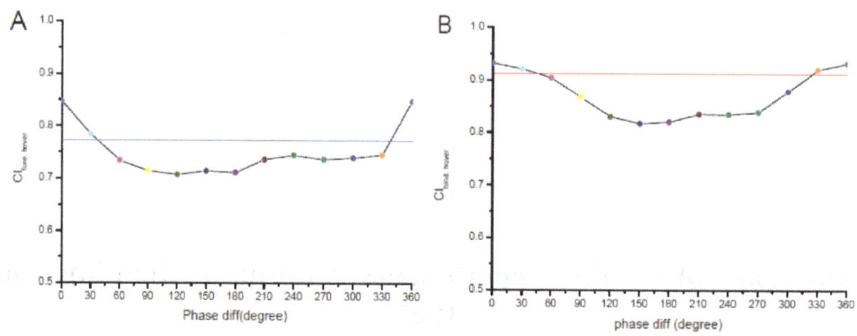

Figure 7. Force coefficients results for hovering flight test. A) Average lift coefficients on forewing; B) average lift coefficients on hindwing; C) total average lift coefficients. The straight lines in each plot indicate the force results without interaction

This section of the study provides a direct proof that inphase flying produces greater aerodynamic forces than anti-phase flight, as well as greater forces than the scenario without wing-wing interaction. This may elucidate the phenomenon of dragonflies flying in-phase during acceleration or maneuvers that necessitate substantial force generation.In order to find out the possible reason that dragonfly uses anti-phase style for hovering mode, we need to compare the time traces of total lift forces generated by γ = 0° with those generated by γ =180° (Fig. 8). It was noted that anti-phase produces uniform flight, whereas flight produced by in-phase stroking is irregular (Rüppell, 1989). Fig. 8 shows the instantaneous lift force coefficients comparison between in-phase and anti-phase flight. As we can see, in-phase brings larger irregularity in the aerodynamic forces than anti-phase does. Observing from the time trace curve for in-phase hovering, there exists a 1/2 cycle period when lift force is closed to zero, while in another 1/2 cycle the peak value is two times of the peak value for anti-phase flight, because forewing and hindwing peak overlap. This irregularity of instantaneous forces increases the body vibration when hovering, while for anti-phase flight the inequality can be compensated to some extent by evenly distributing the peak forces of forewing and hindwing on the whole cycle. Besides minimizing the force irregularities and keeping body posture stable, anti-phase flight can also save energy that might be lost in body vibration (Wang and Russell, 2007). Moreover, Usherwood and Lehmann showed that dragonfly can also save aerodynamic power during anti-phase hovering (Usherwood and Lehmann, 2008). Thus, it is reasonable that dragonflies would rather lose 15% force production for improving flight stability and vibration suppression as well as power efficiency.

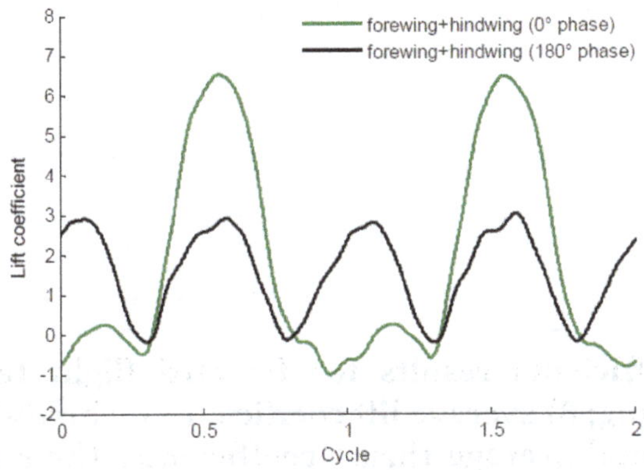

Figure 8. Comparison between total lift forces generated when hovering with = γ 0° and = γ **180° respectively**

7. IMPACT OF PHASE DISPARITY DURING FORWARD FLIGHT

Similarly to the hovering case, we tested the wing aerodynamics by varying phase difference systematically in steps of 30°. The average lift coefficients and average thrust coefficients of both wings are plotted in Fig. 9.

As seen in Fig. 9, during forward flight, interaction patterns for forewing and hindwing differ a lot. Note that the interaction always enhances the forewing's force generation no matter how much γ is, while the hindwing always loses force production because of the interaction. Lift on forewing is enhanced by at most 23% when flapping in-phase and at least 4% when phase difference falls into 120°~330°; Lift on hindwing reaches maximum on 60° and minimum on 270°. Hindwing is subjected to a severe loss on force production up to 38% due to the interaction with forewing. The total force on the two wings does not lose so much force as the hindwing does, due to the considerable enhancement on forewing lift.

Fig. 10 compares the time traces of hindwing lift among the cases of single hindwing, γ =90° and γ =270°. We note that the lift in case of γ =270° was considerably reduced when compared with other cases, and the reduction mainly occurred when hindwing was in the midway of downstroke. This is reasonable because a large portion of lift is produce during this period, since wings reach the highest flapping speed and largest angle of attack around the midway of downstroke.

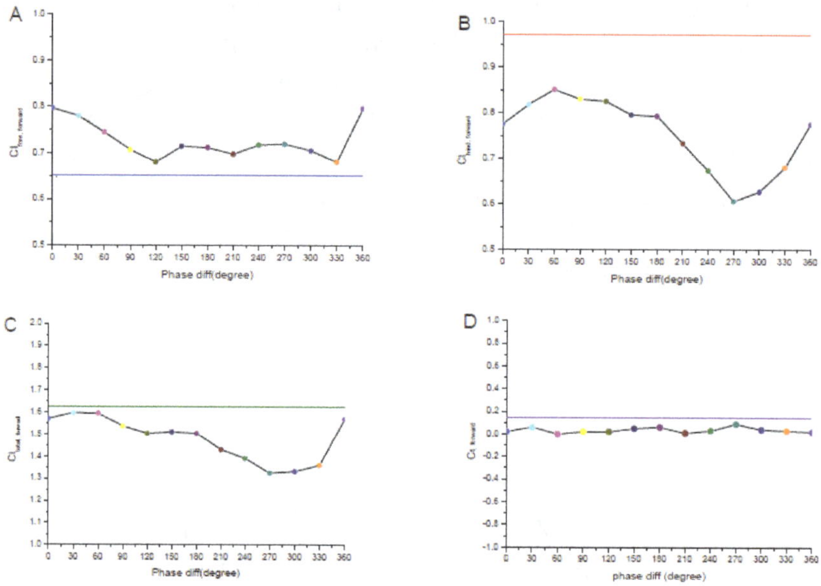

Figure 9. Force coefficients results for forward flight test. A) Average lift coefficients on forewing; B) average lift coefficients on hindwing; C) total average lift coefficients; D) total average thrust coefficients. The straight lines in each plot indicate the force results without interaction

One may explain the reduction in the following way: forewing generates a strong downwash when it finishes a downstroke. If $\gamma = 270°$, that is, forewing leads hindwing by 1/4 cycle, then hindwing is just in the midway of downstroke as forewing finishes downstroke and generates a downwash. In the mean time, because of forward flight, hindwing enters into the downwash area and its lift production decreases due to the downwash effect.

Now we can distinguish the case $\gamma = 90°$ and the case $\gamma = 270°$: the former one can offer dragonfly an 18% higher force than the latter one. On the other hand, 270° offers similar vibration and stability properties as 90°. This may explain why dragonfly never favors the 270° phase difference.

The above results agree qualitatively with the CFD results from (Wang and Sun, 2005) and (Huang and Sun, 2007) to some extent. Their conclusion is that the forewing is only slightly influenced by the wing-wing interaction, but the hindwing lift is greatly reduced by 20~60% during forward flight with a 180°~360° phase difference, compared with that of a single hindwing. In our results, furthermore, there are obvious lift enhancements on the forewing. For thrust force measurement, (Warkentin and DeLaurier, 2007) conducted a systematic series of wind-tunnel tests on an ornithopter configuration consisting of two sets of symmetrically flapping wings, located one behind the other in tandem. It was discovered that the tandem arrangement can increase thrust for certain relative phase differences and longitudinal spacing between the wing sets. In particular, close spacing on the order of one chord length is generally best, and phase differences of approximately -50° to 50° give the highest thrusts and propulsive efficiencies. However, this conclusion is not applicable to dragonfly flight. The thrust curve above demonstrates a reduction in thrust force due to interaction, irrespective of the phase difference. This would be justifiable if we see that the gap between the forewing and hindwing of the dragonfly is significantly lower than the chord separation between the ornithopter wing sets.

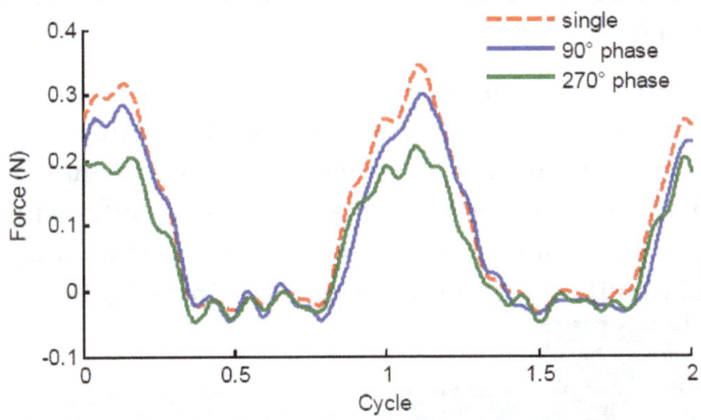

Figure 10. Comparisons of instantaneous hindwing lift in forward flight among the cases of single hindwing, 9 = γ0° and = γ 270°. Note: the vertical axis indicates the forces generated on the flapper experiment and are unscaled

8. DRAGONFLY-INSPIRED M.A.V

At last we introduce several dragonfly robots we have developed in our lab. Fig. 11 shows the latest designs. This model has two pairs of wings driven by a double crank rocker mechanism.

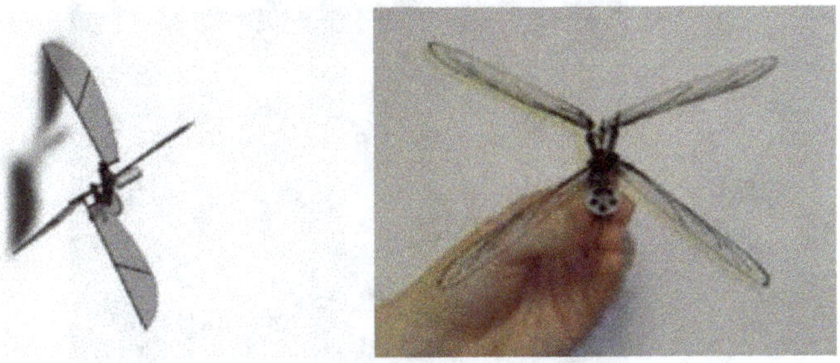

Figure 11. Dragonfly M.A.V prototype

The current prototype weighs 4g including battery and electronics. The following table shows some specifications.

Table 1. Dragonfly prototype specifications

Motor	Battery & Control	Gear
Torque: 44mNm	4V	Maximize Torque
Power: 176mW	0.4A	Ratio: 1:7
16k rpm	Infrared Chip	Precision Molded
Weight: 1.4g	Weight:1.1+1.3g	Weight:0.23g

This prototype was able to generate 28 Hz flapping frequency with four carbon fiber leading edges only and 9 Hz after adding (gluing) the polymer wing onto two leading edges. With all four wings the frequency reduced to 7 Hz. The total weight is 7 grams and the total length is 2 inches.

In order to test different wing performance and property, we have developed a mechanical wing tester. It can be used to study the fatigue cycles of each wing developed and can be also used to visualize the wing kinematics by a high speed camera. Fig.12 shows the wing tester and a sample dragonfly wing made of carbon fiber and polymers. Fig.13 shows the camera images of test wing and Fig.14 shows the frequency plot of different robotic wings under investigation.

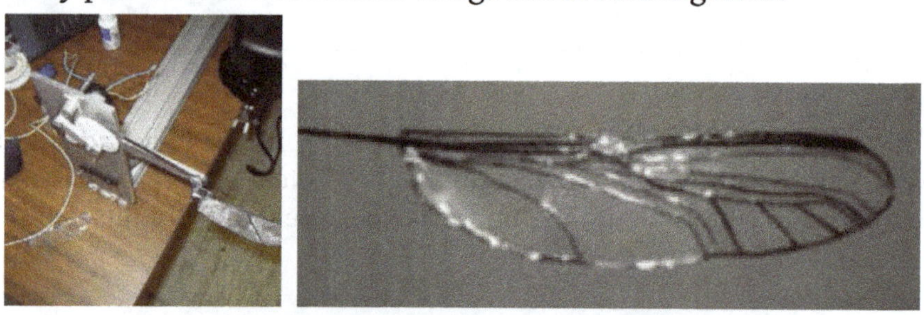

Figure 12 & 13. Wing tester and sample test wing

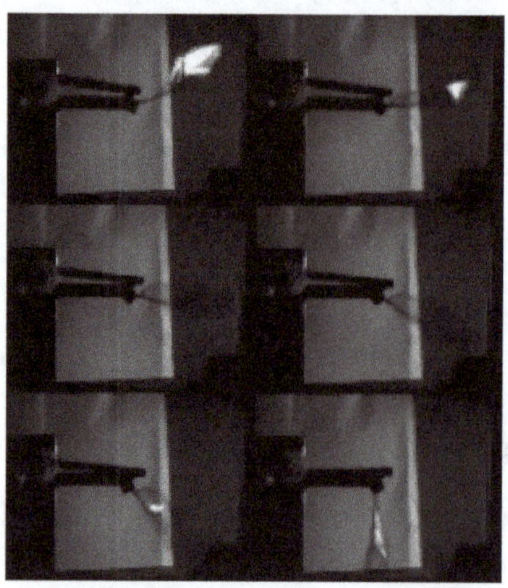

Figure 14. Dragonfly shaped wing flapping at 20Hz

The average wing size is larger than that of the actual insects under examination. Veins are incorporated to enhance rigidity and produce greater force. The reduction in length and size results in an increased frequency. This procedure necessitates much trial and error, as minor variations in wing structure might cause it to disintegrate at elevated frequencies. Dragonflies inherently possess numerous traits integrated into their wings. Subsequent testing will focus on determining the factors that contribute to the robustness and delicacy of their wings under high frequencies.

9. CONCISE SUMMARY

This study examines the impact of forewing-hindwing interactions during hovering and onward flying in dragonflies. In summary, wing-wing interaction adversely affects the total lift force production. The lift of the hindwing was markedly diminished during forward flight as a result of the downwash generated by the forewing. In-phase flight produces greater lift than other phase discrepancies, however a 270° phase difference yields the least lift. During hovering, dragonflies employ anti-phase flight to produce a consistent lift force that enhances stability and minimizes vibrations. A prototype for dragonfly-inspired M.A.V has been shown.

A multitude of tasks must be accomplished in the future to develop an advanced M.A.V. The mechanical design requires enhancement to attain a greater flapping frequency and enhanced efficiency, hence reducing weight and maximizing lift force. Controlling the phase difference in a four-wing Micro Air Vehicle presents a significant challenge for engineers.

REFERENCES

Alexander, D. E. (1984). Unusual phase relationships between the forewings and hindwings in flying dragonflies. J Exp Biol 109, 379-383.

Appleton, F. M. (1974). Dragonflies and flight. Nature Canada 3(3), 25-29.

Azuma, A., Azuma, S., Watanabe, I. and Furuta, T. (1985). Flight mechanics of a dragonfly. J Exp Biol 116, 79-107.

Azuma, A. and Watanabe, T. (1988). Flight performance of a dragonfly. J Exp Biol 137, 221- 252.

Fry, S. N., Sayaman, R. and Dickinson, M. H. (2005). The aerodynamics of hovering flight in Drosophila. J Exp Biol 208, 2303-18.

Huang, H. and Sun, M. (2007). Dragonfly forewing-hindwing interaction at various flight speeds and wing phasing. AIAA J. 45, 508-511.

Lan, C. E. (1979). The unsteady quasi-vortex-lattice method with applications to animal propulsion. J. Fluid Mech. 93, 747-765.

Maybury, W. J. and Lehmann, F. O. (2004). The fluid dynamics of flight control by kinematic phase lag variation between two robotic insect wings. J Exp Biol 207, 4707-26.

Norberg, R. A. (1972). The pterostigma of insect wings and inertial regulator of wing pitch. J. Comp. Physiol 81, 9-22.

Norberg, R. A. (1975). Hovering flight of the dragonfly: Aeschna juncea L., kinematics and aerodynamics. Swimming and Flying in Nature 2, 763-780.

Reavis, M. A. and Luttges, M. W. (1988). Aerodynamic forces produced by a dragonfly. AIAA J. 88-0330, 1-13.

Rüppell, G. (1989). Kinematic analysis of symmetrical flight manoeuvres of odonata. J Exp Biol 144, 13-42.

Thomas, A. L., Taylor, G. K., Srygley, R. B., Nudds, R. L. and Bomphrey, R. J. (2004). Dragonfly flight: free-flight and tethered flow visualizations reveal a diverse array of unsteady lift-generating mechanisms, controlled primarily via angle of attack. J Exp Biol 207, 4299-323.

Usherwood, J. R. and Lehmann, F. O. (2008). Phasing of dragonfly wings can improve aerodynamic efficiency by removing swirl. J R Soc Interface.

Wang, H., Zeng, L., Liu, H. and Yin, C. (2003). Measuring wing kinematics, flight trajectory and body attitude during forward flight and turning maneuvers in dragonflies. J Exp Biol 206, 745-57.

Wang, J. K. and Sun, M. (2005). A computational study of the aerodynamics and forewing- hindwing interaction of a model dragonfly in forward flight. J Exp Biol 208, 3785- 804.

Wang, Z. J. (2004). The role of drag in insect hovering. J Exp Biol 207, 4147-55.

Wang, Z. J. and Russell, D. (2007). Effect of forewing and hindwing interactions on aerodynamic forces and power in hovering dragonfly flight. Physical Review Letters 99.

Warkentin, J. and DeLaurier, J. (2007). Experimental aerodynamic study of tandem flapping membrane wings. AIAA J. 44, 1653-1661.

Whitehouse, F. C. (1941). British Columbia dragonflies (Odonata), with notes on distribution and habits. The American Midland Naturalist 26(3), 488.

10 Chapter

DC SUPPLY SYSTEM DETECTOR OF UNMANNED AERIAL VEHICLE (UAV)

1. INTRODUCTION: AN OVERVIEW

This chapter presents an integrated smart detector within the DC supply system of an unmanned aerial vehicle (UAV) utilizing the AT89C2051, capable of measuring load insulation failures in the direct current supply system to ensure the UAV operates under stable conditions. This chapter initially analyzes the measuring concepts and subsequently presents the system design approaches for both software and hardware. The measurement error has been examined and quantified. In comparison to the Hall instrument, this detector exhibits attributes of cost-effectiveness and minimal power usage. This instrument exhibits exceptional measurement precision and dependability. Empirical investigations demonstrate that the sensor system, with a measurement precision of 0.01mA, satisfies the requirements for detection accuracy, fulfills the anticipated design criteria, and is suitable for dissemination. The development of Unmanned Air Vehicles (UAVs) has accelerated significantly since the Middle Eastern conflicts of the last century, leading to their application in an increasing number of disciplines, while numerous scholars have become involved in this area of study. This research focuses on a smart sensor designed for detecting load insulation failures in the DC supply system of UAVs, highlighting the significance of diverse sensors as a crucial technology in UAV development.In the present of DC supply system[2]-[3], Hall instrument is commonly used to detect unbalance status in the load current, which has characteristics of high detection precision, but high expensive price accordingly. So, in one system, simultaneous using two or more hall instruments are very costly and, further, measurement data can not be calculated on- line, and also increase the burden of the detection and computation of the upper PC. Therefore, a smart sensor based on AT89C2051 is presented in this paper, which can measure load insulation failure in direct current supply system and possess characteristics of low-cost, low-power consumption, and easy to use. This kind of sensor will be pulled- on the positive-negative polar line of each directly current branch circuit. When insulation level of every loop is normal, the current passes through the sensor will have the equal value, and the opposite direction (notice that the current which flow across these coils terminal is defined I+, the negative current is I-, and the composed current is ΔI. Then the zero flux condition will be tested in the synthetic magnetic field of direct current of the sensor, otherwise the output will be nonzero due to ΔI = (I+ - I-) . Consequently, by creating a circuit of the sensor's output state, one can ascertain whether a ground fault occurs in the direct current supply system. Upon fault recognition, the sensor will detect data, enabling the processing of fault information and the transmission of an alert signal.

2. HARDWARE CIRCUIT AND MEASUREMENT PRINCIPLE

2.1 Measure Principle

This smart sensor is mainly used to measure unbalance current in each branch circuit, and check circuit if over-current occurs due to overload or else reasons. Coil is the key testing component of this smart sensor, and is also an important unit in the oscillatory circuit. Branch wire that was measured crossed through the center of coil. When synthetic direct current in the loop fluctuate ΔI (magnetic flux produce ΦΔ changed, and at the same time coil current produce ΔLx changed), frequency loop of the oscillatory circuit will change consequent, and then test the variation of frequency Δf to compute current value crossed.

The relational expression can be described as follows in an ideal case:

$$\Phi = \mu \frac{NI}{2R} a^2 \tag{1}$$

Where

μ — core magnetic conductivity;

a — radius of transverse section of endless solenoid core;

R — radius of annular core;

N — turns.

$$L = \frac{N\Phi}{I} \tag{2}$$

$$f = \frac{2R^2C - 3L}{4RLC\ln(2R^2C/L)} \tag{3}$$

Where R = 5.1 KΩ; C = 1NF.

However, the nonlinear relation exists between coil current and oscillation frequency, which can be detected by conducting experiments, and it is difficult to test current directly. Methods are taken to solve the above problem: firstly, we have a testing experiment and obtain the relation curve between the frequency of the prototype and the corresponding current, and then obtain linear equations of several intervals by piecewise linearization of the curve. Secondly, obtain computing equations best suited for this device by analyzing and comparing the frequency calculated with the data in the table. Finally, actual current value can be calculated using the equation.

2.2 System Design

The whole system architecture of sensor is consisted of several modules: CPU circuit, watchdog circuit, communication circuit, power module and measurement circuit, which is shown in Fig.1.

1) CPU circuit

The CPU chip of smart sensor used is ATMEL AT89C2051 single chip[4], which provides the following features: high performance, ease to use, and high function-price ratio. This chip has 2K flash ROM built-in and be compatible with MCS-51™ Products. The CPU performs the function that records the frequency produced by the oscillatory circuit, and computes the current value according to the relation curve of unbalance current (ΔI) and the oscillatory frequency (Δf). In addition, the CPU is also in charge of the task as communication interface to the Upper PC.

2) Watchdog circuit

MAXIMUM X25045 programmable chip[5] is used in the watchdog circuit, which initialization program is stored in the AT89C2051 chip.

3) Communication circuit

The communication circuit function is mainly performed by MAX485, which is low-power transceiver for RS-485 communication.

Compared with Rs-232 communication, the RS-485 communication features high transmission speed and far transmission distance, which is commonly the first-chosen to the lower system. The communication task is controlled by the single chip.

4) Power module

Operating voltages of this system are +5V and +6.5V, which are converted by the voltage regulator chip TL750L05C [6], +6.5V voltage regulator tube and voltage regulator diode from ±8V supply power.

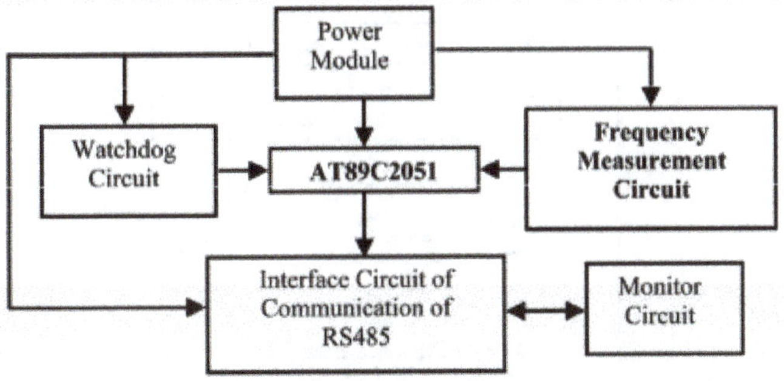

Figure 1. Block diagram of this sensor system

5) Oscillatory circuit

The oscillatory circuit is the key component of frequency measurement circuit, which is consisted of operational amplifier TLE2141CP[7], coils, resistor - condenser oscillatory circuits and high speed diodes, and which operate principle is that the oscillatory frequency of the whole oscillatory circuit will change as the unbalance current flowing across coils

center work a change, then the unbalance current value can be calculated by MCU by measuring the varying quantity of the input voltage frequency. Structure diagram is shown in Fig.2 (the direct current resistance value is 6.22Ω in this sensor, and the inductance value is about 100mH as no current flowing through this sensor).

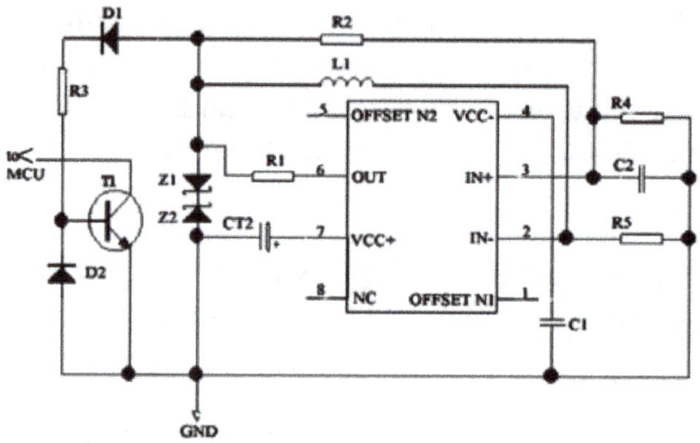

Figure 2. Structure diagram of the oscillatory circuit

3. SOFTWARE FLOWCHART AND METHODS OF MEASUREMENT

3.1 Measure Principle

Software flowchart is shown in Fig.3.

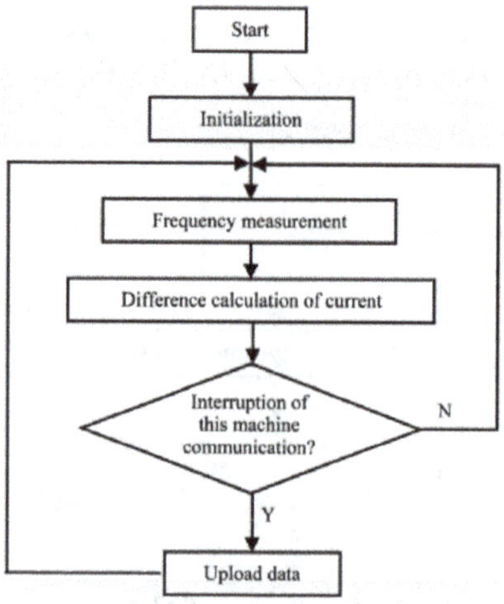

Figure 3. Software flowchart

3.2 Measurement Methods

Equations (1-3) show that it is complicated to compute the frequency directly, however, in view of frequency range (about 75HZ ~1000HZ), the relation between the value of the machine oscillation cycles tested and current calculated is linear in the smallest possible interval. So, a curve can be obtained by taking the method of measuring oscillation cycle in programs[8], which is shown in Fig.5, and then process piecewise linearization. The lateral coordinate represents machine cycles value tested (the machine cycles value tested obtain by converting from frequency in the table 1, and the vertical ordinate represents current calculated (unit: mA). To improve the measurement precision and extend the measuring range[9], the equation coefficient is verified by taking different approaches, such as multi- byte addition, subtraction, double byte multiplication, multi-byte cycle shift division. As practical situation stand, the precision of the current calculated, which achieves 0.01mA, can satisfy the request of measurement precision.

The oscillation cycle value begins to be tested by the MCU at the t1 time (the waveform shown in Fig.4, which close to the square wave is commutated by the diode, is produced by the Wien bridge oscillatory circuit.). However, the formal measurement sets to work at the T2 time when the wave between the t1~T2 time slot was removed. The T2~T3 time slot is a complete cycle and the data tested will be stored in the MCU after the measurement is completed.

By the same above method, the MCU test 30 groups of cycle data in total and make digital filtering respectively at the same time, then calculate the average value of cycles and look for the interval of current tested by the given data table. Finally, according to the corresponding interpolation computation in the locked intervals, the change value of the current of a cycle will be obtained.

4. RESULT ANALYSIS AND METHODS OF IMPROVING PRECISION

4.1 Result Analysis

During the process of software debugging, part of current values, which are shown in Table 1, are obtained by the actual measurement, and then the subsection curve, shown in Fig.5, will be obtained due to the value. As is shown in Table 1, the current error rate is small, which usually in about 0%-0.8%, therefore the current precision is 0.01mA, which can meet the requirement of the actual project.

Table 1. Part of the fault current data Measured by the smart sensor

Actual current value flowed through coils (mA)	Current value tested by the sensor(mA)	Absolute error rate ×100%
0	0	0%
83.14	83.3	0.2%

Actual current value flowed through coils (mA)	Current value tested by the sensor(mA)	Absolute error rate ×100%
96.85	96.98	0.14%
144.69	143.78	0.62%
160.77	161.44	0.45%
192.93	193.63	0.36%
225.08	226.25	0.52%
241.16	246.51	0.59%
305.47	304.05	0.46%
353.7	351.12	0.73%

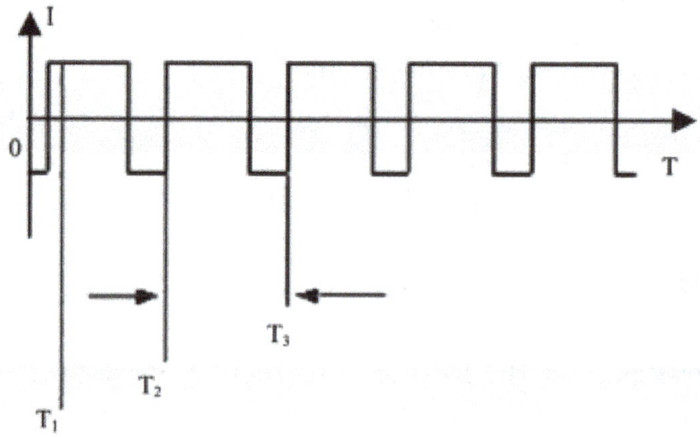

Figure 4. Computing model of a complete cycle

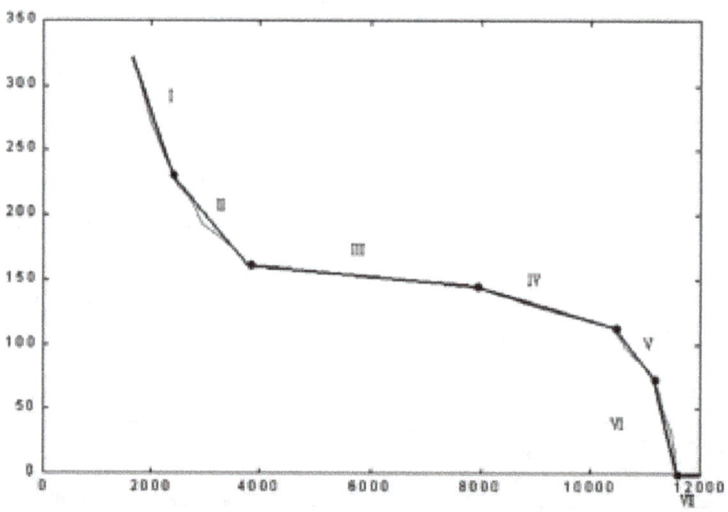

Figure 5. Relation curve between coil current (y-axis) and oscillation cycle (x-axis)

4.2 Methods of Improving Precision

The measurement errors originated from the following sources in this sensor:

Relate to the coil sensitivity $\Delta L_X / \Delta I$: The coil inductance changed ΔL_X proportion to the coil current changed ΔI within a certain range, namely $\Delta L_X \propto K \cdot \Delta I$, in direct where K is the proportional coefficient. Coils aging (mainly the heat loss of the coils) would lead to not only decline sensitivity, but also reduce the proportional coefficient K. The resolve method is to replace coils at fixed periods.

Relate to the oscillation cycle of coils measured: It would produced errors by measuring the oscillation cycle by calculating the machine cycle, and the maximum frequency value tested is 1KHZ due to the actual measurement, so the biggest error produced (the crystal oscillator of AT89C2051 is 11.0592MHZ) is (12 /11.0592µ) /1000µ 100% 0.185% that can satisfy the request of the actual project.

Refer to the subsection on linear fitting through curve measurement. It is evident that the lineal fitting has a more favorable impact within a specific range; however, the lineal fitting errors would be generated as a result of the minimal bias. The resolve technique employed is either the direct interpolation method or the increase of the number of measured points.

5. CONCISE SUMMARY

This sensor exhibits excellent accuracy and real-world performance due to its small size, high reliability, and high performance-to-price ratio.The findings indicate that this design scheme is viable, has the potential to meet the anticipated design criteria, and has the potential to be widely implemented.

REFERENCES

Major William K Lewis, "UCAV-The Next Generation Air-Superiority Fighter? School of Advanced Airpowers Studies, June 2002. Available: http://www.au.af.mil/au/awc/awcgate/saas/lewis.pdf [1]

X.P. Meng, Y. Gao, Electric Power System Analysis, Higher Education Press, Beijing, 2004, pp. 100–135. [2]

T.X. Chen, J.D. Duan, T.T. Chen, G. Chen, A new approach to on-line insulation monitoring of electric power cables based on capacitive-current measurement and its criteria, High Voltage Apparatus, vol. 40, no. 3, 16, June, 2004, pp. 183–185. [3]

8-bit Microcontroller with 2K Bytes Flash AT89C2051. Available: www.atmel.com/atmel/acrobat/doc0368.pdf [4]

X25045 datasheet. Available: http://tu.cndzz.com/down/soft/2206.htm [5] TL750L05C datasheet. Available: www.compel.ru/pdf/TI/TL750LXX. [6]

TLE2141CP datasheet. Available: http://www.alldatasheet.com/datasheet- pdf/pdf/TI/TLE2141CP.html [7]

W.J. Dong, The Methods for Software Anti-interference Applied to MCS-51 Monolithic Processor Operation System, Journal of Luoyang University, vol. 15, no. 4, Dec. 2000, pp. 56–58. [8]

X.R. Chen, P. Cai, H.Q. Zhou, Several practical methods of frequency measurement based on a single chip computer, Industrial Instrumentation & Automation, no.3, 2003, pp. 40–43. [9]

Chapter 11

FORMATION FLIGHT OF UNMANNED AERIAL VEHICLES UTILIZING SLIDING MODE DISTURBANCE OBSERVERS

1. INTRODUCTION: AN OVERVIEW

The purpose of this chapter is to develop a controller that is capable of achieving exceptional formation flight for unmanned aerial vehicles and accurately tracking the desired path in the presence of unknown disturbances. In other words, the objective is to establish the autonomous laws in a manner that enforces the desired formation by adhering to the guidance commands. A smooth controller composed of three time-scaled nested controllers is developed using sliding mode control, which is driven by sliding mode disturbance observer techniques based on the inversion of rotorcraft dynamics and quaternions mathematical representation, given a non-linear mathematical model of a quadrotor (X-shaped rotorcraft). Various control techniques, including the Proportional Integral Derivative (PID) controller, Linear Quadratic Regulator (LQR) feedback controller, adaptive controller, and control Lyapunov functions, are available as a result of existing research efforts in multiple aircraft formation control (Innocenti et al., 2000). Nevertheless, these methods have been unable to resolve the robustness necessary for the tight formation to withstand real-world effects, such as external disturbances, environmental aerodynamic uncertainties, end-effector malfunctions, and simulation model inaccuracies. The proposed techniques are primarily constrained by the combination of robust formation controllers, adapted guidance laws for extensive formation structure capabilities, and accurate path tracking. The absence of the latter is essential for maintaining a close formation in the presence of unknown disturbances and leader maneuvers.Sliding Mode (SM) control is a well-known technique used to achieve robustness and to guarantee the stability of a system (C. Edwards, & S. Spurgeon, 1998). However it presents the inconvenience of control signal high frequency switching which is not appropriate for autopilots. Recent results in Higher Order

Sliding Mode (HOSM) control theory (A. Levant, 2003) present the ability to design a continuous and even smooth control, robust to external disturbances, parametric uncertainties, end effectors failures and modeling errors, which is chattering free and provides for enhanced accuracy. Robust sliding mode control using sliding mode disturbance observer techniques allow us to consider tight formation flight where the separation distances between flying aerial vehicles are minimized. The sliding mode control technique provides the unmanned rotorcrafts with bounded, continuous output tracking for the three controllers: translation tracking, attitude tracking and speed controller for the four propellers.

The structure of this chapter is as follows: The unmanned vehicle is initially shown, followed by several instructions for building the non-linear model using quaternion representation. The formation strategies are devised in a three-dimensional setting utilizing triangle formation techniques. The controller is designed for a single rotorcraft, assuming for simplification that every rotorcraft has identical properties, despite the fact that variations in mass and efficiency might be readily incorporated. Ultimately, simulation outcomes will be exhibited within the Matlab/Simulink environment.

2. THE VEHICLE: X-SHAPED ROTORCRAFT

An example of a X-Shaped rotorcraft is the X-4 Flyer→. It is propelled by four independent, fixed-pitch rotors located on the extremities of an 'X' shaped frame. This specific configuration allows a balanced distribution of thrusts around the center of mass of the vehicle. Motion is provided by independently varying the angular speed and thus the thrust of each rotor. Advantages of such a vehicle are numerous, especially for robust formation control, since soft collisions can be tolerated during flight. The major issue of the X-4-Flyer is that it is unstable by design. This makes it entirely inoperable by a human being, without the help of a fast on-board controller and motion sensors.

Figure 1. X-4 Flyer→

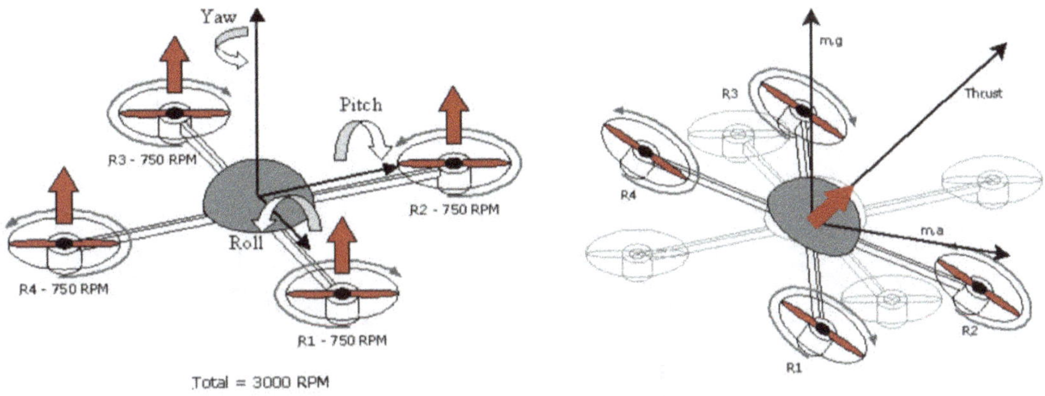

Figure 2. Hover State and Translation

The "X" configuration allows rotations and translations along all three axes, but only four degrees of freedom can be controlled independently because of the coupling between these motions. Like conventional helicopters, pitch and roll rotations are coupled with longitudinal and transversal translations, respectively.

Figure 2 illustrates instances of rotor speed arrangements. In the hover condition, the push generated by the four propellers precisely neutralizes the force of gravity. Yaw is regulated by modifying the torque exerted on each motor. The global reaction torque will induce rotation of the entire airframe around its local yaw axis. Pitch and roll revolutions can be attained by modifying the differential rotor speeds of a pair of rotors revolving in unison. Translation is accomplished by orienting the entire airframe through Pitch/Roll rotations towards the intended acceleration vector and modulating the overall thrust.

2.1 Mathematical Model

The frames of reference are: a fixed-to-earth frame and a local moving frame attached to the center of mass O of the vehicle. G is assumed to be Galilean. Orientation of frame A relative to frame G is expressed with the rotation matrix, equation (1). The airframe orientation in space is given by the Euler angles: roll (φ), pitch (θ) and yaw (ψ) with angular velocities respectively (p, q, r).

$$R_{A/G} = \begin{bmatrix} \cos(\psi)\cos(\theta) & \sin(\phi)\sin(\theta)\cos(\psi) - \cos(\phi)\sin(\psi) & \cos(\phi)\sin(\theta)\cos(\psi) + \sin(\phi)\sin(\psi) \\ \sin(\psi)\cos(\theta) & \sin(\phi)\sin(\theta)\sin(\psi) + \cos(\phi)\sin(\psi) & \cos(\phi)\sin(\theta)\sin(\psi) - \sin(\phi)\cos(\psi) \\ -\sin(\theta) & \cos(\phi)\cos(\theta) & \cos(\phi)\cos(\theta) \end{bmatrix} \quad (1)$$

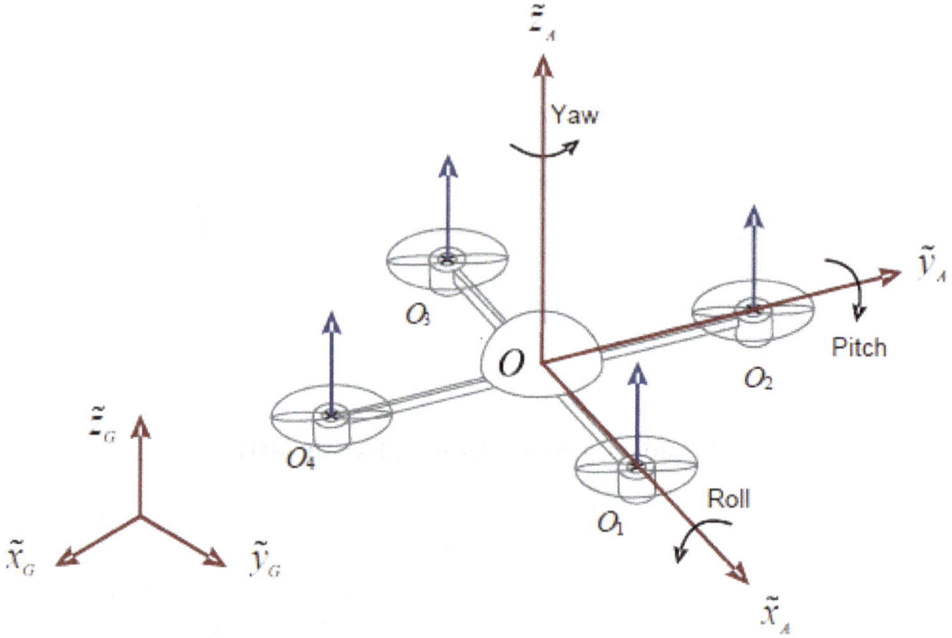

Figure 3. Frame of Reference

The following model represents an ideal case, where all variables are known or measured exactly with no measurement noise and no external disturbances. However, the model does include important realistic parameters such as aerodynamic forces and torques on airframe, propellers, viscous friction on rotor axles, and the full inertia matrix of the airframe.

Using conservation of linear and angular momentum of the airframe, conservation of angular momentum for one rotor and simplifications due to mechanical constraints of Rotor-Airframe coupling, a full realistic dynamical model of a quadrotor is proposed:

Airframe linear acceleration:

$$\begin{pmatrix} \ddot{x} \\ \ddot{y} \\ \ddot{z} \end{pmatrix}_{A/G}^{A} = \dot{\tilde{v}}_{A/G}^{G} = -g \cdot \tilde{z}^{G} + \frac{\rho_{air}}{2m} \cdot \left[R_{G/A}\right] \cdot \begin{pmatrix} \sum_{i=1}^{4} \left(c_{rdf_i} \cdot \omega_i \cdot |\omega_i| \cdot \tilde{f}_i^{A}\right) \\ -\left[C_{ADF}^{A}\right] \cdot \left[R_{A/G}\right] \cdot \tilde{v}_{A/G}^{G} \cdot \left|\tilde{v}_{A/G}^{G\,T}\right| \end{pmatrix} \tag{2}$$

Airframe angular acceleration:

$$\begin{pmatrix} \dot{p} \\ \dot{q} \\ \dot{r} \end{pmatrix}_{A/G}^{A} = \dot{\tilde{\omega}}_{A/G}^{A} = \left[I_{frame/O}^{A}\right]^{-1} \cdot \begin{pmatrix} -sk(\tilde{\omega}_{A/G}^{A}) \cdot \left[\left[I_{frame/O}^{A}\right] \cdot \tilde{\omega}_{A/G}^{A} + \sum_{i=1}^{4} \left(j_{z_i} \cdot \omega_i \cdot \tilde{r}_i^{A}\right)\right] \\ -\sum_{i=1}^{4} \left(\tau_i - c_v \cdot \omega_i\right) \cdot \tilde{r}_i^{A} \\ +\frac{1}{2} \cdot \rho_{air} \cdot \left[-\left[C_{ADF}^{A}\right] \cdot \tilde{\omega}_{A/G}^{A} \cdot \left|\tilde{\omega}_{A/G}^{A\,T}\right| + \sum_{i=1}^{4} \left(c_{rdf_i} \cdot \omega_i \cdot |\omega_i| \cdot \tilde{f}_i^{A} \times \tilde{L}_i^{A}\right)\right] \end{pmatrix} \tag{3}$$

Rotor angular acceleration:

$$\dot{\omega}_i = \frac{1}{j_{z_i}} \cdot \left(\tau_i - \frac{1}{2} \cdot \rho_{air} \cdot c_{rdt_i} \cdot \omega_i \cdot |\omega_i| - c_{v_i} \cdot \omega_i \right) \tag{4}$$

Rotation matrix derivative:

$$[\dot{R}_{G/A}] = [R_{G/A}] \cdot sk(\tilde{\omega}^A_{A/G}) \tag{5}$$

2.2 Quaternions Representation

A quaternion can be thought of as a complex number with 4 dimensions: 3 imaginary dimensions i, j and k; and one real dimension, all orthogonal to one another. Each of the imaginary dimensions has a unit value of the square root of -1. A quaternion is represented as follows:

$$Q = q_1 \cdot i + q_2 \cdot j + q_3 \cdot k + q_4 \tag{6}$$

or

$$Q = \begin{bmatrix} q_1 \\ q_2 \\ q_3 \\ q_4 \end{bmatrix} \tag{7}$$

When used to describe 3D attitude or orientation, the following rule exists between quaternion coefficients and instantaneous axis ($\tilde{u}$) and angle (θ) of rotation:

$$Q = \begin{bmatrix} q_1 \\ q_2 \\ q_3 \\ q_4 \end{bmatrix} = \begin{bmatrix} u_x \cdot \sin(\theta/2) \\ u_y \cdot \sin(\theta/2) \\ u_z \cdot \sin(\theta/2) \\ \cos(\theta/2) \end{bmatrix} \tag{8}$$

It is possible to rotate vectors in 3D space using quaternion through the same process used with rotation matrices. The following formula shows how to rotate vector V = [v, v ,v]T using rotation axis and angle given by quaternion Q

$$Q_{vrot} = Q \otimes Q_v \otimes Q' \tag{9}$$

$$Q_v = v_x i + v_y j + v_z k + 0 \tag{10}$$

3. FORMATION STRATEGIES

Formation strategies are developed to manage any size formation according to three specific schemes: "Tracking," "Placement" and "Anti-Collision." The shape of the formation is derived using "triangular" patterns depending on the number of aerial vehicle (AV).

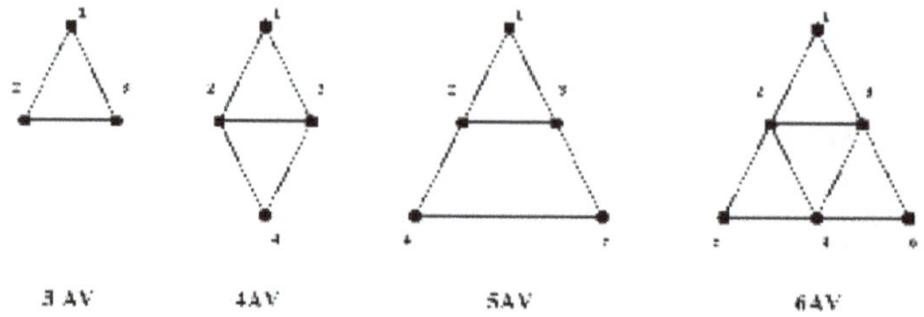

Figure 4. Formations Triangular Patterns

The formation strategies are designed as follows:

- The "Tracking" strategy is specific to the leader of the formation which is in this case the agent at the tip of the triangle. The task of the leader is to track the predefined path.
- The "Placement" strategy consists in positioning an agent in regards to the leader only.
- The "Anti-Collision" strategy consists in positioning an agent in regards to its lateral neighbors (set of safety distances) and to the leader.

Let us consider a formation of three agents in a Cartesian frame centered on the leader (Figure 5). The leader receives the path to follow as a function of time, while the followers receive the distances to keep from each other and to the leader. For an equilateral triangular formation pattern where the distances to keep is d from each vehicle with a vertical separation of h, the formation commands for the followers are:

$$\text{"Placement"}: x_p = x_{leader} - \sqrt{\frac{3}{4}d^2 - h^2}\,,\; y_p = y_{leader} - \frac{d}{2}\,,\; z_p = z_{leader} - h$$

$$\text{"Anti-Collision"}: x_A = x_{leader} - \sqrt{\frac{3}{4}d^2 - h^2}\,,\; y_A = y_P + d\,,\; z_A = z_{leader} - h$$

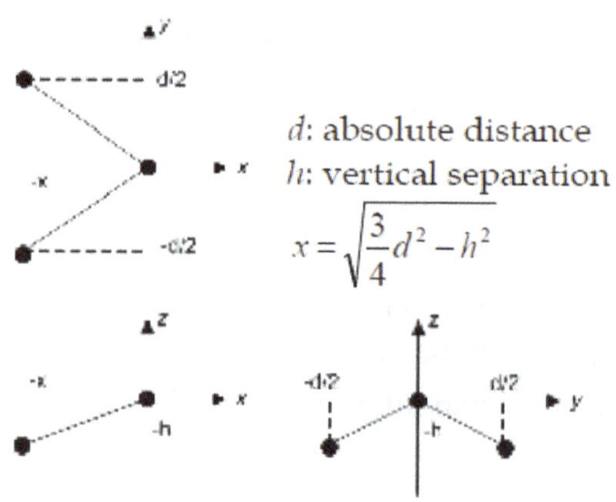

Figure 5. Three Aerial Vehicles 3D Triangular Formation Pattern

4. CONTROLLER ARCHITECTURE

The controller's design adheres to the model's natural time constants, which are optimally distinct. A three-time scale controller employing three stacked feedback control loops, driven by sliding mode observer techniques, is described. This includes a translation tracking controller, an attitude tracking controller, and a speed controller for the four propellers. Figure 6 illustrates the control approach, whereas figure 7 presents the design for each controller, which incorporates three feedback loops and sensors to assess the real-time rotorcraft status.In loop 1 of figure 7, the translation controller is designed to compute the instantaneous acceleration vector needed to track the reference trajectory by comparing the position commands and the actual measured positions from a Global Positioning System.

Loop 2 works in conjunction with the attitude controller to compute the body torques needed to align the rotorcraft in the direction of the desired acceleration. The desired instantaneous acceleration vector is compared to the angular velocities about the three axis of the airframe measured by gyroscopes. Attitude is derived by integration.

The appropriate propeller velocity rates are calculated from the specified torque using inverse kinematics (Loop 3). In loop 4, the propeller controller regulates the propeller velocity to align with the commanded value, in relation to the observed angular rates of each rotor. The closed-loop time constants of the three outer loop controllers must be meticulously selected to ensure the convergence of each loop. Sliding mode control, utilizing sliding mode observer techniques, is employed to stabilize system dynamics and ensure robustness; errors will converge to the origin despite external and internal disturbances.

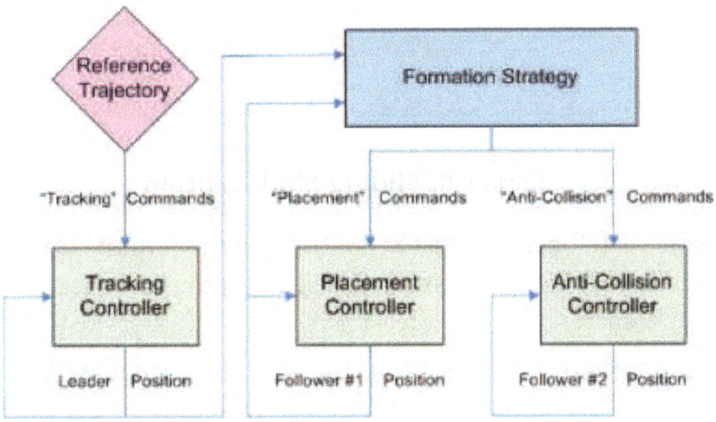

Figure 6. Formation Control Strategy

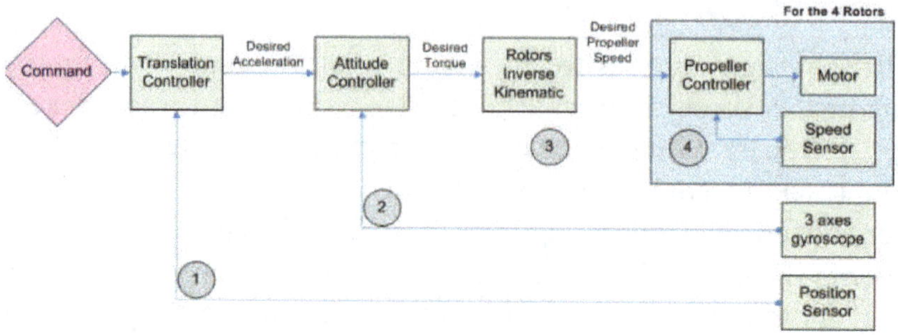

Figure 7. Flight Controller Architecture

5. SLIDING MODE

5.1 Sliding Mode: An Overview

Sliding Mode Control is a technique that enables the user to direct a system to a specified condition, known as the sliding surface, and comprises two stages: the reaching phase and the sliding phase, as seen in figure 8. The reaching phase emphasizes directing the system state towards the sliding surface. The sliding phase involves reducing the sliding variable to zero and is marked by high-frequency, discontinuous control switching.

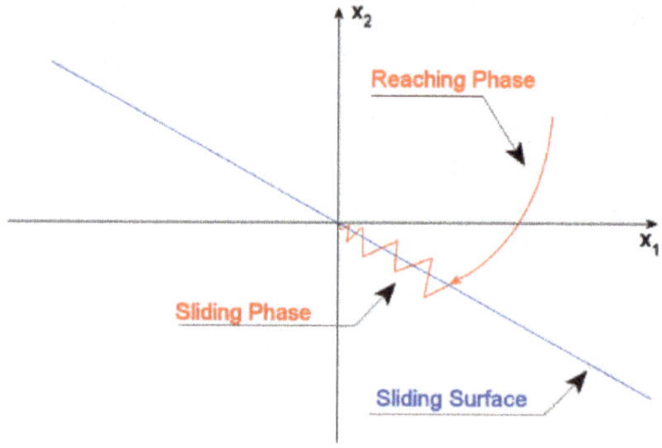

Figure 8. Sliding Mode Motion

The sliding surface is to be designed so that the system exhibits the desired behavior. The control is to be designed to assure that the system reaches the sliding surface and stays on it thereafter

Output tracking problems consider Multi Input Multi Output (MIMO) system such as the following

$$\begin{cases} \dot{x} = f(x,t) + G(x,t)u \\ y = h(x,t) \end{cases} \tag{11}$$

where $x \in \Re^n$, $f(x,t) \in \Re^n$, $y \in \Re^m$, $u \in \Re^m$ and $h(x,t) = [h_1, h_2, \ldots, h_m]^T \in \Re^m$,

$G(x,t)=[g_1,g_2,...,g_m]^T \in \Re^{n\times m}$ are analytic vector and matrix functions.

The system can be expressed in normal form assuming the system is completely feedback linearizable (Isidori, 1995) in a reasonable compact domain $x\in\Gamma(x)$. $L_f^{r_i-1}h_i$ and $L_{g_i}\left(L_f^{r_i-1}h_i\right)$ $\forall i=\overline{1,m}$ are corresponding Lie derivatives, $|E(x,t)|\neq 0$, $\forall x\in\Gamma$ and $\bar{r}=[r_1,r_2,...,r_m]^T$ is a vector relative degree.

$$\begin{bmatrix} y_1^{(r_1)} \\ y_2^{(r_2)} \\ \cdots \\ y_m^{(r_m)} \end{bmatrix} = \begin{bmatrix} L_f^{r_1}h_1(x,t) \\ L_f^{r_2}h_2(x,t) \\ \cdots \\ L_f^{r_m}h_m(x,t) \end{bmatrix} + E(x,t)u \tag{12}$$

$$E(x,t)=\begin{bmatrix} L_{g_1}\left(L_f^{r_1-1}h_1\right) & L_{g_2}\left(L_f^{r_1-1}h_1\right) & \cdots & L_{g_m}\left(L_f^{r_1-1}h_1\right) \\ L_{g_1}\left(L_f^{r_2-1}h_2\right) & L_{g_2}\left(L_f^{r_2-1}h_2\right) & \cdots & L_{g_m}\left(L_f^{r_2-1}h_2\right) \\ \cdots & \cdots & \cdots & \cdots \\ L_{g_1}\left(L_f^{r_m-1}h_m\right) & L_{g_2}\left(L_f^{r_m-1}h_m\right) & \cdots & L_{g_m}\left(L_f^{r_m-1}h_m\right) \end{bmatrix} \tag{13}$$

The goal of output tracking techniques is to derive the control u to drive the system output y to the desired trajectory y_c (command). In other words, the goal is to drive the tracking error $e = y - y_c$ to zero in a finite time. The design problem is in choosing a sliding variable σ that will drive the tracking error to zero and maintain it thereafter.

Choose the following sliding variable

$$\sigma_i = e_i^{(r_i-1)} + c_{i,r_i-2}e_i^{(r_i-2)} + \ldots + c_{i,1}e_i^{(1)} + c_{i,0}e_i + c_{i,-1}e_{i,-1} \tag{14}$$

where $\sigma=[\sigma_1,\sigma_2,...,\sigma_m]^T$, $e_i = y_{i,c}-y_i$, $e_i^{(k)}=\dfrac{d^k e_i}{dt^k}$, $e_{i,-1}=\int e_i dt$ and $c_{i,k}>0\in\Re$.

Taking the derivatives on both sides,

$$\dot{\sigma}_i = e_i^{(r_i)} + c_{i,r_i-2}e_i^{(r_i-1)} + \ldots + c_{i,1}e_i^{(2)} + c_{i,0}e_i^{(1)} + c_{i,-1}e_i \tag{15}$$

From equation (12),

$$y_i^{(r_i)} = L_f^{r_i}h_i(x,t) + (Eu)_i \tag{16}$$

Introducing $\tilde{u}_i = (Eu)_i$,

$$e_i^{(r_i)} = y_{i,c}^{(r_i)} - L_f^{r_i} h_i(x,t) - \tilde{u}_i \tag{17}$$

and substituting (17) into (15),

$$\dot{\sigma}_i = y_{i,c}^{(r_i)} - L_f^{r_i} h_i(x,t) - \tilde{u}_i + c_{i,r_i-2} e_i^{(r_i-1)} + \ldots + c_{i,1} e_i^{(2)} + c_{i,0} e_i^{(1)} + c_{i,-1} \tag{18}$$

Let us use the following notation

$$\begin{cases} \Psi_i = y_{i,c}^{(r_i)} - L_f^{r_i} h_i(x,t) + c_{i,r_i-2} e_i^{(r_i-1)} + \ldots + c_{i,1} e_i^{(2)} + c_{i,0} e_i^{(1)} + c_{i,-1} \\ \Psi_i = \psi_i^0 + \Delta\psi_i \end{cases} \tag{19}$$

where ψ_i^0 represents the known part of Ψi and Δψ I represents unknown bounded disturbances such that $\|\Delta\psi_i\| \le L_i$. Introducing (19) into (18),

$$\dot{\sigma}_i = \psi_i^0 + \Delta\psi_i - \tilde{u}_i \tag{20}$$

Let us design the control as

$$\tilde{u}_i = \tilde{u}_{i,0} + \tilde{u}_{i,1} \tag{21}$$

where $\tilde{u}_{i,0}$ is designed to compensate for the known terms

$$\tilde{u}_{i,0} = \psi_i^0 \tag{22}$$

Therefore, becomes,

$$\dot{\sigma}_i = \Delta\psi_i - \tilde{u}_{i,1} \tag{23}$$

The control problem is to design the control $\tilde{u}_{i,1}$ that will drive $\dot{\sigma}_i$ to zero.

5.2 Traditional Sliding Mode Control

Traditional sliding mode control is based on the analysis of Lyapunov functions to insure asymptotic stability with finite time convergence. Consider the following Lyapunov function

$$V_i = \frac{1}{2}\sigma_i^2 \tag{24}$$

Taking the time derivative, we obtain

$$\dot{V}_i = \dot{\sigma}_i \sigma_i \tag{25}$$

From Lyapunov theory, we derive the following finite-time asymptotic convergence condition

$$\dot{V}_i \le -\rho_i \sqrt{2V_i} \tag{26}$$

with $\rho_i > 0$, $\rho_i \in \Re$, which guarantees a reaching time of $t_{r,i} \le \frac{\sqrt{2V_i(0)}}{\rho_i}$.

Substituting (24) and (25) into (26) we obtain

$$\sigma_i \dot{\sigma}_i \leq -\rho_i \sqrt{\sigma_i^2} \tag{27}$$

$$\sigma_i \dot{\sigma}_i \leq -\rho_i |\sigma_i| \tag{28}$$

From (23), we also have

$$\sigma_i \dot{\sigma}_i = \sigma_i \left(\Delta\psi_i - \tilde{u}_{i,1} \right) \tag{29}$$

$$\sigma_i \dot{\sigma}_i = \sigma_i \Delta\psi_i - \sigma_i \tilde{u}_{i,1} \tag{30}$$

which can be rewritten as

$$\sigma_i \dot{\sigma}_i \leq |\sigma_i| |\Delta\psi_i| - \sigma_i \tilde{u}_{i,1} \tag{31}$$

The disturbances being bounded, $\|\psi_i\| \leq L_i$, (31) becomes

$$\sigma_i \dot{\sigma}_i \leq L_i |\sigma_i| - \sigma_i \tilde{u}_{i,1} \tag{32}$$

or, using the definition of the sign function

$$\sigma_i \dot{\sigma}_i \leq |\sigma_i| \left[L_i - \tilde{u}_{i,1} sign(\sigma_i) \right] \tag{33}$$

The control $\tilde{u}_{i,1}$ has to satisfy the sliding mode existence condition expressed in (28). Therefore, from (33) and (28) we obtain

$$L_i - \tilde{u}_{i,1} sign(\sigma_i) = -\rho_i \tag{34}$$

$$\tilde{u}_{i,1} sign(\sigma_i) = \rho_i + L_i \tag{35}$$

Therefore,

$$u_{i,1} = (\rho_i + L_i) sign(\sigma_i) \tag{36}$$

The final sliding mode control is expressed using (22), (36) and recalling that $\tilde{u}_i = (Eu)_i$

$$u = E^{-1} \left[\psi^0 + (\rho + L)\ sign(\sigma) \right] \tag{37}$$

This conventional sliding mode control will stabilize the sliding variable on the sliding surface but is marked by high-frequency switching. Advancements in higher order sliding mode control and sliding mode control utilizing sliding mode observers are implemented to minimize switching actions and precisely estimate disturbances.

5.3 Sliding Mode Controlled by Sliding Mode Disturbance Observer

This technique aims to develop a sliding mode observer to precisely estimate the unknown disturbance for compensation purposes. Introduce the auxiliary sliding variables S_i and Z_i such that

$$\begin{cases} s_i = \sigma_i + z_i \\ \dot{z}_i = \tilde{u}_{i,1} - v_i \end{cases} \quad (38)$$

Therefore, the dynamic of the auxiliary sliding variable is

$$\dot{s}_i = \left(\Delta\psi_i - \tilde{u}_{i,1}\right) + \left(\tilde{u}_{i,1} - v_i\right) \quad (39)$$

$$\dot{s}_i = \Delta\psi_i - v_i \quad (40)$$

As seen previously in (36), the control that will drive Si to zero and maintain it thereafter is

$$v_i = \left(\rho_i + L_i\right) sign\left(s_i\right) \quad (41)$$

To avoid the high frequency switching motion of the control, a Low-Pass filter is introduced

$$\hat{v}_i = \frac{1}{1+\beta_i s} v_i \quad (42)$$

Where βi is the time constant to be chosen according to design specifications. Therefore, when the auxiliary sliding variable Si stabilizes at zero, (26) becomes

$$\Delta\psi_i - \hat{v}_i = 0 \quad (43)$$

As a result, the control $\hat{v}_i$ is an exact estimate of the unknown disturbances $\Delta\psi_i$.

Let us design the control $\tilde{u}_{i,1}$ to stabilize σ_i to zero in (23) of the form:

$$\tilde{u}_{i,1} = \hat{v}_i + K_{0,i}\sigma_i \quad (44)$$

Where $K_{0,i} > 0$ is chosen to ensure fast reaching of the sliding surface by σ, as soon as the disturbance is estimated and compensated for (i.e., the convergence of σ_i cannot be faster than that of Si). Therefore, the dynamic of the sliding variable is

$$\dot{\sigma}_i = -K_{0,i}\sigma_i \quad (45)$$

which satisfies the sliding mode existence condition: $\sigma_i\dot{\sigma}_i < 0$.

The final sliding mode control driven by sliding mode observer is

$$u = E^{-1}[\psi^0 + \hat{v} + K_0\sigma] \quad (46)$$

6. CONTROLLER DESIGN

6.1 Translation Controller

The following equation expresses the airframe linear position, velocity and acceleration errors, respectively

$$\begin{aligned} \tilde{\varepsilon}_p &= \tilde{p}_c - \tilde{p} \\ \tilde{\varepsilon}_v &= \dot{\tilde{\varepsilon}}_p = \dot{\tilde{p}}_c - \dot{\tilde{p}} \\ \tilde{\varepsilon}_\gamma &= \ddot{\tilde{\varepsilon}}_p = \ddot{\tilde{p}}_c - \ddot{\tilde{p}} \end{aligned} \tag{47}$$

Where $\ddot{\tilde{p}}_c, \dot{\tilde{p}}_c, \tilde{p}_c$ are given by the desired trajectory commands in position, velocity and acceleration, respectively, and $\dot{\tilde{p}}, \tilde{p}$ are measured or estimated in real time by on-board sensors.

Following sliding mode observer techniques, a sliding variable of second order is designed

$$\tilde{\sigma}_p = k_{v1}\dot{\tilde{\varepsilon}}_p + k_{p1}\tilde{\varepsilon}_p + k_{i1}\int \tilde{\varepsilon}_p dt \tag{48}$$

A third order pole placement is used to compute values for the coefficients kv1, kp1 and ki1. In this case, a triangular left half plane pattern pole placement is used to compute values for the coefficients, using the desired response time for position tracking.

The dynamics of the sliding variable can be expressed as

$$\dot{\tilde{\sigma}}_p = \tilde{\psi}_p + \Delta\tilde{\psi}_p - k_{v1}\tilde{\gamma} \tag{49}$$

where $\tilde{\psi}_p$ represents the known terms,

$$\tilde{\psi}_p = k_{v1}(\ddot{\tilde{p}}_c + \tilde{g}) + k_{p1}\dot{\tilde{\varepsilon}}_p + k_{i1}\tilde{\varepsilon}_p \tag{50}$$

And $\tilde{\psi}_p$ the unknown bounded disturbances, $|\Delta\tilde{\psi}_p| \le L_p$, where L_p is a positive constant. Therefore, the control can be expressed in a sum of control terms compensating respectively for the known and unknown terms

$$\tilde{\gamma} = (\tilde{u}_{0p} + \tilde{u}_{1p}) / k_{v1} \tag{51}$$

The disturbance observer is designed considering the following auxiliary sliding variable

$$\tilde{\psi}_p = k_{v1}(\ddot{\tilde{p}}_c + \tilde{g}) + k_{p1}\dot{\tilde{\varepsilon}}_p + k_{i1}\tilde{\varepsilon}_p \tag{52}$$

$$\begin{cases} \tilde{s}_p = \tilde{\sigma}_p + \tilde{z}_p \\ \dot{\tilde{z}}_p = \tilde{u}_{1p} - \tilde{v}_p \end{cases} \tag{53}$$

The resulting dynamics are

$$\dot{\tilde{s}}_p = \Delta\tilde{\psi}_p - \tilde{v}_p \tag{54}$$

The dynamics of the auxiliary variable are completely compensated with a control of the form

$$\tilde{v}_p = \rho_p \, sign(\tilde{s}_p) \, , \, \rho_p >> L_p \tag{55}$$

The high frequency switching is filtered out using the equivalent control

$$\hat{\tilde{v}}_p = \frac{\tilde{v}_p}{\tau_p s + 1} \tag{56}$$

where τ p is the filter time constant to be tuned.

Finally, the resulting designs of the control terms are

$$\begin{aligned} \tilde{u}_{0p} &= \tilde{\psi}_p \\ \tilde{u}_{1p} &= \hat{\tilde{v}}_p + K_p \tilde{\sigma}_p \end{aligned} \tag{57}$$

where $K_p > 0$ assures exponential convergence of the sliding variable to zero.

6.2 Attitude Controller

Applying inverse rotation kinematics, the desired instantaneous axis and angle of rotation are defined by computing the cross product between current thrust force vector $\tilde{F}$ and the desired acceleration vector $\tilde{\gamma}$ normalized to unit length.

$$\tilde{u}_d = \frac{\tilde{F}}{\|\tilde{F}\|} \times \frac{\tilde{\gamma}}{\|\tilde{\gamma}\|} \tag{58}$$

The direction of unit vector $\tilde{u}_d$ gives the desired axis of rotation. Therefore, the actual attitude error angle is

$$\tilde{\varepsilon}_\theta = \cos^{-1}\left(\frac{\tilde{F}}{\|\tilde{F}\|} \times \frac{\tilde{\gamma}}{\|\tilde{\gamma}\|} \right) \tag{59}$$

Equations (60) express the attitude, angular velocity and angular acceleration errors, respectively

$$\begin{aligned} \tilde{\varepsilon}_\theta &= \tilde{\theta}_c - \tilde{\theta} = \cos^{-1}\left(\frac{\tilde{F}}{\|\tilde{F}\|} \times \frac{\tilde{\gamma}}{\|\tilde{\gamma}_d\|} \right) \cdot \frac{\tilde{u}_u}{\|\tilde{u}_d\|} \\ \tilde{\varepsilon}_w &= \dot{\tilde{\varepsilon}}_\theta = \tilde{\omega}_c - \tilde{\omega} = -\tilde{\omega} \\ \dot{\tilde{\varepsilon}}_w &= \ddot{\tilde{\varepsilon}}_\theta = \dot{\tilde{\omega}}_c - \dot{\tilde{\omega}} = -\dot{\tilde{\omega}} \end{aligned} \tag{60}$$

Following similar sliding mode observer techniques as presented in Section 2.D, a sliding variable of second order is designed

$$\tilde{\sigma}_\theta = k_{v2} \dot{\tilde{\varepsilon}}_\theta + k_{p2} \tilde{\varepsilon}_\theta + k_{i2} \int \tilde{\varepsilon}_\theta dt \tag{61}$$

A third order pole placement is used to compute values for the coefficients k_{v2}, k_{p2} and k_{i2}. In this case, a Butterworth optimal pole placement as presented in previous work [45] is used to compute values for the coefficients, using the desired response time for attitude tracking.

Equation (62) expresses the dynamics of the sliding variable

$$\dot{\tilde{\sigma}}_\theta = \tilde{\psi}_\theta + \Delta\tilde{\psi}_\theta - k_{v2}\dot{\tilde{\omega}} \tag{62}$$

where $\tilde{\psi}_\theta$ represents the known terms,

$$\tilde{\psi}_\theta = k_{p2}\dot{\tilde{\varepsilon}}_\theta + k_{i2}\tilde{\varepsilon}_\theta \tag{63}$$

and $\Delta\tilde{\psi}_\theta$ the unknown bounded disturbances, $|\Delta\tilde{\psi}_\theta| < L_\theta$, where Lθ is a positive constant.

Once again, the control is expressed in terms of a sum of control terms compensating respectively for the known and unknown terms

$$\dot{\tilde{\omega}} = (\tilde{u}_{0\theta} + \tilde{u}_{1\theta}) / k_{v2} \tag{64}$$

The disturbance observer is designed considering the following auxiliary sliding variable

$$\begin{cases} \tilde{s}_\theta = \tilde{\sigma}_\theta + \tilde{z}_\theta \\ \dot{\tilde{z}}_\theta = \tilde{u}_{1\theta} - \tilde{v}_\theta \end{cases} \tag{65}$$

The resulting dynamics are

$$\dot{\tilde{s}}_\theta = \Delta\tilde{\psi}_\theta - \tilde{v}_\theta \tag{66}$$

The dynamics of the auxiliary variable are completely compensated for with control of the form

$$\tilde{v}_\theta = \rho_\theta\, sign(\tilde{s}_\theta)\,,\ \rho_\theta >> L_\theta \tag{67}$$

The high frequency switching is filtered out using the equivalent control

$$\hat{\tilde{v}}_\theta = \frac{\tilde{v}_\theta}{\tau_\theta s + 1} \tag{68}$$

where τ_θ is the filter time constant to be tuned. Finally, the resulting design of the control terms is

$$\begin{aligned} \tilde{u}_{0\theta} &= \tilde{\psi}_\theta \\ \tilde{u}_{1\theta} &= \hat{\tilde{v}}_\theta + K_\theta\tilde{\sigma}_\theta \end{aligned} \tag{69}$$

where $K_\theta > 0$ assure exponential convergence of the sliding variable to zero.

6.3 Propeller Controller

The speed of each rotor (w_i, i = 1...4) is computed using inverse dynamics as presented in and compared to the desired speed (w_{id}, i = 1...4) corresponding to the desired linear ($\ddot{\tilde{\gamma}}$) and angular ($\dot{\tilde{\omega}}$) accelerations. The necessary torques to apply to each electric motor are then computed in order to achieve these desired speeds.

The angular velocity error of each rotor is defined as

$$\begin{aligned} \varepsilon_{\omega i} &= \omega_{id} - \omega_i \\ \dot{\varepsilon}_{\omega i} &= \dot{\omega}_{id} - \dot{\omega}_i = -\dot{\omega}_i \end{aligned} \tag{70}$$

Using the same sliding mode observer techniques as for the previous two controllers, the sliding variable

$$\sigma_{\omega i} = k_{p3}\varepsilon_{\omega i} + k_{i3}\int \varepsilon_{\omega i} dt \tag{71}$$

A classical second order placement is used, with the desired response time, to define kp3 and ki3. The dynamics of the sliding variable are expressed as

$$\dot{\sigma}_{\omega i} = \psi_{\omega i} + \Delta\psi_{\omega i} - k_{p3}\dot{\omega}_i \tag{72}$$

where $\psi_{\omega i}$ represents the known terms

$$\psi_{\omega i} = k_{i3}\varepsilon_{\omega i} \tag{73}$$

and $\Delta\psi_{\omega i}$ the unknown bounded disturbances, $\Delta\psi_{\omega i} < L_{\omega i}$, where $L_{\omega i}$ is a positive constant. Once again, the control is expressed in terms of a sum of control terms compensating respectively for the known and unknown terms

$$\dot{\omega}_i = (u_{0\omega i} + u_{1\omega i}) / k_{p3} \tag{74}$$

The disturbance observer is designed considering the auxiliary sliding variable

$$\begin{cases} s_{\omega i} = \sigma_{\omega i} + z_{\omega i} \\ \dot{z}_{\omega i} = u_{1\omega i} - v_{\omega i} \end{cases} \tag{75}$$

The resulting dynamics are

$$\dot{s}_{\omega i} = \Delta\psi_{\omega i} - v_{\omega i} \tag{76}$$

The dynamics of the auxiliary variable are completely compensated for with control of the form

$$v_{\omega i} = \rho_{\omega i}\, sign(s_{\omega i}),\ \rho_{\omega i} >> L_{\omega i} \tag{77}$$

The high frequency switching is filtered out using the equivalent control

$$\hat{v}_{\omega i} = \frac{v_{\omega i}}{\tau_{\omega i} s + 1} \tag{78}$$

where τ_{wi} is the filter time constant to be tuned. Finally, the resulting design of the control terms is

$$\begin{aligned} u_{0\omega i} &= \psi_{\omega i} \\ u_{1\omega i} &= \hat{v}_{\omega i} + K_{\omega i}\sigma_{\omega i} \end{aligned} \tag{79}$$

where $K_{wi} > 0$ assures exponential convergence of the sliding variable to zero.

7. SIMULATION RESULTS

The following figures are the simulation results for the formation of three X-4 Flyer flying in a close formation in the face of simulated wind gusts acting upon each rotorcraft (modeled as offset on the velocity vector in frame G). The formation is to fly in a triangular pattern with an absolute distance of three meters (d = 3m) between each rotorcraft and a vertical separation of one meter to the leader. The trajectory to follow is a sinusoid at a constant elevation of 2 meters and a constant longitudinal velocity of 2m/s. Note that the first follower will stay at the initial position for one second following take off of the leader and then follow the "Placement" commands. The second follower will stay at the initial position for two seconds following the take off of the leader (or one second following the take off of the first follower) and then follow the "Anti-Collision" commands. The simulation time is 10sec with a computing frequency of a 1000Hz.

7.1 Leader "Path Tracking"

The characteristics of the X-4 Flyer, such as position, velocity, angular velocity and propeller angular velocity are illustrated, as well as the control torques and the HOSM controller variables.

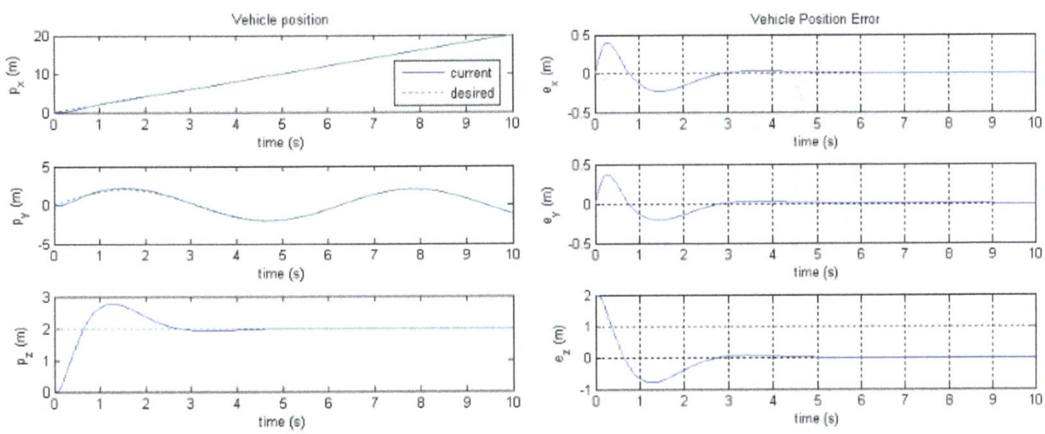

Figure 9. Leader Position and Position Error versus Time

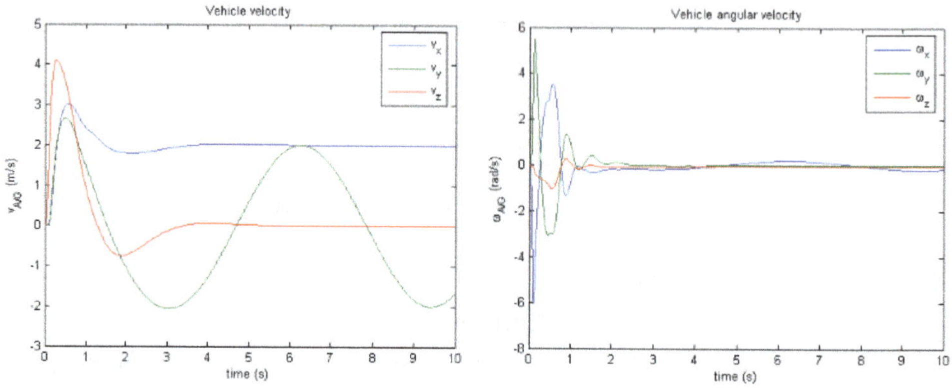

Figure 10. Leader Velocity and Angular Velocity versus Time

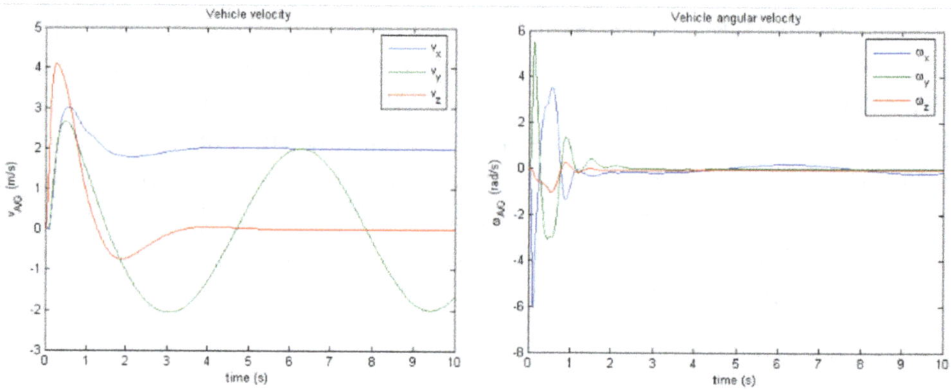

Figure 11. Leader Propeller Angular Velocity and Control Torques versus Time

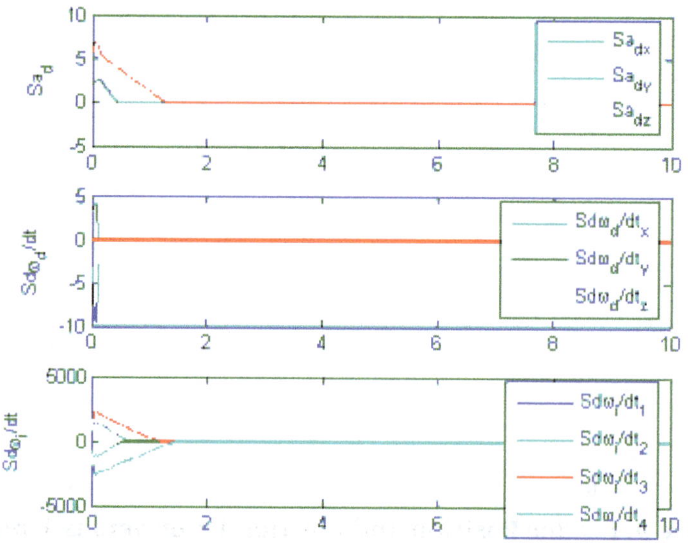

Figure 12. Leader Auxiliary Sliding Variables versus Time

The leader achieves perfect tracking of the desired path in a smooth and robust manner. The position errors in 3 reach zero in finite time. The unknown disturbances are estimated and completely compensated for using robust HOSM techniques.

First Follower "Placement"

The following figures represent the "Placement" simulation results for the first follower. The characteristics of the X-4 Flyer, such as position, velocity, angular velocity and propeller angular velocity are illustrated, as well as the control torques and the HOSM controller variables.

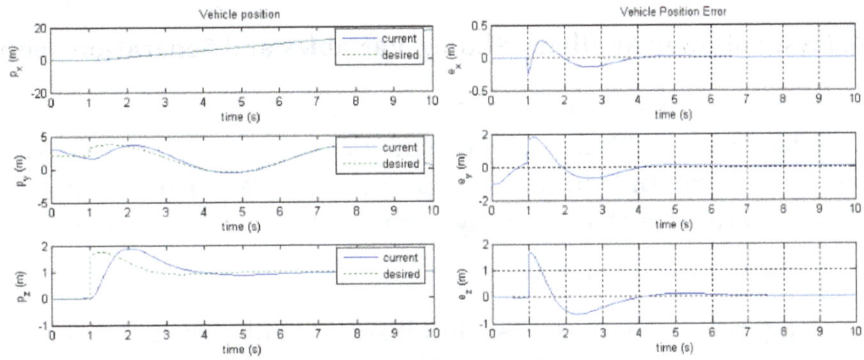

Figure 13. First Follower Position and Position Error versus Time

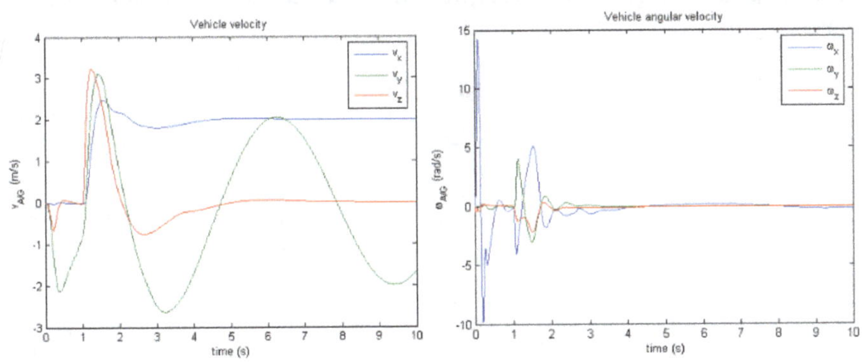

Figure 14. First Follower Velocity and Angular Velocity versus Time

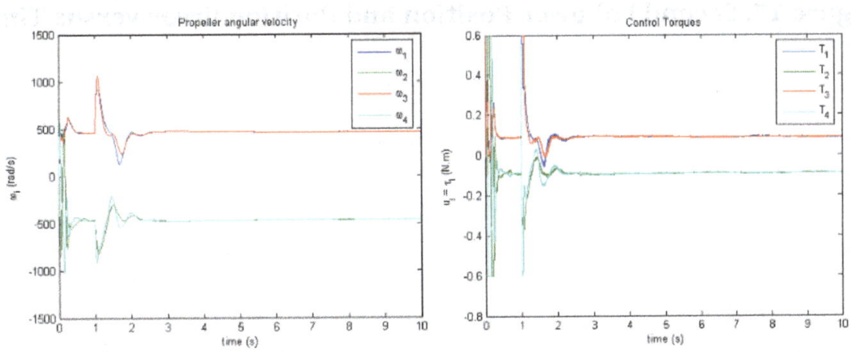

Figure 15. First Follower Propeller Angular Velocity and Control Torques versus Time

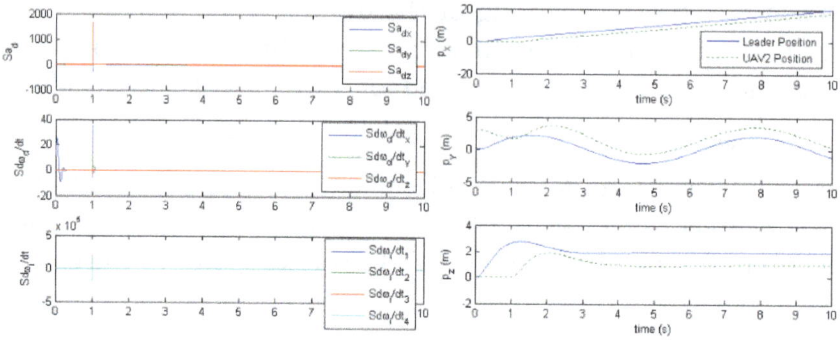

Figure 16. First Follower Auxiliary Sliding Variables and Separation versus Time

After staying at the initial position for the first second, the rotorcraft initiates the formation following the "Placement" commands. The rotorcraft achieves perfect positioning after a smooth transient, the position errors reach zero in finite time. The disturbances are estimated and fully compensated for, via robust HOSM.

7.3 Second Follower "Anti-Collision"

The following figures represent the "Anti-Collision" simulation results for the second follower. The characteristics of the X-4 Flyer, such as position, velocity, angular velocity and propeller angular velocity are illustrated, as well as the control torques and the HOSM controller variables.

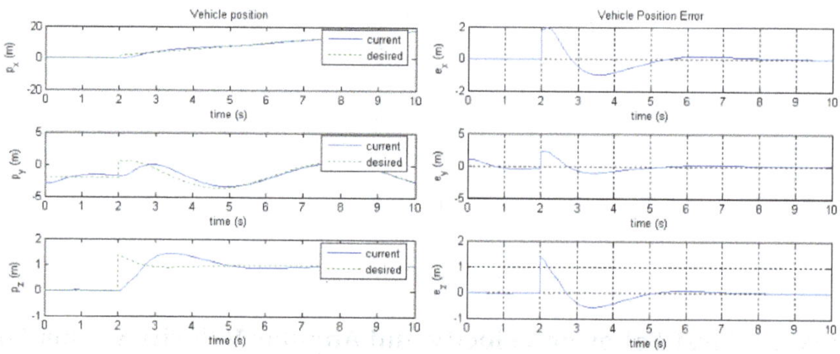

Figure 17. Second Follower Position and Position Error versus Time

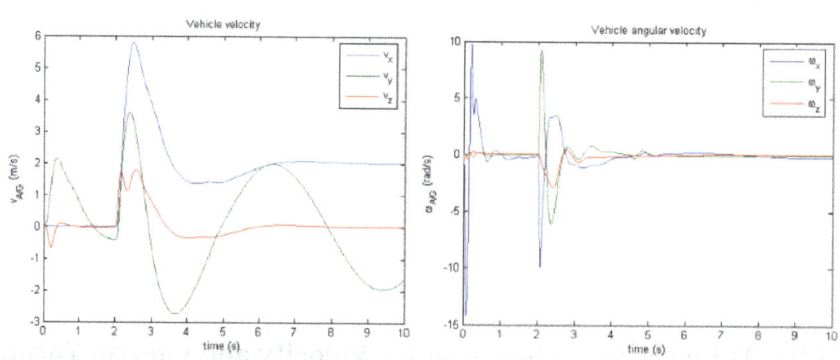

Figure 18. Second Follower Velocity and Angular Velocity versus Time

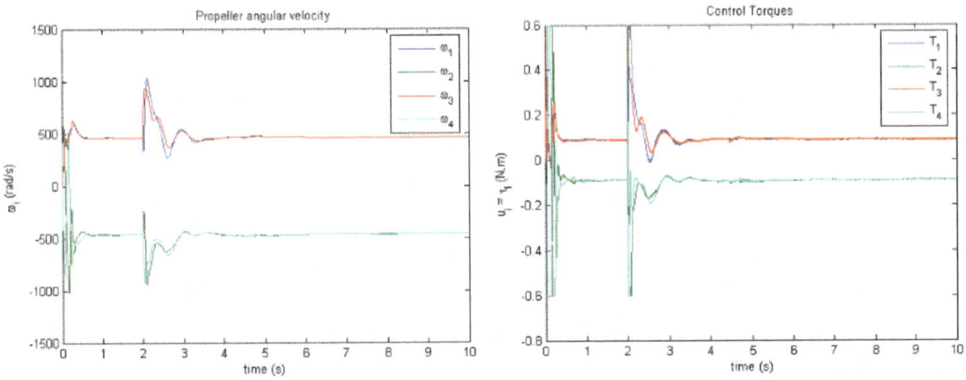

Figure 19. Second Follower Propeller Angular Velocity and Control Torques versus Time

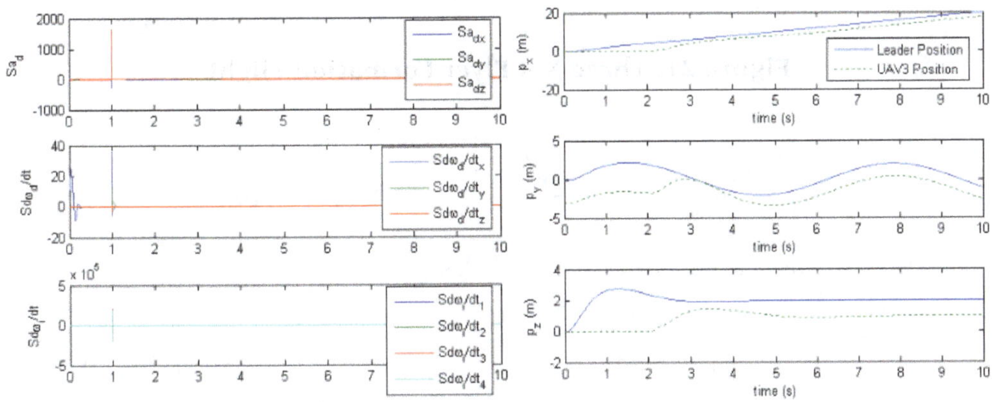

Figure 20. Second Follower Auxiliary Sliding Variables and Separation versus Time

The second follower commences the "Anti-Collision" technique after a two-second delay from the beginning condition. The rotorcraft attains optimal placement following a brief transitory period. The rotorcraft must accelerate to align with the formation, necessitating the full utilization of available resources to swiftly attain the target position, as evidenced by the saturation of the torques. The disruptions are assessed and completely mitigated by robust HOSM.

7.4 Formation tracking

The rotorcrafts establish and maintain the intended configuration along the correct trajectory following a brief transitory period. The precise distance between rotorcrafts attains the specified spacing. The intended triangle configuration along the specified sinusoidal trajectory is illustrated in the subsequent illustrations.

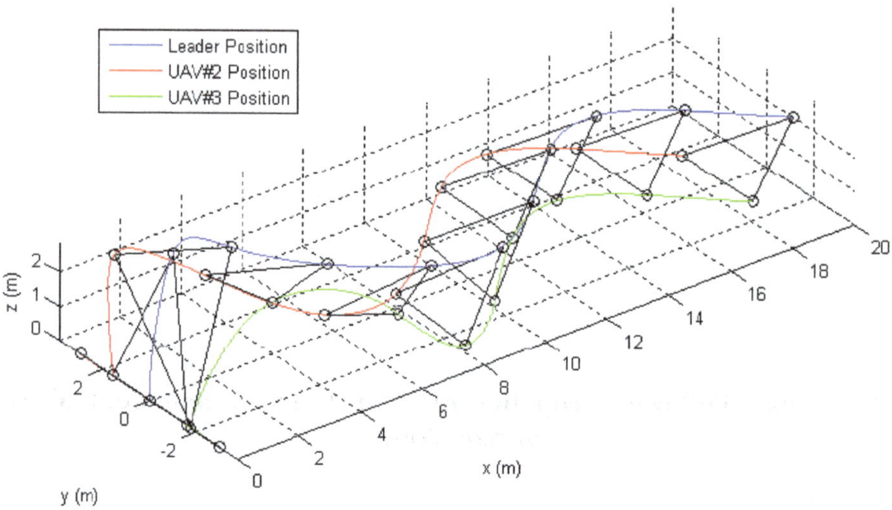

Figure 21. Three X-4 Flyer Formation Flight

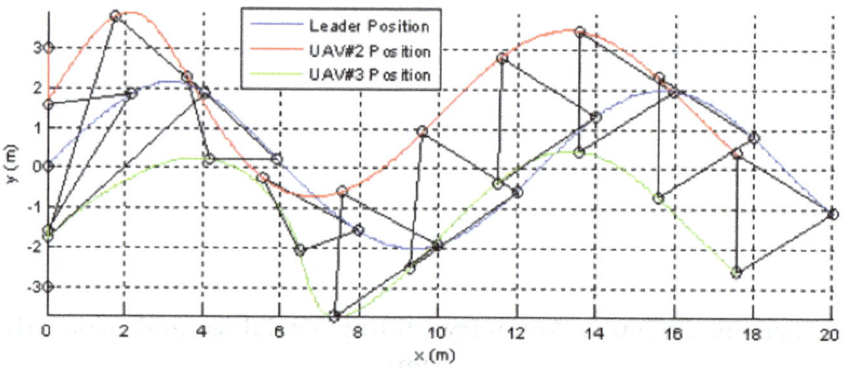

Figure 22. Three X-4 Flyer Formation Flight X-Y Projection

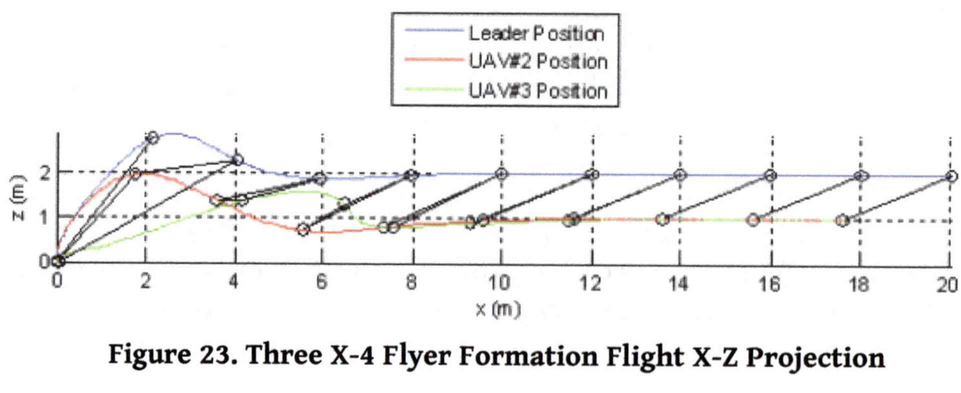

Figure 23. Three X-4 Flyer Formation Flight X-Z Projection

7.5 Actuator Malfunction

One of the propellers of the first follower is malfunctioning; four seconds into the simulation, the propeller's effectiveness diminishes, and the motor torque decreases by 50%. To maintain rotorcraft equilibrium, the propeller's angular velocity must be rapidly recovered by modulating the applied torque. The controller doubles the requisite motor torques to swiftly compensate for the efficiency loss, so recovering the angular velocity. The flying formation remains unaffected by the actuator malfunction, so demonstrating the resilience of the proposed HOSM approaches.

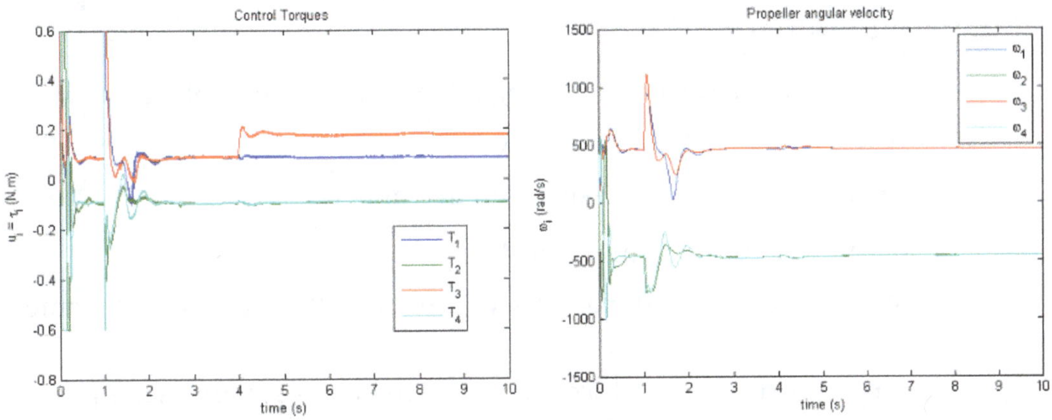

Figure 24. Compensating Torques and Restored Angular Velocity

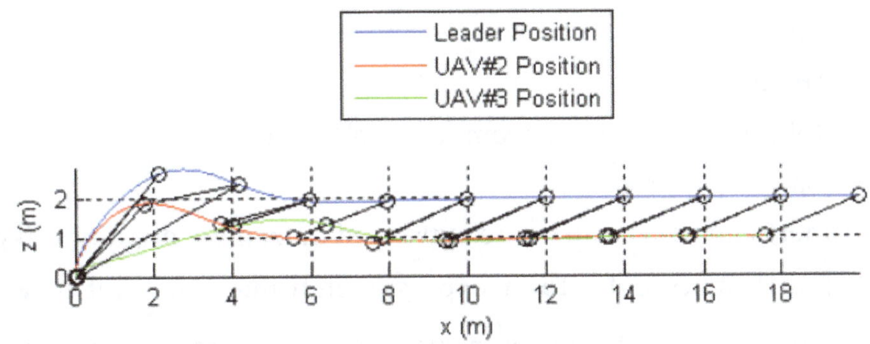

Figure 25. Formation Flying with Actuator Malfunction X-Z projection

8. CONCISE SUMMARY

This chapter applies formation techniques to the close formation flight of three X-shaped rotorcraft in a fully three-dimensional environment. Robust formation tracking is accomplished by sliding mode approaches informed by sliding mode disturbance observers. Simulation studies indicate that SMDO controllers perform seamless and continuous control despite unknown boundary disturbances and actuator failures. The controls ensure that the tracking errors converge to zero within a finite time following a brief transient period associated with the startup of the formation. Optimal close formation is attained despite leader maneuvers and unforeseen interruptions, demonstrating the efficacy of the recommended tactics. Potential extensions of the research encompass the execution of real formation flying scenarios involving leader reassignment and formation reconfiguration. An inbuilt collision avoidance system could be incorporated into each aerial vehicle within the formation. Outcomes utilizing prospective field methodologies are noteworthy and documented in the literature.

9. LIST OF SYMBOLS

$\tilde{x}_A, \tilde{y}_A, \tilde{z}_A$	Unit vectors respectively along XA, YA and ZA axes of frame A
$\tilde{V}^A$	Vector V expressed in frame A
$\dot{\tilde{V}}^A$	Derivative of vector V expressed in frame A
$[M^A]$	Matrix M expressed in frame A
$\tilde{F}^A_{B_1 \to B_2}$	Force vector exerted from body B1 to body B2 expressed in frame A
$\tilde{\tau}^A_{B_1 \to B_2}$	Torque vector exerted from body B1 to body B2 expressed in frame A
$sk(\tilde{V}^A)$	Skew matrix of vector V expressed in frame A (used for cross product)
(ϕ, θ, ψ)	Euler angles
(p, q, r)	Angular velocities
$[R_{G/A}]$	Rotation matrix to go from frame A to frame G
D	Diameter of propeller (m)
$[I^A_{frame/O}]$	Inertia matrix of airframe expressed in frame A at point O (kg.m)
$[I^A_{rotor/O_i}]$	Inertia matrix of rotor 'i' expressed in frame A at point Oi (kg.m)
$\tilde{V}^G_{A/G}$	Linear velocity vector of frame A relative to frame G (ground) (m.s-1)
$\tilde{\omega}^A_{A/G} = (p, q, r)^T$	Angular velocity vector of frame A relative to frame G (rad.s-1)
$\tilde{\omega}^A_{R_i/A}$	Angular velocity vector of rotor 'i' relative to frame A (rad.s-1)
ω_i	Angular velocity of rotor 'i' with respect to frame A (rad.s-1)
$\tilde{r}^A_i$	Axis of rotation (unit vector) of rotor 'i' expressed in frame A.
$\tilde{f}^A_i$	Direction of thrust produced by rotor 'i' expressed in frame A.

τ_i	Torque produced by motor 'i' on rotor 'i' (N.m = kg.m2.s-2)
F_i	Thrust force produced by rotor 'i' expressed in frame A (N = kg.m.s-2)
$\tilde{L}_i^A$	Vector $\overline{O_iO}$ between rotor and center of mass expressed in frame A (m)
$\tilde{\tau}_{F_i}^A$	Torque produced by rotor 'i' on airframe expressed in frame A (N.m)
$[C_{ADF}^A]$	Drag force coefficient matrix of airframe expressed in frame A (m2)
$[C_{ADT}^A]$	Drag torque coefficient matrix of airframe expressed in frame A (m5)
$[C_{RDT_i}^A]$	Drag torque coefficient matrix of rotor 'i' expressed in frame A (m)
$[C_{V_i}^A]$	Viscous coefficient matrix of rotor 'i' expressed in frame A (kg. m2.s-1)
c_{rdf_i}	Drag torque coefficient of rotor 'i' along its axis of rotation (m5)
c_{v_i}	Viscous friction coefficient of rotor 'i' (kg.m2.s-2)
$\tilde{p}_c$, $\dot{\tilde{p}}_c$, $\ddot{\tilde{p}}_c$	Commanded trajectory position, velocity and acceleration
$\tilde{\varepsilon}_p$, $\tilde{\varepsilon}_v$, $\tilde{\varepsilon}_\gamma$	Position, velocity and acceleration error vectors respectively
k_{v1}, k_{p1}, k_{i1}	Position controller feedback coefficients
k_{v2}, k_{p2}, k_{i2}	Attitude controller feedback coefficients
k_{p3}, k_{i3}	Propeller controller feedback coefficients
$\tilde{F}_i$	Thrust force produced by each rotor expressed in frame A (N)
$\tilde{F}$	Current total thrust force expressed in frame G (N)

REFERENCES

C. Edwards, & S. Spurgeon, Sliding Mode Control, Taylor & Francis, Bristol, PA, 1998.

M. Innocenti, L. Pollini, & F. Guilietti, Autonomous Formation Flight, IEEE Controls System Magazine, Vol. 20, No. 6, 2000, pp. 34-44.

A. Isidori, Nonlinear Control Systems, Springer-Verlag, London, 3rd Ed., 1995.

A. Levant, Higher-Order Sliding Modes, Differentiation and Output-Feedback Control, International journal of Control, 76 (9/10), pp. 924-941, 2003.

12 Chapter

DESIGN AND EXPERIMENTS OF AUTONOMOUS FORMATION FLIGHT

1. INTRODUCTION: AN OVERVIEW

Formation flight has been utilized by numerous avian species for centuries due to its aerodynamic and social advantages. The traditional "V" shape formation that birds used not only facilitated communication between individuals, but also reduced the induced drag for each trailing bird, thereby reducing the energy required for flight (Weimerskirch et al. 2001). Additionally, the advantages of formation flight have been assessed for manned aircraft. Nevertheless, the aerodynamic advantages of human-piloted close formations are scarcely realized for an extended period of time due to the elevated risk factors. Consequently, autonomous formation control that is dependable may be a desirable attribute for both human-piloted aircraft and Unmanned Aerial Vehicles (UAVs). In recent years, the formation control problem has been the subject of extensive discussion, with a variety of applications in the field of ground mobile robotics, aircraft systems, and other space vehicles. (Scharf et al. 2004) categorized the spacecraft formation flight control algorithms into five architectures in their survey paper:

- "Multiple-Input Multiple-Output, in which the formation is treated as a single multiple- input, multiple-output plan;
- Leader/Follower, in which individual spacecraft controllers are connected hierarchically;
- Virtual Structure, in which spacecraft are treated as rigid bodies embedded in an overall virtual structure;
- Cyclic, in which individual spacecraft controllers are connected nonhierarchically;
- Behavioral, in which multiple controllers for achieving different (and possibly competing) objectives are combined."

Similar classifications can be extended to the formation control of other types of vehicles. However, due to the complexity and non-linearity associated with the aircraft dynamics, the 'leader-follower' approach was by far the most popular method for aircraft formation flight control. The advantage of the 'leader-follower' approach lies in its conceptual simplicity, where the formation flight problem is reduced to a set of tracking problems that can be analyzed and solved using standard control techniques.

In the early 1990s, a series of publications from D'Azzo and his colleagues (Dargan et al. 1992) (Buzogany et al. 1993) (Reyna et al. 1994) (Veth et al. 1995) outlined the foundation for the control of the 'leader-follower' formation flight using compensation-type controllers.

Since then, a variety of control techniques has been evaluated includeing optimal control (McCammish et al. 1996) (Dogan et al. 2005), adaptive control (Boskovic & Mehra 2003), fuzzy control (Li et al. 2005), robust control (Li et al. 2006), feedback linearization (Singh et al. 2000) (Venkataramanan & Dogan 2003), and sliding mode (Schumacher & Singh 2000). (Allen et al. 2002) performed a string stability analysis of an autonomous formation for measuring of how position errors propagate from one vehicle to another in a cascaded system. (Giulietti et al. 2000) simulated the scenario with the presence of a failure of one of the nodes, such as the loss of an aircraft.

Experimental studies for evaluating the aerodynamic effects of formation flight and for validating formation control laws have been conducted with both wind tunnel experiments (Gingras 1999) (Fowler & D'Andrea 2003) (Kutay et al. 2005) and flight-testing (Napolitano 2005) (Gu et al. 2006) (Lavretsky 2002) (Hanson et al. 2002) (How et al. 2004) (Johnson et al. 2004). Related to flight-testing efforts, the NASA Dryden Flight Research Center Autonomous Formation Flight (AFF) project in 2001 demonstrated formation control in the lateral and vertical channels with a pair of F/A-18 aircraft (Hanson et al. 2002). (How et al. 2004) at MIT performed a 2-aircraft formation flight using timing control. (Johnson et al. 2004) at Georgia Tech have been performing a series of vision-based formation flight since 2004.

This chapter details the research initiative culminating in the flight demonstration of autonomous formation flying with three YF-22 research aircraft developed, constructed, equipped, and evaluated at West Virginia University (WVU). The remainder of the chapter is structured as follows. Section two delineates the formation geometry and the design of the formation controller. Section three delineates the test-bed aircraft and the avionics equipment. Section four presents the identification of both the linear and nonlinear mathematical models of the aircraft. Section five delineates the formation flight simulation environment and the onboard software. Section six presents several flight-testing phases and the final experimental results. The final portion presents a concise conclusion.

2. FORMATION CONTROLLER DESIGN

In the selected 'leader-follower' formation flight configuration, a Radio Control (R/C) pilot manually controls the 'leader' throughout the flight. Each 'follower' executes the formation control laws to maintain a pre-defined position and orientation with respect to the 'leader'. A main objective of the controller design is to maintain the formation geometry under maneuvered flight conditions. In addition, a minimum amount of information exchange is also desired between the 'leader' and 'follower' aircraft.

Since the trajectory dynamics generally have a much larger time constant than the attitude dynamics, the formation controller can be designed with an inner/outer-loop structure. In this configuration, the outer-loop controller minimizes the lateral, forward, and vertical distance error while the inner-loop controller performs disturbance attenuation and attitude tracking. The definition of formation geometry and designs of the inner-loop and outer-loop controllers are described next.

2.1 Formation Geometry

Consider that the flight path typically lies in a horizontal plane, the formation flight control can be simplified as two decoupled horizontal and vertical tracking problems. The general formation geometry is shown in Fig. 1. For navigation purposes, the position and velocity of both 'leader' and 'follower' aircraft are expressed with respect to a pre-defined Local Tangent Plane (LTP) and are measured by the on-board GPS receivers.

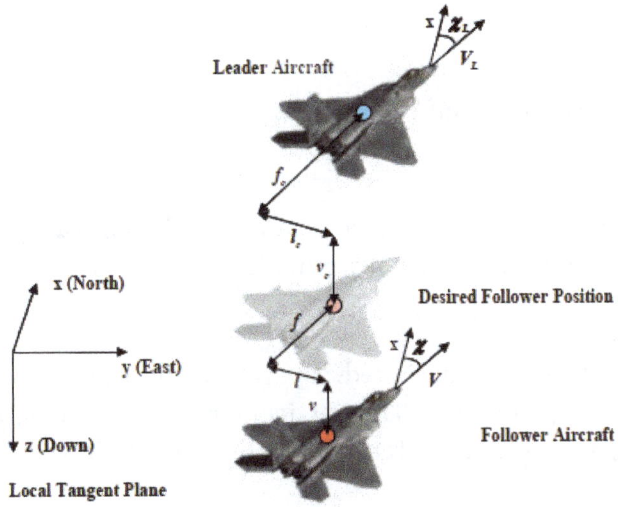

Figure 1. Formation Geometry

A. Horizontal Geometry

The horizontal formation geometric parameters include the desired forward clearance f_c and the desired lateral clearance l_c, as defined in Fig. 1. The forward distance error f and lateral distance error l can be calculated using the following relationship:

$$\begin{bmatrix} l \\ f \end{bmatrix} = \begin{bmatrix} \sin(\chi_L) & -\cos(\chi_L) \\ \cos(\chi_L) & \sin(\chi_L) \end{bmatrix} \begin{bmatrix} x_L - x \\ y_L - y \end{bmatrix} - \begin{bmatrix} l_c \\ f_c \end{bmatrix} \tag{1}$$

, where χ is the aircraft azimuth angle, x and y are the aircraft position along the x and y axis, and a subscript 'L' is used for all 'leader' parameters. A trigonometric expression for the azimuth angle is given as:

$$\sin(\chi) = \frac{V_y}{\sqrt{V_x^2 + V_y^2}} \tag{2}$$

, where V_x and V_y are the projections of the velocity along the x and y axes of the LTP. Equation (1) transforms the position error from an LTP reference frame to a reference frame oriented along the velocity of the 'leader'. In addition, the derivatives of f and l can be described as:

$$\begin{bmatrix} \dot{l} \\ \dot{f} \end{bmatrix} = \begin{bmatrix} V_{xy}\sin(\chi - \chi_L) \\ V_{Lxy} - V_{xy}\cos(\chi - \chi_L) \end{bmatrix} + \Omega_L \begin{bmatrix} f \\ -l \end{bmatrix} \tag{3}$$

, where Ω is the aircraft angular turn rate and V_{xy} is the aircraft velocity in the horizontal plane. It is obvious that the lateral and forward distance controls are coupled.

B. Vertical Geometry

The vertical distance error v can be obtained as:

$$v = z_L - z - v_c \tag{4}$$

, where v_c is the desired vertical clearance and z is the aircraft position along the z axis. The vertical geometry is also shown in Fig. 1.

2.2 Outer-Loop Controller

The outer-loop controller is designed separately for the decoupled vertical and horizontal formation geometries. The vertical controller is a simple linear altitude tracker providing the desired pitch angle θ_d to be followed by the inner-loop controller,

$$\theta_d = K_v v + K_{vs}\dot{v} \tag{5}$$

where K is the feedback gain to be selected later.

The outer-loop horizontal controller calculates the desired throttle position Δt and the desired roll angle command φd for the inner-loop control laws to follow:

$$\begin{bmatrix} \delta_T \\ \phi_d \end{bmatrix} = f(\chi - \chi_L, \begin{bmatrix} f \\ l \end{bmatrix}) \tag{6}$$

where f() is a nonlinear function to be determined next. The design of the horizontal controller is based on a Non-Linear Dynamic Inversion (NLDI) approach, which algebraically transforms the nonlinear system into a linear one (Isidori 1995) (Slotine & Li 1991) (Calise & Rysdyk 1998) so that standard linear control techniques can be applied. The use of nonlinear technique provides an effective way for controlling the aircraft under a wide range of maneuvering conditions.

In this specific problem, to minimize the forward and lateral distance errors f and l, the desired bank angle φ_d and throttle command δt are used as outer-loop control inputs. From equation (3), the second derivatives of f and l can be calculated as:

$$\begin{bmatrix} \ddot{\ell} \\ \ddot{f} \end{bmatrix} = \begin{bmatrix} \sin(\chi - \chi_L) \\ -\cos(\chi - \chi_L) \end{bmatrix} \dot{V}_{xy} + \begin{bmatrix} \cos(\chi - \chi_L) \\ \sin(\chi - \chi_L) \end{bmatrix} V_{xy}(\Omega - \Omega_L) + \begin{bmatrix} f \\ -\ell \end{bmatrix} \dot{\Omega}_L + \begin{bmatrix} \dot{f} \\ -\dot{\ell} \end{bmatrix} \Omega_L \tag{7}$$

To establish the relationship between this equation and control inputs [δ_t, φ_d], let's first look at the aircraft forward translational acceleration equation (Stevens & Lewis 1992):

$$\begin{aligned} \dot{V} &= \frac{1}{m}(Y \sin\beta - D\cos\beta + T\cos\alpha\cos\beta) - g\sin\gamma \\ &= \left(\frac{\cos\alpha\cos\beta}{m}\right)T - \left(\frac{\bar{q}S(C_D\cos\beta - C_Y\sin\beta)}{m} + g\sin\gamma\right) \triangleq \omega_1 T - \omega_2 \end{aligned} \tag{8}$$

, where D, Y, and T are the drag, side force, and thrust,, respectively, m is the aircraft mass, g is the acceleration due to gravity, α, β, and γ are the angle of attack, angle of sideslip, and flight path angle, respectively, with $\cos\gamma = V_{xy}/V$, $\bar{q}$ is the dynamic pressure, S is the wing area, and CD and CY are the dimensionless drag and side force coefficients.

The projection of V onto the level plan is given by (assuming a quasi steady state condition with $\dot{\gamma}$ = 0):

$$\dot{V}_{xy} = \dot{V}\cos\gamma = T\omega_1\cos\gamma - \omega_2\cos\gamma = \frac{V_{xy}}{V}\omega_1 T - \frac{V_{xy}}{V}\omega_2 = \frac{V_{xy}}{V}\omega_1(T_b + K_T\delta_T) - \frac{V_{xy}}{V}\omega_2 \tag{9}$$

, where Tb and KT are constants to be provided by the aircraft propulsion model. Assuming a coordinated turn condition for both the 'leader' and 'follower' aircraft:

$$\Omega = \dot{\chi} \cong \dot{\psi} \cong \frac{g}{V}\tan\phi \tag{10}$$

, where ψ is the aircraft heading angle and φ is the roll angle. Also assuming a steady wings-level or steady turning flight condition for the 'leader':

$$\dot{\Omega}_L = 0 \tag{11}$$

equation (7) becomes

$$\begin{bmatrix}\ddot{\ell}\\ \ddot{f}\end{bmatrix} = \begin{bmatrix} V_{xy}\cos(\chi-\chi_L) & \frac{V_{xy}}{V}\omega_1\sin(\chi-\chi_L) \\ V_{xy}\sin(\chi-\chi_L) & -\frac{V_{xy}}{V}\omega_1\cos(\chi-\chi_L)\end{bmatrix}\begin{bmatrix}\frac{g}{V}\tan(\phi_d)\\ T_b+K_T\delta_T\end{bmatrix} + \frac{V_{xy}}{V}\omega_2\begin{bmatrix}-\sin(\chi-\chi_L)\\ \cos(\chi-\chi_L)\end{bmatrix}$$
$$- \Omega_L V_{xy}\begin{bmatrix}\cos(\chi-\chi_L)\\ \sin(\chi-\chi_L)\end{bmatrix} + \Omega_L\begin{bmatrix}\dot{f}\\ -\dot{\ell}\end{bmatrix} \qquad (12)$$

Since the (2 x 2) matrix relating inputs and second derivatives of the output from (12) is invertible, the resulting inversed relationship is given by:

$$\begin{bmatrix}\frac{g}{V}\tan(\phi_d)\\ T_b+K_T\delta_T\end{bmatrix} = \frac{1}{V_{xy}}\begin{bmatrix}\cos(\chi-\chi_L) & \sin(\chi-\chi_L)\\ \frac{V}{\omega_1}\sin(\chi-\chi_L) & -\frac{V}{\omega_1}\cos(\chi-\chi_L)\end{bmatrix}\begin{bmatrix}\ddot{\ell}_d\\ \ddot{f}_d\end{bmatrix}$$
$$+\begin{bmatrix}\Omega_L\\ \frac{\omega_2}{\omega_1}\end{bmatrix} + \begin{bmatrix}\dot{\ell}\sin(\chi-\chi_L)-\dot{f}\cos(\chi-\chi_L)\\ -\frac{V}{\omega_1}\dot{\ell}\cos(\chi-\chi_L)-\frac{V}{\omega_1}\dot{f}\sin(\chi-\chi_L)\end{bmatrix}\frac{\Omega_L}{V_{xy}} \qquad (13)$$

By imposing α = α0, β = 0, the lateral NLDI control law is:

$$\phi_d = \arctan\left\{\frac{1}{g\cos\gamma}\left[\ddot{\ell}_d\cos(\chi-\chi_L)+\ddot{f}_d\sin(\chi-\chi_L)\right] + \frac{V}{g}\Omega_L + \left[\dot{\ell}\sin(\chi-\chi_L)-\dot{f}\cos(\chi-\chi_L)\right]\frac{\Omega_L}{g\cos\gamma}\right\} \qquad (14)$$

and the forward control law is:

$$\delta_T = \frac{m}{K_T\cos\gamma}\left[\ddot{\ell}_d\sin(\chi-\chi_L)-\ddot{f}_d\cos(\chi-\chi_L)\right] + \frac{1}{K_T}\left(\frac{1}{2}\rho_0 V^2 S(C_{D0}+C_{D\alpha}\alpha_0)+m\sin\gamma-T_b\right) - \frac{m}{K_T\cos\gamma}\Omega_L\left[\dot{\ell}\cos(\chi-\chi_L)+\dot{f}\sin(\chi-\chi_L)\right] \qquad (15)$$

The application of the control inputs (14) and (15) to the system described by (12) cancels the non-linearities, leading to the linear relationship:

$$\begin{bmatrix}\ddot{\ell}\\ \ddot{f}\end{bmatrix} = \begin{bmatrix}\ddot{\ell}_d\\ \ddot{f}_d\end{bmatrix} \qquad (16)$$

, which can be controlled with compensator-type linear control laws:

$$\ddot{\ell}_d = -K_{\ell s}\dot{\ell} - K_\ell \ell$$
$$\ddot{f}_d = -K_{fs}\dot{f} - K_f f \qquad (17)$$

2.3 Inner-Loop Controller

The objective of the inner-loop controller design is to achieve desirable disturbance attenuation and tracking capabilities while maintaining a reasonable stability margin and damping ratio. The longitudinal inner-loop control law tracks the desired pitch angle, as supplied by the outer-loop controller, using the following relationship:

$$i_H = K_q q + K_\theta(\theta - \theta_d) \tag{18}$$

, where iH is the aircraft stabilator deflection, q is the roll rate, and θ is the roll angle.

The lateral-directional inner-loop control laws track a desired bank angle, supplied by the outer-loop controller, while augmenting the lateral-directional stability of the aircraft:

$$\delta_A = K_p p + K_\phi(\phi - \phi_d) \tag{19}$$

$$\delta_R(s) = K_r \frac{s}{s + \omega_0} r \tag{20}$$

where δ_A and δ_R are the aileron and rudder deflections, p and r are the pitch and yaw rates, φ is the pitch angle, and ω_0 is the washout filter constant to be selected.

3. TEST-BED DEVELOPMENT

3.1 Aircraft Platform

A set of three YF-22 research aircraft were designed and developed for the formation flight experiments. Although these aircraft feature similarities with the Lockheed's YF-22 aircraft, they are not dynamically scaled models. Instead, the research team focused on designing aircraft that could provide desirable handling quality and payload capacity. The WVU YF- 22 fleet is shown in Fig. 2. Additional information about the design and manufacturing of the WVU YF-22 research aircraft is available at (Napolitano 2005).

Figure 2. WVU YF-22 Research Aircraft (Formation Flight Fleet) An overview of the aircraft specifications is provided in Table 1.

Table 1. Specifications of the test-bed aircraft

Wingspan	1.96 m
Length	3.05 m with probe
Height	0.61 m
Wing Area	1.37 m2
Weight	23 Kg
Fuel Capability	3.5 L
Maximum Flight Duration	12 minutes
Cruise Airspeed	42 m/s
Takeoff Speed	30 m/s
Engine / Thrust	RAM1000 / 125 N
T/W Ratio (fully fueled)	0.55
W/S Ratio (fully fueled)	16.5 Kg/ m2

During takeoff/landing and part of the flight, the aircraft is under manual control with a 10- channel Pulse Code Modulation (PCM) R/C system. The ground pilot has control of the aircraft primary control surfaces (stabilators, ailerons, and rudders), secondary control surfaces (flaps), engine throttle, brakes, and a 'controller switch' to activate/deactivate the on-board autonomous flight control.

The turbine propulsion system generates up to 125 N of thrust. An Electronic Control Unit (ECU) monitors the exhaust gas temperature and engine compressor pressure, and, in turn, controls the turbine RPM by regulating the fuel supplied to the turbine. The fuel consumption is rated at approximately 0.35 liter/minute for a maximum RPM (127,000) setting. Throughout the flight experiment, with exception of the takeoff phase, the throttle setting is typically within the (½ - ¾) range, with fuel consumption in the range of (0.15-0.3) liter/minute.

3.2 Avionics System

The avionics system is structured as a modulated system to fulfill the demands of several research areas, including formation flight control, fault-tolerant flight control, and vision-based navigation. Each 'follower' aircraft is outfitted with a comprehensive suite of avionics for data collecting, communication, and flight control. It receives the 'leader' position information at a 50Hz update rate through a 900 Hz RF modem. The 'leader' avionics is a stripped down version of the 'follower' avionics with main objectives as data acquisition and communication. Fig. 3 shows the formation configuration and capabilities of the 'leader' and 'follower' avionics systems.

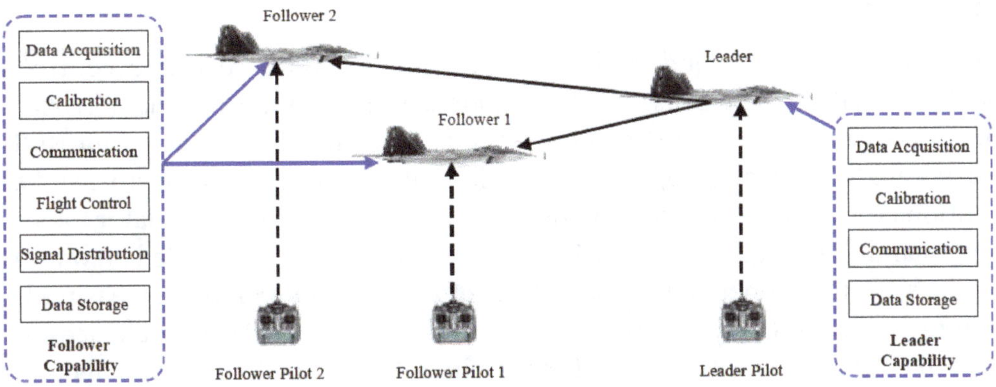

Figure 3. Formation Configuration and Capabilities

A view of the installed 'follower' avionics system is shown in Fig. 4. In general, the avionics receives pilot commands, monitors aircraft states, performs data communication, generates formation control commands, and distributes control signals to primary control surfaces and the propulsion system. A description of major avionics sub-systems is provided next.

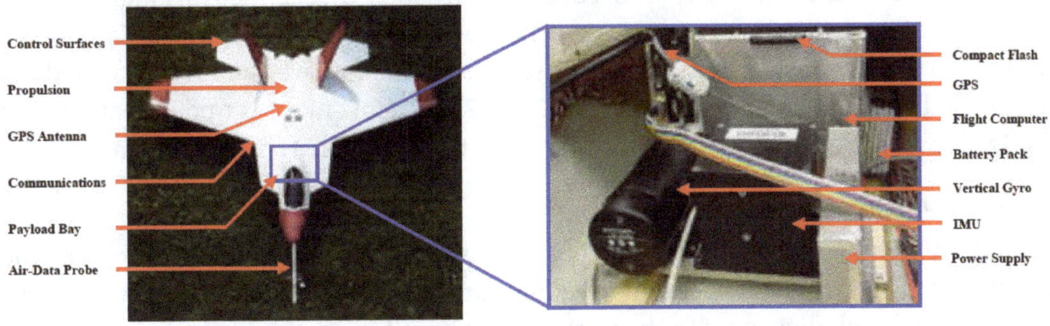

Figure 4. 'Follower' Aircraft and Avionics System

A. Flight Computer

The flight computer is based on a PC-104 format computer stack with a 300Mhz CPU module, a 32-channel 16-bit data acquisition module, a power supply/communication module, and an IDE compact flash adapter. In addition, a set of custom Printed Circuit Board (PCB) was developed for interfacing sensor components, generating Pulse-Width Modulation (PWM) control signals, and distributing signals to each control actuator. The PC-104 format is selected because of its compact size and expandability. An 8 MB compact flash card stores the operation system, the flight control software, and the collected flight data. A 14.8v 3300mAh Li-Poly battery pack can power the avionics system for more than an hour, providing sufficient ground testing and flight mission time.

B. Sensor Suite

Flight data is collected and calibrated on-board for both real-time control and post-flight analysis. The sensor suite include a SpaceAge mini air-data probe, two SenSym pressure sensors, a Crossbow IMU400 Inertial Measurement Unit (IMU), a Goodrich VG34 vertical gyro, a Novatel OEM4 GPS receiver, a thermistor, and eight potentiometers measuring primary control surfaces deflections (stabilators, ailerons, rudders) and flow angles (α, β). A digital video camera is also installed on one of the 'followers' for flight documentation. A total of 22 analog channels are measured with a 16-bit resolution. The sampling rate was initially set at 100 Hz for data acquisition flights and later reduced to 50 Hz for matching the control command update rate (limited by the R/C system). Consider the aircraft short period mode of 7.7 rad/sec (1.2 Hz), a 50 Hz sampling rate provides a substantial amount of oversampling.

The analog signals measured on-board include absolute pressure (0-103.5 kPa), dynamic pressure (0-6.9 kPa), angle of attack (±25º), sideslip angle (±25º), air temperature (-10-70ºC), roll angle (±90º), pitch angle (±60º), 3-axis accelerations (±10g), 3-axis angular rates (±200º/sec), 6-channel primary control surfaces deflections (±15º), and several avionics health indicators. A GPS receiver provides direct measurements of the aircraft 3-axis position and velocity with respect to an Earth-Centered-Earth-Fixed (ECEF) Cartesian coordinate system. These measurements are then transformed into a LTP used by the formation controller. The GPS measurement is updated at a rate of 20 Hz, providing a substantial advantage over the low-cost 1Hz GPS system.

C. Control Signal Distribution System

A Control Signal Distribution System (CSDS) is designed to give the 'follower' pilot the freedom to switch between manual and autonomous modes at any time during the flight. A block diagram for the CSDS is shown in Fig. 5.

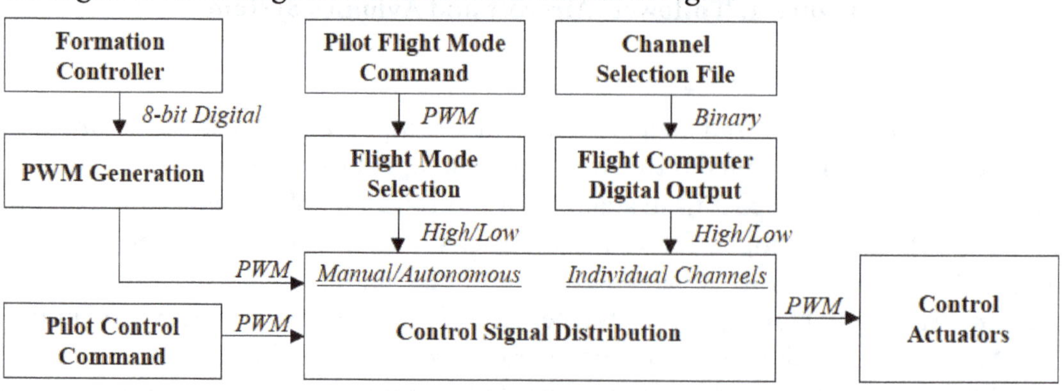

Figure 5. Control Signal Distribution System

In autonomous mode, the flight computer may operate all or a subset of six control channels: the left stabilator, right stabilator, left aileron, right aileron, dual rudders, and engine throttle. Two switching mechanisms are implemented to guarantee the safety of the aircraft: 'Hardware Switching' and 'Software Switching.' 'Hardware Switching' enables the pilot to revert to manual control immediately under any condition. In the event of avionics power failure, manual control is automatically activated. 'Software Switching' provides the flight computer with the capability to manage any combination of the aircraft's major control surfaces and propulsion through pre-programmed options. Software Switching is executed by a combination of hardware and software components. The on-board software retrieves pre-defined channel selection data from a log file during the initialization phase of execution. Upon activation of the 'controller switch,' the program transmits the channel selection signal via the digital output port of the data acquisition card. The signal is thereafter transmitted to a controller board to determine the pilot or on-board control. This capability allows for the independent testing of various components inside the flight control system. This subsequently enhances the flexibility and elevates the safety of the flight-testing activity.

D. Electro-Magnetic Interference

Electromagnetic Interference (EMI) can present substantial risks to airplane safety. This is particularly applicable to small UAVs, where several electronic components are restricted within a constrained area. The most susceptible component of the avionics system is frequently the R/C link between the ground pilot and the aircraft, which directly influences the safety of both the aircraft and the ground personnel. The proximity to several interference sources, including the CPU, vertical gyro, RF modem, and any connecting cables functioning as antennas, can significantly diminish the range of the R/C system. Given that prevention is recognized as the most effective method against EMI, meticulous attention is applied to the selection of 'commercial-off-the-shelf' products, as well as the design and installation of customized components. Low pass filters are specifically engineered for the power system; all power and signal lines are insulated and adequately grounded; and aluminum enclosures are constructed and sealed with copper or aluminum tape to protect the hardware components. Upon integration of the avionics equipment into the airframe, ferrite chokes are affixed to certain cables according to the noise levels assessed using a spectrum analyzer. Despite the significance of comprehensive laboratory EMI testing, the erratic nature of EMI issues necessitates adherence to stringent R/C ground range testing protocols prior to each takeoff to guarantee aircraft operation safety.

4. MODELING AND PARAMETER IDENTIFICATION

The existence of a precise mathematical model of the test-bed is essential for choosing formation control parameters and creating a high-fidelity simulation environment. The modeling approach mostly relies on empirical data obtained from ground tests and flight-testing experiments.

4.1 Identification of the Aircraft Linear Mathematical Model

The decoupled linear aircraft model is determined through a Parameter IDentification (PID) effort. A series of initial test flights are performed to collect data used for the identification process. Typical pilot-injected maneuvers, including stabilator doublets, aileron doublets, rudder doublets, and aileron/rudder doublets, are performed with various magnitudes to excite the aircraft longitudinal and lateral-directional dynamics. Fig. 7 represents a typical aileron/rudder doublets maneuver, where a rudder doublet is performed immediately after an aileron doublet.

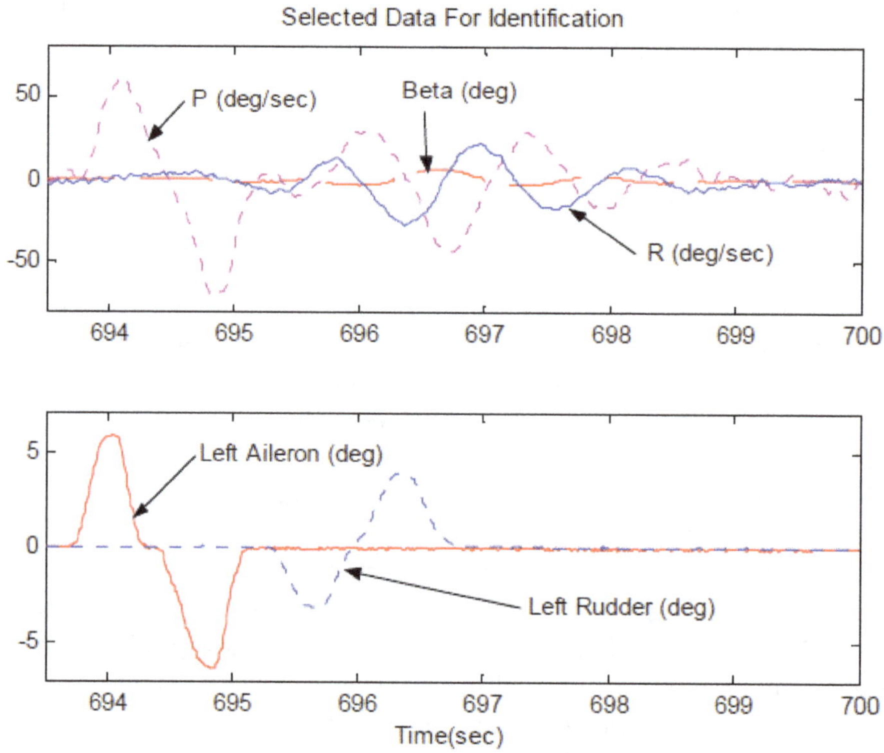

Figure 6. Flight Data for Linear Model Identification

The identification of the linear model is performed using a 3-step process. First, after a detailed examination of the flight data, two data segments with the best quality for each class of maneuvers are selected. Next, a subspace-based identification method (Ljung 1999) is used to perform the parameter identification with one set of data. Finally, the identified linear model is validated through comparing the simulated aircraft response with the remaining unused data set. This identification process is repeated until a satisfactory agreement is achieved. Following the identification study, the estimated linear longitudinal and lateral-directional aerodynamic model in continuous time are found to be:

$$\begin{bmatrix} \dot{V}_T \\ \dot{\alpha} \\ \dot{q} \\ \dot{\theta} \end{bmatrix} = \begin{bmatrix} -0.284 & -23.096 & 0 & -0.171 \\ 0 & -4.117 & 0.778 & 0 \\ 0 & -33.884 & -3.573 & 0 \\ 0 & 0 & 1 & 0 \end{bmatrix} \begin{bmatrix} V_T \\ \alpha \\ q \\ \theta \end{bmatrix} + \begin{bmatrix} 20.168 \\ 0.544 \\ -39.085 \\ 0 \end{bmatrix} i_H \tag{21}$$

$$\begin{bmatrix} \dot{\beta} \\ \dot{p} \\ \dot{r} \\ \dot{\phi} \end{bmatrix} = \begin{bmatrix} 0.430 & 0.094 & -1.030 & 0.237 \\ -67.334 & -7.949 & 5.640 & 0 \\ 20.533 & -0.655 & -1.996 & 0 \\ 0 & 1 & 0 & 0 \end{bmatrix} \begin{bmatrix} \beta \\ p \\ r \\ \phi \end{bmatrix} + \begin{bmatrix} 0.272 & -0.771 \\ -101.845 & 33.474 \\ -6.261 & -24.363 \\ 0 & 0 \end{bmatrix} \begin{bmatrix} \delta_A \\ \delta_R \end{bmatrix} \tag{22}$$

, where V_T is the true airspeed. This model represents the aircraft in a steady and level flight at V_T= 42 m/s, H= 310 m above the sea level, at trimmed condition with α = 3 deg, with inputs i_H = -1°, $\delta_A = \delta_R$ =0° and a thrust force along the x body axis of the aircraft T = 54.62 N. The decoupled linear model is used later for the formation controller design.

4.2 Identification of the Non-Linear Mathematical Model

A more detailed non-linear mathematical model is identified for the development of a formation flight simulator. The identification process for a non-linear dynamic system relies on detailed knowledge of the system dynamics along with the application of minimization algorithms (Maine & Iliff 1986). In general, the non-linear model of an aircraft system can be described using the following general form (Stevens & Lewis 1992), (Roskam 1995):

$$\begin{aligned} \dot{x} &= f(x,\delta,G,F_A(x,\delta),M_A(x,\delta)); \\ y &= g(x,\delta,G,F_A(x,\delta),M_A(x,\delta)); \end{aligned} \tag{23}$$

where x is the state vector, y is the output vector, δ is the input vector, G is a vector of geometric parameters and inertia coefficients, and F_A and M_A are aerodynamic forces and moments acting on the aircraft. The functions f and g are known as analytic functions modeling the dynamics of a rigid-body system. The aerodynamic forces and moments are expressed using the aerodynamic coefficients (Roskam 1995), including drag coefficient C_D, side force coefficient C_Y, lift coefficient C_L, rolling moment coefficient Cl, pitching moment coefficient C_m, and yawing moment coefficient C_n:

$$F_A = \bar{q}S \begin{bmatrix} C_D(x,\delta) \\ C_Y(x,\delta) \\ C_L(x,\delta) \end{bmatrix}, \quad M_A = \bar{q}S \begin{bmatrix} bC_l(x,\delta) \\ \bar{c}C_m(x,\delta) \\ bC_n(x,\delta) \end{bmatrix} \tag{24}$$

The moments of inertia of the aircraft are experimentally evaluated with a 'swing pendulum' experimental set-up (Soule & Miller 1934), as shown in Fig. 7

Figure 7 Experimental Setup for Measuring Aircraft Moments of Inertia

The product of inertia Ixz could not be evaluated using the pendulum-based method. Thus, the remaining issue is to determine Ixz along with the values of the aerodynamic derivatives of the aircraft. The relationship from the coefficients of the linear models (21) and (22) to the values of the aerodynamic derivatives and geometric-inertial parameters are known (Stevens & Lewis 1992). After inverting these relationships and using the experimental values of the geometric and inertial parameters, initial values for each of the aerodynamic stability derivatives are calculated. A Sequential Quadratic Programming (SQP) technique (Hock & Schittowski 1983) is then used to iteratively minimize the Root Mean Square (RMS) of the difference between the actual and simulated aircraft outputs [Campa et al. 2007]. The resulting non-linear mathematical model is given by:

Geometric and inertial:

$\overline{c}$ = 0.76 m, b = 1.96 m, S = 1.37 m^2
I_{xx} = 1.61 Kg m^2, I_{yy} = 7.51 Kg m^2, I_{zz} = 7.18 Kg m^2, I_{xz} = -0.24 Kg m^2
M = 20.64 Kg, T = 54.62 N

Longitudinal aerodynamic derivatives:

$C_{D0} = 0.0085$, $C_{D\alpha} = 0.5079$, $C_{Dq} = 0.0000$, $C_{DiH} = -0.0339$
$C_{L0} = -0.0492$, $C_{L\alpha} = 3.2580$, $C_{Lq} = -0.0006$, $C_{LiH} = 0.1898$
$C_{m0} = 0.0226$, $C_{m\alpha} = -0.4739$, $C_{mq} = -3.4490$, $C_{miH} = -0.3644$

Lateral-Directional aerodynamic derivatives:

C_{Y0} = 0.0156, $C_{Y\beta}$ = 0.2725, C_{Yp} = 1.2151,
C_{Yr} = -1.1618, $C_{Y\delta A}$ = 0.1836, $C_{Y\delta R}$ = -0.4592
C_{l0} = -0.0011, $C_{l\beta}$ = -0.0380, C_{lp} = -0.2134,
C_{lr} = 0.1147, $C_{l\delta A}$ = -0.0559, $C_{l\delta R}$ = 0.0141
C_{n0} = -0.0006, $C_{n\beta}$ = 0.0361, C_{np} = -0.1513,
C_{nr} = -0.1958, $C_{n\delta A}$ = -0.0358, $C_{n\delta R}$ = -0.0555

where c is the mean aerodynamic chord, b is the wing span, S is the wing area, and m is the aircraft mass with a 60% fuel capacity.

A final validation of the non-linear model is then conducted using the validation flight data set, as it was performed for the linear mathematical model. Figure 8 shows a substantial agreement between the measured and the simulated data with the non-linear model.

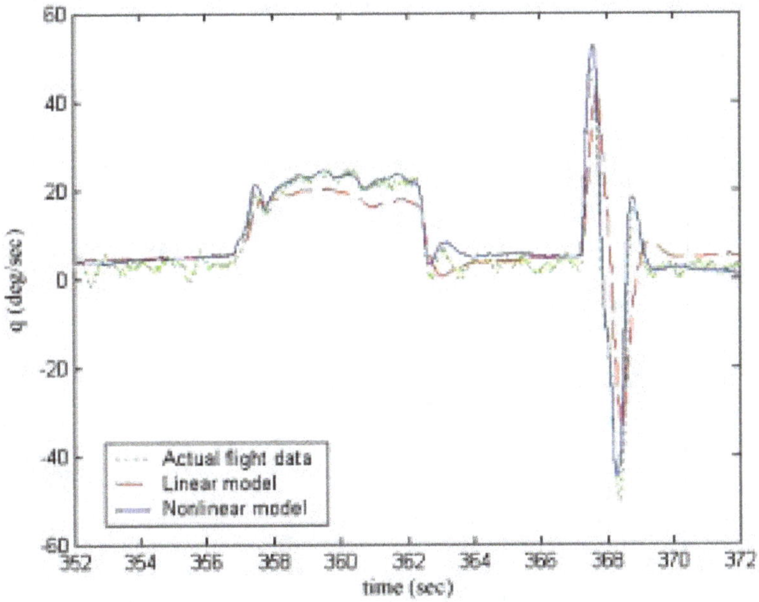

Figure 8. Linear and Non-linear Model Simulations Compared to Actual Flight Data

4.3 Engine and Actuator Models

The engine mathematical model is defined as the transfer function from the throttle command to the actual engine thrust output. The evaluation of this model is important as the jet propulsion system has a substantially lower bandwidth compared with rest of the control system. Fig. 9 provides a photo and a schematic drawing of the experimental set-up used for the identification process.

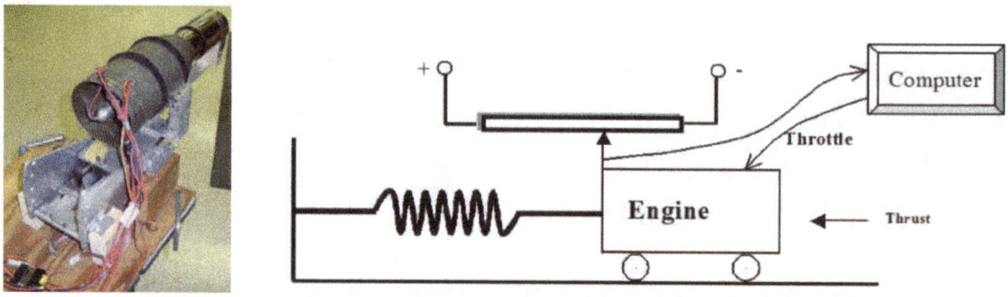

Figure 9. Engine Ground Test Setup and Schematic

The turbine is mounted on a customized engine test stand where the motion is limited to be only along the thrust force (x) direction. The thrust is then measured by reading the displacement of a linear potentiometer. The throttle control is based on 8-bit PWM signal generated by the computer with a throttle range between 0 and 255. During the test, a sequence of throttle commands is sent to the turbine and the corresponding thrust is measured with the data acquisition system, as shown in Fig. 10.

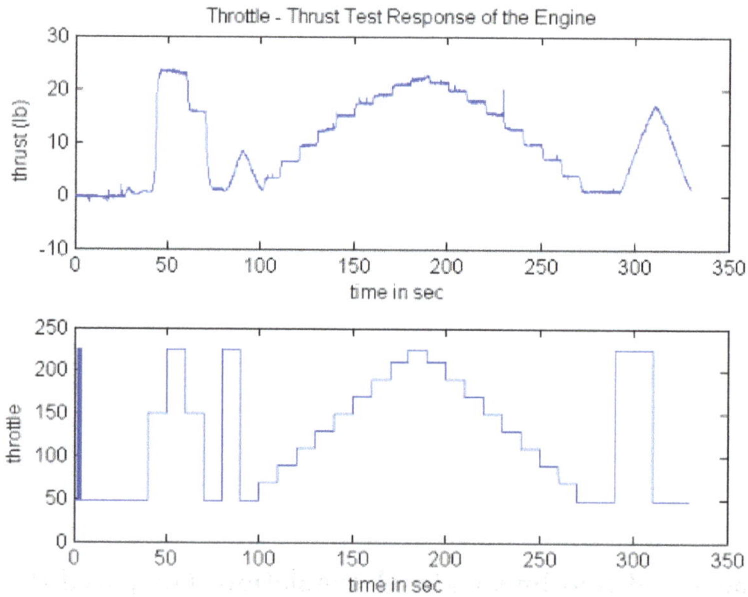

Figure 10. Throttle Thrust Response in Test Time Sequence

The first step of the engine model identification is to identify the static gain of the engine response from the throttle position to the thrust output. For simplicity purposes, a linear fitting is used. The linearized input-output relationship under steady-state condition is found to be:

$$T\,(N) = T_b + K_T\delta_T \tag{25}$$

, with $K_T = 0.624$ and T_b – =25.86.

To quantify the transient response of the engine dynamics, a standard prediction error method is applied to selected data segments where the throttle input consists of a series of step-like signals, as shown in Fig. 10. The identification result shows that the engine dynamic response can be approximated with a 1st order system and a pure time delay:

$$G_T(s) = \frac{T(s) - T_b}{\delta_T(s)} = \frac{K_T}{1 + \tau_T s} e^{-\tau_d s} \tag{26}$$

, with $\tau_T = 0.25$ sec, and $\tau_d = 0.26$ sec . As indicated by the values of τ_T and τ_d, the low bandwidth of the turbine propulsion system poses a fundamental limitation of the achievable formation flight performance under maneuvered flight conditions.

Digital R/C servos are used as actuators for the aircraft primary control surfaces. The actuator dynamics is defined as the transfer function from the 8-bit digital command to the actuator's actual position. During ground experiments, a set of step inputs is sent to the actuator. Both the control command and aircraft surface deflection are then recorded. The procedure is repeated for all six actuators on each of the primary control surfaces. From data analysis it is found that the actuator model could be approximated by the following transfer function:

$$G_{Act}(s) = \frac{1}{1 + \tau_a s} e^{-\tau_{ad} s} \tag{27}$$

where the actuator time constant τa and the time delay constant τ_{ad} were identified to be-0.04 sec and 0.02 sec respectively.

5. CONTROLLER IMPLEMENTATIONS AND SIMULATION

5.1 Controller Parameters

Once a complete set of aircraft mathematical model is available, controller parameters are designed based on the classic root-locus method. The time delays in the engine model (26) and actuator model (27) are replaced by 1st order Pade approximations to facilitate the controller design. The final selections of controller parameters are listed in Table 2.

Table 2. Formation Controller Parameters

Inner-Loop Controller			Outer-Loop Controller		
Longitudinal	Lateral	Directional	Forward	Lateral	Vertical
Kq = 0.12	Kp = 0.04	Kr = 0.16	Kf = 0.24	K_ℓ = 0.20	Kv = 3.23
Kθ = 0.50	Kφ = 0.35	ω0 = 1.80	Kfs = 2.06	$K_{\ell s}$ = 0.89	Kvs = 1.76

5.2 Simulation Environment

A Simulink®-based formation flight simulation environment is developed using the mathematical model and the formation control laws described in previous sections. This environment provides a platform for validating and refining the formation control laws prior to performing the actual flight tests. The simulation schemes are interfaced with the Matlab® Virtual Reality Toolbox (VRT), where objects and events of a virtual world can be driven by signals from the simulation. The collected flight data can also be played back 'side-by-side' with the simulated aircraft response. This provides an important tool for validating the accuracy of the identified nonlinear aircraft model. In addition, the ability for VRT to visualize the entire formation flight operation, especially with the freedom of selecting different viewpoints, provides a substantial amount of intuition during the controller design and flight planning process. Figure 11 shows the formation flight simulation environment.

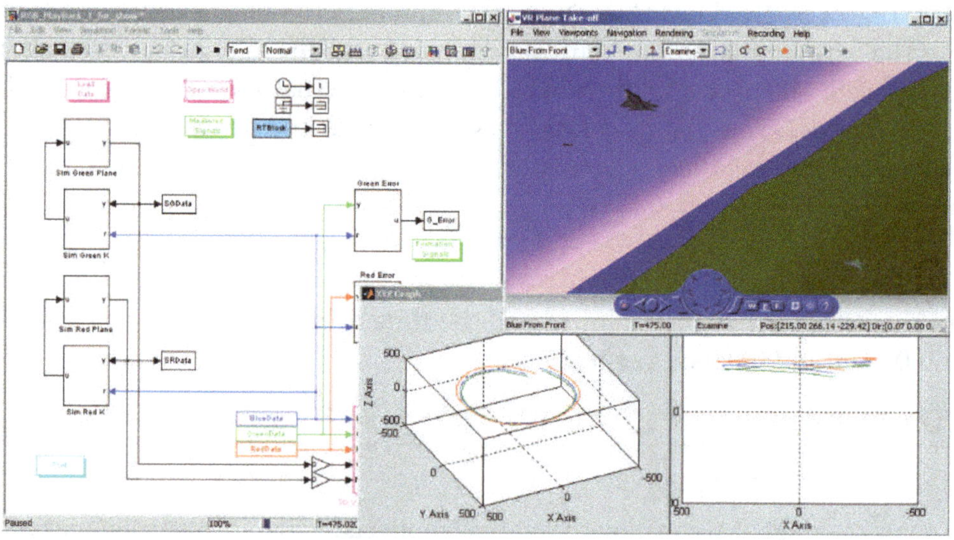

Figure 11. Formation Flight Simulation Environment

5.3 Robustness Assessment

The resilience of the formation controller is examined using a Monte Carlo method, conducting a series of simulation simulations to assess the deterioration of closed-loop stability and tracking performance due to measurement noise and modeling error. The subsequent two formation configurations are examined:

$$\text{Configuration \#1:} \quad l_c = -20m, \quad f_c = 20m, \quad v_c = 20m \tag{28}$$

$$\text{Configuration \#2:} \quad l_c = 20m, \quad f_c = 20m, \quad v_c = -20m \tag{29}$$

These two configurations are later used during the flight-testing program. To simulate the effects of the measurement error/noise, a set of random noise is applied on all inputs of the formation controller. Specifically, random values following Gaussian distributions with zero means are added to the simulation parameters using the following standard deviation values:

- 2 deg/sec for angular rates (p, q, r);
- 2 deg for Euler angles (θ, φ);
- 4 m for horizontal position components (x,y);
- 8 m for vertical position component (z);
- 2 m/sec for horizontal velocity components (Vx, Vy);
- 4 m/s for vertical velocity component (Vz).

These values are substantially higher than the typical measurement noise observed in the actual flight data. Simulation studies reveal that the average tracking error increased by 6% and 20% for configurations #1 and #2, respectively, compared to the ideal conditions without measurement error.

An assessment of the closed-loop stability with the existence of multiplicative modeling uncertainties is also performed with a 2-step process. First, a ±10% variation is applied on each of the 30 longitudinal and lateral-directional aerodynamic derivatives, one at a time. Simulation studies show no unstable conditions for all configurations with the three most sensitive coefficients found to be $C_{D\alpha}$, $C_{L\alpha}$, and $C_{m\alpha}$. The second step is to vary the value of these three coefficients along with seven additional parameters by ±5% and perform a simulation for each possible combination. The selected parameters are C_{D0}, C_{miH}, C_{Y0}, $C_{l\beta}$, C_{lp}, $C_{l\delta A}$, $C_{n\beta}$, $C_{n\delta R}$. Therefore, a batch set of simulations (2048 total) by varying combinations of parameter changes are performed using both formation configuration #1 and #2. Again, no unstable conditions are observed in this analysis. The worst-case degradation of tracking performance is found to be 1.98 m for the lateral distance error, 1.05 m for the forward distance error, and 4.41 m for the vertical distance error. Overall, the simulation result indicates that the designed formation controller has adequate robustness characteristics with respect to modeling errors.

5.4 On-Board Software

The formation controller software module, once validated through simulation studies, is integrated with other software components to perform real-time data acquisition, communication, and control. The on-board software is implemented as a Simulink scheme with each component written in C-language as a Matlab 'S-function'. An executable file is compiled using Matlab Real-Time Workshop (RTW) as a real-time extended DOS target for flight test experiments. The modulated software design provides flexibility for quick on-site reconfigurations to meet various flight-testing objectives.

The main tasks for 'leader' on-board software is to perform data acquisition, communication, and data storage. The 'follower' software also executes the formation control laws, selects the operational mode of the aircraft, decides primary control channels to be controlled on-board, and calibrates the flight control commands. Figure 12 shows a sample of the 'follower' on-board software scheme.

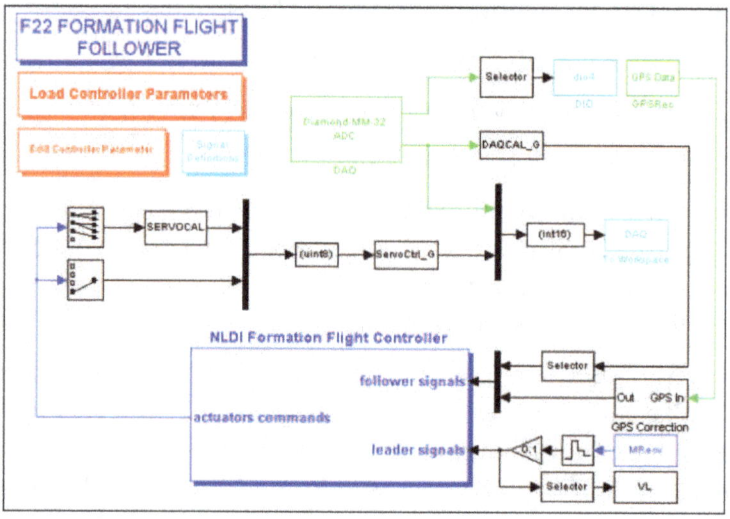

Figure12. Simulink Diagram of the 'Follower' Aircraft Software

6. FLIGHT-TESTING OF FORMATION CONTROL LAWS

Flight testing is the most authentic phase for controller validation and is, by far, the most perilous. Although the utilization of UAVs significantly mitigates risks linked to control system validation, meticulous preparation is crucial for assuring safe operations and identifying potential issues. This section offers a summary of several flight-testing phases and the results of the autonomous formation flight studies.

6.1 Flight Testing Phases

The flight-testing program is divided into six major phases with increasing complexity and risk level:

A. Flight for Assessment of Handling Qualities

This initial phase is for evaluating the handling qualities and dynamic characteristics of the test-bed aircraft. After a few satisfactory test flights, 'artificial' payloads of incremental weight are installed to test the structural integrity and the handling qualities under a full payload configuration.

B. Data Acquisition Flights

The avionics equipment has been installed, and flight data has been gathered for PID analysis. A series of specialized PID maneuvers is executed across several flights to stimulate the aircraft dynamics. Common PID maneuvers encompass stabilator doublets, aileron-rudder doublets, and various engine throttle inputs.

C. Inner-Loop Controller Validation

The stability and tracking performance of the designed inner-loop linear controller is validated during this phase. Both the longitudinal and lateral-directional inner-loop control laws are tested. The flight control hardware is also validated during this phase.

D. Outer-Loop Controller Sub-System Validation

Individual sub-systems of the outer-loop controller are tested. Experiments are performed to test the altitude-hold, heading-hold, and velocity-hold control and their combinations. Sample flight data in Fig. 13 shows the result from a heading-hold control experiment.

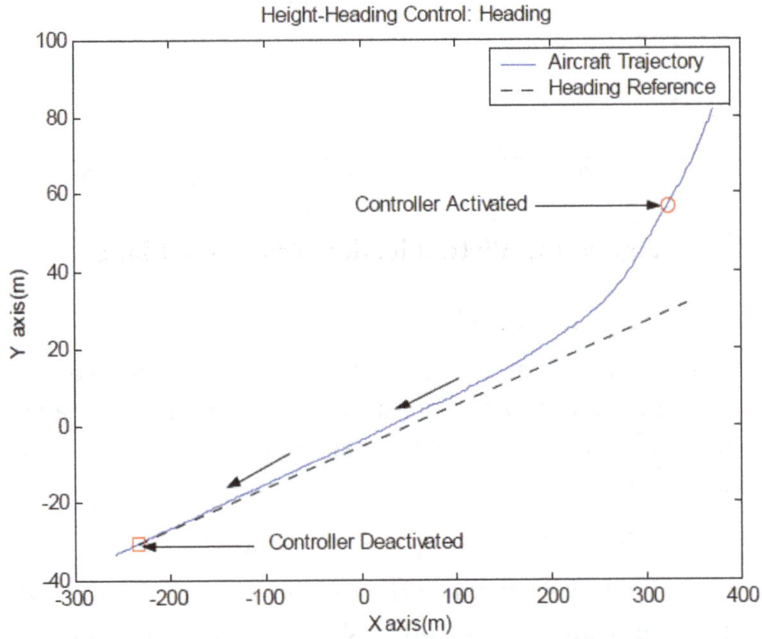

Figure 13. Heading Control Experiment

E. 'Virtual Leader' Flights

A 'Virtual Leader' (VL) approach is implemented as an alternative method for testing the formation controller without the risk and logistic issues associated with a full-blown multiple-aircraft experiment. The VL experiment consists of a single aircraft tracking a previously recorded flight trajectory. This trajectory is initially stored on-board the 'follower' aircraft and later moved to a ground station to test the performance of the communication link. The VL flights are proven to be invaluable for the validation and fine- tuning of formation control laws. A total of 12 VL flights are performed using various formation parameters. Fig. 14 is a sample flight data demonstrating the ability for the formation controller to reduce a large initial error and maintain the formation flight.

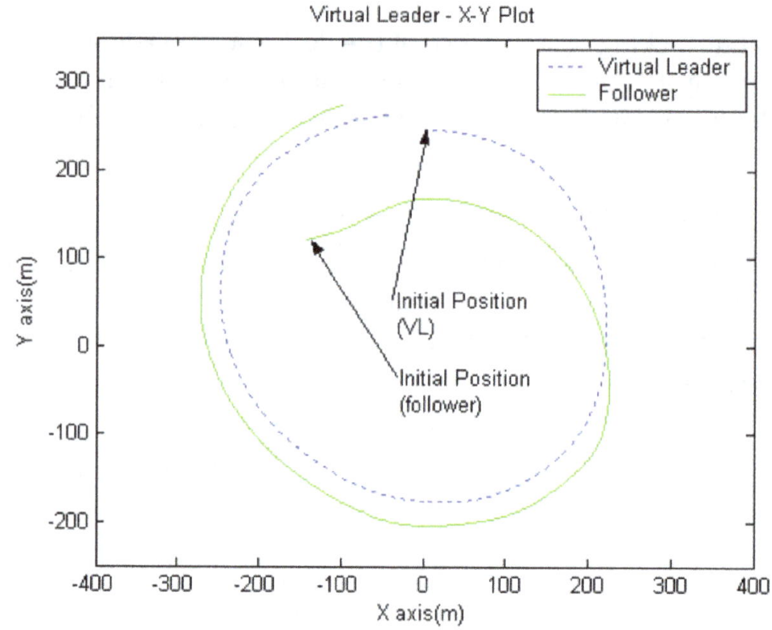

Figure 14. 'Virtual leader' Test - X-Y Plane

F. Multiple Aircraft Formation Flights

After various formation geometries and initial conditions are explored with the VL experiments, the flight-testing program proceeds to the multiple aircraft testing. A total of four 2-aircraft formation flight experiments are performed along with one 3-aircraft formation demonstration.

6.2 Three-Aircraft Formation Flight Experiment

The procedure for the 3-aircraft formation experiment is the following. The 'blue' aircraft, acting as the 'leader', takes off first while the 'red' aircraft ('follower #1') takes off approximately 35 seconds later. After the 'red' aircraft reaches a pre-defined 'rendezvous' area behind the 'leader', the ground pilot engages the on-board formation control. Once the 2-aircraft formation is stabilized for approximately 50 seconds, the 'green' aircraft ('follower #2') takes off and approaches a 'rendezvous' area behind the 2-aircraft already in formation. After the 'green' pilot engages the autonomous control, the trajectory of the 3-aircraft formation is solely controlled by the 'leader' R/C pilot. Fig. 15 shows a ground photo of the 3-aircraft formation experiment.

Figure 15. 3-Aircraft Formation Flight Test

The pre-selected formation geometries include configuration #1 (Equation 28) for the 'red' aircraft and configuration #2 (Equation 29) for the 'green' aircraft. Fig. 16 represents a 40- second portion of flight trajectory during the formation flight. The 3-aircraft formation configuration is engaged for approximately 275 seconds. Fig. 17 shows the aircraft altitude during the formation flight.

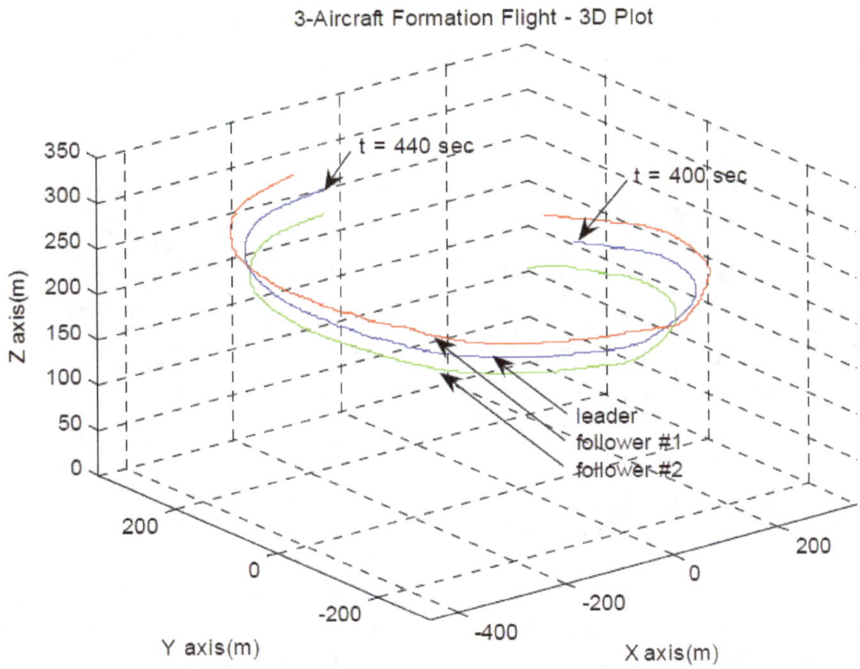

Figure 16. 3- Aircraft Formation Test - 3D Trajectory (Blue='leader', Red=Outside 'follower', Green=Inside 'follower')

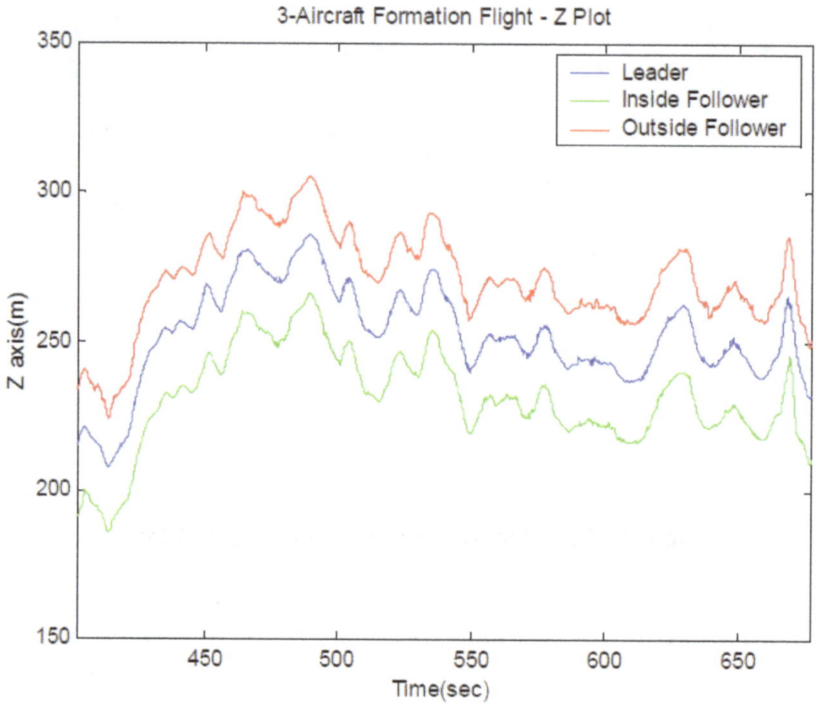

Figure 17. 3-Aircraft Formation Test – Altitude

The mean and standard deviation of the steady state tracking error for the flight test are shown in Table 3. The simulation results calculated with the same 'leader' trajectory are also supplied for comparison purposes.

Table 3. 3-Aircraft Formation Test – Error Analysis

3-aircraft formation Experiment		fc (m)	lc (m)	vc (m) Mean	Forward Distance Error (m)		Lateral Distance Error (m)		Vertical Distance Error (m)	
					Std. Dev	Mean	Std. Dev	Mean	Std. Dev	
Green Aircraft	Flight Data	20	-20	20	-2.49	2.46	13.24	3.45	1.15	1.04
	Simulation	20	-20	20	-3.59	2.50	14.31	3.30	1.34	0.71
Red Aircraft	Flight Data	20	20	-20	27.28	3.73	-2.59	2.29	1.15	0.95
	Simulation	20	20	-20	25.30	3.82	-0.46	1.98	1.19	0.66

The experiment including a three-plane formation confirms the efficacy of the formation controller, test-bed aircraft, and onboard avionics system design. The statistical analysis shows that the 'outside' aircraft - 'follower #2' - achieves desirable lateral tracking performance but with a larger forward tracking error. Conversely, the 'inside' aircraft exhibits favorable forward tracking and a marginally diminished lateral tracking capability. Both 'follower' aircraft have exceptional tracking capability in the vertical channel. The standard deviation of all tracking errors is relatively low, with a maximum value of 3.73 m, indicating a smooth trajectory following performance. A significant correlation between the simulation results and actual flight data is seen, demonstrating an accurate nonlinear mathematical model of the aircraft.

7. CONCISE SUMMARY

This chapter encapsulates the findings of an initiative aimed at demonstrating closed-loop formation flight with research UAVs. A 'leader-follower' method is employed in the design of the formation controller. A dual time-scale methodology is employed, featuring a nonlinear outer-loop and a linear inner-loop controller. The flight-testing program was executed across three flight seasons from 2002 to 2004, encompassing roughly 100 flight sessions. The successive flight-testing phases validate the overall design of the formation control laws and the performance of the testbed aircraft and avionics components. The implementation of a 'virtual leader' methodology is an invaluable and secure method for the preliminary evaluation of formation control laws. In the concluding flight sessions, five formation flight experiments are successfully conducted, comprising four 2-aircraft formations and one 3-aircraft formation. Flight data verifies the effective performance of the established 'leader-follower' formation control laws.

REFERENCES

Weimerskirch, H.; Martin, J.; Clerquin, Y.; Alexandre, P. & Jiraskova, S. (2001). Energy savings in flight formation. Nature. Vol. 413, pp. 697-698.

Scharf, D. P.; Hadaegh, F. Y. & Ploen, S. R. (2004). A survey of spacecraft formation flying guidance and control (part II): control. American Control Conference, Boston, MA, June 2004, pp. 2976–2985.

Dargan, J.L.; Patcher, M. & D'Azzo, J.J. (1992). Automatic formation flight control. Proc. AIAA Guidance, Navigation and Control Conf.. Hilton Head, SC, Aug. 1992, pp. 838- 857.

Buzogany, L.E.; Patcher, M. & D'Azzo, J.J. (1993). Automated control of aircraft in formation flight. Proc. AIAA Guidance, Navigation and Control Conf.. Monterey, CA, Aug. 1993, pp. 1349-1370.

Veth, M.; Pachter, M. & D'Azzo, J.J. (1995). Energy preserving formation flight control. AIAA-1995-335, Aerospace Sciences Meeting and Exhibit, 33rd, Reno, NV, Jan 9-12, 1995

Reyna, V.P.; Pachter, M. & D'Azzo, J.J. (1994). Formation Flight. Control Automation. Proc. AIAA Guidance, Navigation, and Control Conference, pp 1379-1404,

McCammish, S.; Pachter, M.; D'Azzo, J. J. & Reyna, V. (1996) Optimal Formation Flight Control. AIAA Guidance, Navigation, and Control Conference, San Diego, CA, Jul. 1996.

Dogan, A.; Sato, S.; & Blake W. (2005). Flight Control and Simulation for Aerial Refueling. AIAA paper 2005-6264, Proc. AIAA Guidance, Navigation, and Control Conference, San Francisco, CA, Aug 15-18, 2005

Boskovic, J.D. & Mehra, R.K. (2003). An adaptive reconfigurable formation flight control design, Proc. American Control Conference. June 2003 Vol 1, pp. 284-289

Li, Y.; Li, B.; Sun, Z. & Song, Y.D. (2005). Fuzzy technique based close formation flight control. Industrial Electronics Society, 2005. IECON 2005. 32nd Annual Conference of IEEE, Nov. 2005

Li, B.; Liao, X.H.; Sun, Z.; Li, Y.H. & Song, Y.D. (2006). Robust Autopilot for Close Formation Flight of Multi-UAVs, System Theory. Proceeding of the Thrity-Eighth Southeastern Symposium, March 2006, pp. 258- 262

Singh, S.N.; Chandler, P.; Schumacher, C.; Banda, S. & Pachter M. (2000). Adaptive feedback linearizing nonlinear close formation control of UAVs, Proc. American Control Conference, 2000. vol.2, pp.854-858

Venkataramanan, S. & Dogan, A. (2003). Nonlinear Control for Reconfiguration of UAV Formation. AIAA paper 2003-5725, Proc. AIAA Guidance, Navigation, and Control Conference, Austin, TX, Aug. 2003

Schumacher, C.J. & Singh, S.N. (2000). Nonlinear Control of Multiple UAVs in Close-Coupled Formation Flight. AIAA Paper 2000-4373, Aug. 2000

Allen, M.J.; Ryan, J.; Hanson, C.E. & Parle, J.F. (2002). String Stability of a Linear Formation Flight Control System. AIAA Paper 2002-4756, Aug. 2002

Giulietti, F.; Pollini, L. & Innocenti, M. (2000). Autonomous formation flight. IEEE Control Systems Magazine, Vol. 20, No. 6, pp. 34-44, Dec. 2000

Gingras, R.D. (1999). Experimental Investigation of a Multi-Aircraft Formation, AIAA. Paper- 99-3143, 1999

Fowler, J. M. & D'Andrea, R. (2003). A formation flight experiment: Constructing a test-bed for research in control of interconnected systems. Control Systems Magazine, 23(5) pp.35-43, 2003

Kutay, A.T.; Fowler, J.M.; Calise, A.J. & D'Andrea, R. (2005). Distributed Adaptive Output Feedback Control Design and Application to a Formation Flight Experiment. AIAA Guidance, Navigation and Control Conference, August 2005

Napolitano, M.R. (2005). Development of formation flight control algorithms using 3 YF-22 flying models. AFOSR Report A994434, Apr. 2005, Available: (http://www.stormingmedia.us/99/9944/A994434.html)

Gu, Y.; Seanor, B.; Campa, G.; Napolitano, M.R.; Rowe, L.; Gururajan, S.; Perhinschi, M.G. & Wan, S. (2006). Design and Flight Testing Evaluation of Formation Control Laws. IEEE Transactions on Control Systems Technology, November 2006

Lavretsky, E. (2002). F/A-18 autonomous formation flight control system design. AIAA Guidance, Navigation and Control Conference, AIAA Paper 2002-4757, Monterey, CA, Aug. 2002.

Hanson, C.E.; Ryan, J.; Allen, M.J. & Jacobson, S.R. (2002). An Overview of Flight Test Results for a Formation Flight Autopilot. AIAA Paper 2002-4755, Aug. 2002

How, J. P.; King, E. & Kuwata, Y. (2004). Flight Demonstrations of Cooperative Control for UAV Teams. Proc. AIAA 3rd Unmanned Unlimited Technical Conference, Workshop and Exhibit, Sept. 2004. AIAA-2004-6490

Johnson, E.N.; Calise, A.J.; Sattigeri, R. & Watanabe, Y. (2004) Approaches to Vision-Based Formation Control. Proc. IEEE Conference on Decision and Control, December 2004.

Isidori, A. (1995). Nonlinear Control Systems, Springer-Verlag, London, Third edition, 1995.

Slotine, J.-J. E. & Li, W. (1991). Applied Nonlinear Control, Prentice Hall, New Jersey, 1991.

Calise, A.J. & Rysdyk, R.T. (1998). Nonlinear adaptive flight control using neural networks.

IEEE Control Systems Magazine, Vol. 18, No. 6, pp. 14-25, Dec. 1998

Stevens, B. & Lewis, F. (1992). Aircraft Control and Simulation, John Wiley & Sons, NY, 1992.

Ljung, L. (1999). System Identification: Theory for the User, 2nd Ed., PTR Prentice Hall, Upper

Saddle River, Englewood Cliffs, NJ, 1999.

Maine, R.E. & Iliff, K.W. (1986). Identification of dynamic systems: theory and formulation.

NASA RF 1168, June 1986.

Roskam, J. (1995) Airplane Flight Dynamics and Automatic Flight controls, DARcorporation, KS, 1995.

Soule, H.A. & Miller, M. P. (1934). The experimental determination of the moments of inertia of airplanes. NACA Report 467, 1934. Available: (http://naca.larc.nasa.gov/reports/1934/).

Hock, W. & Schittowski, K. (1983). A Comparative performance evaluation of 27 nonlinear programming codes. Computing, Vol. 30, p. 335, 1983.

Campa, G.; Gu, Y.; Seanor, B.; Napolitano, M. R.; Pollini, L. & Fravolini, M. L. (2007) Design And Flight Testing Of Nonlinear Formation Control Laws. Control Engineering Practice, pp 1077-1092, Vol. 15, Issue 9, September 2007.

13 Chapter

VIBRATION-INDUCED PARTICULATE MATTER OSCILLATOR NOISE AND ITS MITIGATION

1. INTRODUCTION: AN OVERVIEW

High-precision oscillators are crucial in contemporary communication and navigation systems, as well as in radars and sensors utilized in unmanned aerial vehicles, helicopters, missiles, and other dynamic platforms. These systems must deliver their requisite performance even under mild to severe dynamic environmental circumstances. Oscillators frequently exhibit sufficiently low inherent phase modulation (PM) noise to meet specific system requirements in a static environment. These oscillators are susceptible to acceleration, which may manifest as steady acceleration, vibration, shock, or audio interference. In many applications, the acceleration encountered by an oscillator manifests as vibration, potentially causing mechanical deformations that compromise the oscillator›s typically low phase noise. (Filler, 1988; Vig et al., 1992; Howe et al., 2005). This diminishes the performance of the entire electronic system reliant on the oscillator's low phase noise. For instance, when radars and sensors affixed to helicopters experience intense low- and medium-frequency vibrations, the resultant noise in the system's reference oscillator leads to target blurring and potential false detections. This sensitivity to vibration mostly arises from phase changes within the oscillator's positive-feedback loop, typically caused by physical deformations in the frequency-determining part, the resonator. Factors contributing to heightened acceleration sensitivity of the resonator encompass nonlinear or sensitive mechanical coupling effects and an absence of mechanical symmetry that mitigates frequency variations in the resonator. Vibration induces mechanical deformations in non-frequency-determining electrical components, resulting in phase variations. (Steinberg, 2000). Because these fluctuations are inside the oscillator feedback loop and are integrated according to Leeson's model (Leeson, 1966), they can become large at Fourier, or offset, frequencies close to carrier frequency. An oscillator's sensitivity to vibration is characterized traditionally by acceleration sensitivity, which is the normalized frequency change per g (1 g is the acceleration of gravity near the earth's surface, approximately 9.8 m/s^2). Typically, frequency shifts in oscillators are on the order of

10^{-8} to 10^{-10} per g, primarily because of the physical deformations.

Vibration-induced noise can be suppressed by physical means and further by electronic means if a suitably low-cost way of measuring and correcting the vibration-induced noise from an oscillator is implemented. Passive mechanical isolation systems consist of elastic and damping materials that translate vibration energy to different frequencies where they are less troublesome and/or damped (Renoult et al., 1989). Active mechanical systems use accelerometers and mechanical actuators to measure and cancel motion induced by the vibration. Hybrid active-passive systems allow higher degrees of vibration isolation to be achieved, but such systems are not easily miniaturized, are somewhat complex, and are power-consuming (Weglein, 1989). In principle, atom-based frequency-determining elements such as those used in atomic frequency standards have extremely low acceleration sensitivity (Thieme et al., 2004). However, the state-selection, RF interrogation, and detection electronics are more complex than in oscillators, and the corresponding large volume of atomic standards make them equally vulnerable to mechanical deformation under vibration. Some method of suppressing induced frequency shifts is often required to even approach 10-10 per g (Kwon & Hahn, 1983). More compact atomic standards allow for simpler mechanical vibration isolation to be incorporated (Riley, 1992).

Strategies for electronically reducing acceleration sensitivity have traditionally relied on accurately detecting this vibration with sensors (Healy et al., 1983) and even using the resonator itself as a vibration sensor (Watts et al., 1988). Suppression at one vibration frequency along one axis in quartz oscillators by electronic means has been explored with success (Rosati & Filler, 1981). More recently, significant advances have been made in which this electronic vibration suppression is effective over a wide range of vibration frequencies from a few hertz to 200 Hz. This is accomplished by fabricating high-Q quartz resonators in which the "cross" g-sensitivities of the three orthogonal axes are decoupled to a high degree (Bloch et al., 2006).

This chapter aims to introduce the topic of vibration-induced PM noise by examining the method for assessing acceleration sensitivity and presenting the characterization results for a sample of devices functioning at microwave frequencies. Strategies for mitigating vibration-induced noise are also examined.

2. DEFINING ACCELERATION SENSITIVITY

If the vibration frequency from mechanical shock or other external processes is f_v, the vibration-induced phase fluctuations cause the carrier frequency to deviates from its nominal frequency, f_0, by an amount $\pm \Delta f$, at a rate of f_v. Spurious sidebands, a highly undesirable type of noise in many applications, will appear at $f_0 \pm f_v$. The red curve in Figure 1 shows the PM noise of one test oscillator that is subjected to 100 Hz vibration along one axis. Note that the intrinsic random electronic noise is degraded by additional noise due to this vibration (shown as the noise pedestal on both sides of an ideal carrier signal). Also, the blue curve indicates that as the vibration increases, so do the sidebands, eventually exceeding the carrier power.

Low acceleration sensitivity at one frequency such as 100 Hz does not necessarily mean that phase noise due to acoustic and structure-borne vibration is suppressed. While vibration- induced noise modulation on an oscillator may be proportional to overall acceleration sensitivity, the proportionality as a function of f_v can be complicated in the range of audio frequencies of concern here (from a few Hertz to 2 kHz). Resonator deformations that affect its center frequency depend on designs of mounting, elastic properties of materials, acoustic resonances, sound and vibration isolation, orientation, etc. Therefore, suppression of only "dc or time-independent" acceleration sensitivity due to what is commonly called 2 g- tipover (Vig et al., 1992) or steady acceleration has limitations and is insufficient to solve the larger problem of "ac or time dependent" acceleration sensitivity due to vibration. Acceleration sensitivity and vibration sensitivity are often used interchangeably for time- dependent accelerations. The acceleration sensitivity is characterized more fully as a function of f_v, as discussed next.

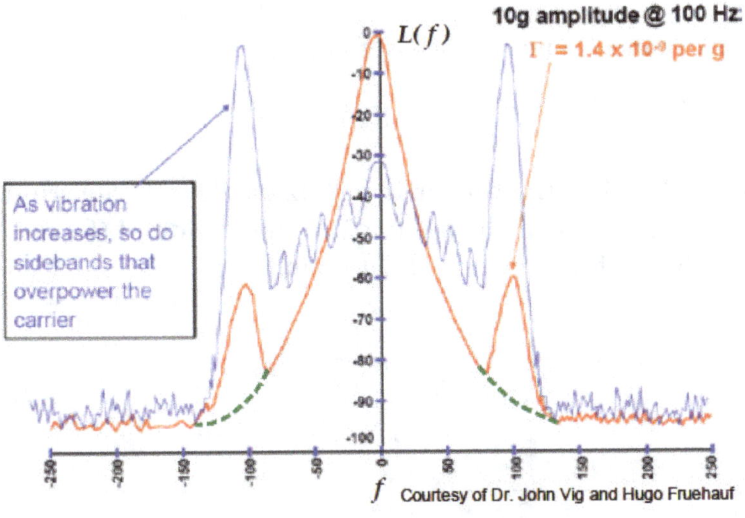

Figure 1. Phase noise of an oscillator that is subjected to vibration at fv = 100 Hz. f is the offset frequency from the carrier

Acceleration sensitivity of an oscillator is explained in detail by Filler [Filler, 1988]. When an oscillator is subjected to acceleration, its resonant frequency shifts. The peak frequency shift Δf_{peak}, which is proportional to magnitude of the acceleration and dependent on the direction of acceleration, is given by a peak fractional-frequency change y peak as

$$y_{peak} = \frac{\Delta f_{peak}}{f_0} = \bar{\Gamma} \bullet \bar{a}, \tag{1}$$

where f_0 is the frequency of the oscillator with no acceleration, Γ is the acceleration sensitivity vector and $\vec{a}$ is the peak applied acceleration vector. The magnitude of acceleration is expressed in units of g. When the direction of applied acceleration is parallel to the axis of acceleration sensitivity vector, it will have the greatest effect on Δf_{peak}.

By definition, $S_y(f)$ is the power spectral density of root-mean-square (rms) fractional-frequency change, y_{rms} (Sullivan et al., 1990), and is given by

$$S_y(f) = \frac{|y_{rms}|^2}{BW} = \left|\frac{\Delta f_{rms}}{f_0}\right|^2 \frac{1}{BW} = \left|\frac{\Delta f_{peak}}{\sqrt{2} f_0}\right|^2 \frac{1}{BW} \quad [1/\text{Hz}], \tag{2}$$

where f is the offset, or Fourier, frequency away from the carrier and BW is the bandwidth of the spectral-density measurement. Also, Sy(f) is related to power spectral density of phase fluctuations as

$$S_\phi(f) = S_y(f)\left(\frac{f_0}{f}\right)^2 \text{.....}[\text{rad}^2/\text{Hz}], \tag{3}$$

and the single sideband PM noise L(f) is defined as

$$L(f) \equiv \frac{1}{2} S_\phi(f) \quad [\text{dBc/Hz}], \tag{4}$$

where dBc/Hz is dB below the carrier in a 1 Hz bandwidth. Substituting f = f_v the vibration frequency, and normalizing to a 1 Hz bandwidth, L(f_v) can be related to acceleration sensitivity for a small modulation index as

$$\begin{aligned} L(f_v) &= \frac{1}{2}\left(\frac{\Delta f_{peak}}{\sqrt{2} f_0}\right)^2 \left(\frac{f_0}{f_v}\right)^2 \\ &= \left(\frac{\vec{\Gamma} \bullet \vec{a}}{2 f_v} f_0\right)^2. \end{aligned} \tag{5}$$

Or, when expressed in dB,

$$L(f_v) = 20 \log\left(\frac{\vec{\Gamma} \bullet \vec{a}}{2 f_v} f_0\right). \tag{6}$$

Equation 6 may be rearranged to obtain

$$\Gamma_i = \frac{2f_V}{a_i f_0} 10^{\left(\frac{L(f_v)}{20}\right)}, \tag{7}$$

where Γ_i is the component of acceleration sensitivity vector in the i(i = x, y and z) direction. For a sinusoidal vibration, $|\vec{a}|$ is the peak applied vibration level in units of g, and $L(f_v)$ is expressed in units of dBc. In most cases, vibration experienced by an oscillator is random instead of sinusoidal. Under random vibration the power is randomly distributed over a range of frequencies, phases, and amplitudes, and the acceleration is represented by its power spectral density (PSD). For random vibration, $|\vec{a}| = \sqrt{2PSD}$, and its unit is g/√Hz. Also, for a random vibration, L(fv) is expressed in units of dBc/Hz. The sum of acceleration sensitivity squared in all three axes gives the total acceleration sensitivity, or gamma ($\vec{\Gamma}$), and its magnitude is defined as

$$|\vec{\Gamma}| = \sqrt{\Gamma_x^2 + \Gamma_y^2 + \Gamma_z^2}. \tag{8}$$

$\vec{\Gamma}$ of an oscillator can be calculated from equation 8 once the PM noise of the oscillator is measured for all three axes. Noteworthy to this discussion, the sidebands generated by oscillators under vibration are a more serious issue, as the signal frequency increases due either to frequency multiplication or direct frequency generation at higher frequency. Systems are in place that require ultralow PM noise from reference oscillators operating in the range of 6 to 18 GHz. The vibration-induced PM noise of an oscillator with frequency f0 upon frequency multiplication by a factor of N is given by

$$L(f_v) = 20\log\left[\frac{\vec{\Gamma} \bullet \vec{a}}{2f_v}(Nf_0)\right]. \tag{9}$$

The vibration frequency fv is unaffected because it is an external influence. It is clear from equation 9 that, given a nominal $|\vec{\Gamma}| \sim 1 \times 10^{-9}/g$, the level of vibration sidebands in phase- noise plots of L(f) can become excessively large at X-band and higher ranges. For example, a 10 MHz oscillator with a vibration sensitivity of 1 × 10-9/g when experiencing an acceleration of 5 g produces a sideband level of -72 dBc at f_v = 100Hz. For the same $|\vec{\Gamma}|$ and under identical vibration conditions, a 10 GHz oscillator will produce a sideband of -12 dBc, a factor of 20 log (N =1000) higher. Often the sidebands are larger than the carrier, and there are also conditions where the carrier disappears and all of the power appears in the sidebands. This seriously affects or even prohibits the use of microwave systems that employ phase-locked loops, because large sidebands due to vibration cause large phase excursions and unlock the loops (Filler, 1988; Wallin et al., 2003).

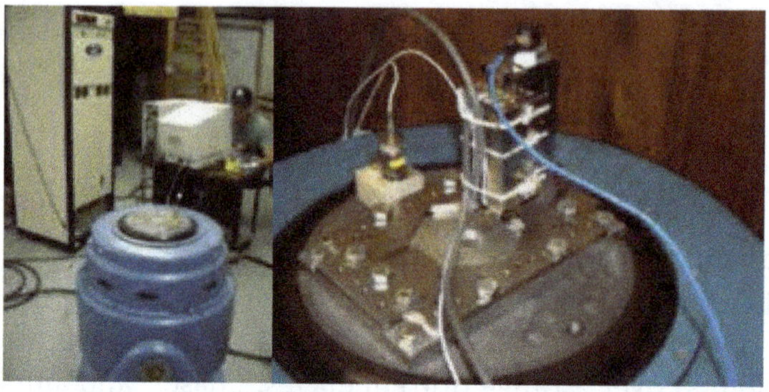

Figure 2. The picture on the left shows the vibration actuator (shown in blue); the amplifier for driving the actuator (the vertical rack-mount system); and the controlling computer. The picture on the right is a device under test (DUT) mounted on the actuator

3. MEASUREMENT TECHNIQUES

In order to measure the acceleration sensitivity of different microwave oscillators and components, the device needs to be characterized while subjected to vibration. The equipment needed to vibrate the device consists of a mechanical actuator or "shaker," its associated power amplifier, an accelerometer, and computer control system. The computer uses the accelerometer to sense the vibration of the actuator and generates the desired vibration profile using closed-loop feedback. The single axis actuator used for these tests has the capability to vibrate either in a random vibration pattern or in various sinusoidal patterns, including continuous wave and swept. When the actuator vibrates in one axis, the cross-axis leakage is low, as shown in Figure 3. These data are taken with a 3-axis accelerometer mounted on the actuator.

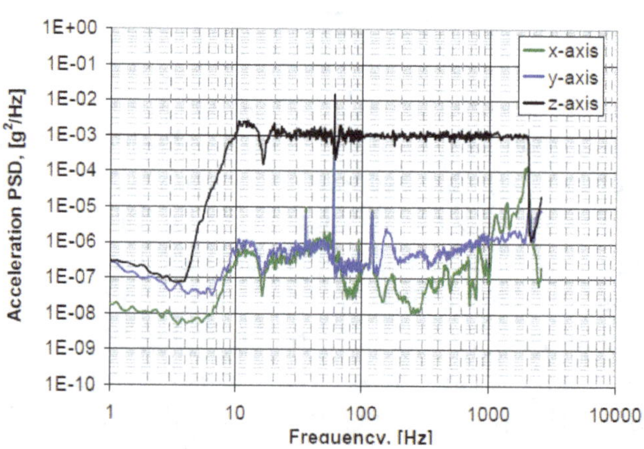

Figure 3. Plot showing the acceleration power spectral density (PSD) along x, y and z axes when vibration is along the z-axis, showing that cross-axis leakage is small. A random vibration profile of acceleration PSD = 1.0 mg²/Hz (rms) is used for 10 Hz ≤ fv ≤ 2000 Hz; fv is the vibration frequency

For all the vibration tests discussed later, both sinusoidal and random vibration testing are chosen. First, a random vibration pattern is used, vibrating at frequencies between 10 and 2000 Hz, followed by a sinusoidal vibration at 10, 20, 30, 50, 70, 90, 100, 200, 300, 500, 700, 900, 1000, and 2000 Hz. This frequency range is chosen because it is the full range for the available vibration table, adequately covering smaller ranges associated with most applications (Section 5).

3.1 Experimental Setup for Residual PM Noise Measurement

Residual noise is the noise that is added to a signal by its passage through a two-port device. Figure 4 shows the block diagram of a PM noise measurement system used to measure the residual noise of a two-port or a non-oscillatory device such as a bandpass filter or amplifier as well as a cable and connector under vibration (Walls & Ferre-Pikal, 1999). The output power of a reference oscillator is split into two paths. One path is used to drive the device under test (DUT), and the other path is connected to a delay line. The delay is chosen so that the delay introduced in one path is equal to the delay in the other path. A phase shifter is used to set phase quadrature or 90-degrees between two paths, and the resulting signals are fed to a double-balanced mixer, acting as a phase detector. The baseband signal at the output of the phase detector is amplified and measured on a fast Fourier transform (FFT) analyzer. The output voltage V0(t) is given by

$$V_0(t) = k_d G \Delta\phi(t) \qquad \text{for } \Delta\phi(t) << 1, \tag{10}$$

where k_d is the mixer sensitivity, G is the gain of the baseband IF amplifier, and $\Delta\varphi(t)$ is the difference in phase fluctuations between two inputs to the phase detector. The PM noise is obtained from

$$S_\phi(f) = \frac{PSD\left[V_0(t)\right]}{(k_d G)^2}. \tag{11}$$

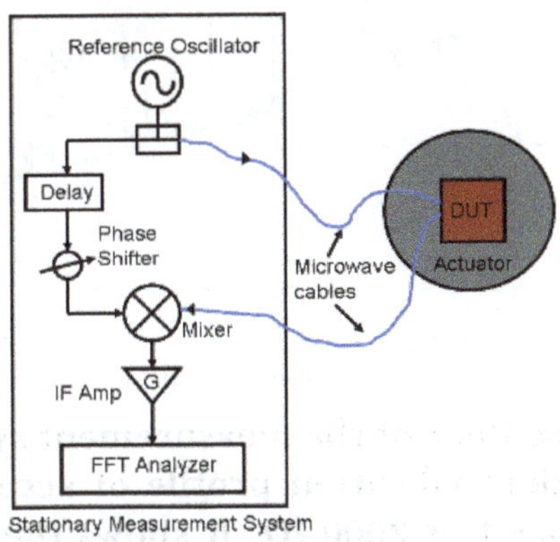

Figure 4. Block diagram of an experimental setup for residual PM noise measurement of components under vibration. DUT – Device Under Test; IF Amp – Intermediate Frequency Amplifier

Because the delays in the two signal paths are equal, the PM noise from the reference oscillator is equal and correlated in each path and thus cancels. At the output of the mixer, the noise from the vibrating DUT and connecting cables appears because it is not correlated between the two inputs of the mixer. A low noise phase detector and IF amplifier are chosen for this measurement and their noise contributions are much lower than the dominating vibration-induced noise of DUT and cables.

In order to accurately measure the vibration sensitivity of a DUT, it is very important to know the vibration sensitivity noise floor first. For the noise floor measurement, the DUT is replaced with an appropriate length of rigid coaxial cable. Compared to all other experimental components, it is the microwave cables, blue in color (Figure 4) and connected between the measurement system and the DUT mounted on the actuator that generally set the vibration sensitivity noise floor. When the DUT is under vibration, the cables flex between the vibrating and stationary (measurement system) reference frames. The flexure of the cables causes the relative position between the different segments of the outer conductor, the dielectric, and the inner conductor of the cable to vary and thus changes the electrical characteristics of the cables. The main challenge is to obtain a reproducible low noise floor at close to carrier offset frequencies. The noise floor is very dependent on the configuration and tension of cables the running between the vibrating and stationary reference frames; a slight change may cause the noise to vary anywhere from 10 to 30 dB, as shown in Figure 5.

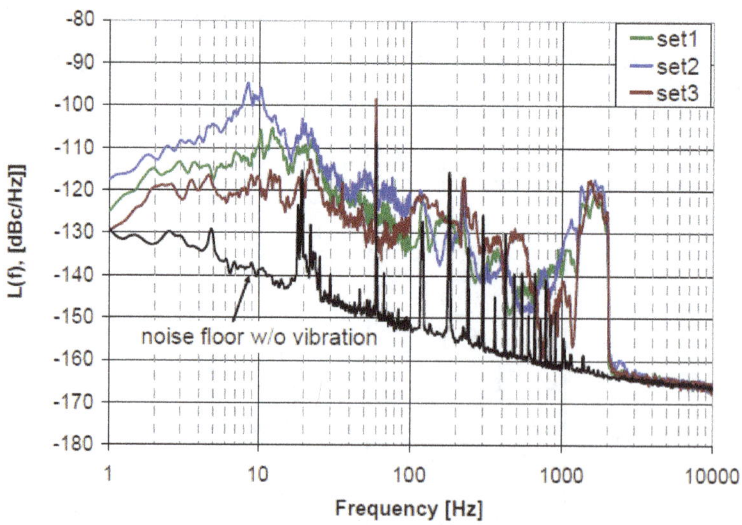

Figure 5. Residual PM noise floor of the measurement system measured with and without vibration. A random vibration profile of acceleration PSD = 1.0 mg2/Hz (rms) is used for 10 Hz ≤ fv ≤ 2000 Hz. It shows the variation in the close to carrier noise for three different sets of measurement cable configurations. The bottom curve shows the noise floor measured under no vibration. Narrow spurs are power line EMI pick-up and should be ignored

Sometimes the noise floor is so high that it is impossible to accurately characterize low- vibration-sensitive components. In order to measure the acceleration sensitivity of a component accurately, the following precautionary measures should be taken:

- Rigidly mount the DUT on the vibration table to avoid any mechanical resonance inside the frequency range of interest.
- Experiment with different amount of cable slack or tension between the stationary and vibrating reference frames to obtain the best noise floor.
- Properly secure the cables to minimize flexing and strain due to vibration. It is also important to properly secure the power leads for the DUT.
- Reduce the acoustic noise and external vibration in the test area.
- The vibration actuator often has cooling fans; prevent this airflow from disturbing the cables.
- No other components except the DUT and accelerometer should be mounted on the vibration table.
- If possible, use 1 to 3 dB attenuators at the connector interfaces to minimize the effect of voltage-standing-wave-ratio (VSWR) induced mechanical and multipath phase fluctuations.

Check the noise floor in between the measurements by replacing the DUT with a short cable.

3.2 Experimental Setup for Absolute PM Noise Measurement

Absolute noise is the noise measured on a one-port device such as an oscillator. Several measurement techniques can be used to measure the PM noise and thus the acceleration sensitivity of an oscillator. The most commonly used techniques are the heterodyne (two- oscillator) PM noise measurement system and the homodyne (delay-line discriminator) measurement system (Lance et al., 1984; Sullivan et al., 1990; Walls & Ferre-Pikal, 1999). In this section, these measurement techniques are briefly described. The block diagram of a single-channel heterodyne (two-oscillator) PM noise measurement system is shown in Figure 6. In this system, the signals from the DUT mounted on the vibration actuator and a stationary reference oscillator of the same frequency are fed into a double-balanced mixer. A lowpass filter (LPF) is used after the mixer to filter the higher-order harmonics, and a phase- locked loop (PLL) is used to lock the reference frequency to the DUT frequency and to maintain quadrature between the two input signals to the mixer. The output voltage of the mixer is proportional to the difference between the phase fluctuations of the two sources. This signal is amplified, and its power spectral density is measured on a FFT analyzer.

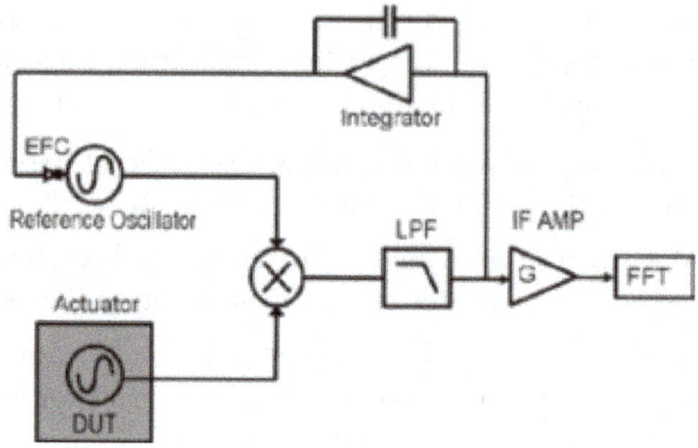

Figure 6. Block diagram for a heterodyne PM noise measurement system. DUT-Device Under Test; LPF – Low Pass Filter; EFC – Electrical Frequency Control; IF AMP – Intermediate Frequency Amplifier

The advantage of the heterodyne measurement system is its high sensitivity; however, there are notable disadvantages. It requires a reference oscillator of the same frequency, a PLL, and calibration of the mixer sensitivity (kd). Among these, the main measurement uncertainty comes from the PLL. In order to measure the vibration induced PM noise at close to carrier offset frequencies, a low loop bandwidth is required. Furthermore, when the DUT experiences a large level of vibration, it may be difficult to keep the DUT and reference oscillator phase locked due to large phase excursions. This situation requires the PLL bandwidth to be increased, which complicates the calibration inside the PLL bandwidth.

A PM noise measurement system that needs neither a second source nor a phase-locked loop is a delay-line discriminator system, shown in Figure 7. As discussed in Section 3.1, in a residual PM noise measurement of a device, it is important to keep the delay in both signal paths as equal as possible so that the source noise is correlated and cancels at the phase detector. However, in a delay-line discriminator technique, introduction of a long delay in one path intentionally un-correlates the noise by increasing the delay time so that one can measure the PM noise of the source. The differential delay introduced in one path creates a frequency dependent phase shift between the two inputs of the phase detector. This frequency dependent phase shift converts frequency fluctuations of the source into differential phase fluctuations between the inputs of the phase detector. When these two signals are adjusted for phase quadrature, the phase detector converts the phase fluctuations into their voltage equivalent for measurement and analysis. The voltage at the output of the IF amplifier is given by

$$V_0(t) \cong (2\pi\nu_0 k_d \tau) G y_{rms}(t) = \nu_0 k_\nu G y_{rms}(t), \tag{12}$$

where ν_0, τ, k_d, G, y_{rms}(t), and kν are respectively the carrier frequency, the time delay introduced by the delay line, the mixer sensitivity to phase fluctuations, the voltage gain of the baseband IF amplifier, the rms fractional frequency fluctuation and mixer sensitivity to frequency fluctuations. Equation 12 indicates that the output voltage of the mixer is proportional to the frequency fluctuation in the source; thus, this system measures the frequency modulated (FM) noise of the DUT. The FM and PM noises can be calculated from the following relations respectively,

$$\left.\begin{aligned} S_y(f) &= \frac{PSD\left[V_0(t)\right]}{(v_0 k_V G)^2} \\ S_\phi(f) &= \left(\frac{1}{f^2}\right)^2 \frac{PSD\left[V_0(t)\right]}{(k_V G)^2} \end{aligned}\right\} \text{ for } f \ll 1/\tau. \quad (13)$$

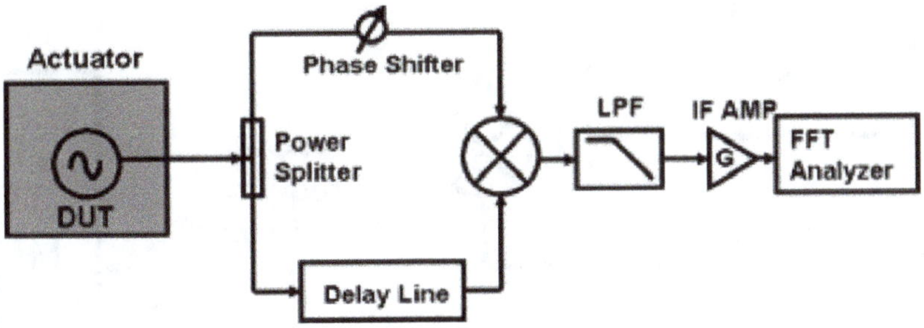

Figure 7. Block diagram of a measurement system that uses a delay-line to measure PM noise of an oscillator under vibration

The close to carrier noise floor of a delay line discriminator measurement is inversely proportional to the time delay, τ. By increasing the time delay, the noise floor can be lowered; however, longer delay causes higher insertion loss of the RF signal thus reducing measurement sensitivity, which effectively increases the noise floor. This problem can be overcome by use of a low loss delay, such as a photonic delay line (Rubiola et al., 2005). The long fiber delay provides a low-loss carrier of RF signals. A 2 km fiber typically provides 0.4 dB loss and about 10 μs delay whereas insertion loss due to a 25 m semi-rigid cable at 10 GHz is approximately 20 dB with a delay of about 100 ns. Using a 10 μs photonic delay instead of a 100 ns RF delay, almost 40 dB of improvement in the noise floor can be achieved.

There is another technique known as direct digital PM noise measurement (Grove et al., 2004), which requires two sources similar to a heterodyne technique. However, like a delay line discriminator technique it needs no phase-locked loop. This PM noise measurement system uses fast analog-to-digital converters to digitize the input RF signal and performs all down-conversion and phase detection functions by digital signal processing. It provides a low noise floor, require no calibration of

k_d or k_v, and can compare oscillators at different frequencies. High dynamic range with no phase-locked loop involved means the PM noise of a vibrating noisy source can be measured accurately for offset frequencies very close to the carrier. Figure 8 shows the setup used to measure acceleration sensitivity of different microwave oscillators discussed in Section 5.2 using this measurement system. The oscillator under test is mounted on a vibration table with the output going to one input of a mixer. This output is then mixed with a very-low-PM-noise oscillator to generate a beat frequency anywhere between 1 MHz to 30 MHz for the most straightforward digitizers. A low-noise 10 MHz quartz crystal oscillator serves nicely as the digitizer's reference clock.

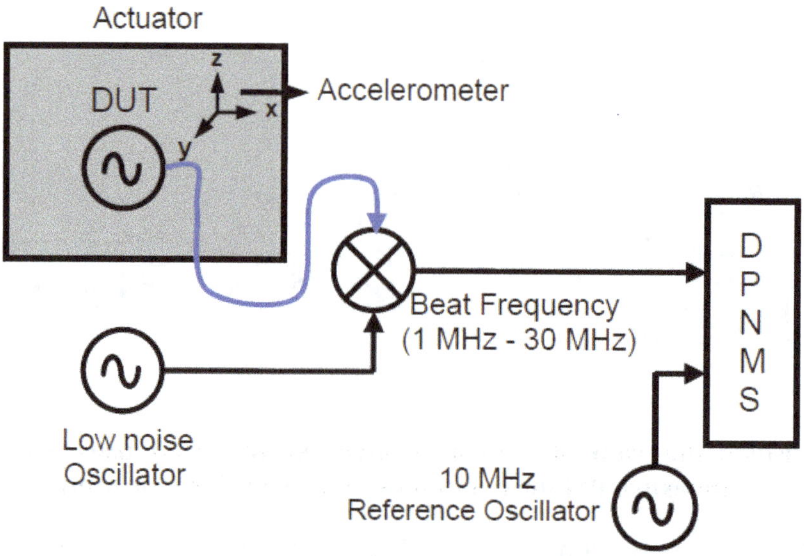

Figure 8. Block diagram of an experimental setup for measuring acceleration sensitivity of an oscillator. DPNMS – Direct digital PM noise measurement system

4. RESULTS - ACCELERATION SENSITIVITY OF NON-OSCILLATORY DEVICES

Most microwave components are sensitive to vibrations to some extent. Among them, microwave cables, bandpass filters, mechanical phase shifters and connectors are the most sensitive. Therefore, it is very important to select vibration-tolerant components in order to build a system with low vibration sensitivity. The local oscillator is one of the prime performance-limiting components in an RF receiver. However, with increasing demand for low acceleration sensitivity of oscillators in various applications, the sensitivity to vibration of these non-oscillatory components will soon be the limiting factor and thus cannot be ignored (Driscoll & Donovan, 2007). In this section, the performance of some of the non- oscillatory microwave and optical components under vibration is discussed. The components used for test are commercially available and shown in Figure 9.

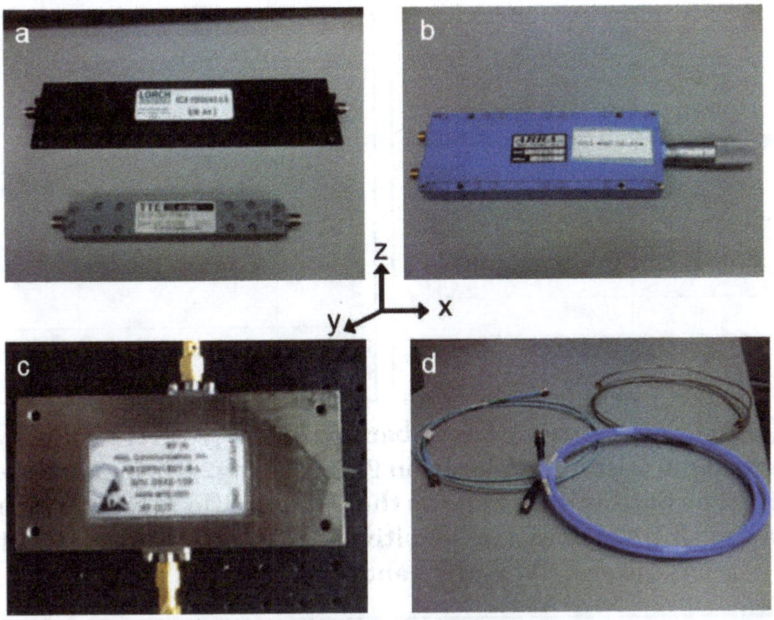

Figure 9. Picture of the non-oscillatory components used for vibration test. (a) Bandpass filter, (b) Mechanical phase shifter, (c) Amplifier and (d) Microwave cables. Arbitrary x and y axes are chosen in the plane of the page, and the z-axis is normal to the page

4.1 Microwave Components

First, two 10 GHz bandpass cavity filters of different quality factors (Q) are tested under random vibration. A random vibration profile of acceleration PSD 1.0 mg^2/Hz (rms) for offset frequencies 10 Hz to 2000 Hz is used. Figure 10(a) shows the PM noise floor of the measurement system as well as the PM noise of the filters under vibration. A comparison of acceleration sensitivity of these filters in rad/g is shown in Figure 10(b). The result is obtained by converting L(f) to $S_\varphi(f)$ using equation 4 and dividing it by $|\vec{a}| = \sqrt{2PSD}$. The result shows that the filter with higher Q (narrow bandwidth) is more sensitive to vibration. This is due to the fact that the transfer function phase of a band pass filter has its steepest slope at the center frequency. Any vibration that modulates the resonant structure of the filter also modulates the center frequency and thus the phase shift through the filter. The phase slope is proportional to the filter Q; this causes the high-Q filter to be more sensitive to small mechanical distortions under vibration. Figure 10(b) also indicates that the microwave cables used for the measurement set a noise floor that provides an acceleration sensitivity of 10^{-4} to 10^{-6} rad/g. For perspective, this means that in order to measure a 10 GHz oscillator with an acceleration sensitivity of 10^{-12}/g experiencing a random noise of 0.05 g^2/Hz (rms) between 10 Hz $\leq f_v \leq$ 1000 Hz, the cable should have vibration sensitivity better than 7×10^{-4} rad/g and 7×10^{-6} at offset frequencies 10 Hz and 1000 Hz, respectively.

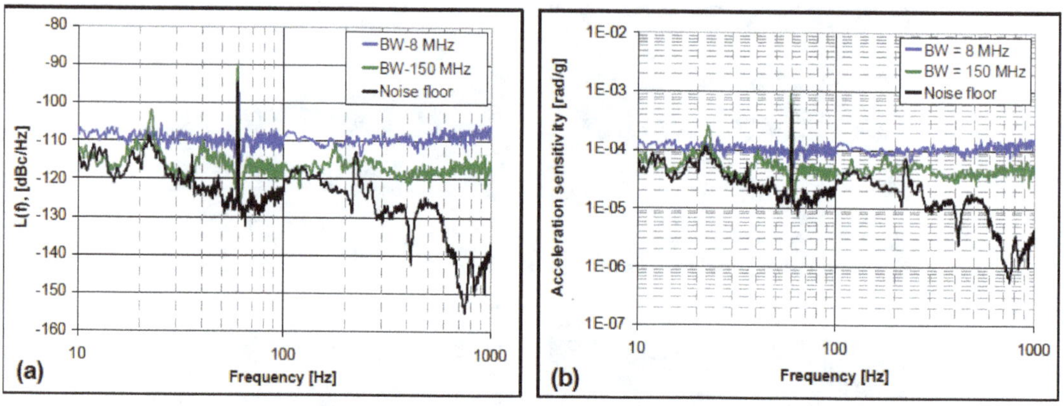

Figure 10. (a) PM noise of two 10 GHz bandpass cavity filters under vibration. A random vibration profile of acceleration PSD = 1.0 mg2/Hz (rms) is used for 10 Hz ≤ fv ≤ 2000 Hz. The bottom curve shows the PM noise floor set by flexing of cables under vibration. (b) Acceleration sensitivity of the same two 10 GHz bandpass cavity filters. BW is the bandwidth of band pass filter

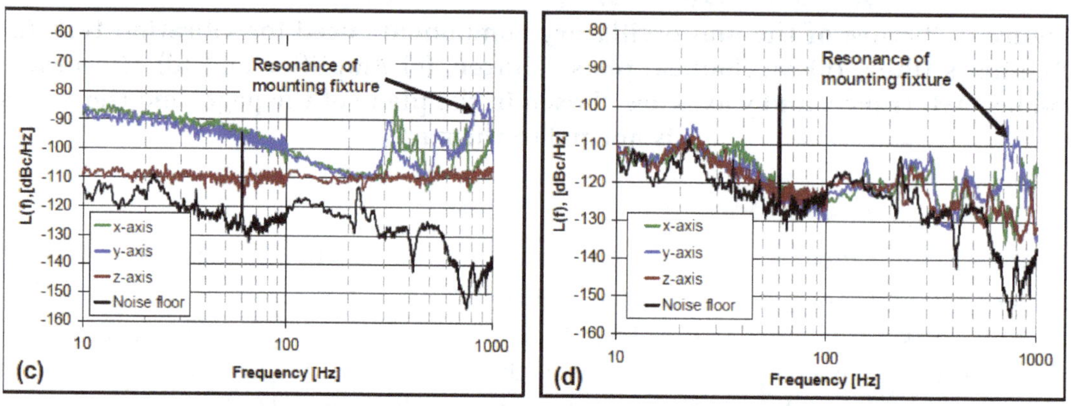

Figure 10. (c) PM noise of a 10 GHz high-Q band pass filter along x, y and z axes. (d) PM noise of a mechanical phase shifter along x, y and z axes. A random vibration profile of acceleration PSD = 1.0 mg2/Hz (rms) is used for 10 Hz ≤ fv ≤ 2000 Hz

The PM noise of a high-Q bandpass filter and a mechanical phase shifter is measured along all three axes. See Figure 9 for the orientation of the axes. Figure 10(c) shows the PM noise of a bandpass filter under vibration along all three axes. Close to the carrier, the filter is more sensitive to vibration along the x- and y-axis than along the z-axis. It is also important to note that sometimes the mechanical resonances of the fixture used to mount the component on the actuator can appear. This effect is visible for the bandpass filter above 300 Hz. Therefore, it is important to either minimize this effect, if possible, or separate it from the results of actual vibration induced noise. In the case of the phase shifter, the noise is limited mostly by the noise floor of the measurement system, as shown in Figure 10(d).

Next, three different types of cables, viz. flexible, hand formable and semi-rigid cables, each 12 feet long, are chosen for the noise floor measurement. The PM noise floors set by each of these cables under vibration are comparable, similar to that shown in Figure 5. Finally, the PM noises of a few amplifiers at 10 GHz are also measured under vibration (Hati et al. 2007). The acceleration sensitivity of these components is much lower than the acceleration sensitivity noise floor provided by the cables. As a result, an accurate measurement is not possible. However, it can be concluded from the experimental results that the acceleration sensitivity of the phase shifter and amplifier under test is no greater than 10-4 rad/g.

4.2 Optical Components

Optical fibers are widely used in fiber-optic communication, ring-laser gyroscopes, optoelectronic oscillators (OEOs), and delay line discriminators, among a number of other applications (Yao & Maleki, 1996; Römisch et al., 2000; Kadiwar & Giles, 1989; Rubiola et al., 2005; Minacian, 2006). Mechanical distortions due to vibration induce phase fluctuations in the fiber. In this section, the effect of vibration on the phase fluctuations in the optical fiber wound on a spool is discussed. Letting l be the length of the fiber (assumed to be uniformly wound on a cylindrical spool) and vg the group velocity of amplitude modulated electromagnetic waves down the fiber; then the group delay through the fiber is $\tau_d = l/vg$. In fiber, stretching can occur either as a result of temperature changes of the spool on which the fiber is wound or as a result of axial vibrations that accelerate, and hence, deform the spool. The fiber length change (δl) due to deformation of spool, which is mounted at the bottom, is given by (Ashby et al., 2007)

$$\frac{\delta l}{l} = \frac{v\rho a(t)h}{2E}, \tag{14}$$

where ν, ρ, a(t), h and E are, respectively, Poisson's ratio of the spool material, spool density, time-dependent acceleration, height of the spool, and Young's modulus (describing the spool's stiffness). Equation 14 shows that the stiffer the spool, the smaller the length change, and the denser the spool, the larger the length change, because the force on the fiber is larger. It also shows a linear dependence on the time-dependent acceleration. Further, a change in length due to stretching will also result in stresses within the fiber that can change the optical properties of the fiber, resulting in a change in the group velocity through its stress-optic coefficients. When these changes are small, the fractional frequency shift is a superposition of these two contributions:

$$\frac{\delta f}{f_0} \approx \frac{1}{v_g}\left(\frac{\delta v_g}{\delta\left(\frac{\delta l}{l}\right)} - v_g\right)\frac{\delta l}{l}. \tag{15}$$

The changes in the group velocity for a stretched fiber are quite small, and we can neglect them in a first approximation so that the fractional frequency shift due to vibration-induced fiber length changes can be adequately approximated by

$$\frac{\delta f}{f_0} \approx -\frac{\delta l}{l}. \tag{16}$$

Thus, when used in an optoelectronic oscillator, which is described in Section 5.2, the main effect on a desired frequency comes from the change of length itself rather than from the change in group velocity. The small time-delay fluctuations due to fiber length fluctuations inside the loop directly map to larger frequency fluctuations at the output of the oscillator (Leeson, 1966), which is principally why suppressing fiber vibration sensitivity is so important.

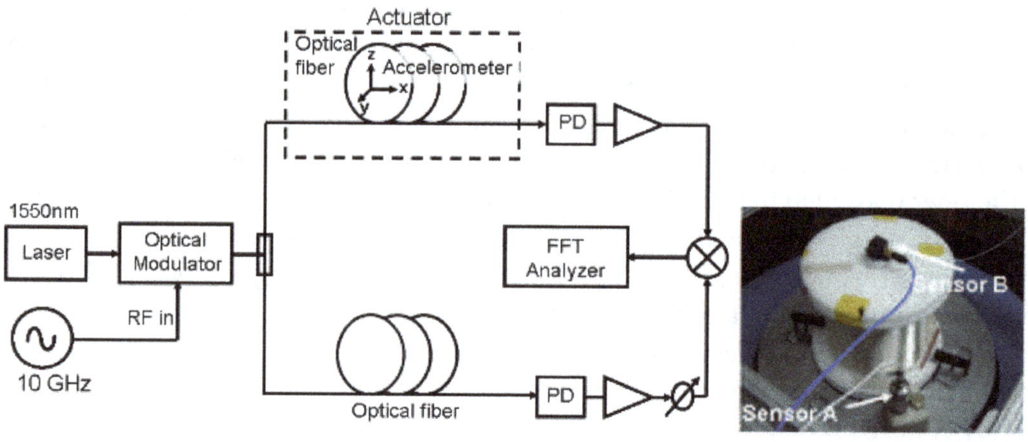

a) b)

Figure 11. (a) Experimental setup to study vibration-induced PM noise in a fiber delay line mounted on a vibration table. (b) Optical fiber of length 3 km wound on a ceramic spool and mounted on a vibration table. Orientation as shown presents the vibration (acceleration) along the z-axis of spool

Figure 11(a) shows the setup used to measure the effect of vibration on a fiber delay line using a residual PM noise measurement. It consists of a 1550 nm communications-grade laser whose output is sent into an optical modulator that is amplitude-modulated by a 10 GHz RF signal. This RF-modulated optical signal is then split into two signal paths, each composed of a 3 km length of single mode fiber (SMF-28) wound on a cylindrical spool, a photo-detector, an RF amplifier, and a phase shifter. As discussed earlier, a mechanical phase shifter is used to set phase quadrature between the two signals so that the output voltage noise of the mixer is proportional to the PM noise. This signal is then measured on a FFT analyzer. Because the delays in the two signal paths are equal, this technique cancels the PM noise introduced by the laser, optical modulator, and the reference oscillator.

In order to test the vibration properties of the fiber, one of the 3 km fibers wound on a stiff cylindrical ceramic spool is secured to a vibration table. The cylindrical spool is approximately 11.5 cm in diameter with a length between end caps of 10 cm. An accelerometer mounted to the actuator, as shown in Figure 11(b) (sensor A), provides closed-loop feedback to a computer for control of the vibration profile. Sensor B is a triaxial accelerometer that is used to sense vibrations affecting the fiber spool. Its z-axis is aligned with the axis of the spool, and both the x and y axes are in the radial directions. For this test, the z-axis sensitivity is the subject of consideration because the vibration sensitivity of a fiber-on-spool is greatest along this axis (Ashby et al. 2007; Huang et al. 2000). The spool is subjected to a random vibration of acceleration PSD equal to 0.5 mg2/Hz (rms) along the z- axis, and the PM noise is measured. In Figure 12, the bottom curve is the noise floor, and the topmost curve is the same measurement of PM noise under vibration. The dotted brown line shows that the acceleration sensitivity of this fiber is 2.5×10-3 rad/g, which is calculated for L(f) = -85 dBc/Hz. The resonance at 600 Hz is a characteristic of the hollow cylindrical spool geometry. The ceramic spool is chosen due to its stiffness because it is less sensitive to vibration than other materials such as plastics (Taylor et al., 2008). The measurement of acceleration sensitivity of other optical components such as optical connectors, fiber laser, optical modulator, and photo-detector are also important; however, the largest sensitivity to vibration is usually from the fiber.

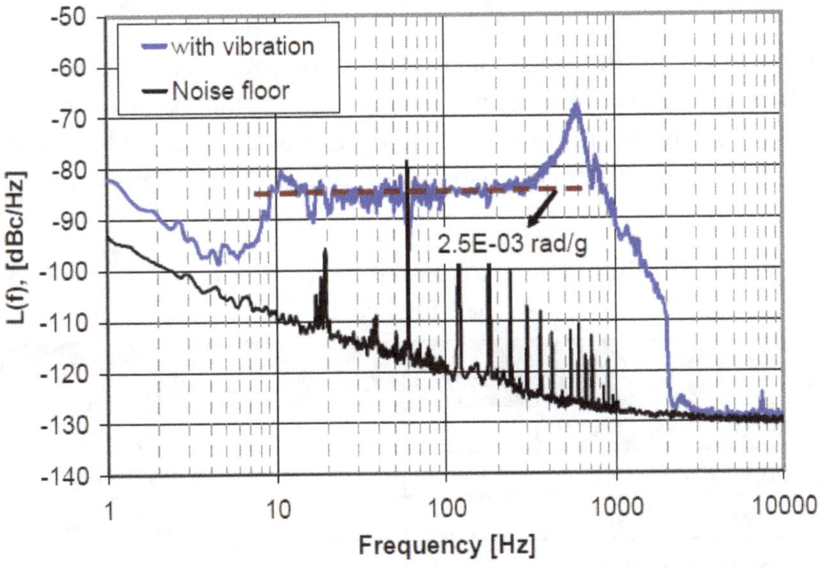

Figure 12. The residual PM noise of a 3 km fiber under random vibration along the z-axis is shown. A random vibration profile of acceleration PSD of 0.5 mg2/ Hz (rms) is used for 10 Hz ≤ fv ≤ 2000 Hz. The dotted brown curve shows the acceleration sensitivity corresponding to L(f) = -85 dBc/Hz. The bottom curve is the noise floor measured under no vibration

5. RESULTS - ACCELERATION SENSITIVITY OF DIFFERENT CLASSES OF OSCILLATORS

Low-noise oscillators must operate under "real world" vibration/acceleration conditions. This is crucial for high-speed communications systems, advanced surveillance systems, weapons detection, and many other applications. The PM noise requirement for each application is different, and a single oscillator cannot meet the PM noise requirement over a broad range of offset frequencies. For example, a good low noise 10 MHz quartz oscillator typically has a PM noise L(10 Hz) = -130 dBc/Hz and a white PM noise level of -170 dBc/Hz. The PM noise of this 10 MHz signal at 10 Hz offset, when multiplied to generate a signal at 10 GHz, becomes -70 dBc/Hz, which is reasonably low and suitable for a 10 GHz radar to detect slow-moving objects (Fruehauf, 2007). However, the white PM noise after ideal multiplication becomes -110 dBc/Hz and is too high to detect fast-moving targets such as supersonic aircraft. There are several X-band oscillators topologies such as a YIG (Yttrium, Iron and Garnet) oscillator, dielectric resonator oscillator (DRO), surface transverse wave (STW) oscillator, sapphire loaded cavity oscillator, or a simple air-dielectric cavity resonator oscillator that can provide very low phase noise at higher offset frequencies. A combination of any of these oscillators with a quartz oscillator can achieve a low PM noise performance over a broad frequency range.

Oscillators that are capable of satisfying specific system requirements in quiet environments are readily available. However, in the vibrating environments of airborne platforms, the PM noise of oscillators degrades significantly. Table 1 and Figure 13 show the typical level of acceleration for certain operating environments (Mancini, 2004). Acceleration levels at the oscillator depend on how and where the oscillator is mounted. Platform resonances can greatly amplify the acceleration levels.

Table 1. Typical level of acceleration for certain operating environments. * See Figure 13

Environment	Acceleration typical levels, in units of g
Tractor-trailer (3-80 Hz)	0.2 peak
Armored personnel carrier	0.5 to 3 rms
Ship – calm seas	0.02 to 0.1 peak
Ship – rough seas	0.8 peak
Propeller aircraft	0.3 to 5 rms
Helicopter*	0.1 to 7 rms
Missile – boost phase	15 peak
Jet aircraft*	0.02 to 2 rms

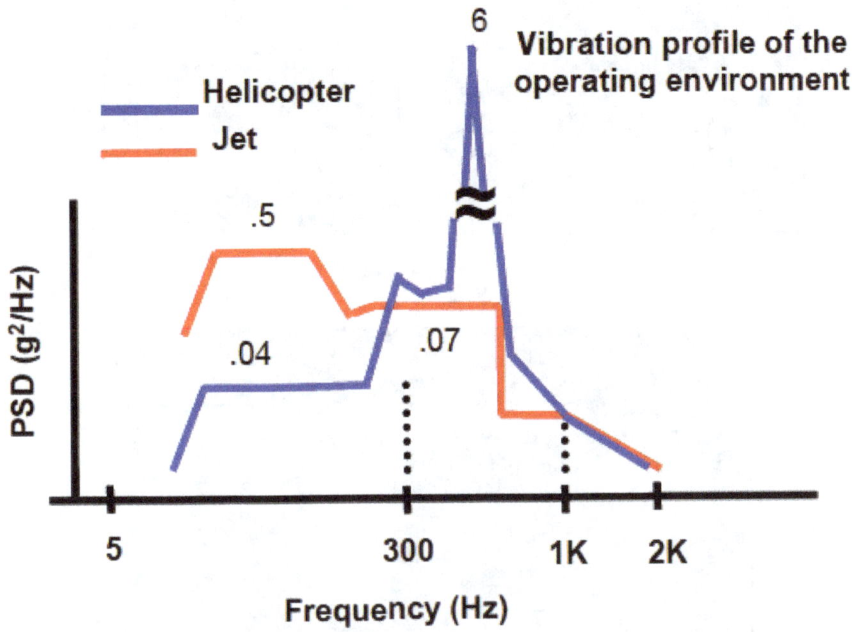

Figure 13. Typical aircraft random vibration profile

Section 5 is primarily focused on studying the effect of vibration on the phase noise of electronic and opto-electronic oscillators at 10 GHz and also comparing their acceleration sensitivity. The acceleration sensitivity of a quartz oscillator is not discussed here because enormous amounts of work have already been done and the data are readily available in the literature (Vig, 1992; Driscoll, 1993; Filler, 1983; Kosinski, 2000). A good quartz oscillator typically has an acceleration sensitivity of 1 × 10-9/g, and the lowest known acceleration sensitivity for a quartz oscillator is 2 × 10-11/g for 10 Hz ≤ fv ≤ 200 Hz (Bloch et al., 2006).

5.1 Electronic Oscillators

Figure 14 shows three different types of oscillators chosen for the vibration test, viz., a DRO at 10 GHz, a silicon germanium (SiGe) amplifier-based STW oscillator at 2.5 GHz (Hay, et al., 2004), and a TE023 mode air-dielectric ceramic-cavity resonator oscillator (ACCRO) at 10 GHz (Hati et al., 2006). A STW oscillator is chosen because these are the best performing oscillators in the frequency range 1 to 3 GHz. Below 1 GHz surface acoustic wave (SAW) oscillators perform well (Parker & Montress, 1988), and above 3 GHz DROs provide the best compromise between performance and cost. There are several other commercially available oscillators above 3 GHz that have extremely low phase noise but are large, specialized and expensive by comparison. At first, the PM noise of these oscillators is measured with no vibration; the PM noise plots are shown in Figure 15.

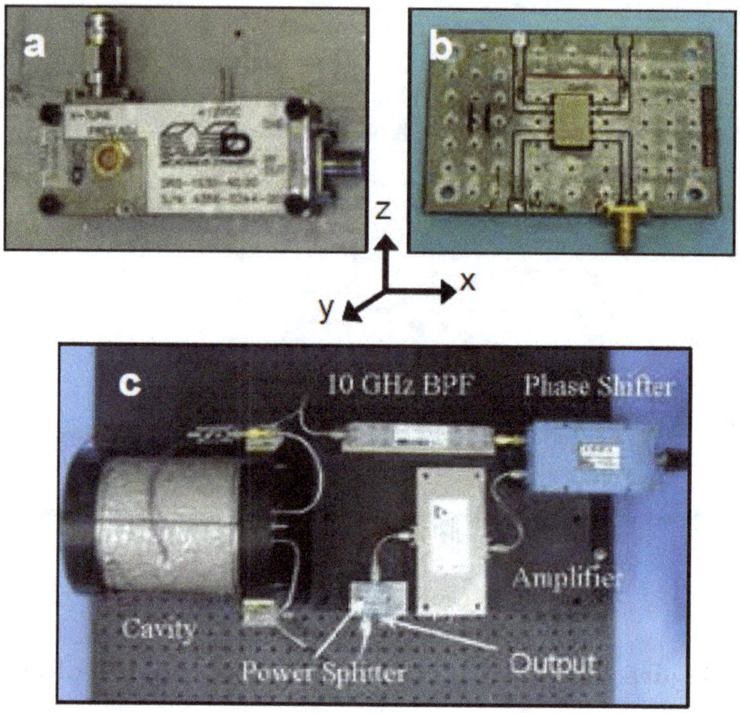

Figure 14. Pictures of three different types of oscillators used for vibration test. (a) DRO, (b) STW oscillator, (c) ACCRO. Arbitrary x and y axes are chosen in the plane of the page, and the z-axis is normal to the page

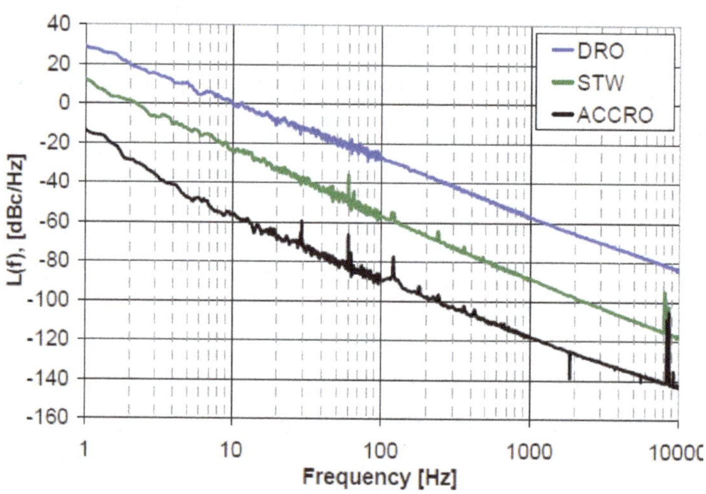

Figure 15. PM noise of three different oscillators at 10 GHz without vibration. For straight comparison, the PM noise of 2.5 GHz STW oscillator is normalized to 10 GHz

A commercial DRO at 10 GHz is subjected to a random vibration along three axes independently. The degradation in PM noise due to vibration in the z-axis is shown in Figure 16(a). The effect of random vibration in the x and y axes is not noticeable because the PM noise of the stationary DRO is much higher than the noise induced by random vibration.

In order to measure the acceleration sensitivity in all three axes, the oscillator is subjected to sinusoidal vibration with higher g-levels at different frequencies. The results are shown in Figure 16(b). This particular DRO is more sensitive to vibration along the z-axis than along the other two axes. Further, the acceleration sensitivities of two DROs of comparable size and weight but different Q are compared. Figure 17(a) and 17(b) respectively show the PM noise and acceleration sensitivity of these DROs. The DRO with higher Q is more sensitive to vibration for the same reason as discussed in Section 4.1 for a high-Q bandpass filter.

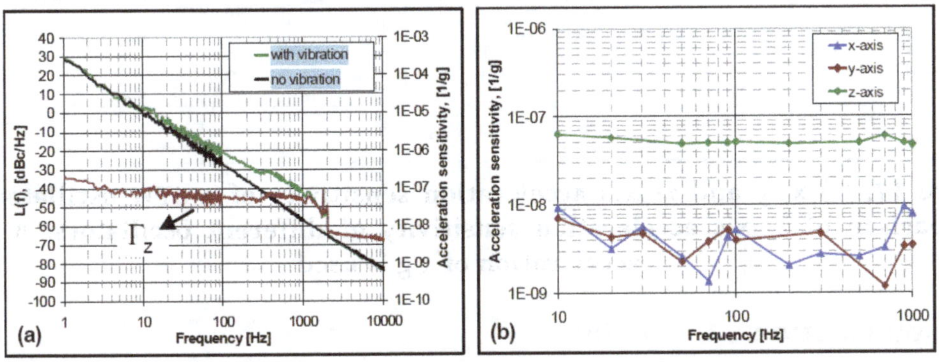

Figure 16. (a) PM noise of the DRO with and without vibration along the z-axis. The brown curve indicates the z-axis acceleration sensitivity of this particular DRO. (b) x, y, and z-axis gamma. An acceleration PSD of 0.5 mg2/Hz (rms) is used for random vibration (Fig. 16a) and a peak acceleration of 1 g is used for sinusoidal vibration (Fig. 16b)

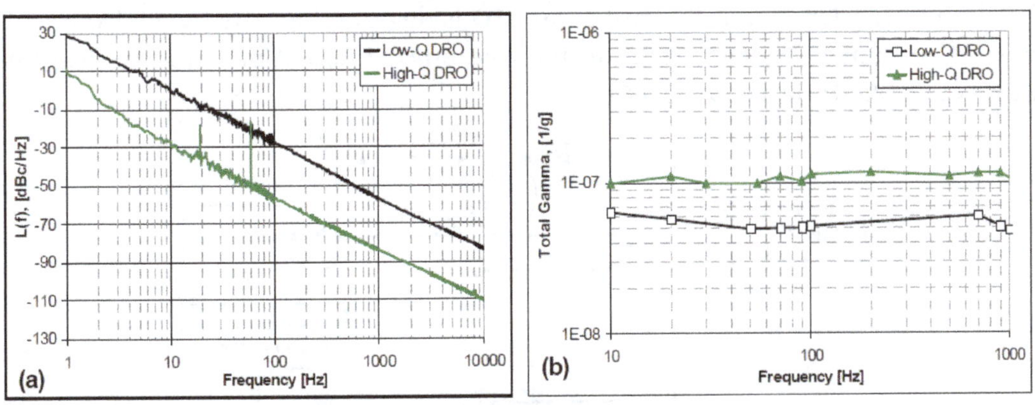

Figure 17. (a) PM noise of two DROs of different Q without vibration. (b) Plot of total gamma for the DROs. An acceleration PSD = 0.5 mg2/Hz (rms) is used for 10 Hz ≤ fv ≤ 2000 Hz. The lower PM noise oscillator has correspondingly higher acceleration sensitivity

Vibration-sensitivity experiments are also performed for a STW oscillator in all three axes, and the results are shown in Figure 17(c). For the ACCRO, the vibration measurement is confined to a single axis normal to the mounting plate, as shown in Figure 10. Figure 17(d) shows the z-axis acceleration sensitivity of three oscillators; the acceleration sensitivity of the STW oscillator is two orders of magnitude lower than that of the DRO.

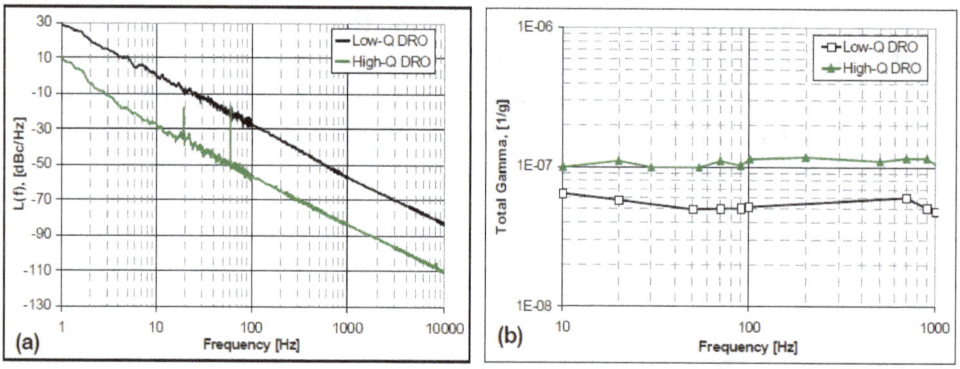

Figure 17. (c) x, y and z-axis acceleration sensitivity of a STW oscillator. (d) Comparison of z-axis acceleration sensitivity of different oscillators. A peak acceleration of 1 g is used

5.2 Optoelectronic Oscillator

Low-noise, microwave-frequency oscillators are key components of systems that require high spectral purity. An optoelectronic oscillator (OEO) is an example that has emerged as a low-noise source in recent years (Yao & Maleki, 1996; Römisch et al., 2000; Eliyahu et al., 2008). The high spectral purity signal of an OEO is achieved by using a long optical fiber that provides a very high quality factor (Q). However, the close to carrier spectral purity of an OEO is degraded mostly by environmental sensitivities, one being the vibration-induced phase fluctuations in its optical fiber (Howe et al., 2007; Hati et al., 2008).

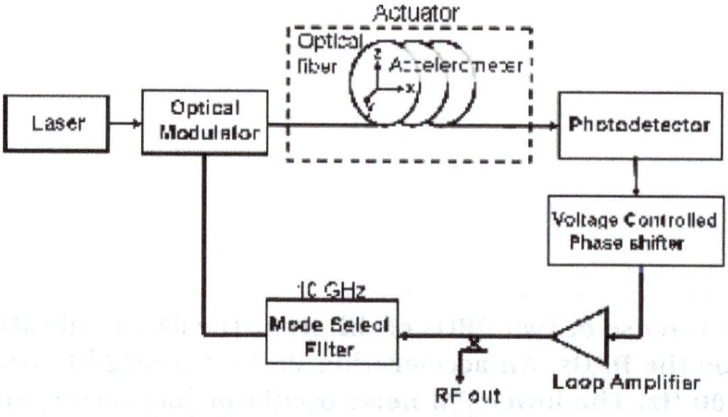

Figure 18. Block diagram of an optoelectronic oscillator

In a typical optoelectronic device as shown in Figure 18, light from a laser passes through an electro-optic amplitude modulator, the output of which is fed to a long optical fiber and detected with a photodetector. The output of the photodetector is then amplified, filtered, and fed back to the modulator port, which amplitude-modulates the laser light. When loop gain is greater than 1, this configuration leads to self-sustained oscillations. A typical OEO gives a large number of modes with frequencies given by (Yao & Maleki, 1996; Römisch et al., 2000)

$$f_0 = \frac{(K+1/2)}{\tau_d}. \tag{17}$$

where K is an integer whose value is selected by the filter and $\tau_d = l/v_g = nl/c$ is the group delay through the fiber with index of refraction n and length l and c is the velocity of light. The quality factor of an OEO is proportional to length l and is given by Q = πτdf0 = πnlf0/c.

An OEO at 10 GHz is designed by using the same 3 km (SMF-28) fiber as mentioned in Section 4.2 to study the effect of vibration on the overall PM noise performance of the oscillator. The PM noise of this OEO subjected to random vibration is shown in Figure 19. The PM noise under vibration degrades almost 30 to 40 dB from its normal stationary PM noise performance. This is due to the fact that the phase perturbations due to vibration in the fiber, which is the most vibration sensitive component in the loop, translate to frequency fluctuations inside the OEO's resonator bandwidth. The z-axis acceleration sensitivity of this OEO is approximately 5 × 10-9/g, as shown by the brown plot in Figure 19.

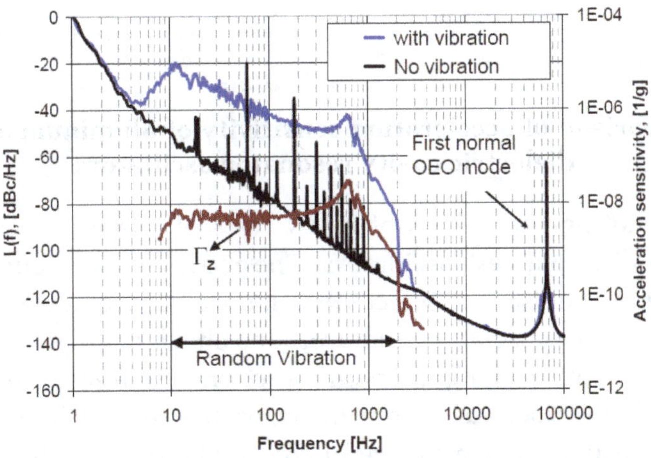

Figure 19. Plot comparing the PM noise of an optoelectronic oscillator with and without vibration. A random vibration profile of acceleration PSD = 0.5 mg2/Hz (rms) is used for 10 Hz ≤ fv ≤ 2000 Hz. The brown curve shows the z-axis acceleration sensitivity of this OEO, which is approximately 5 × 10-9/g. The first mode for a 3 km fiber, which is approximately 67 kHz from the carrier, is also shown

6. ACCELERATION SENSITIVITY REDUCTION

In this section, a few methods of reducing vibration-induced noise from vibration-sensitive components are discussed. The most common approach for reducing vibration-induced PM noise is to select low-vibration-sensitive materials. A comparison is made for two air- dielectric cavity oscillators at 10 GHz, one cavity made of aluminum and another one made of ceramic. The two cavities are chosen so that they have comparable volume, almost identical loaded Q's of 22,000 (TE023 mode) and insertion losses of 6 dB. All the other components of the two oscillators are identical. These two oscillators are tested for different sinusoidal vibration frequencies. The acceleration sensitivity of the ceramic cavity oscillator is found to be almost one sixth that of the aluminum cavity oscillator, as shown in Figure 20. This is because ceramic is stiffer than aluminum and thus less sensitive to vibration.

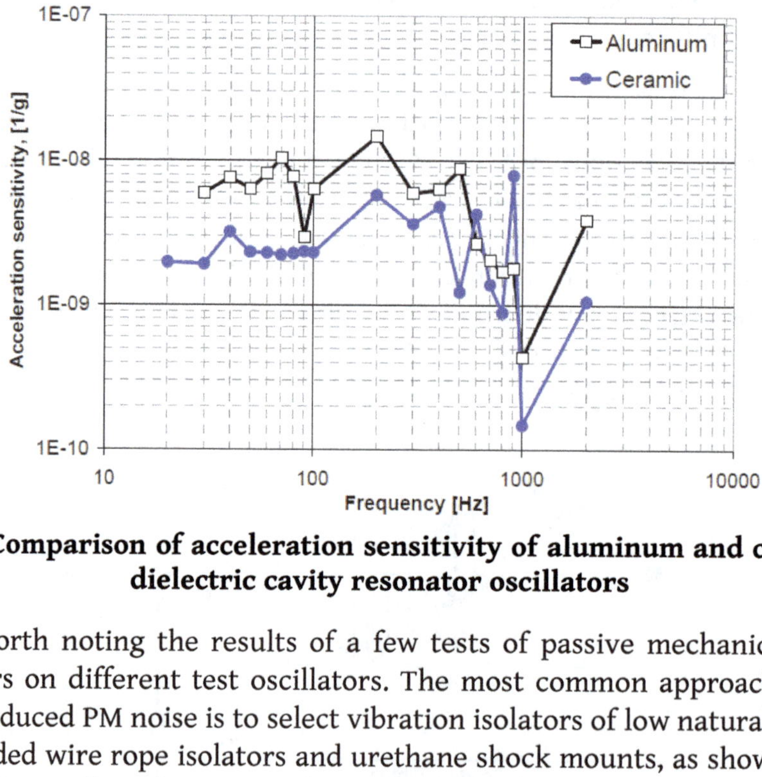

Figure 20. Comparison of acceleration sensitivity of aluminum and ceramic air-dielectric cavity resonator oscillators

It is also worth noting the results of a few tests of passive mechanical dampers and isolators on different test oscillators. The most common approach to reduce vibration-induced PM noise is to select vibration isolators of low natural frequency. Small stranded wire rope isolators and urethane shock mounts, as shown in Figure 21(a), provide excellent damping and omnidirectional isolation. When size is critical, these small shock absorbers are often incorporated inside the oscillator package to improve the PM noise performance of the oscillator in high vibration induced environments. Further, when space is not an issue, an external vibration isolation platform can be used. In this case, the whole system including the oscillator can be mounted on this vibration-free platform to achieve the precision needed for the particular application. There are several commercially available vibration isolation platforms with extremely good performance and very low natural resonance frequencies, less than 1 Hz. A significant improvement in the acceleration sensitivity of the DRO and STW oscillators is observed when tested under vibration with these passive dampers as shown in Figure 21(b).

Finally, an active electronic vibration cancellation technique can be sometimes used to reduce the vibration sensitivity. A 3 km long optical fiber and an optoelectronic oscillator are chosen to illustrate this scheme. Work on control of environmental noise in optical fiber has previously been implemented in systems where either a portion of the system undergoes vibration or a stable reference is available to measure the vibration-induced noise (Foreman et al., 2007). It is very important to assess the degree to which vibration induced phase fluctuations φv(t) can be correlated with and predicted by an accelerometer so that it becomes possible to electronically cancel the effect of vibration-induced noise in the fiber (Hati et al., 2008).

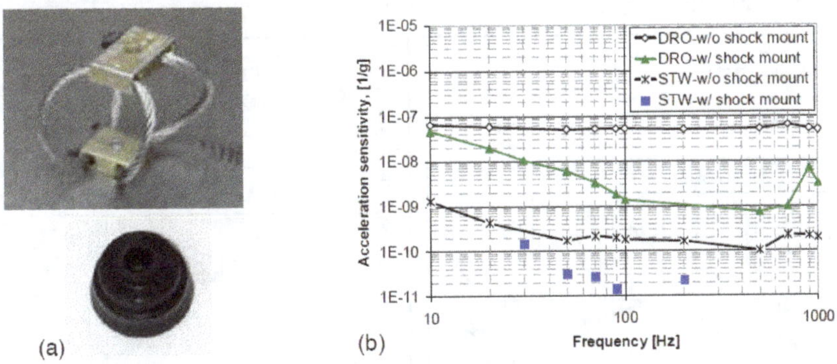

Figure 21. (a) Wire-rope and urethane passive shock mounts. (b) Improvement in acceleration sensitivity resulting from use of shock mounts

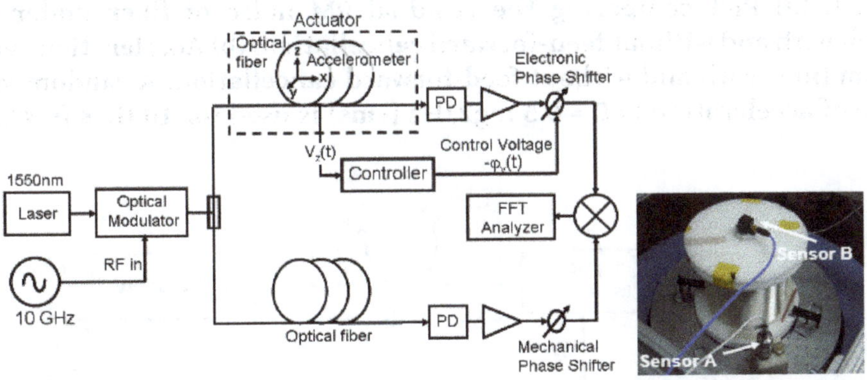

Figure 22. Experimental setup to study vibration-induced PM noise and its suppression in a fiber delay line mounted on a vibration table. PD is a photodetector that converts 10 GHz optical modulation to RF

The 3 km optical fiber (SMF-28) wound on a ceramic spool is subjected to vibration. While the fiber is under vibration, an estimate of the opposite phase of the φv(t) signal is generated based on vibration sensors, in this case, a z-axis accelerometer (sensor B) mounted on the top of the spool. The z-axis is the most sensitive axis, by an order of magnitude or more compared to the x and y axes (Ashby et al. 2007, Huang et al. 2000). An electronic phase shifter, as shown in Figure 22, corrects the phase perturbations of the demodulated 10 GHz signal sensed by the accelerometer by feed-forward correction. Figure 23(a) shows preliminary results and proof-of-concept of active noise control applied to the ceramic spool of optical fiber. The bottom curve is the noise floor, and the topmost curve is the same measurement of PM noise while the spool is subjected to a random vibration. The middle curve is the residual PM noise with the noise control on. Particularly noteworthy is that the residual PM noise through the spool of fiber is reduced by 15 dB to 25 dB. For this experiment, a flat frequency response is used for the feed-forward phase correction. The acceleration sensitivity of 3 km fiber with and without feed-forward cancellation is also shown in Figure 23(b).

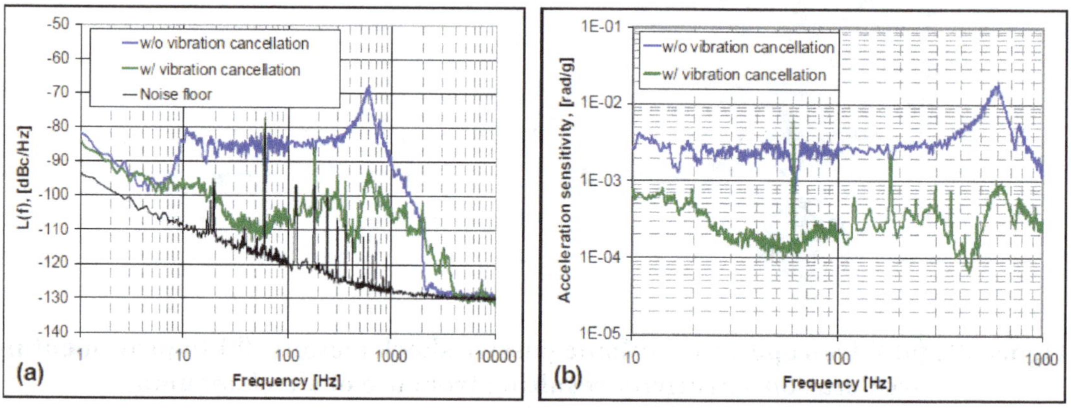

Figure 23. (a) Plot comparing the residual PM noise of fiber under random vibration with and without feed-forward cancellation. (b) Acceleration sensitivity of a 3 km fiber with and without feed-forward cancellation. A random vibration profile of acceleration PSD = 0.5 mg2/Hz (rms) is used for 10 Hz ≤ fv ≤ 2000 Hz

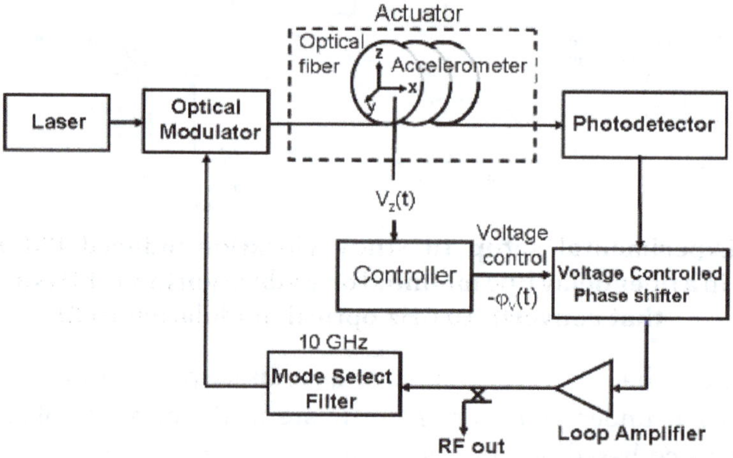

Figure 24. Block diagram of an OEO with active vibration-induced noise control

The same technique is implemented in an OEO to cancel the vibration-induced noise. In an OEO under vibration, an accelerometer signal is used to accurately estimate the complex conjugate of the vibration-induced phase modulation, as depicted in Figure 24. This estimate modulates the oscillator's output frequency by virtue of Leeson's model (Leeson, 1966) in such a way as to correct the induced in-loop phase perturbations. The PM noise of the 10 GHz OEO with and without vibration, as well as with vibration cancellation, is shown in Figure 25(a). Finally, the z-axis acceleration sensitivity of the OEO at 10 GHz is calculated by use of equation 7 and Figure 25(a) and is shown in Figure 25(b). There is an improvement by almost an order of magnitude in acceleration sensitivity over the full range of vibration frequencies tested. A flat frequency response is used for the feed-forward phase correction in this case. However, a custom-tailored frequency response can be used to achieve better cancellation at different vibration frequencies. In order to verify this, the fiber spool is subjected to a sinusoidal acceleration of 1 g at 10 Hz, 50 Hz, 100 Hz, 200 Hz and 1 kHz. The bottom curve of Figure 25(b) shows that further improvement in z-axis sensitivity is achieved by optimizing the phase and amplitude of the feed-forward phase correction at these specific sinusoidal frequencies.

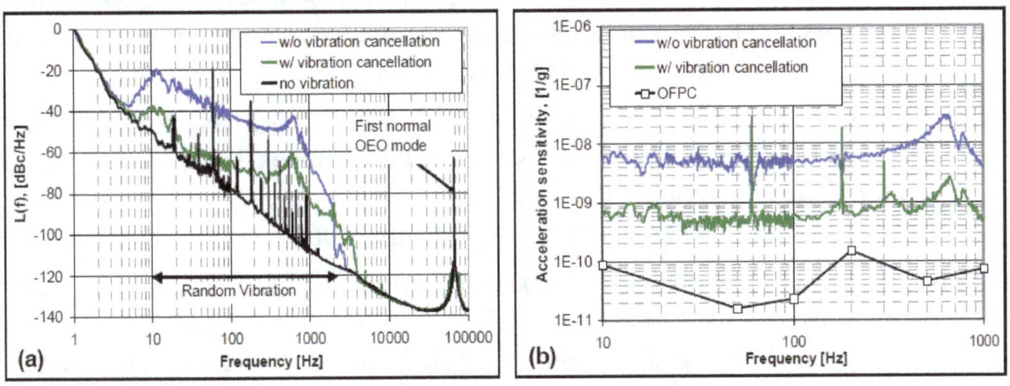

Figure 25. (a) PM noise of a 10 GHz OEO with and without vibration cancellation. A random vibration profile of acceleration PSD of 0.5 mg2/Hz (rms) is used for 10 Hz ≤ fv ≤ 2000 Hz. (b) Plot of z-axis acceleration sensitivity with and without vibration cancellation. The bottom curve corresponds to a single-frequency sinusoidal peak acceleration of 1 g. For each sinusoidal frequency, an optimized feed-forward phase correction (OFPC) is used

7. CONCISE SUMMARY

Structure-borne vibration is commonplace in numerous applications, resulting in elevated PM noise in oscillators that impairs various systems. Consequently, it is imperative to use components that exhibit minimal phase alterations under vibration to construct a system with reduced vibration sensitivity. This chapter derives the relationship between acceleration sensitivity and phase modulation noise of an oscillator. Various methodologies for quantifying the PM noise level of components subjected to vibration are also examined. The acceleration sensitivity of several advanced microwave and optical components is assessed. Findings indicate that high-Q components exhibit greater sensitivity to vibration. Finally, various passive and active methods to mitigate or eliminate vibration-induced noise in these oscillators are also examined. The findings indicate that the effective implementation of passive and active cancellation techniques can enhance the acceleration sensitivity of an oscillator by many orders of magnitude.

REFERENCES

Ashby, N.; Howe, D. A.; Taylor, J.; Hati, A. & Nelson, C. (2007). Optical fiber vibration and acceleration model, 2007 IEEE International Frequency Control Symposium Jointly with the 21st European Frequency and Time Forum, pp. 547-551, Geneva, Switzerland, 29 May–1 June, 2007

Bloch, M; Mancini, O. & Stone, C. (2005). Method for achieving highly reproducible acceleration insensitive quartz crystal oscillators, U.S. Patent 07106143, 2006

Driscoll, M.M. (1993). Reduction of Quartz Oscillator flicker-of-frequency and white phase noise (floor) levels of acceleration sensitivity via use of multiple resonators. IEEE Transactions on Ultrasonics, Ferroelectrics, and Frequency Control, Vol. 40, No.4, pp. 427-430, July 1993

Driscoll, M.M. & Donovan, J.B. (2007). Vibration-Induced Phase Noise: It Isn't Just About the Oscillator, Proceedings of 2007 IEEE International Frequency Control Symposium Jointly with the 21st European Frequency and Time Forum, pp. 535-540, Geneva, Switzerland, 29 May–1 June, 2007

Eliyahu, D.; Seidel, D. & Maleki, L. (2008). RF amplitude and phase-noise reduction of an optical link and an opto-electronic oscillator. IEEE Transactions on Microwave Theory and Techniques, Vol. 56, No. 2, pp. 449-456, February 2008

Filler, R.L.; Kosinski, J.A. & Vig, J.R. (1983). Further Studies on the Acceleration Sensitivity of Quartz Resonators, Proceedings of 37th Annual Symposium on Frequency Control, pp. 265 – 271, 1983

Filler, R.L. (1988). The acceleration sensitivity of quartz crystal oscillators: A review. IEEE Transactions on Ultrasonics, Ferroelectrics, and Frequency Control, Vol. 35, No. 3, pp. 297-305, May 1988

Foreman, S.M; Ludlow, A.D; Miranda, M.H.G.; Stalnaker, J. E.; Diddams, S.A. & Ye, J. (2007). Coherent optical phase transfer over a 32-km fiber with 1 s instability at 10-17. Physical Review Letter, Vol. 99, pp. 153601 (4), October 2007

Fruehauf, H. (2007). "g"-Compensated, Miniature, High Performance Quartz Crystal Oscillators, http://www.freqelec.com/tech_lit.html, April 2007.

Grove, J.; Hein, J.; Retta, J.; Schweiger, P.; Solbrig, W. & Stein, S.R. (2004). Direct-digital phase-noise measurement, Proceedings of the 2004 IEEE International Frequency Control Symposium and Exposition, pp. 287 - 291, 23-27, August 2004

Hati, A.; Howe, D.A. & Nelson, C.W. (2006). Comparison of AM noise in commercial amplifiers and oscillators at X-band, Proceedings of 2006 IEEE International Frequency Control Symposium, pp. 740-744, June 2006

Hati, A.; Nelson, C.W; Howe, D.A.; Ashby, N.; Taylor, J., Hudek, K.M.; Hay, C.; Seidel, D.J. & Eliyahu, D. (2007). Vibration sensitivity of microwave components, Proceedings of 2007 IEEE International Frequency Control Symposium Jointly with the 21st European Frequency and Time Forum, pp. 541-546, Geneva, Switzerland, 29 May–1 June, 2007

Hati, A.; Nelson, C.W.; Taylor, J.; Ashby, N. & Howe, D.A. (2008). Cancellation of Vibration- Induced Phase Noise in Optical Fibers. IEEE Photonics Technology Letters, Accepted for future publication, 2008

Hay, C.E.; Harrell, M.E. & Kansy, R.J. (2004). 2.4 and 2.5 GHz miniature, low-noise oscillators using surface transverse wave resonators and a SiGe sustaining amplifier, Proceedings of the 2004 IEEE International Frequency Control Symposium and Exposition, pp. 174 – 179, 23-27 August 2004

Healy, D.J.; Hahn, H. & Powell, S. (1983). A measurement technique for determination of frequency vs. acceleration characteristics of quartz crystal units, Proceedings of the 37th Annual Symposium on Frequency Control, pp. 284-289, 1983

Howe, D.A; Lanfranchi, J.; Cutsinger, L.; Hati, A. & Nelson, C.W. (2005). Vibration-induced PM noise in oscillators and measurements of correlation with vibration sensors, Proceedings of the 2005 IEEE International Frequency Control Symposium and Exposition, pp. 494-498, 29-31 August, 2005

Howe, D.A.; Hati, A.; Nelson, C.W.; Taylor, J. & Ashby, N. (2007). Active vibration-induced PM noise control in optical fibers: preliminary studies, 2007 IEEE International Frequency Control Symposium Jointly with the 21st European Frequency and Time Forum, pp. 552-556, Geneva, Switzerland, 29 May–1 June, 2007

Huang, S.; Tu, M.; Yao, S. & Maleki, L. (2000). A turnkey optoelectronic oscillator with low acceleration sensitivity, Proceedings of the 2000 IEEE/EIA International Frequency Control Symposium and Exhibition, pp. 269-279, 2000

Kadiwar R.K & Giles, I.P. (1989). Optical fibre Brillouin ring laser gyroscope. Electronics Letters, Vol. 25, No. 25, pp. 1729 – 1731, 1989

Kosinski, J. A. (2000), Theory and design of crystal oscillators immune to acceleration: Present state of the art, Proceedings of the 2000 IEEE/EIA International Frequency Control Symposium and Exhibition, pp. 260-268, 2000.

Kwon, T.M. & Hahn, T. (1983). Improved vibration performance in passive atomic frequency standards by servo-loop control, Proceedings of the 37th Annual Symposium on frequency control, pp. 28-20, 1983

Lance, A.L.; Seal, W.D. & Labaar, F. (1984). Phase noise and AM noise measurements in the frequency domain. Infrared and Millimeter Waves, Vol. 11, pp. 239-289, 1984

Leeson, D.B. (1966). A simple model of feed back oscillator noise spectrum, Proceeding of the IEEE, Vol. 54, No. 2, pp. 329-330, 1966

Mancini, O. (2004). Tutorial on Precision Frequency Generation, Microwave Theory & Techniques Society of the IEEE Long Island Section, 18 March 2004. http://www.ieee.li/pdf/viewgraphs_freq.pdf

Minasian, R. (2006). Photonic signal processing of microwave signals. IEEE Transactions on Microwave Theory and Technology, Vol. 54, No. 2, pp. 832–846, February 2006

Parker, T.E. & Montress, G.K. (1988). Precision surface-acoustic-wave (SAW) oscillators. IEEE Transactions on Ultrasonics, Ferroelectrics, and Frequency Control, Vol. 35, No. 3, pp. 342 - 364, May 1988

Renoult, P.; Girardet, E. & Bidart, L. (1989). Mechanical and acoustic effects in low phase noise piezoelectric oscillators, Proceeding of the 43rd Annual Symposium on Frequency Control, pp. 439-446, 31 May-2 June, 1989

Riley, W. (1992). The Physics of the environmental sensitivity of rubidium gas cell atomic Frequency standards. IEEE Transactions on Ultrasonics, Ferroelectrics, and Frequency Control, Vol. 39, pp. 232-240, 1992

Römisch, S.; Kitching, J.; Ferre-Pikal, E.S.; Hollberg L. & Walls, F.L. (2000). Performance evaluation of an optoelectronic oscillator. IEEE Transactions on Ultrasonics, Ferroelectrics, and Frequency Control, Vol. 47, pp. 1159-1165, September 2000

Rosati V.J. & Filler, R. L. (1981). Reduction of the effects of vibration on SC-cut quartz crystal oscillators, Proceedings of the 35th Annual Symposium on Frequency Control, pp. 117-121, 1981

Rubiola, E.; Salik, E.; Huang, S.; Yu, N. & Maleki, L. (2005). Photonic delay technique for PM noise measurement of microwave oscillators. Journal Optical Society of America B: Optical Physics, Vol. 22, No.5, pp. 987-997, May 2005

Steinberg, D.S. (2000). Vibration Analysis for Electronic Equipment, John Wiley & Sons, Inc. 9780471376859, NY, USA

Sullivan, D.B; Allan, D.W.; Howe, D.A. & Walls F. L. (1990). Characterization of clocks and oscillators. National Institute of Standards and Technology Technical Note 1337, Section A-6, March 1990

Taylor, J.; Nelson, C.W.; Hati, A.; Ashby, N. & Howe, D.A. (2008). Vibration-induced PM noise measurements of a rigid optical fiber spool, Proceedings of the 2008 IEEE International Frequency Control Symposium, pp. 807-810, Hawaii, May 2008

Thieme, B., Zoschg, D. & Baister, G. (2004). Space worthy electronics package for the 35kg space active hydrogen maser on ACES, 18th European Frequency and Time Forum, April 2004

Vig, J.R.; Audoin, C.; Cutler, L.S.; Driscoll, M.M.; EerNisse, E.P.; Filler, R.L.; Garvey, R.M.; Riley, W.L.; Smythe, R.C. & Weglein, R.D. (1992). Acceleration, vibration and shock effects-IEEE standards project P1193, Proceedings of the 46rd Annual. Symposium on Frequency Control, pp. 763-781, 1992

Wallin, T.; Josefsson, L. & Lofter, B. (2003). Phase noise performance of sapphire microwave oscillators in airborne radar systems, GigaHertz 2003. Proceedings from the Seventh Symposium, Linköping, Sweden, November 4–5, 2003

Walls, F.L. & Ferre-Pikal, E.S. (1999). Measurement of frequency, phase noise and amplitude noise. Wiley Encyclopedia of Electrical and Electronics Engineering, Vol. 12, pp. 459-473, June 1999

Watts, M.H.; EerNisse, E.P.; Ward, R.W. & Wiggins, R.B. (1988). Technique for measuring the acceleration sensitivity of SC-cut quartz resonators, Proceedings of the 42rd Annual Symposium on Frequency Control, pp. 442-446, 1988

Weglein, R.D. (1989). The vibration-induced phase noise of a visco-elastically supported crystal resonator, Proceedings of the 43rd Annual Symposium on Frequency Control, pp. 433-438, 1989

Yao, X.S. & Maleki, L. (1996). Optoelectronic microwave oscillator. Journal Optical Society of America B, Vol. 13, No.8, pp. 1725-1735, 1996

14 Chapter

LOCALIZATION OF QUADROTOR UAV FORMATIONS UTILIZING NEURAL NETWORK CONTROL AND WIRELESS SENSOR NETWORKS

1. INTRODUCTION: AN OVERVIEW

Quadrotor helicopters have emerged as a prominent unmanned aerial vehicle (UAV) platform in recent years, with numerous researchers engaging in their control (Dierks & Jagannathan, 2008).A collaborative team of UAVs is frequently more efficient than an individual UAV in situations such as surveillance, search and rescue, and perimeter security. Consequently, the literature has proposed the formation control of UAVs.

Saffarian and Fahimi present a modified leader-follower framework and propose a model predictive nonlinear control algorithm to achieve the formation (Saffarian & Fahimi, 2008). Although the approach is verified via numerical simulations, proof of convergence and stability is not provided. In the work of Fierro et al., cylindrical coordinates and contributions from wheeled mobile robot formation control (Desai et al., 1998) are considered in the development of a leader-follower based formation control scheme for aircrafts whereas the complete dynamics are assumed to be known (Fierro et al., 2001). The work by Gu et al. proposes a solution to the leader-follower formation control problem involving a linear inner loop and nonlinear outer-loop control structure, and experimental results are provided (Gu et al., 2006). The associated drawbacks are the need for a dynamic model and the measured position and velocity of the leader has to be communicated to its followers. Xie et al. present two nonlinear robust formation controllers for UAV's where the UAV's are assumed to be flying at a constant altitude. The first approach assumes that the velocities and accelerations of the leader UAV are known while the second approach relaxes this assumption (Xie et al., 2005). In both the designs, the dynamics of the UAV's are assumed to be available. Then, Galzi and Shtessel propose a robust formation controller based on higher order sliding mode controllers in the presence of bounded disturbances (Galzi & Shtessel, 2006).

This study presents a novel leader-follower formation control framework for quadrotor UAVs utilizing spherical coordinates, wherein the desired position of a follower UAV is defined by a specific spacing, sd, and desired angles of incidence, αd, and bearing, βd. A novel control law for leader-follower formation control is formulated using neural networks (NN) to acquire the comprehensive dynamics of the UAV in real-time, encompassing unmodeled dynamics such as aerodynamic drag amid constrained perturbations. Despite the aquadrotor UAV being underactuated, a unique neural network virtual control input system for leader-follower formation control is proposed, enabling the regulation of all six degrees of freedom of the UAV with only four control inputs. We propose a graph theory-based framework for discovery, localization, and cooperative control. Discovery enables the UAVs to establish an ad hoc mobile sensor network, while localization permits each UAV to ascertain its position and orientation in relation to its neighbors, hence determining the formation shape. This chapter is structured as follows. Initially, Section 2 introduces the leader-follower formation control problem for UAVs and provides the necessary background knowledge. Subsequently, the neural network control law is formulated for both the follower UAVs and the formation leader, with the stability of the entire formation discussed in Section 3. Section 4 introduces the localization and routing system for UAV formation control, Section 5 shows numerical simulations, and Section 6 offers concluding observations.

2. BACKGROUND

2.1 Quadrotor UAV Dynamics

Consider a quadrotor UAV with six DOF defined in the inertial coordinate frame, Ea, as $[x, y, z, \phi, \theta, \psi]^T \in E^a$ where $\rho = [x, y, z]^T \in E^a$ are the position coordinates of the UAV and $\Theta = [\phi, \theta, \psi]^T \in E^a$ describe its orientation referred to as roll, pitch, and yaw, respectively. The translational and angular velocities are expressed in the body fixed frame attached to the center of mass of the UAV, Eb, and the dynamics of the UAV in the body fixed frame can be written as (Dierks & Jagannathan, 2008)

$$M\begin{bmatrix} \dot{v} \\ \dot{\omega} \end{bmatrix} = \bar{S}(\omega)\begin{bmatrix} v \\ \omega \end{bmatrix} + \begin{bmatrix} N_1(v) \\ N_2(\omega) \end{bmatrix} + \begin{bmatrix} G(R) \\ 0_{3x1} \end{bmatrix} + U + \tau_d \tag{1}$$

where $U = \begin{bmatrix} 0 & 0 & u_1 & u_2^T \end{bmatrix}^T \in \Re^6$,

$$M = \begin{bmatrix} mI_{3x3} & 0_{3x3} \\ 0_{3x3} & J \end{bmatrix} \in \Re^{6x6}, \quad \overline{S}(\omega) = \begin{bmatrix} -mS(\omega) & 0_{3x3} \\ 0_{3x3} & S(J\omega) \end{bmatrix} \in \Re^{6x6}$$

and m is a positive scalar that represents the total mass of the UAV, $J \in \Re^{3x3}$ represents the positive definite inertia matrix, $v(t) = [v_{xb}, v_{yb}, v_{zb}]^T \in \Re^3$ represents the translational velocity, $\omega(t) = [\omega_{xb}, \omega_{yb}, \omega_{zb}]^T \in \Re^3$ represents the angular velocity, $N_i(\bullet) \in \Re^{3x1}, i = 1,2$, are the nonlinear aerodynamic effects, $u_1 \in \Re^1$ provides the thrust along the z-direction, $u_2 \in \Re^3$ provides the rotational torques, $\tau_d = [\tau_{d1}^T, \tau_{d2}^T]^T \in \Re^6$ and $\tau_{di} \in \Re^3$, $i = 1,2$ represents unknown, but bounded disturbances such that $\|\tau_d\| < \tau_M$ for all time t, with τ_M being a known positive constant, $I_{nxn} \in \Re^{nxn}$ is an nxn identity matrix, and $0_{mxl} \in \Re^{mxl}$ represents an mxl matrix of all zeros. Furthermore, $G(R) \in \Re^3$ represents the gravity vector defined as $G(R) = mgR^T(\Theta)E_z$ where $E_z = [0,0,1]^T$ is a unit vector in the inertial coordinate frame, g = 9.81m /s2 , and $S(\bullet) \in \Re^{3x3}$ is the general form of a skew symmetric matrix defined as in (Dierks & Jagannathan, 2008). It is important to highlight $w^T S(\gamma) w = 0$ for any vector $w \in \Re^3$, and this property is commonly referred to as the skew symmetric property (Lewis et al., 1999).

The matrix $R(\Theta) \in \Re^{3x3}$ is the translational rotation matrix which is used to relate a vector in the body fixed frame to the inertial coordinate frame defined as (Dierks & Jagannathan, 2008)

$$R(\Theta) = R = \begin{bmatrix} c_\theta c_\psi & s_\phi s_\theta c_\psi - c_\phi s_\psi & c_\phi s_\theta c_\psi + s_\phi s_\psi \\ c_\theta s_\psi & s_\phi s_\theta s_\psi + c_\phi c_\psi & c_\phi s_\theta s_\psi - s_\phi c_\psi \\ -s_\theta & s_\phi c_\theta & c_\phi c_\theta \end{bmatrix} \tag{2}$$

where the abbreviations s(•) and c(•) have been used for sin(•) and cos(•), respectively. It is important to note that $R^{-1} = R^T$, $\dot{R} = RS(\omega)$ and $\dot{R}^T = -S(\omega)R^T$. It is also necessary to define a rotational transformation matrix from the fixed body to the inertial coordinate frame as (Dierks & Jagannathan, 2008)

$$T(\Theta) = T = \begin{bmatrix} 1 & s_\phi t_\theta & c_\phi t_\theta \\ 0 & c_\phi & -s_\phi \\ 0 & s_\phi / c_\theta & c_\phi / c_\theta \end{bmatrix} \tag{3}$$

where the abbreviation t(•) has been used for tan(•). The transformation matrices R and T are nonsingular as long as $-(\pi/2) < \varphi < (\pi/2)$, $-(\pi/2) < \theta < (\pi/2)$ and $-\pi \le \psi \le \pi$. These regions will be assumed throughout the development of this work, and will be referred to as the stable operation regions of the UAV. Under these flight conditions, it is observed that $\|R\|_F = R_{max}$ and $\|T\|_F < T_{max}$ for known constants Rmax and Tmax (Neff et al., 2007).

Finally, the kinematics of the UAV can be written as

$$\begin{aligned} \dot{\rho} &= Rv \\ \dot{\Theta} &= T\omega \end{aligned} \tag{4}$$

2.2 Neural Networks

In this work, two-layer NN's are considered consisting of one layer of randomly assigned constant weights $V_N \in \Re^{axL}$ in the first layer and one layer of tunable weights $W_N \in \Re^{Lxb}$ in the second with a inputs, b outputs, and L hidden neurons. A compromise is made here between tuning the number of layered weights with computational complexity. The universal approximation property for NN›s (Lewis et al., 1999) states that for any smooth function fN(xN), there exists a NN such that $f_N(x_N) = W_N^T \sigma(V_N^T x_N) + \varepsilon_N$ where εN is the bounded NN functional approximation error such that $\|\varepsilon_N\| < \varepsilon_M$, for a known constant ε_M and $\sigma(\cdot): \Re^a \to \Re^L$ is the activation function in the hidden layers. It has been shown that by randomly selecting the input layer weights V_N, the activation function $\sigma(\bar{x}_N) = \sigma(V_N^T x_N)$ forms a stochastic basis, and thus the approximation property holds for all inputs, $x_N \in \Re^a$, in the compact set S . The sigmoid activation function is considered here. Furthermore, on any compact subset of $\Re^n$, the target NN weights are bounded by a known positive value, WM, such that $\|W_N\|_F \le W_M$. For complete details of the NN and its properties, see (Lewis et al., 1999).

2.3 Three Dimensional Leader-Follower Formation Control

Throughout the development, the follower UAV's will be denoted with a subscript 'j' while the formation leader will be denoted by the subscript 'i'. To begin the development, an alternate reference frame is defined by rotating the inertial coordinate frame about the z-axis by the yaw angle of follower j, ψ_j, and denoted by E_j^a, In order to relate a vector in E^a to E_j^a the transformation matrix is given by.

$$R_{aj} = \begin{bmatrix} \cos\psi_j & \sin\psi_j & 0 \\ -\sin\psi_j & \cos\psi_j & 0 \\ 0 & 0 & 1 \end{bmatrix}, \tag{5}$$

where $R_{aj}^T = R_{aj}^{-1}$.

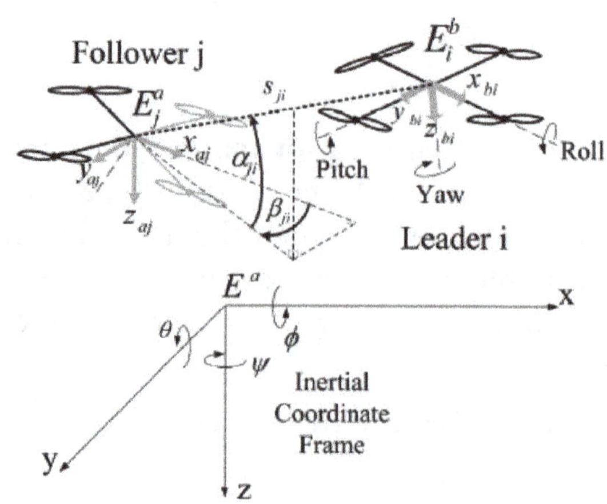

Figure 1. UAV leader-follower formation control

The objective of the proposed leader-follower formation control approach is for the follower UAV to maintain a desired separation, sjid, at a desired angle of incidence, $\alpha_{jid} \in E_j^a$, and bearing, $\beta_{jid} \in E_j^a$, with respect to its leader. The incidence angle is measured from the $x_{aj}-y_{aj}$ plane of follower j while the bearing angle is measured from the positive x_{aj} - axis as shown in Figure 1. It is important to observe that each quantity is defined relative to the follower j instead of the leader i (Fierro et al., 2001), (Desai et al., 1998). Additionally, in order to specify a unique configuration of follower j with respect to its leader, the desired yaw of follower j is selected to be the yaw angle of leader i, $\psi_i \in E^a$ as in (Saffarian & Fahimi, 2008). Using this approach, the measured separation between follower j and leader i is written as

$$\rho_i - \rho_j = R_{aj}^T s_{ji} \Xi_{ji}, \tag{6}$$

where

$$\Xi_{ji} = \begin{bmatrix} \cos\alpha_{ji} \cos\beta_{ji} \\ \cos\alpha_{ji} \sin\beta_{ji} \\ \sin\alpha_{ji} \end{bmatrix}. \tag{7}$$

Thus, to solve the leader-follower formation control problem in the proposed framework, a control velocity must be derived to ensure

$$\left.\begin{matrix} \lim_{t\to\infty}(s_{jid} - s_{ji}) = 0, \ \lim_{t\to\infty}(\beta_{jid} - \beta_{ji}) = 0, \\ \lim_{t\to\infty}(\alpha_{jid} - \alpha_{ji}) = 0, \ \lim_{t\to\infty}(\psi_{jd} - \psi_j) = 0 \end{matrix}\right\}. \tag{8}$$

Throughout the development, s_{jid}, α_{jid} and β_{jid} will be taken as constants, while the constant total mass, m_j, is assumed to be known. Additionally, it will be assumed that reliable communication between the leader and its followers is available, and the leader communicates its measured orientation, Θ_i, and its desired states, $\psi_{id}, \dot{\psi}_{id}, \ddot{\psi}_{id}$, $v_{id}, \dot{v}_{id}$. This is a far less stringent assumption than assuming the leader communicates all of its measured states to its followers (Gu et al., 2006). Additionally, future work will relax this assumption. In the following section, contributions from single UAV control will be considered and extended to the leader-follower formation control of UAV's.

3. LEADER-FOLLOWER FORMATION TRACKING CONTROL

In single UAV control literature, the overall control objective UAV j is often to track a desired trajectory, $\rho_{jd} = [x_{jd}, y_{jd}, z_{jd}]^T$, and a desired yaw ψ_{jd} while maintaining a stable flight configuration (Dierks & Jagannathan, 2008). The velocity v_{jzb} is directly controllable with the thrust input. However, in order to control the translational velocities v_{jxb} and v_{jyb}, the pitch and roll must be controlled, respectively, thus redirecting the thrust. With these objectives in mind, the frameworks for single UAV control are extended to UAV formation control as follows.

3.1 Follower UAV Control Law

Given a leader i subject to the dynamics and kinematics (1) and (4), respectively, define a reference trajectory at a desired separation s_{jid}, at a desired angle of incidence, α_{jid}, and bearing, β_{jid} for follower j given by

$$\rho_{jd} = \rho_i - R_{ajd}^T s_{jid} \Xi_{jid} \tag{9}$$

where Rajd is defined as in (5) and written in terms of ψ_{jd} , and Ξ_{jid} is written in terms of the desired angle of incidence and bearing, α_{jid}, βjid,respectively, similarly to (7). Next, using (6) and (9), define the position tracking error as

$$e_{j\rho} = \rho_{jd} - \rho_j = R_{aj}^T s_{ji} \Xi_{ji} - R_{ajd}^T s_{jid} \Xi_{jid} \in E^a \tag{10}$$

which can be measured using local sensor information. To form the position tracking error dynamics, it is convenient to rewrite (10) as $e_{j\rho} = \rho_i - \rho_j - R_{ajd}^T s_{jid} \Xi_{jid}$ revealing

$$\dot{e}_{j\rho} = R_i v_i - R_j v_j - \dot{R}_{ajd}^T s_{jid} \Xi_{jid} \,. \tag{11}$$

Next, select the desired translational velocity of follower j to stabilize (11)

$$v_{jd} = [v_{jdx} \;\; v_{jdy} \;\; v_{jdz}]^T = R_j^T \left(R_i v_{id} - \dot{R}_{ajd}^T s_{jid} \Xi_{jid} + K_{j\rho} e_{j\rho} \right) \in E^b \tag{12}$$

where $K_{j\rho} = diag\{k_{j\rho x}, k_{j\rho y}, k_{j\rho z}\} \in \Re^{3x3}$ is a diagonal positive definite design matrix of positive design constants and vid is the desired translational velocity of leader i. Next, the translational velocity tracking error system is defined as

$$e_{jv} = \begin{bmatrix} e_{jvx} \\ e_{jvy} \\ e_{jvz} \end{bmatrix} = \begin{bmatrix} v_{jdx} \\ v_{jdy} \\ v_{jdz} \end{bmatrix} - \begin{bmatrix} v_{jxb} \\ v_{jyb} \\ v_{jzb} \end{bmatrix} = v_{jd} - v_j \,. \tag{13}$$

Applying (12) to (11) while observing $v_j = v_{jd} - e_{jv}$ and similarly $e_{iv} = v_{id} - v_i$, reveals the closed loop position error dynamics to be rewritten as

$$\dot{e}_{j\rho} = -K_{j\rho} e_{j\rho} + R_j e_{jv} - R_i e_{iv} \,. \tag{14}$$

Next, the translational velocity tracking error dynamics are developed. Differentiating (13), observing

$$\dot{v}_{jd} = -S(\omega_j) v_{jd} + R_j^T \left(R_i S(\omega_i) v_{id} + R_i \dot{v}_{id} - \ddot{R}_{ajd}^T s_{jid} \Xi_{jid} \right) + R_j^T K_{j\rho} \left(R_i v_i - R_j v_j - \dot{R}_{ajd}^T s_{jid} \Xi_{jid} \right),$$

substituting the translational velocity dynamics in (1), and adding and subtracting $R_j^T (K_{j\rho} (R_i v_{id} + R_j v_{jd}))$ reveals

z

$$\begin{aligned} \dot{e}_{jv} = \dot{v}_{jd} - \dot{v}_j = &-N_{j1}(v_j)/m_j - S(\omega_j) e_{jv} - G(R_j)/m_j - u_{j1} E_{jz}/m_j - \bar{\tau}_{jd1} \\ &+ R_j^T \left(R_i S(\omega_i) v_{id} + R_i \dot{v}_{id} - \ddot{R}_{ajd} s_{jid} \Xi_{jid} + K_{j\rho} \left(R_j e_{jv} - K_{j\rho} e_{j\rho} \right) \right) - R_j^T K_{j\rho} (R_i e_{iv} - R_j e_{jv}) \end{aligned} \,. \tag{15}$$

Next, we rewrite (2) in terms of the scaled desired orientation vector $\bar{\Theta}_{jd} = [\bar{\theta}_{jd}\ \bar{\phi}_{jd}\ \psi_{jd}]^T$, where $\bar{\theta}_{jd} = \pi\theta_{jd}/(2\theta_{d\max}), \bar{\phi}_{jd} = \pi\phi_{jd}/(2\phi_{d\max})$, and $\theta_{d\max} \in (0,\pi/2)$ and $\theta_{d\max} \in (0,\pi/2)$ are the maximum desired roll and pitch, respectively, define $R_{jd}=R_j(\bar{\Theta}_{jd})$, and add and subtract $G(R_{jd})/m_j$ and $R_{jd}^T\Lambda_j$ with $\Lambda_j = R_i\dot{v}_{id} - \ddot{R}_{ajd}^T s_{jid}\Xi_{jid} + K_{j\rho}R_j e_{jv} - K_{j\rho}e_{j\rho}$ to $\dot{e}_{jv}$ to yield

$$\dot{e}_{jv} = -G(R_{jd})/m_j + R_{jd}^T\Lambda_j + A_{jc1}f_{cj1}(x_{cj1}) - u_{j1}E_{jz}/m_j - K_{j\rho}R_i e_{iv} - \bar{\tau}_{jd1} \tag{16}$$

where $A_{jc1} = diag\{\cos(\bar{\theta}_{jd}), \cos(\bar{\phi}_{jd}), 1\} \in \Re^{3x3}$ and

$$\begin{aligned} f_{cj1}(x_{cj1}) = {} & A_{jc1}^{-1}\left(G(R_{jd})/m_j - G(R_j)/m_j + (R_j^T - R_{jd}^T)\Lambda_j\right) + \\ & A_{jc1}^{-1}\left(K_{j\rho}R_j e_{jv} - S(\omega_j)e_{jv} - N_{j1}(v_j)/m_j + R_j^T R_i S(\omega_i)v_{id} + R_j^T K_{j\rho}(1-K_{j\rho})e_{j\rho}\right) \end{aligned} \tag{17}$$

is an unknown function which can be rewritten as $f_{jc1}(x_{jc1}) = [f_{jc11}\ f_{jc12}\ f_{jc13}]^T \in \Re^3$. In the forthcoming development, the approximation properties of NN will be utilized to estimate the unknown function $f_{jc1}(x_{jc1})$ by bounded ideal weights W_{jc1}^T, V_{jc1}^T such that $\|W_{jc1}\|_F \le W_{Mc1}$ for an unknown constant W_{Mc1}, and written as where $\varepsilon_{jc1} \le \varepsilon_{Mc1}$ is the bounded NN approximation error where ε_{Mc1} is a known constant. The NN estimate of f_{jc1} is written as $\hat{f}_{jc1} = \hat{W}_{jc1}^T\sigma(V_{jc1}^T\hat{x}_{jc1}) = \hat{W}_{jc1}^T\hat{\sigma}_{jc1} = [\hat{W}_{jc11}^T\hat{\sigma}_{jc1}\ \hat{W}_{jc12}^T\hat{\sigma}_{jc1}\ \hat{W}_{jc13}^T\hat{\sigma}_{jc1}]^T$ where $\hat{W}_{jc1}^T$ is the NN estimate of W_{jc1}^T, $\hat{W}_{jc1i}^T, i = 1,2,3$ is the ith row of $\hat{W}_{jc1}^T$, and $\hat{x}_{jc1}$ is the NN input defined as

$$\hat{x}_{jc1} = [1\ \Theta_j^T\ \Theta_i^T\ \Lambda_j^T\ v_{jd}^T\ v_{id}^T\ \dot{v}_{id}^T\ \psi_{jd}\ \dot{\psi}_{jd}\ \ddot{\psi}_{jd}\ \omega_j^T\ v_j^T\ e_{jv}^T\ e_{j\rho}^T]^T.$$

Note that $\hat{x}_{jc1}$ is an estimate of x_{jc1} since the follower does not know ωi. However, Θ_i is directly related to ω_i; therefore, it is included instead.

Remark 1: In the development of (16), the scaled desired orientation vector was utilized as a design tool to specify the desired pitch and roll angles. If the un-scaled desired orientation vector was used instead, the maximum desired pitch and roll would remain within the stable operating regions. However, it is desirable to saturate the desired pitch and roll before they reach the boundaries of the stable operating region.

Next, the virtual control inputs θ_{jd} and φ_{jd} are identified to control the translational velocities v_{jxb} and v_{jyb}, respectively. The key step in the development is identifying the desired closed loop velocity tracking error dynamics. For convenience, the desired translational velocity closed loop system is selected as

$$\dot{e}_{jv} = -K_{jv}e_{jv} - \bar{\tau}_{jd1} - K_{j\rho}R_i e_{iv} \tag{18}$$

where $K_{jv} = diag\{k_{jv1}\cos(\bar{\theta}_{jd}), k_{jv2}\cos(\bar{\phi}_{jd}), k_{v3}\}$ is a diagonal positive definite design matrix with each $k_{vi} > 0$, $i = 1,2,3$, and $\bar{\tau}_{jd1} = \tau_{jd1}/m_j$. In the following development, it will be shown that $\theta_d \in (-\pi/2, \pi/2)$ and $\phi_d \in (-\pi/2, \pi/2)$; therefore, it is clear that Kv > 0 . Then, equating (16) and (18) while considering only the first two velocity error states reveals

$$-g\begin{bmatrix} -s_{\bar{\theta}jd} \\ c_{\bar{\theta}jd}s_{\bar{\phi}jd} \end{bmatrix} + \begin{bmatrix} c_{\bar{\theta}jd}(k_{jv1}e_{jvx} + f_{jc11}) \\ c_{\bar{\phi}jd}(k_{jv2}e_{jvy} + f_{jc12}) \end{bmatrix} + \begin{bmatrix} c_{\bar{\theta}jd}c_{\psi jd} & c_{\bar{\theta}jd}s_{\psi jd} & -s_{\bar{\theta}jd} \\ s_{\bar{\phi}jd}s_{\bar{\theta}jd}c_{\psi jd} - c_{\bar{\phi}jd}s_{\psi jd} & s_{\bar{\phi}jd}s_{\bar{\theta}jd}s_{\psi jd} + c_{\bar{\phi}jd}c_{\psi jd} & s_{\bar{\phi}jd}c_{\bar{\theta}jd} \end{bmatrix} \begin{bmatrix} \Lambda_{j1} \\ \Lambda_{j2} \\ \Lambda_{j3} \end{bmatrix} = \begin{bmatrix} 0 \\ 0 \end{bmatrix} \quad (19)$$

where $\Lambda_j = [\Lambda_{j1} \ \Lambda_{j2} \ \Lambda_{j3}]^T$ was utilized. Then, applying basic math operations, the first line of (19) can be solved for the desired pitchθ_{jd} while the second line reveals the desired roll φjd. Using the NN estimates, $\hat{f}_{cj1}$, The desired pitch θjd can be written as

$$\theta_{jd} = \frac{2\theta_{max}}{\pi} a\tan\left(\frac{N_{\theta jd}}{D_{\theta jd}}\right) \quad (20)$$

where $N_{\theta jd} = c_{\psi jd}\Lambda_{j1} + s_{\psi jd}\Lambda_{j2} + k_{jv1}e_{jvx} + \hat{f}_{jc11}$ and $D_{\theta jd} = \Lambda_{j3} - g$. Similarly, the desired roll angle, φjd, is found to be

$$\phi_{jd} = \frac{2\phi_{max}}{\pi} a\tan\left(\frac{N_{\phi jd}}{D_{\phi jd}}\right) \quad (21)$$

where $N_{\phi jd} = s_{\psi jd}\Lambda_{j1} - c_{\psi jd}\Lambda_{j2} - k_{jv2}e_{jvy} + \hat{f}_{jc12}$ and $D_{\phi jd} = c_{\bar{\theta}jd}(\Lambda_{j3} - g) + s_{\bar{\theta}jd}c_{\psi jd}\Lambda_{j1} + s_{\bar{\theta}jd}s_{\psi jd}\Lambda_{j2}$.

Remark 2: The expressions for the desired pitch and roll in (20) and (21) lend themselves very well to the control of a quadrotor UAV. The expressions will always produce desired values in the stable operation regions of the UAV. It is observed that a tan(•) approaches ± π 2 as its argument increases. Thus, introducing the scaling factors in $\bar{\theta}_{jd}$ and $\bar{\phi}_{jd}$ results in $\theta_{jd} \in (-\theta_{max}, \theta_{max})$ and $\phi_{jd} \in (-\phi_{max}, \phi_{max})$, and the aggressiveness of the UAV's maneuvers can be managed.

Now that the desired orientation has been found, next define the attitude tracking error as

$$e_{j\Theta} = \Theta_{jd} - \Theta_j \in E^a \quad (22)$$

where the dynamics are found using (4) to be $\dot{e}_{j\Theta} = \dot{\Theta}_{jd} - T_j\omega_j$. In order to drive the orientation errors (22) to zero, the desired angular velocity, ω_{jd}, is selected as

$$\omega_{jd} = T_j^{-1}(\dot{\Theta}_{jd} + K_{j\Theta}e_{j\Theta}) \quad (23)$$

where $K_{j\Theta} = diag\{k_{j\Theta 1}, k_{j\Theta 2}, k_{j\Theta 3}\} \in \Re^{3x3}$ is a diagonal positive definite design matrix all with positive design constants. Define the angular velocity tracking error as

$$e_{j\omega} = \omega_{jd} - \omega_j \quad (24)$$

and observing $\omega_j = \omega_{jd} - e_{j\omega}$, the closed loop orientation tracking error system can be written as

$$\dot{e}_{j\Theta} = -K_{j\Theta}e_{j\Theta} + T_j e_{j\omega} \quad (25)$$

Examining (23), calculation of the desired angular velocity requires knowledge of $\dot{\Theta}_{jd}$; however, $\dot{\Theta}_{jd}$ is not known in view of the fact $\dot{\Lambda}_j$ and $\dot{\hat{f}}_{jc1}$ are not available. Further, development of u_{j2} in the following section will reveal $\dot{\omega}_{jd}$ is required which in turn implies $\ddot{\Lambda}_j$ and $\ddot{f}_{jc1}$ must be known. Since these requirements are not practical, the universal approximation property of NN is invoked to estimate ω_{jd} and $\dot{\omega}_{jd}$ (Dierks and Jagannathan, 2008).

To aid in the NN virtual control development, the desired orientation, $\Theta_{jd} \in E^a$, is reconsidered in the fixed body frame, E^b, using the relation $\dot{\Theta}^b_{jd} = T_j^{-1}\dot{\Theta}_{jd}$. Rearranging (23), the dynamics of the proposed virtual controller when the all dynamics are known are revealed to be

$$\begin{aligned}\dot{\Theta}^b_{jd} &= \omega_{jd} - T_j^{-1}K_{j\Theta}e_{j\Theta} \\ \dot{\omega}_{jd} &= \dot{T}_j^{-1}(\dot{\Theta}_{jd} + K_{j\Theta}e_{j\Theta}) + T_j^{-1}(\ddot{\Theta}_{jd} + K_{j\Theta}\dot{e}_{j\Theta})\end{aligned} \tag{26}$$

For convenience, we define a change of variable as $\Omega_d = \omega_d - T^{-1}K_\Theta e_\Theta$, and the dynamics (26) become

$$\begin{aligned}\dot{\Theta}^b_{jd} &= \Omega_{jd} \\ \dot{\Omega}_{jd} &= \dot{T}_j^{-1}\dot{\Theta}_{jd} + T_j^{-1}\ddot{\Theta}_{jd} = f_{j\Omega}(x_{j\Omega}) = f_{j\Omega}\end{aligned} \tag{27}$$

Defining the estimates of Θ^b_{jd} and Ω_{jd} to be $\hat{\Theta}^b_{jd}$ and $\hat{\Omega}_{jd}$, respectively, and the estimation error $\tilde{\Theta}^b_{jd} = \Theta^b_{jd} - \hat{\Theta}^b_{jd}$, the dynamics of the proposed NN virtual control inputs become

$$\begin{aligned}\dot{\hat{\Theta}}^b_{jd} &= \hat{\Omega}_{jd} + K_{j\Omega1}\tilde{\Theta}^b_{jd} \\ \dot{\hat{\Omega}}_{jd} &= \hat{f}_{j\Omega} + K_{j\Omega2}\tilde{\Theta}^b_{jd}\end{aligned} \tag{28}$$

where $K_{j\Omega1}$ and $K_{j\Omega2}$ are positive constants. The estimate $\hat{\omega}_{jd}$ is then written as

$$\hat{\omega}_{jd} = \hat{\Omega}_{jd} + K_{j\Omega3}\tilde{\Theta}^b_{jd} + T_j^{-1}K_{j\Theta}e_{j\Theta} \tag{29}$$

where $K_{j\Omega3}$ is a positive constant.

In (28), universal approximation property of NN has been utilized to estimate the unknown function $f_{j\Omega}(x_{j\Omega})$ by bounded ideal weights $W^T_{j\Omega}, V^T_{j\Omega}$ such that $\|W_{j\Omega}\|_F \le W_{M\Omega}$ for a known constant $W_{M\Omega}$, and written as $f_{j\Omega}(x_{j\Omega}) = W^T_{j\Omega}\sigma(V^T_{j\Omega}x_{j\Omega}) + \varepsilon_{j\Omega}$ where $\varepsilon_{j\Omega}$ is the bounded NN approximation error such that $\|\varepsilon_{j\Omega}\| \le \varepsilon_{\Omega M}$ for a known constant $\varepsilon_{\Omega M}$. The NN estimate of $f_{j\Omega}$ is written as $\hat{f}_{j\Omega} = \hat{W}^T_{j\Omega}\sigma(V^T_{j\Omega}\hat{x}_{j\Omega}) = \hat{W}^T_{j\Omega}\hat{\sigma}_{j\Omega}$ where $\hat{W}^T_{j\Omega}$ is the NN estimate of $W^T_{j\Omega}$ and $\hat{x}_{j\Omega}$ is the NN input written in terms of the virtual control estimates, desired trajectory, and the UAV velocity. The NN input is chosen to take the form of $\hat{x}_\Omega = [1\ \Lambda^T_j\ (\Theta^b_{jd})^T\ \hat{\Omega}^T_{jd}\ v^T_j\ \omega^T_j]^T$. Observing $\tilde{\omega}_{jd} = \omega_{jd} - \hat{\omega}_{jd} = \tilde{\Omega}_{jd} - K_{j\Omega3}\tilde{\Theta}^b_{jd}$, subtracting (28) from (27) and adding and subtracting $W^T_{j\Omega}\hat{\sigma}_{j\Omega}$, the virtual controller estimation error dynamics are found to be

$$\begin{aligned}\dot{\tilde{\Theta}}^b_{jd} &= \tilde{\omega}_{jd} - (K_{j\Omega1} - K_{j\Omega3})\tilde{\Theta}^b_{jd} \\ \dot{\tilde{\Omega}}_{jd} &= \tilde{f}_{j\Omega} - K_{j\Omega2}\tilde{\Theta}^b_{jd} + \xi_{j\Omega}\end{aligned} \tag{30}$$

where $\tilde{\Omega}_{jd} = \Omega_{jd} - \hat{\Omega}_{jd}$, $\tilde{f}_{j\Omega} = \tilde{W}^T_{j\Omega}\hat{\sigma}_{j\Omega}$, $\tilde{W}^T_{j\Omega} = W^T_{j\Omega} - \hat{W}^T_{j\Omega}$, $\xi_{j\Omega} = \varepsilon_{j\Omega} + W^T_{j\Omega}\tilde{\sigma}_{j\Omega}$, and $\tilde{\sigma}_{j\Omega} = \sigma_{j\Omega} - \hat{\sigma}_{j\Omega}$. Furthermore, $\|\xi_{j\Omega}\| \le \xi_{\Omega M}$ with $\xi_{\Omega M} = \varepsilon_{\Omega M} + 2W_{M\Omega}\sqrt{N_\Omega}$ a computable constant with NΩ the constant number of hidden layer neurons in the virtual control NN. Similarly, the estimation error dynamics of (29) are found to be

$$\dot{\tilde{\omega}}_{jd} = -K_{j\Omega3}\tilde{\omega}_{jd} + \tilde{f}_{j\Omega} - \bar{K}_{j\Omega}\tilde{\Theta}^{b}_{jd} + \xi_{j\Omega} \tag{31}$$

where $\bar{K}_{j\Omega} = K_{j\Omega2} - K_{j\Omega3}(K_{j\Omega1} - K_{j\Omega3})$. Examination of (30) and (31) reveals $\tilde{\Theta}^{b}_{jd}, \tilde{\omega}_{jd}$, and $\tilde{f}_{j\Omega}$ to be equilibrium points of the estimation error dynamics when $\|\xi_{j\Omega}\| = 0$.

To this point, the desired translational velocity for follower j has been identified to ensure the leader follower objective (8) is achieved. Then, the desired pitch and roll were derived to drive $v_{jxb} \to v_{jdx}$ and $v_{jyb} \to v_{jdy}$, respectively. Then, the desired angular velocity was found to ensure $\Theta_j \to \Theta_{jd}$. What remains is to identify the UAV thrust to guarantee $v_{jzb} \to v_{jdz}$ and rotational torque vector to ensure $\omega_j \to \omega_{jd}$. First, the thrust is derived.

Consider again the translational velocity tracking error dynamics (16), as well as the desired velocity tracking error dynamics (18). Equating (16) and (18) and manipulating the third error state, the required thrust is found to be

$$\begin{aligned} u_{j1} &= m_j c_{\bar{\theta}jd} c_{\bar{\phi}jd}(\Lambda_{j3} - g) + m_j(c_{\bar{\phi}jd} s_{\bar{\theta}jd} c_{\psi jd} + s_{\bar{\phi}jd} s_{\psi jd})\Lambda_{j1} \\ &+ m_j(c_{\bar{\phi}jd} s_{\bar{\theta}jd} s_{\psi jd} - s_{\bar{\phi}jd} c_{\psi jd})\Lambda_{j2} + m_j k_{jvz} e_{vj3} + m_j \hat{f}_{jc13} \end{aligned} \tag{32}$$

where $\hat{f}_{jc13}$ is the NN estimate in (17) previously defined. Substituting the desired pitch (20), roll (21), and the thrust (32) into the translational velocity tracking error dynamics (16) yields

$$\dot{e}_{jv} = -K_{jv}e_{jv} + A_{jc1}(W^T_{jc1}\sigma_{jc1} + \varepsilon_{jc}) - A_{jc1}\hat{W}^T_{jc1}\hat{\sigma}_{jc1} - K_{j\rho}R_i e_{iv} - \bar{\tau}_{jd1}',$$

and adding and subtracting $A_{jc1}W^T_{jc1}\hat{\sigma}^T_{jc1}$ reveals

$$\dot{e}_{jv} = -K_{jv}e_{jv} + A_{jc1}\tilde{W}^T_{jc1}\hat{\sigma}_{jc1} - K_{j\rho}R_i e_{iv} + \xi_{jc1} \tag{33}$$

with $\xi_{jc1} = A_{jc1}W^T_{jc1}\tilde{\sigma}^T_{jc1} + A_{jc1}\varepsilon_{jc1} - \bar{\tau}_{jd1}$, $\tilde{W}_{jc1} = W_{jc1} - \hat{W}_{jc1}$ and $\tilde{\sigma}_{jc1} = \sigma_{jc1} - \hat{\sigma}_{jc1}$. Further, $\|A_{jc1}\|_F = A_{c1\max}$ max for a known constant $A_{c1\max}$, and $\|\xi_{jc1}\| \le \xi_{Mc1}$ for a computable constant $\xi_{Mc1} = A_{c1\max}\varepsilon_{Mc1} + 2A_{c1\max}W_{Mc1}\sqrt{N_c} + \tau_M / m_j$.

Next, the rotational torque vector, u_{j2}, will be addressed. First, multiply the angular velocity tracking error (24) by the inertial matrix J_j, take the first derivative with respect to time, and substitute the UAV dynamics (1) to reveal

$$J_j\dot{e}_{j\omega} = f_{jc2}(x_{jc2}) - u_{j2} - \tau_{jd2} \tag{34}$$

with $f_{jc2}(x_{jc2}) = J_j\dot{\omega}_{jd} - S(J_j\omega_j)\omega_j - N_{j2}(\omega_j)$. Examining $f_{jc2}(x_{jc2})$, it is clear that the function is nonlinear and contains unknown terms; therefore, the universal approximation property of NN is utilized to estimate the function $f_{jc2}(x_{jc2})$ by bounded ideal weights W^T_{jc2}, V^T_{jc2} such that $\|W_{jc2}\|_F \le W_{Mc2}$ for a known constant W_{Mc2} and written as $f_{jc2}(x_{jc2}) = W^T_{jc2}\sigma(V^T_{jc2}x_{jc2}) + \varepsilon_{jc2}$ where ε_{jc2} is the bounded NN functional reconstruction error such that $\|\varepsilon_{jc2}\| \le \varepsilon_{Mc2}$ for a known constant ε_{Mc2}. The NN estimate of f_{jc2} is given by

$\hat{f}_{jc2}=\hat{W}_{jc2}^{T}\sigma(V_{jc2}^{T}\hat{x}_{jc2})=\hat{W}_{jc2}^{T}\hat{\sigma}_{jc2}$ where $\hat{W}_{jc2}^{T}$ is the NN estimate of $\hat{W}_{jc2}^{T}$ and $\hat{x}_{jc2}=[1\ \omega_j^T\ \dot{\hat{\Omega}}_{jd}^T\ \tilde{\Theta}_{jd}^{b\ T}\ e_{j\Theta}^T]^T$ is the input to the NN written in terms of the virtual controller estimates. By the construction of the virtual controller, $\dot{\hat{\omega}}_{jd}$ is not directly available; therefore, observing (29), the terms $\dot{\hat{\Omega}}_{jd}^T$, $\tilde{\Theta}_{jd}^{b\ T}$, and $e_{j\Theta}^T$ have been included instead.

Using the NN estimate $\hat{f}_{jc2}$ and the estimated desired angular velocity tracking error $\hat{e}_{j\omega}=\hat{\omega}_{jd}-\omega_j$, the rotational torque control input is written as

$$u_{j2}=\hat{f}_{jc2}+K_{j\omega}\hat{e}_{j\omega}, \tag{35}$$

and substituting the control input (35) into the angular velocity dynamics (34) as well as adding and subtracting $W_{jc2}^{T}\hat{\sigma}_{jc}$, the closed loop dynamics become

$$J_j\dot{e}_{j\omega}=-K_{j\omega}e_{j\omega}+\tilde{W}_{jc2}^{T}\hat{\sigma}_{jc2}+K_{j\omega}\tilde{\omega}_{jd}+\xi_{jc2}, \tag{36}$$

where $\tilde{W}_{jc2}^{T}=W_{jc2}^{T}-\hat{W}_{jc2}^{T}$, $\xi_{jc2}=\varepsilon_{jc2}+W_{jc2}^{T}\tilde{\sigma}_{jc}-\tau_{jd2}$, and $\tilde{\sigma}_{jc2}=\sigma_{jc2}-\hat{\sigma}_{jc2}$. Further, for a computable constant $\xi_{Mc2}=\varepsilon_{Mc2}+2W_{Mc2}\sqrt{N_{c2}}+\tau_{dM}$ where Nc2 is the number of hidden layer neurons.

As a final step, we define $\tilde{W}_{jc}=[\tilde{W}_{jc1}\ 0;\ 0\ \tilde{W}_{jc2}]$ and $\hat{\sigma}_{jc}=[\hat{\sigma}_{jc1}^{T}\ \ \hat{\sigma}_{jc2}^{T}]^{T}$ so that a single NN can be utilized with N_c hidden layer neurons to represent $\hat{f}_{jc}=[\hat{f}_{jc1}^{T}\ \ \hat{f}_{jc2}^{T}]^{T}\in\Re^{6}$. In the following theorem, the stability of the follower j is shown while considering $e_{iv}=0$. In other words, the position, orientation, and velocity tracking errors are considered along with the estimation errors of the virtual controller and the NN weight estimation errors of each NN for follower j while ignoring the interconnection errors between the leader and its followers. This assumption will be relaxed in the following section.

Theorem 3.1.1: (Follower UAV System Stability) Given the dynamic system of follower j in the form of (1), let the desired translational velocity for follower j to track be defined by (12) with the desired pitch and roll defined by (20) and (21), respectively. Let the NN virtual controller be defined by (28) and (29), respectively, with the NN update law given by

$$\dot{\hat{W}}_{j\Omega}=F_{j\Omega}\hat{\sigma}_{j\Omega}\left(\tilde{\Theta}_{jd}^{b}\right)^{T}-\kappa_{j\Omega}F_{j\Omega}\hat{W}_{j\Omega}, \tag{37}$$

where $F_{j\Omega}=F_{j\Omega}^{T}>0$ and $\kappa_{j\Omega}>0$ are design parameters. Let the dynamic NN controller for follower j be defined by (32) and (35), respectively, with the NN update given by

$$\dot{\hat{W}}_{jc}=F_{jc}\hat{\sigma}_{jc}\left(A_{jc}\hat{e}_{jS}\right)^{T}-\kappa_{jc}F_{jc}\hat{W}_{jc}, \tag{38}$$

where $A_{jc}=[A_{jc1}\ 0_{3x3};0_{3x3}\ I_{3x3}]\in\Re^{6x6}$, $\hat{e}_{jS}=\left[e_{jv}^{T}\ \ \hat{e}_{j\omega}^{T}\right]^{T}$, $F_{jc}=F_{jc}^{T}>0$ and $\kappa_{jc}>0$ are constant design parameters. Then there exists positive design constants $K_{j\Omega1}, K_{j\Omega2}, K_{j\Omega3}$, and positive definite design matrices $K_{j\rho}, K_{j\Theta}, K_{jv}, K_{j\omega}$, such that the virtual controller estimation errors $\tilde{\Theta}_{jd}^{b}, \tilde{\omega}_{jd}$ and the virtual control NN weight estimation errors, $\tilde{W}_{j\Omega}$, the position, orientation, and translational and angular velocity tracking errors, $e_{j\rho}, e_{j\Theta}, e_{jv}, e_{j\omega}$, respectively, and the dynamic controller NN weight estimation errors, $\tilde{W}_{jc}$, are all SGUUB. Proof: Consider the following positive definite Lyapunov candidate

$$V_j = V_{j\Omega} + V_{jc}, \tag{39}$$

where

$$V_{j\Omega} = \frac{1}{2}\tilde{\Theta}_{jd}^{b\ T}\overline{K}_{j\Omega}\tilde{\Theta}_{jd}^{b} + \frac{1}{2}\tilde{\omega}_{jd}^{\ T}\tilde{\omega}_{jd} + \frac{1}{2}tr\{\tilde{W}_{j\Omega}^{T}F_{j\Omega}^{-1}\tilde{W}_{j\Omega}\}$$

$$V_{jc} = \frac{1}{2}e_{j\rho}^{T}e_{j\rho} + \frac{1}{2}e_{j\Theta}^{T}e_{j\Theta} + \frac{1}{2}e_{jv}^{T}e_{jv} + \frac{1}{2}e_{j\omega}^{T}J_{j}e_{j\omega} + \frac{1}{2}tr\left\{\tilde{W}_{jc}^{T}F_{jc}^{-1}\tilde{W}_{jc}\right\}$$

whose first derivative with respect to time is given by $\dot{V}_j = \dot{V}_{j\Omega} + \dot{V}_{jc}$. Considering first $\dot{V}_{j\Omega}$, and substituting the closed loop virtual control estimation error dynamics (30) and (31) as well as the NN tuning law (37) , reveals

$$\dot{V}_{j\Omega} = -\overline{K}_{j\Omega 2}\tilde{\Theta}_{jd}^{b\ T}\tilde{\Theta}_{jd}^{b} - K_{j\Omega 3}\tilde{\omega}_{jd}^{\ T}\tilde{\omega}_{jd} + \tilde{\omega}_{jd}^{\ T}\xi_{j\Omega} + tr\left\{\tilde{W}_{j\Omega}^{T}\left(\kappa_{j\Omega}\hat{W}_{j\Omega} - \hat{\sigma}_{j\Omega}\left(\tilde{\Theta}_{jd}^{b}\right)^{T} + \hat{\sigma}_{j\Omega}\tilde{\omega}_{jd}^{\ T}\right)\right\}$$

where $\overline{K}_{j\Omega 2} = (K_{j\Omega 1} - K_{j\Omega 3})(K_{j\Omega 2} - K_{j\Omega 3}(K_{j\Omega 1} - K_{j\Omega 3}))$ and $\overline{K}_{j\Omega 2} > 0$ provided $K_{j\Omega 1} > K_{j\Omega 3}$ and $K_{j\Omega 2} > K_{j\Omega 3}(K_{j\Omega 1} - K_{j\Omega 3})$. Observing $\|\hat{\sigma}_{j\Omega}\| \le \sqrt{N_{j\Omega}}$, $\|W_{\Omega j}\|_F \le W_{M\Omega}$ for a known constant, $W_{M\Omega}$, and $tr\{\tilde{W}_{j\Omega}^{T}(W_{j\Omega} - \tilde{W}_{j\Omega})\} \le \|\tilde{W}_{j\Omega}\|_F W_{M\Omega} - \|\tilde{W}_{j\Omega}\|_F^2$, $\dot{V}_{j\Omega}$ can then be rewritten as

$$\dot{V}_{\Omega} \le -\overline{K}_{j\Omega 2}\|\tilde{\Theta}_{jd}^{b}\|^2 - K_{j\Omega 3}\|\tilde{\omega}_{jd}\|^2 - \kappa_{j\Omega}\|\tilde{W}_{j\Omega}\|_F^2 + \|\tilde{\omega}_{jd}\|\xi_{\Omega M} + \|\tilde{\Theta}_{jd}^{b}\|\|\tilde{W}_{j\Omega}\|_F\sqrt{N_{j\Omega}} + \|\tilde{\omega}_{jd}\|\|\tilde{W}_{j\Omega}\|_F\sqrt{N_{j\Omega}} + \kappa_{j\Omega}\|\tilde{W}_{j\Omega}\|_F W_{M\Omega}.$$

Now, completing the squares with respect to $\|\tilde{W}_{j\Omega}\|_F$, $\|\tilde{\Theta}_{jd}^{b}\|$, and $\|\tilde{\omega}_{jd}\|$, an upper bound for $\dot{V}_{j\Omega}$ is found to be

$$\dot{V}_{j\Omega} \le -\left(\overline{K}_{j\Omega 2} - \frac{N_{j\Omega}}{\kappa_{j\Omega}}\right)\|\tilde{\Theta}_{jd}^{b}\|^2 - \left(\frac{K_{j\Omega 3}}{2} - \frac{N_{j\Omega}}{\kappa_{j\Omega}}\right)\|\tilde{\omega}_{jd}\|^2 - \frac{\kappa_{j\Omega}}{4}\|\tilde{W}_{j\Omega}\|_F^2 + \eta_{j\Omega} \tag{40}$$

where $\eta_{j\Omega} = \kappa_{j\Omega}W_{M\Omega}^2 + \xi_{\Omega M}^2/(2K_{j\Omega 3})$. Next, considering $\dot{V}_{jc}$ and substituting the closed loop kinematics (14) and (25), dynamics (33) and (36), and NN tuning law (38) while considering $e_{jv} = 0$ reveals

$$\dot{V}_{jc} = -e_{j\rho}^{T}K_{j\rho}e_{j\rho} - e_{j\Theta}^{T}K_{j\Theta}e_{j\Theta} - e_{jv}^{T}K_{jv}e_{jv} - e_{j\omega}^{T}K_{j\omega}e_{j\omega} + e_{j\rho}^{T}R_{j}e_{jv} + e_{j\Theta}^{T}T_{j}e_{j\omega} + e_{jv}^{T}\xi_{jc1} + e_{j\omega}^{T}K_{j\omega}\tilde{\omega}_{jd} + e_{j\omega}^{T}\xi_{jc2}$$
$$+ \kappa_{jc}tr\left\{\tilde{W}_{jc}^{T}(W_{jc} - \tilde{W}_{jc})\right\} + tr\left\{\tilde{W}_{jc}^{T}\hat{\sigma}_{jc2}(e_{j\omega}^{T} - \hat{e}_{j\omega}^{T})\right\}$$

Then, observing $\tilde{\omega}_{jd} = e_{j\omega} - \hat{e}_{j\omega}$ and completing the squares with respect to $e_{j\rho}, e_{j\Theta}, e_{jv}, e_{j\omega}$ and $\tilde{W}_{jc}$, and upper bound for $\dot{V}_{jc}$ is found to be

$$\dot{V}_{jc} \le -\frac{K_{j\rho\min}}{2}\|e_{j\rho}\|^2 - \frac{K_{j\Theta\min}}{2}\|e_{j\Theta}\|^2 - \left(\frac{K_{jv\min}}{2} - \frac{R_{\max}^2}{2K_{\rho\min}}\right)\|e_{jv}\|^2 - \frac{\kappa_{jc}}{3}\|\tilde{W}_{jc}\|_F^2 - \left(\frac{K_{j\omega\min}}{3} - \frac{T_{\max}^2}{2K_{j\Theta\min}}\right)\|e_{j\omega}\|^2 \tag{41}$$
$$+ \frac{3K_{j\omega\min}}{4}\|\tilde{\omega}_{jd}\|^2 + \frac{3N_{jc}}{4\kappa_{jc}}\|\tilde{\omega}_{jd}\|^2 + \eta_{jc}$$

where $K_{j\rho\min}$, $K_{j\Theta\min}$, $K_{jv\min}$, and $K_{j\omega\min}$ are the minimum singular values of $K_{j\rho}$, $K_{j\Theta}$, K_{jv}, and $K_{j\omega}$, respectively, and $\eta_{jc} = \xi_{c1M}^2/(2K_{jv\min}) + \xi_{c2M}^2/(2K_{j\omega\min}) + 3W_{Mc}\kappa_{jc}/4$. Now, combining (40) and (41), an upper bound for $\dot{V}_j$ is written as

$$\dot{V}_j \le -\left(\overline{K}_{j\Omega} - \frac{N_{j\Omega}}{\kappa_{j\Omega}}\right)\left\|\tilde{\Theta}_{jd}^b\right\|^2 - \left(\frac{K_{j\Omega 3}}{2} - \frac{N_{j\Omega}}{\kappa_{j\Omega}} - \frac{3K_{j\omega\min}}{4} - \frac{3N_{jc}}{4\kappa_{jc}}\right)\left\|\tilde{\omega}_{jd}\right\|^2 - \frac{K_{j\rho\min}}{2}\left\|e_{j\rho}\right\|^2 - \frac{K_{j\Theta\min}}{2}\left\|e_{j\Theta}\right\|^2 \tag{42}$$

$$-\left(\frac{K_{jv\min}}{2} - \frac{R_{\max}^2}{2K_{\rho\min}}\right)\left\|e_{jv}\right\|^2 - \left(\frac{K_{j\omega\min}}{3} - \frac{T_{\max}^2}{2K_{j\Theta\min}}\right)\left\|e_{j\omega}\right\|^2 - \frac{\kappa_{j\Omega}}{4}\left\|\tilde{W}_{j\Omega}\right\|_F^2 - \frac{\kappa_{jc}}{3}\left\|\tilde{W}_{jc}\right\|_F^2 + \eta_{j\Omega} + \eta_{jc}$$

Finally, (42) is less than zero provided

$$\overline{K}_{j\Omega 2} > \frac{N_{j\Omega}}{\kappa_{j\Omega}},\quad K_{j\Omega 3} > \frac{2N_{j\Omega}}{\kappa_{j\Omega}} + \frac{3K_{j\omega\min}}{2} + \frac{3N_{jc}}{2\kappa_{jc}},\quad K_{jv\min} > \frac{R_{\max}^2}{K_{\rho\min}},\quad K_{j\omega\min} > \frac{3T_{\max}^2}{2K_{j\Theta\min}} \tag{43}$$

and the following inequalities hold:

$$\left\|\tilde{\Theta}_{jd}^b\right\| > \sqrt{\frac{\eta_{j\Omega} + \eta_{jc}}{\overline{K}_{j\Omega 2} - N_{j\Omega}/\kappa_{j\Omega}}} \quad or \quad \left\|\tilde{W}_{jc}\right\|_F > \sqrt{\frac{3(\eta_{j\Omega} + \eta_{jc})}{\kappa_{jc}}} \quad or \quad \left\|e_{jv}\right\| > \sqrt{\frac{2(\eta_{j\Omega} + \eta_{jc})}{K_{jv\min} - R_{\max}^2/K_{\rho\min}}} \tag{44}$$

$$or \quad \left\|e_{j\rho}\right\| > \sqrt{\frac{2(\eta_{j\Omega} + \eta_{jc})}{K_{j\rho\min}}} \quad or \quad \left\|e_{j\Theta}\right\| > \sqrt{\frac{2(\eta_{j\Omega} + \eta_{jc})}{K_{j\Theta\min}}} \quad or \quad \left\|\tilde{W}_{j\Omega}\right\|_F > \sqrt{\frac{4(\eta_{j\Omega} + \eta_{jc})}{\kappa_{j\Omega}}}$$

$$or \quad \left\|e_{j\omega}\right\| > \sqrt{\frac{\eta_{j\Omega} + \eta_{jc}}{K_{j\omega\min}/3 - T_{\max}^2/(2K_{j\Theta\min})}} \quad or \quad \left\|\tilde{\omega}_{jd}\right\| > \sqrt{\frac{\eta_{j\Omega} + \eta_{jc}}{\frac{K_{j\Omega 3}}{2} - \frac{N_{j\Omega}}{\kappa_{j\Omega}} - \frac{3K_{j\omega\min}}{4} - \frac{3N_{jc}}{4\kappa_{jc}}}}$$

Therefore, it can be concluded using standard extensions of Lyapunov theory (Lewis et al., 1999) that V is less than zero outside of a compact set, revealing the virtual controller estimation errors, $\tilde{\Theta}_{jd}^b, \tilde{\omega}_{jd}$, and the NN weight estimation errors, $\tilde{W}_{j\Omega}$, the position, orientation, and translational and angular velocity tracking errors $e_{j\rho}, e_{j\Theta}, e_{jv}, e_{j\omega}$, respectively, and the dynamic controller NN weight estimation errors, $\tilde{W}_{jc}$, are all SGUUB.

3.2 Formation Leader Control Law

The dynamics and kinematics for the formation leader are defined similarly to (1) and (4), respectively. In our previous work (Dierks and Jagannathan, 2008), an output feedback control law for a single quadrotor UAV was designed to ensure the robot tracks a desired path, $\rho_{id} = [x_{id}, y_{id}, z_{id}]^T$, and desired yaw angle, ψ_{id}. Using a similarly approach to (10)-(14), the state feedback control velocity for leader i is given by (Dierks and Jagannathan, 2008)

$$v_{id} = [v_{idx}\ \ v_{idy}\ \ v_{idz}]^T = R_i^T\left(\dot{\rho}_{id} + K_{ip}e_{ip}\right) \in E^b \tag{45}$$

The closed loop position tracking error then takes the form of

$$\dot{e}_{ip} = -K_{ip}e_{ip} + R_i\, e_{iv} \tag{46}$$

Then, using steps similar to (15)-(21), the desired pitch and roll angles are given by

$$\theta_{id} = \frac{2\theta_{\max}}{\pi} a\tan\left(\frac{N_{i\theta d}}{D_{i\theta d}}\right) \tag{47}$$

where $N_{i\theta d} = c_{i\psi d}\left(\ddot{x}_{id} + k_{ipx}\dot{x}_{id} - v_{iR1}\right) + s_{i\psi d}\left(\ddot{y}_{id} + k_{ipy}\dot{y}_{id} - v_{iR2}\right) + k_{iv1}e_{vx} + \hat{f}_{ic11}$ and $D_{i\theta d} = \ddot{z}_{id} + k_{ipz}\dot{z}_{id} - v_{iR3} - g$ and

$$\phi_{id} = \frac{2\phi_{\max}}{\pi} a\tan\left(\frac{N_{i\phi d}}{D_{i\phi d}}\right) \tag{48}$$

where $N_{i\phi d} = s_{i\psi d}\left(\ddot{x}_{id} + k_{ipx}\dot{x}_{id} - v_{iR1}\right) - c_{i\psi d}\left(\ddot{y}_{id} + k_{ipy}\dot{y}_{id} - v_{iR2}\right) - k_{iv2}e_{vy} + \hat{f}_{ic12}$, $D_{\phi d} = c_{\bar{\theta}di}\left(\ddot{z}_{id} + k_{ipz}\dot{z}_{id} - v_{iR3} - g\right) + s_{\bar{\theta}id}c_{\psi id}\left(\ddot{x}_{id} + k_{ipx}\dot{x}_{id} - v_{iR1}\right) + s_{\bar{\theta}di}s_{\psi id}\left(\ddot{y}_{id} + k_{ipy}\dot{y}_{id} - v_{iR2}\right)$ with $v_{iR} = [v_{iR1}\ v_{iR2}\ v_{iR3}]^T = K_{ip}R_i v_i$ and $\hat{f}_{ic1} = \left[\hat{f}_{ic11}\ \hat{f}_{ic12}\ \hat{f}_{ic13}\right]^T$ is a NN estimate of the unknown function $f_{ic1}(x_{ic1})$ (Dierks and Jagannathan, 2008). The desired angular velocity as well as the NN virtual controller for the formation leader is defined similarly to (23) and (28) and (29), respectively, and finally, the thrust and rotation torque vector are found to be

$$\begin{aligned} u_{i1} &= m_i c_{\bar{\theta}id} c_{\bar{\phi}id}\left(\ddot{z}_{id} + k_{ipz}\dot{z}_{id} - v_{iR3} - g\right) + m_i\left(c_{\bar{\phi}id}s_{\bar{\theta}id}s_{\psi id} - s_{\bar{\phi}id}c_{\psi id}\right)\left(\ddot{y}_{id} + k_{ipy}\dot{y}_{id} - v_{iR2}\right) + m_i k_{iv3}e_{ivz} \\ &+ m_i\left(c_{\bar{\phi}id}s_{\bar{\theta}id}c_{\psi id} + s_{\bar{\phi}id}s_{\psi id}\right)\left(\ddot{x}_{id} + k_{ipx}\dot{x}_{id} - v_{iR1}\right) + m_i\hat{f}_{ic13} \end{aligned} \tag{49}$$

$$u_{i2} = \hat{f}_{ic2} + K_{i\omega}\hat{e}_{i\omega} \tag{50}$$

where $\hat{f}_{ic2} \in \Re^3$ is a NN estimate of the unknown function $f_{ic2}(x_{ic2})$ and $\hat{e}_{i\omega} = \hat{\omega}_{id} - \omega_i$. The closed loop orientation, virtual control, and velocity tracking error dynamics for the formation leader are found to take a form similar to (25), (30) and (31), and (33) and (36), respectively (Dierks and Jagannathan, 2008).

Next, the stability of the entire formation is considered in the following theorem while considering the interconnection errors between the leader and its followers.

3.3 Quadrotor UAV Formation Stability

Before proceeding, it is convenient to define the following augmented error systems consisting of the position and translational velocity tracking errors of leader i and N follower UAV's as

$$e_\rho = \left[e_{i\rho}^T\ e_{j\rho}^T\Big|_{j=1} \cdots e_{j\rho}^T\Big|_{j=N}\right]^T \in \Re^{3(N+1)}$$

$$e_v = \left[e_{iv}^T\ e_{jv}^T\Big|_{j=1} \cdots e_{jv}^T\Big|_{j=N}\right]^T \in \Re^{3(N+1)}.$$

Next, the transformation matrix (2) is augmented as

$$R_F = diag\left\{R_i, R_j\Big|_{j=1},\ldots,R_j\Big|_{j=N}\right\} \in \Re^{3(N+1)x3(N+1)} \tag{51}$$

while the NN weights for the translational velocity error system are augmented as

$$\hat{W}_{c1} = diag\left\{\hat{W}_{ic1}, \hat{W}_{jc1}\Big|_{j=1},\ldots,\hat{W}_{jc1}\Big|_{j=N}\right\} \in \Re^{(N \cdot N_{jc1}+N_{ic1})x3(N+1)}$$

$$\hat{\sigma}_{c1} = \left[\hat{\sigma}_{ic1}^T \ \hat{\sigma}_{jc1}^T\Big|_{j=1},\ldots,\hat{\sigma}_{jc1}^T\Big|_{j=N}\right]^T \in \Re^{(N \cdot N_{jc1}+N_{ic1})}$$

Now, using the augmented variables above, the augmented closed loop position and translational velocity error dynamics for the entire formation are written as

$$\dot{e}_\rho = -K_\rho e_\rho + (I - G_F)R_F \, e_v \tag{52}$$

$$\dot{e}_v = -K_v e_v + A_{cF}\tilde{W}_{c1}^T\hat{\sigma}_{c1} - K_\rho G_F R_F e_v + \xi_{c1} \tag{53}$$

where $A_{cF} = diag\left\{A_{ic}, A_{jc}\Big|_{j=1},\ldots,A_{jc}\Big|_{j=N}\right\}$ with A_{ic} defined similarly to A_{jc} in terms of $\overline{\Theta}_{id}$, $K_\rho = diag\left\{K_{i\rho}, K_{j\rho}\Big|_{j=1},\ldots K_{j\rho}\Big|_{j=N}\right\}$, $K_v = diag\left\{K_{iv}, K_{jv}\Big|_{j=1},\ldots,K_{jv}\Big|_{j=N}\right\}$, G_F is a constant matrix relating to the formation interconnection errors defined as

$$G_F = [0 \ 0; \ F_T \ \ 0] \in \Re^{(N+1)x(N+1)} \tag{54}$$

and $F_T \in \Re^{NxN}$ is constant and dependent on the specific formation topology. For instance, in a string formation where each follower follows the UAV directly in front of it, follower 1 tracks leader i, follower 2 tracks follower 1, etc., and F_T becomes the identity matrix.

In a similar manner, we define augmented error systems for the virtual controller, orientation, and angular velocity tracking systems as

$$\tilde{\Theta}_d^b = \left[(\tilde{\Theta}_{id}^b)^T, (\tilde{\Theta}_{jd}^b)^T\Big|_{j=1}\ldots(\tilde{\Theta}_{jd}^b)^T\Big|_{j=N}\right]^T \in \Re^{3(N+1)}, \quad \tilde{\omega}_d = \left[\tilde{\omega}_{id}^T, \tilde{\omega}_{jd}^T\Big|_{j=1}\ldots\tilde{\omega}_{jd}^T\Big|_{j=N}\right]^T \in \Re^{3(N+1)},$$

$$e_\Theta = \left[e_{i\Theta}^T \ e_{j\Theta}^T\Big|_{j=1}\ldots e_{j\Theta}^T\Big|_{j=N}\right]^T \in \Re^{3(N+1)}, \quad e_\omega = \left[e_{i\omega}^T \ e_{j\omega}^T\Big|_{j=1}\ldots e_{j\omega}^T\Big|_{j=N}\right]^T \in \Re^{3(N+1)}, \tag{55}$$

respectively. It is straight forward to verify that the error dynamics of the augmented variables (55) takes the form of (30), (31), (25), and (36), respectively, but written in terms of the augmented variables (55).

Theorem 3.3.1: (UAV Formation Stability) Given the leader-follower criterion of (8) with 1 leader and N followers, let the hypotheses of Theorem 3.1.1 hold. Let the virtual control system for the leader i be defined similarly to (28) and (29) with the virtual control NN update law defined similarly to (37). Let control velocity and desire pitch and roll for the leader be given by (45), (47), and (48), respectively, along with the thrust and rotation torque vector defined by (49) and (50), respectively, and let the control NN update law be defined identically to (38). Then, the position, orientation, and velocity tracking errors, the virtual control estimation errors, and the NN weights for each NN for the entire formation are all SGUUB.

Proof: Consider the following positive definite Lyapunov candidate

$$V = V_\Omega + V_c, \tag{56}$$

where

$$V_\Omega = \frac{1}{2}\tilde{\Theta}_d^{b\,T}\overline{K}_\Omega\tilde{\Theta}_d^b + \frac{1}{2}\tilde{\omega}_d^{\,T}\tilde{\omega}_d + \frac{1}{2}tr\{\tilde{W}_\Omega^T F_\Omega^{-1}\tilde{W}_\Omega\}$$

$$V_c = \frac{1}{2}e_\rho^T e_\rho + \frac{1}{2}e_\Theta^T e_\Theta + \frac{1}{2}e_v^T e_v + \frac{1}{2}e_\omega^T J e_\omega + \frac{1}{2}tr\{\tilde{W}_c^T F_c^{-1}\tilde{W}_c\}$$

with $\overline{K}_\Omega = diag\{\overline{K}_{i\Omega}I, \overline{K}_{j\Omega}I|_{j=1}..\overline{K}_{j\Omega}I|_{j=N}\}$, $\tilde{W}_\Omega = diag\{\tilde{W}_{i\Omega}, \tilde{W}_{j\Omega}|_{j=1}..\tilde{W}_{j\Omega}|_{j=N}\}$, $J = diag\{J_i, J_j|_{j=1}..J_j|_{j=N}\}$,

and $\tilde{W}_c = diag\{\tilde{W}_{ic}, \tilde{W}_{jc}|_{j=1}..\tilde{W}_{jc}|_{j=N}\}$. The first derivative of (56) with respect to time is given by $\dot{V} = \dot{V}_\Omega + \dot{V}_c$, and performing similar steps as those used in (40)-(41) reveals

$$\dot{V}_\Omega \le -\left(\overline{K}_{\Omega 2\min} - \frac{N_\Omega}{\kappa_{\Omega\min}}\right)\left\|\tilde{\Theta}_d^b\right\|^2 - \left(\frac{K_{\Omega 3\min}}{2} - \frac{N_\Omega}{\kappa_{\Omega\min}}\right)\left\|\tilde{\omega}_d\right\|^2 - \frac{\kappa_\Omega}{4}\left\|\tilde{W}_\Omega\right\|_F^2 + \eta_\Omega, \tag{57}$$

$$\dot{V}_c \le -\frac{K_{\rho\min}}{2}\left\|e_\rho\right\|^2 - \frac{K_{\Theta\min}}{2}\left\|e_\Theta\right\|^2 - \left(\frac{K_{v\min}}{2} - \frac{\eta_1^2}{2K_{\rho\min}} - \eta_2\right)\left\|e_{jv}\right\|^2 - \frac{\kappa_{c\min}}{3}\left\|\tilde{W}_c\right\|_F^2 - \left(\frac{K_{\omega\min}}{3} - \frac{(N+1)T_{\max}^2}{2K_{\Theta\min}}\right)\left\|e_\omega\right\|^2, \tag{58}$$
$$+\frac{3K_{\omega\min}}{4}\left\|\tilde{\omega}_d\right\|^2 + \frac{3N_c}{4\kappa_{c\min}}\left\|\tilde{\omega}_d\right\|^2 + \eta_c$$

where $\overline{K}_{\Omega 2\min}$ min is the minimum singular value of $\overline{K}_{2\Omega}$, $\kappa_{\Omega\min}$ is the minimum singular value of $\kappa_\Omega = diag\{\kappa_{i\Omega}I, \kappa_{j\Omega}I|_{j=1}..\kappa_{j\Omega}I|_{j=N}\}$ with I being the identity matrix, $K_{\Omega 3\min}$ is the minimum singular value of $K_{\Omega 3} = diag\{K_{i\Omega 3}I, K_{j\Omega 3}I|_{j=1}..K_{j\Omega 3}I|_{j=N}\}$, N_Ω is the number of hidden layer neurons in the augmented virtual control system, and η_Ω is a computable constant based on $\eta_{i\Omega}$ and $\eta_{j\Omega}, j = 1,...N$. Similarly, $K_{\rho\min}, K_{\Theta\min}, K_{v\min}, K_{\omega\min}$, and $\kappa_{c\min}$ are the minimum singular values of the augmented gain matrices $K_\rho, K_\Theta, K_v, K_\omega$, and κ_c respectively, where $\eta_1 = R_{F\max}\sqrt{1+2N}$, $\eta_2 = K_{\rho\min}R_{F\max}\sqrt{N}$ are a known computable constants and ηc is a computable constant dependent on ηic and ηjc, j = 1,...N. Now, using (57) and (58), an upper bound for $\dot{V}$ is found to be

$$\dot{V}_\Omega \le -\left(\overline{K}_{\Omega 2\min} - \frac{N_\Omega}{\kappa_{\Omega\min}}\right)\left\|\tilde{\Theta}_d^b\right\|^2 - \left(\frac{K_{\Omega 3\min}}{2} - \frac{N_\Omega}{\kappa_{\Omega\min}} - \frac{3K_{\omega\min}}{4} - \frac{3N_c}{4\kappa_{c\min}}\right)\left\|\tilde{\omega}_d\right\|^2 - \frac{\kappa_\Omega}{4}\left\|\tilde{W}_\Omega\right\|_F^2 - \frac{\kappa_{c\min}}{3}\left\|\tilde{W}_c\right\|_F^2 \tag{59}$$
$$-\frac{K_{\rho\min}}{2}\left\|e_\rho\right\|^2 - \frac{K_{\Theta\min}}{2}\left\|e_\Theta\right\|^2 - \left(\frac{K_{v\min}}{2} - \frac{\eta_1^2}{2K_{\rho\min}} - \eta_2\right)\left\|e_{jv}\right\|^2 - \left(\frac{K_{\omega\min}}{3} - \frac{(N+1)T_{\max}^2}{2K_{\Theta\min}}\right)\left\|e_\omega\right\|^2 + \eta_c + \eta_\Omega$$

Finally, (59) is less than zero provided

$$\overline{K}_{\Omega 2\min} > \frac{N_\Omega}{\kappa_{\Omega\min}},\quad K_{\Omega 3\min} > \frac{2N_\Omega}{\kappa_{\Omega\min}} + \frac{3K_{\omega\min}}{2} + \frac{3N_c}{2\kappa_{c\min}},\quad K_{v\min} > \frac{\eta_1^2}{K_{\rho\min}} + 2\eta_2,\quad K_{\omega\min} > \frac{3(N+1)T_{\max}^2}{2K_{\Theta\min}} \tag{60}$$

and the following inequalities hold:

$$\left\|\tilde{\Theta}_d^b\right\| > \sqrt{\frac{\eta_\Omega + \eta_c}{\overline{K}_{\Omega 2\min} - N_\Omega/\kappa_{\Omega\min}}} \;\; or \;\; \left\|\tilde{W}_c\right\|_F > \sqrt{\frac{3(\eta_\Omega + \eta_{jc})}{\kappa_{c\min}}} \;\; or \;\; \left\|e_{jv}\right\| > \sqrt{\frac{2(\eta_\Omega + \eta_c)}{K_{v\min} - \eta_1^2/K_{\rho\min} - 2\eta_2}} \tag{61}$$

$$or \;\; \left\|e_\rho\right\| > \sqrt{\frac{2(\eta_\Omega + \eta_c)}{K_{\rho\min}}} \;\; or \;\; \left\|e_\Theta\right\| > \sqrt{\frac{2(\eta_\Omega + \eta_c)}{K_{\Theta\min}}} \;\; or \;\; \left\|\tilde{W}_\Omega\right\|_F > \sqrt{\frac{4(\eta_\Omega + \eta_c)}{\kappa_{\Omega\min}}}$$

$$or \;\; \left\|e_\omega\right\| > \sqrt{\frac{\eta_\Omega + \eta_c}{K_{\omega\min}/3 - (N+1)T_{\max}^2/(2K_{\Theta\min})}} \;\; or \;\; \left\|\tilde{\omega}_d\right\| > \sqrt{\frac{\eta_\Omega + \eta_c}{\frac{K_{\Omega 3\min}}{2} - \frac{N_\Omega}{\kappa_{\Omega\min}} - \frac{3K_{\omega\min}}{4} - \frac{3N_c}{4\kappa_{c\min}}}}$$

Therefore, it can be concluded using standard extensions of Lyapunov theory (Lewis et al., 1999) that V is less than zero outside of a compact set, revealing the position, orientation, and velocity tracking errors, the virtual control estimation errors, and the NN weights for each NN for the entire formation are all SGUUB.

Remark 3: The conclusions of Theorem 3.3.1 are independent of any specific formation topology, and the Lyapunov candidate (56) represents the most general form required show the stability of the entire formation. Examining (60) and (61), the minimum controller gains and error bounds are observed to increase with the number of follower UAV's, N. These results are not surprising since increasing number of UAV's increases the sources of errors which can be propagated throughout the formation.

Remark 4: Once a specific formation as been decided and the form of F_T is set, the results of Theorem 3.3.1 can be reformulated more precisely. For this case, the stability of the formation is proven using the sum of the individual Lyapunov candidates of each UAV as opposed to using the augmented error systems (51)-(55).

4. OPTIMIZED ENERGY-DELAY SUB-NETWORK ROUTING (OEDSR) PROTOCOL FOR UAV LOCALIZATION AND DISCOVERY

In the previous section, a group of UAV's was modeled as a nonlinear interconnected system. We have shown that the basic formation is stable and each follower achieves its separation, angle of incidence, and bearing relative to its leader with a bounded error. The controller assignments for the UAV's can be represented as a graph where a directed edge from the leader to the followers denotes a controller for the followers while the leader is trying to track a desired trajectory. The shape vector consists of separations and orientations which in turn determines the relative positions of the follower UAV's with respect to its leader.

Then, a group of N UAV's is built on two networks: a physical network that captures the constraints on the dynamics of the lead UAV and a sensing and communication network, preferably wireless, that describes information flow, sensing and computational aspects across the group. The design of the graph is based on the task in hand. In this graph, nodes and edges represent UAV's and control policies, respectively. Any such graph can be described by its adjacency matrix (Das et al., 2002).

In order to solve the leader-follower criterion (8), ad hoc networks are formed between the leader(s) and the follower(s), and the position of each UAV in the formation must be determined on-line. This network is dependent upon the sensing and communication aspects. As a first step, a leader is elected similar to the case of multi-robot formations (Das et al., 2002) followed by the discovery process in which the sensory information and physical networks are used to establish a wireless network. The outcome of the leader election process must be communicated to the followers in order to construct an appropriate shape. To complete the leader-follower formation control task (8), the controllers developed in the previous section require only a single-hop protocol; however, a multi-hop architecture is useful to relay information throughout the entire formation like the outcome of the leader election process, changing tasks, changing formations, as well alerting the UAV's of approaching moving obstacles that appear in a sudden manner.

The optimal energy-delay sub-network routing (OEDSR) protocol (Jagannathan, 2007) allows the UAV's to communicate information throughout the formation wirelessly using a multi-hop manner where each UAV in the formation is treated as a hop. The energy-delay routing protocol can guarantee information transfer while minimizing energy and delay for real-time control purposes even for mobile ad hoc networks such as the case of UAV formation flying.

We envision four steps to establish the wireless ad hoc network. As mentioned earlier, leader election process is the first step. The discovery process is used as the second step where sensory information and the physical network are used to establish a spanning tree. Since this is a multi-hop routing protocol, the communication network is created on- demand unlike in the literature where a spanning tree is utilized. This on-demand nature would allow the UAV's to be silent when they are not being used in communication and generate a communication path when required. The silent aspect will reduce any inference to others. Once a formation becomes stable, then a tree can be constructed until the shape changes. The third step will be assignment of the controllers online to each UAV based on the location of the UAV. Using the wireless network, localization is used to combine local sensory information along with information obtained via routing from other UAV's in order to calculate relative positions and orientations. Alternatively, range sensors provide relative separations, angles of incidence, and bearings. Finally cooperative control allows the graph obtained from the network to be refined. Using this graph theoretic formulation, a group is modeled by a tuple = Γ (ϑ, P, H) where ϑ is the reference trajectory of the robot, P represents the shape vectors describing the relative positions of each vehicle with respect to the formation reference frame (leader), and H is the control policy represented as a graph where nodes represent UAV and edges represent the control assignments. Next, we describe the OEDSR routing protocol where each UAV will be referred to as a "node."

In OEDSR, sub-networks are formed around a group of nodes due to an activity, and nodes wake up in the sub-networks while the nodes elsewhere in the network are in sleep mode. An appropriate percentage of nodes in the sub-network are elected as cluster heads (CHs) based on a metric composed of available energy and relative location to an event (Jagannathan, 2007) in each sub-network. Once the CHs are identified and the nodes are clustered relative to the distance from the CHs, the routing towards the formation leader (FL) is initiated. First, the CH checks if the FL is within the communication range. In such case, the data is sent directly to the FL. Otherwise, the data from the CHs in the sub-network are sent over a multi-hop route to the FL. The proposed routing algorithm is fully distributed since it requires only local information for constructing routes, and is proactive adapting to changes in the network. The FL is assumed to have sufficient power supply, allowing a high power beacon from the FL to be sent such that all the nodes in the network have knowledge of the distance to the FL. It is assumed that all UAV's in the network can calculate or measure the relative distance to the FL at any time instant using the formation graph information or local sensory data. Though the OEDSR protocol borrows the idea of an energy-delay metric from OEDR (Jagannathan, 2007), selection of relay nodes (RN) does not maximize the number of two hop neighbors. Here, any UAV can be selected as a RN, and the selection of a relay node is set to maximize the link cost factor which includes distance from the FL to the RN.

4.1 Selection of an Optimum Relay-Node-Based link cost factor

Knowing the distance information at each node will allow the UAV to calculate the Link Cost Factor (LCF). The link cost factor from a given node to the next hop node 'k' is given by (62) where D_k represent the delay that will be incurred to reach the next hop node in range, the distance between the next hop node to the F_L is denoted by Δx_k, and the remaining energy, E_k, at the next hop node are used in calculation of the link cost as

$$LCF_k = \frac{E_k}{D_k \cdot \Delta x_k} \tag{62}$$

In equation (62), checking the remaining energy at the next hop node increases network lifetime; the distance to the F_L from the next hop node reduces the number of hops and end- to-end delay; and the delay incurred to reach the next hop node minimizes any channel problems. When multiple RNs are available for routing of the information, the optimal RN is selected based on the highest LCF. These clearly show that the proposed OEDSR protocol is an on demand routing protocol. For detailed discussion of OEDSR refer to (Jagannathan, 2007). The route selection process is illustrated through the following example. This represents sensor data collected by a follower UAV for the task at hand that is to be transmitted to the FL.

4.2 Routing Algorithm through an Example

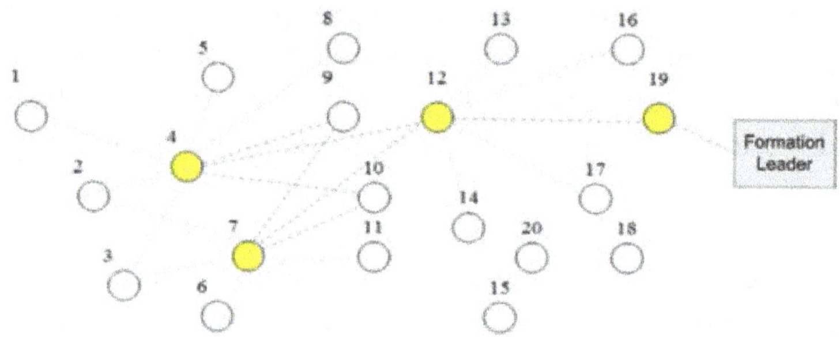

Figure 2. Relay node selection

Consider the formation topology shown in Figure 2. The link cost factors are taken into consideration to route data to the FL. The following steps are implemented to route data using the OEDSR protocol:

1. Start with an empty relay list for source UAV n: Relay(n)={ }. Here UAV n_4 and n_7 are CHs.
2. First, CH n_4 checks with which nodes it is in range with. In this case, CH n_4 is in range with nodes n_1, n_2, n_3, n_5, n_8, n_9, n_{12}, and n_{10}.
3. The nodes n_1, n_2, and n_3 are eliminated as potential RNs because the distance from them to the FL is greater than the distance from CH n_4 to the FL.
4. Now, all the nodes that are in range with CH n_4 transmit RESPONSE packets and CH n_4 makes a list of possible RNs, which in this case are n_5, n_8, n_9, n_{12}, and n_{10}.
5. CH n4 sends this list to CH n7. CH n7 checks if it is range with any of the nodes in the list.
6. Nodes n_9, n_{10}, and n_{12} are the nodes that are in range with both CH n_4 and n_7. They are selected as the potential common RNs.
7. The link cost factors for n_9, n_{10}, and n_{12} are calculated.
8. The node with the maximum value of LCF is selected as the RN and assigned to Relay(n). In this case, Relay(n)={n_{12}}.
9. Now UAV n_{12} checks if it is in direct range with the FL, and if it is, then it directly routes the information to the FL.

Otherwise, n_{12} is assigned as the RN, and all the nodes that are in range with node n_{12} and whose distance to the FL is less than its distance to the FL are taken into consideration. Therefore, UAV's n_{13}, n_{16}, n_{19}, and n_{17} are taken into consideration.

The LCF is calculated for n_{13}, n_{16}, n_{19}, n_{14}, and n_{17}. The node with the maximum LCF is selected as the next RN. In this case Relay(n) = {n19}.

Next the RN n_{19} checks if it is in range with the FL. If it is, then it directly routes the information to the FL. In this case, n_{19} is in direct range, so the information is sent to the FL directly.

4.3 Optimality Analysis for OEDSR

To prove that the proposed route created by OEDSR protocol is optimal in all cases, it is essential to show it analytically.

Assumption 1: It is assumed that all UAV's in the network can calculate or measure the relative distance to the FL at any time instant using the formation graph information or local sensory data.

Theorem 4.3.1: The link cost factor-based routing generates viable RNs to the FL.

Proof: Consider the following two cases

Case I: When the CHs are one hop away from the FL, the CH selects the FL directly. In this case, there is only one path from the CH to the FL. Hence, OEDSR algorithm does not need to be used.

Case II: When the CHs have more than one node to relay information, the OEDSR algorithm selection criteria are taken into account. In Figure 3, there are two CHs, CH_1 and CH_2. Each CH sends signals to all the other nodes in the network that are in range. Here, CH_1 first sends out signals to n_1, n_3, n_4, and n_5 and makes a list of information about the potential RN. The list is then forwarded to CH_2. CH_2 checks if it is in range with any of the nodes in the list. Here, n_4 and n_5 are selected as potential common RNs. A single node must be selected from both n_4 and n_5 based on the OEDSR link cost factor. The cost to reach n from CH is given by (2). So based on the OEDSR link cost factor, n_4 is selected as the RN for the first hop. Next, n_4 sends signals to all the nodes it has in range, and selects a node as RN using the link cost factor. The same procedure is carried on till the data is sent to the FL. Lemma 4.3.2: The intermediate UAV's on the optimal path are selected as RNs by the previous nodes on the path.

Proof: A UAV is selected as a RN only if it has the highest link cost factor and is in range with the previous node on the path. Since OEDSR maximizes the link cost factor, intermediate nodes that satisfy the metric on the optimal path are selected as RNs.

Lemma 4.3.3: A UAV can correctly compute the optimal path (with lower end to end delay and maximum available energy) for the entire network topology.

Proof: When selecting the candidate RNs to the CHs, it is ensured that the distance form the candidate RN to the FL is less than the distance from the CH to the FL. When calculating the link cost factor, available energy is divided by distance and average end-to-end delay to ensure that the selected nodes are in range with the CHs and close to the FL. This helps minimize the number of multi-point RNs in the network.

Theorem 4.3.4: OEDSR protocol results in an optimal route (the path with the maximum energy, minimum average end-to-end delay and minimum distance from the FL) between the CHs and any source destination.

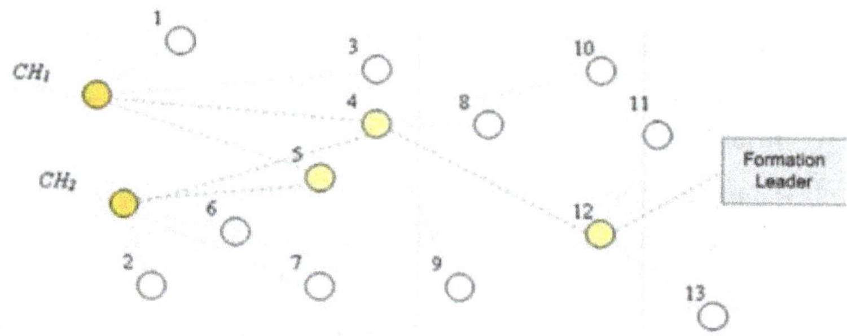

Figure 3. Link cost calculation

5. SIMULATION RESULTS

A wedge formation of three identical quadrotor UAV's is now considered in MATLAB with the formation leader located at the apex of the wedge. In the simulation, follower 1 should track the leader at a desired separation s_{jid} = 2m, desired angle of incidence α_{jid} = 0(rad), and desired bearing $\beta_{jid} = \pi/3$ (rad) while follower 2 tracks the leader at a desired separation s_{jid} = 2m desired angle of incidence, $\alpha_{jid} = -\pi/10$(rad), and desired bearing $\beta_{jid} = -\pi/3$(rad) . The desired yaw angle for the leader and thus the formation is selected to be $\psi_d = \sin(0.3\pi\, t)$. The inertial parameters of the UAV's are taken to be as m = 0.9kg and J = diag{0.32, 0.42, 0.63}kg m^2, and aerodynamic friction is modeled as in (Dierks and Jagannathan, 2008). The parameters outlined in Section 3.3 are communicated from the leader to its followers using single hop communication whereas results for OEDSR in a mobile environment can be found in (Jagannathan, 2007).

Each NN employs 5 hidden layer neurons, and for the leader and each follower, the control gains are selected to be, $K_{\Omega 1}=23$, $K_{\Omega 2}=80$, $K_{\Omega 3}=20$, $K_\rho = \mathrm{diag}\{10,10, 30\}$, $k_{v1}=10$, $k_{v2}=10$, $k_{v3} = 30$, $K_\Theta = \mathrm{diag}\{30,30,30\}$, and $K_\omega = \mathrm{diag}\{25, 25, 25\}$. The NN parameters are selected as, $F_\Omega = 10$, $\kappa_\Omega = 1$, and $F_c = 10$, $\kappa_c = 0.1$, and the maximum desired pitch and roll values are both selected as $2\pi / 5$.

Figure 4 displays the quadrotor UAV formation trajectories while Figures 5-7 show the kinematic and dynamic tracking errors for the leader and its followers. Examining the trajectories in Figure 4, it is important to recall that the bearing angle, β_{ji}, is measured in the inertial reference frame of the follower rotated about its yaw angle. Examining the tracking errors for the leader and its followers in Figures 5-7, it is clear that all states track their desired values with small bounded errors as the results of Theorem 3.3.1 suggest. Initially, errors are observed in each state for each UAV, but these errors quickly vanish as the virtual control NN and the NN in the actual control law learns the nonlinear UAV dynamics. Additionally, the tracking performance of the underactuated states v_x and v_y implies that the desired pitch and roll, respectively, as well as the desired angular velocities generated by the virtual control system are satisfactory.

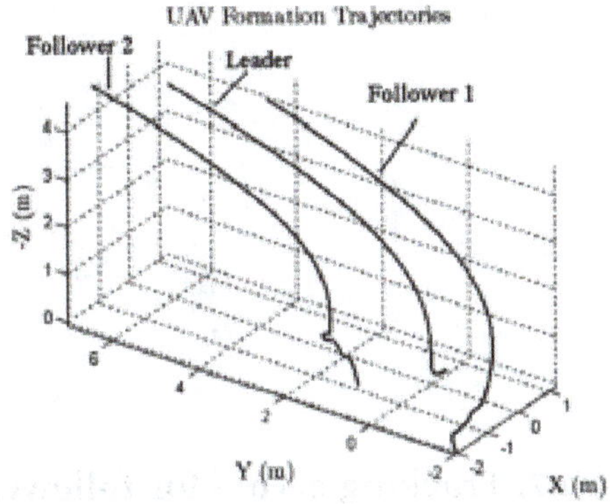

Figure 4. Quadrotor UAV formation trajectories

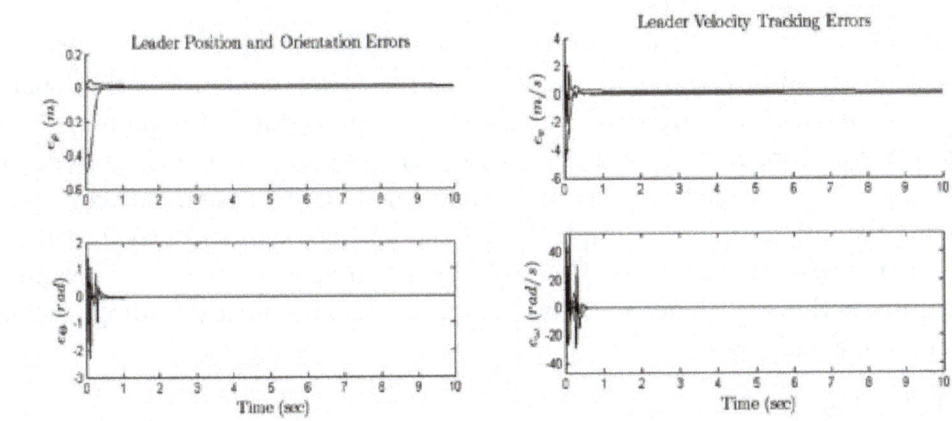

Figure 5. Tracking errors for the formation leader

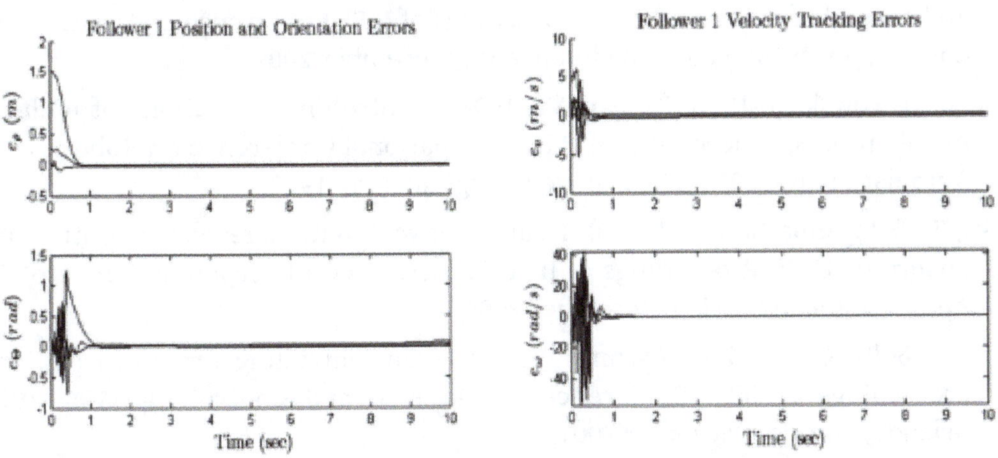

Figure 6. Tracking errors for follower 1

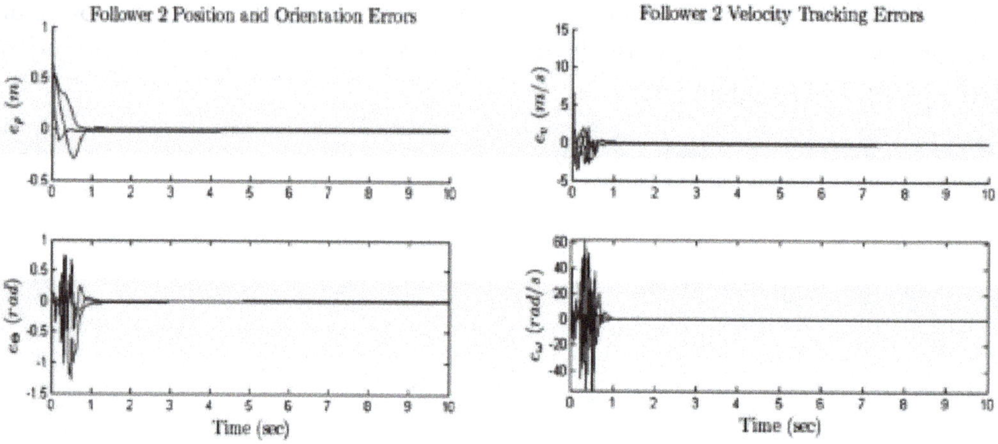

Figure 7. Tracking errors for follower 2

6. CONCISE SUMMARY

A unique framework for quadrotor UAV leader-follower formation control was introduced, using an innovative neural network formation control law that enables each follower to track its leader without requiring knowledge of dynamics. All six degrees of freedom are effectively monitored with merely four control inputs, despite the presence of unmodeled dynamics and constrained disturbances. A discovery and localization technique utilizing graph theory and ad hoc networks was introduced, ensuring maximum utilization of the UAV's communication channels. Lyapunov analysis ensures the SGUUB of the entire formation, and numerical findings validate the theoretical assertions.

REFERENCES

Das, A.; Spletzer, J.; Kumar, V.; & Taylor, C. (2002). Ad hoc networks for localization and control of mobile robots, Proceedings of IEEE Conference on Decision and Control, pp. 2978-2983, Philadelphia, PA, December 2002

Desai, J.; Ostrowski, J. P.; & Kumar, V. (1998). Controlling formations of multiple mobile robots, Proceedings of IEEE International Conference on Robotics and Automation, pp. 2964-2869, Leuven, Belgium, May 1998

Dierks, T. & Jagannathan, S. (2008). Neural network output feedback control of a quadrotor UAV, Proceedings of IEEE Conference on Decision and Control, To Appear, Cancun Mexico, December 2008

Fierro, R.; Belta, C.; Desai, J.; & Kumar, V. (2001). On controlling aircraft formations, Proceedings of IEEE Conference on Decision and Control, pp. 1065-1079, Orlando, Florida, December 2001

Galzi, D. & Shtessel, Y. (2006). UAV formations control using high order sliding modes, Proceedings of IEEE American Control Conference, pp. 4249-4254, Minneapolis, Minnesota, June 2006

Gu, Y.; Seanor, B.; Campa, G.; Napolitano M.; Rowe, L.; Gururajan, S.; & Wan, S. (2006). Design and flight testing evaluation of formation control laws. IEEE Transactions on Control Systems Technology, Vol. 14, No. 6, (November 2006) page numbers (1105-1112)

Jagannathan, S. (2007). Wireless Ad Hoc and Sensor Networks, Taylor & Francis, ISBN 0-8247- 2675-8, Boca Raton, FL

Lewis, F.L.; Jagannathan, S.; & Yesilderek, A. (1999). Neural Network Control of Robot Manipulators and Nonlinear Systems, Taylor & Francis, ISBN 0-7484-0596-8, Philadelphia, PA

Neff, A.E.; DongBin, L.; Chitrakaran, V.K.; Dawson, D.M.; & Burg, T.C. (2007). Velocity control for a quad-rotor uav fly-by-camera interface, Proceedings of the IEEE Southeastern Conference, pp. 273-278, Richmond, VA, March 2007

Saffarian, M. & Fahimi, F. (2008). Control of helicopters' formation using non-iterative nonlinear model predictive approach, Proceedings of IEEE American Control Conference, pp. 3707-3712, Seattle, Washington, June 2008

Xie, F.; Zhang, X.; Fierro, R; & Motter, M. (2005). Autopilot based nonlinear UAV formation controller with extremum-seeking, Proceedings of IEEE Conference on Decision and Control, pp 4933-4938, Seville, Spain, December 2005

15 Chapter

ASYMMETRIC HOVERING FLAPPING FLIGHT: A COMPUTATIONAL ANALYSIS

1. INTRODUCTION: AN OVERVIEW

In the early 1990s, Micro Air Vehicles (MAVs) emerged as a potential alternative for reconnaissance missions in confined areas. The American Defense Advanced Research Projects Agency (DARPA) initiated workshops on the concept and, in 1997, raised funds to conduct a multi-year programme whose objective was to develop a low cost, high autonomy unmanned aircraft, with a largest dimension limited to 15 centimeters (6 inches). In terms of aerodynamics, this typical size specification places the corresponding airflow in the range of low Reynolds number flows, typically between 10^2 and 104. Multiple prototypes were evaluated, grounded in traditional fixed and rotary wing designs. At low Reynolds numbers, supported by research on insect flight, the flapping wing concept emerged as an alternative solution, indicating improved aerodynamic performance (lift, efficiency), flight agility, hovering capability, and reduced noise generation. The latter results from the intricate wing movement characterized by combined translational (downstroke and upstroke) and rotational (supination and pronation) actions. Innovative studies focused on the aerodynamics of flapping wings were conducted by biologists aiming to assess the lift produced by insects. Following numerous attempts utilizing the quasi-steady technique, it was acknowledged that unstable aerodynamic mechanisms are crucial for sustaining an insect in flight, particularly when hovering (Jensen, 1956; Weis-Fogh, 1973; Ellington 1984). Precisely, three phenomena may be distinguished:

1. The presence of a leading edge vortex (LEV or dynamic stall mechanism) during the translating phases, acting as a low pressure suction region on the extrados of the wing. Due to its importance in aeronautics, this phenomenon was extensively studied experimentally (Walker, 1931) and analytically (Polhamus, 1971) before its evidence was demonstrated and analysed in the context of flapping wings (Maxworthy, 1979; Dickinson & Götz, 1993).

2. The Kramer effect, assimilated to the supplementary air circulation resulting from the combined translating and rotating motions (Kramer, 1932; Bennett, 1970; Dickinson et al, 1999).
3. The wake capture mechanism, occurring as the wing encounters and interacts with its own wake generated during previous phases (Dickinson, 1994; Dickinson et al, 1999).

One should keep in mind that the spatial and temporal behaviours of such unsteady phenomena highly depend on the wing kinematics. Thus, when analysing the flow dynamics generated by a flapping wing, it is essential to clearly precise the flight configuration studied. Basically, two main approaches may be distinguished. The first one is the forward-flapping flight configuration for which the wing flaps into a head wind. The resulting aerodynamic force can be divided into two components: the forward (horizontal) and the upward (vertical) components which must be sufficiently strong to counter the body drag force and the weight respectively. The second one, referred to as the hovering-flapping flight configuration, is in fact an extreme mode of flight where the head wind velocity is zero. The resulting mean aerodynamic force is here strictly vertical, the average horizontal force over a flapping period being null. In this particular case, the unsteady effects are preponderant relatively to the quasi steady effects. Considering hovering flapping flight leads to further differentiation whether the wing flaps along a horizontal stroke plane (symmetric or normal hovering) or an inclined stroke plane (asymmetric hovering). The former case has been under much consideration since it appeared to be the most common configuration observed in the world of insects. Specifically, the influence of various kinematic parameters (e.g. angle of attack, position of the centre of rotation, Reynolds number etc) was experimentally (Sane & Dickinson, 2001; Sane & Dickinson, 2002; Kurtulus, 2005) and numerically (Wu & Sun, 2004; Kurtulus, 2005) analysed, giving satisfying agreement. On the other hand, asymmetric hovering studies seemed restricted to biologic configurations (reproducing the wing kinematics of the dragonfly) and very few works reported parametrical results (Wang, 2004) dedicated to MAVs improvement. Introducing asymmetry in hovering flapping flight is an attractive strategy, as the lifting force arises from the interplay of lift and drag, suggesting improved aerodynamic efficiency. This chapter employs two-dimensional numerical simulations to assess the flow dynamics and the consequent aerodynamic loads on a wing executing asymmetric hovering flapping movements at Reynolds number 1000. In contrast to prior studies, the current parametric analysis emphasizes the aerodynamic performance of non-biological structures for Micro Air Vehicle applications. The comparison of asymmetric structures with extensively researched symmetric configurations guarantees reference to a broader context.

2. FLAPPING KINEMATICS AND PARAMETERS

Normal (or symmetric) flapping motions are defined by precisely opposing downstroke and upstroke wing kinematics. The drag produced during the downstroke counteracts the drag generated during the upstroke, as it is an aerodynamic force component collinear to the stroke plane. If the stroke plane is oriented horizontally, the drag can be equated to the horizontal force component, which averages to zero over time, hence maintaining the hovering condition. This study introduces asymmetry by establishing distinct angles of attack for the downstroke and upstroke. The resultant asymmetric movements demonstrate unequal drag throughout the downstroke and upstroke, rendering the horizontal stroke plane insufficient for maintaining the hovering situation. Thus, the latter should be tilted (angle β), resulting in a combined action of both lift and drag as the effective lifting (or vertical) force.

A set of symmetric and asymmetric cases is analysed for a two dimensional NACA0012 profile. The wing kinematics result from the combination of translating and rotating motions as shown in figure 1. Basically, the flapping motions may be decomposed into different regions whether the wing is translating at constant speed and fixed angle of attack (region T) or is submitted to both varying translation speed and rotating motion (regions R). Region T and regions R are respectively 4 and 1 chord long, so that the wing travels along a total course of 6 chords during one stroke. The rotation is applied around a spanwise axis located ¼ chord away from the leading edge. The constant wing velocity V0 reached during the pure translation phases (region T) is calculated with respect to the Reynolds number such that:

$$Re = \frac{cV_0}{\upsilon} \tag{1}$$

where c is the chord of the NACA0012 profile and υ the kinematic viscosity of air. Note that the Reynolds number represents the adimensional ratio between inertial and viscous forces and is here fixed to 1000 such that the flow is considered to be laminar. The flapping period T is defined as:

$$T = \frac{4c}{|V_0|}\left(2 + \frac{\pi}{2}\right) \tag{2}$$

leading to a flapping frequency of approximately 10 Hz. The varying parameters are chosen to be the downstroke and upstroke angles of attack (αd, αu), the difference between the two reflecting the asymmetry of the motion. In order to ensure the continuity of the translating and rotating accelerations as necessary to numerically solve the Navier-Stokes equations, the instantaneous velocities follow 4th order polynomial motion laws as displayed in figure 1.

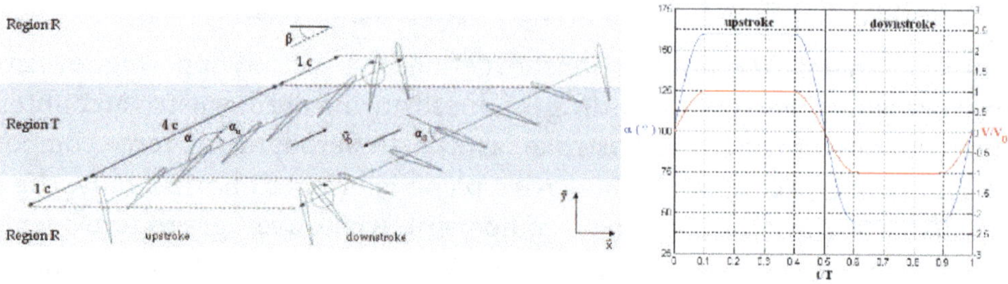

Figure 1. Schematization of the flapping kinematics and time evolution of the angle of attack and translating velocity (αd=45°, αu=20°)

3. INVESTIGATION TOOLS

3.1 Numerical solver

The aerodynamic flow established in the vicinity of the airfoil is computed using a finite element method. The two dimensional time-dependent Navier-Stokes equations (equations (3) and (4) in Cartesian tensor notation for general compressible and incompressible flows) are directly solved (DNS) in the fixed inertial reference frame through a moving mesh method, assuming laminar incompressible flow.

$$\frac{\partial \rho}{\partial t}+\frac{\partial}{\partial x_j}(\rho u_j)=0 \tag{3}$$

$$\frac{\partial \rho u_i}{\partial t}+\frac{\partial}{\partial x_j}(\rho u_j u_i-\tau_{ij})=-\frac{\partial p}{\partial x_i} \tag{4}$$

with the constitutive relation for a laminar Newtonian fluid:

$$\tau_{ij}=2\mu s_{ij}-\frac{2}{3}\mu\frac{\partial u_k}{\partial x_k}\delta_{ij} \tag{5}$$

$$s_{ij}=\frac{1}{2}\left(\frac{\partial u_i}{\partial x_j}+\frac{\partial u_j}{\partial x_i}\right) \tag{6}$$

Symbols are listed in table 1 below.

Table 1. Definition of the Navier-Stokes equations symbols

t	time
xi	Cartesian coordinate
ui	Fluid velocity component
p	Piezometric pressure
ρ	density

μ	Molecular dynamic viscosity
τij	Stress tensor components
sij	Rate of strain tensor
δij	Kronecker delta

The Navier-Stokes equations are solved on a non-conformal O-type grid as shown in figure 2. The latter is composed of 57500 cells, forming a refined rectangular domain close to the profile and a coarser circular domain at the far field. The instantaneous airfoil velocity is computed by means of user-defined subroutines and implemented through a no-slip boundary condition. The pressure is fixed and assumed to be the standard air pressure at the far field, located 15 chords away from the airfoil centre of rotation. The velocities at the corresponding cell faces are linked to the local pressure gradients, allowing the local flow to be inwards or outwards. For a two-dimensional configuration, symmetric boundary conditions are applied on the front and back side of the grid. The spatial and temporal discretisations are performed using an upwind differencing scheme and a PISO scheme respectively. The PISO approach consists of a special implicit scheme, based on the fully implicit Euler scheme and explicit deferred correctors, which results in a formal accuracy lying between first and second order. Setting 1000 time steps per flapping period leads to a maximum and mean Courant number (CFL) of approximately 10 and 0.5 respectively. Note that the results exposed in the present paper concern the 7th flapping period for which the flow is ensured to be periodical.

Higher resolution and larger computational domain tests (keeping a similar CFL) demonstrated that the solution is mesh independent.

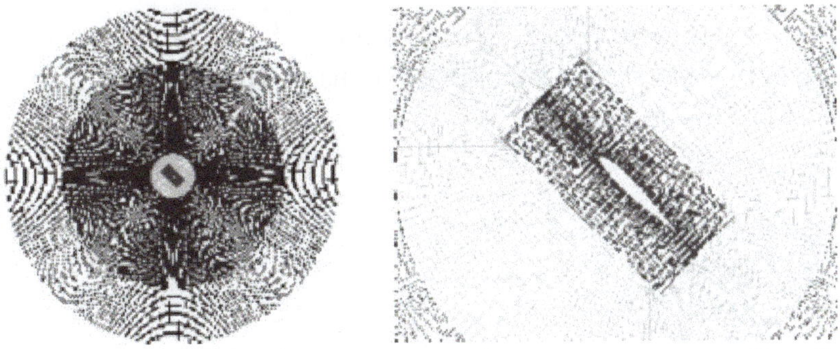

Figure 2. General and zoomed view of the computational grid

3.2 Vortex characterization

The numerical flow solver returns the velocity and the pressure flow fields in the whole computational domain at each time step. In order to characterize the flow dynamics, it is convenient to derive specific criteria which facilitate the vortex cores identification. The present analysis focuses on the λ_2 criteria (Jeong & Hussain, 1995), whose definition for a two-dimensional flow in the xy plane of velocity components uv is:

$$\lambda_2 = \frac{\partial u}{\partial y}\frac{\partial v}{\partial x} - \frac{\partial u}{\partial x}\frac{\partial v}{\partial y} \tag{7}$$

On the contrary to the vorticity, the λ_2 criterion differentiates the strain regions from the vortex core regions. Precisely, λ_2 is the second highest eigenvalue of the tensor Ω_2+S_2, where S and Ω are respectively the symmetric and antisymmetric components of the velocity gradient tensor. Negative values of λ_2 put into evidence local minima of pressure usually illustrating vortex cores.

3.3 Aerodynamic loads

The presence of vortical structures in the vicinity of the flapping airfoil highly influences the generation of aerodynamic loads. As will be exposed in the following section, vortex cores are usually assimilated to low pressure regions which may act as lift enhancers if located on the extra dos of the airfoil. As a consequence, the understanding of flapping airfoil performances imposes to establish a correlation between the vortex behaviours and the resulting aerodynamic force F. The latter is inferred by integrating both pressure and viscous stresses along the wing surface and may be decomposed into the lift F_l and drag F_d components (respectively orthogonal and collinear to the stroke plane) or the vertical F_y and horizontal F_x components, linked by the following relations:

$$F_x = \pm F_d \cos\beta - F_l \sin\beta \tag{8}$$

$$F_y = F_l \cos\beta \pm F_d \sin\beta \tag{9}$$

The basic two-dimensional aerodynamic coefficients are obtained by adimensionalizing the previous components using the constant translating velocity V0 and the chord of the airfoil c:

$$C_i = \frac{2F_i}{\rho c V_0^2} \tag{10}$$

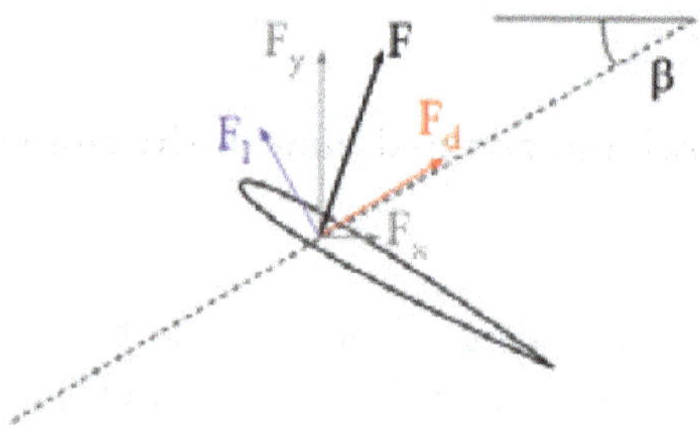

Figure 3. Relation between the resulting aerodynamic force F and the various components Fl, Fd, Fy and Fx

As a first approach, it is of interest to evaluate the mean values of these coefficients by averaging them over a flapping period T:

$$\overline{C_i} = \frac{1}{T}\int_0^T C_i(t)dt \tag{11}$$

Further coefficients are introduced to evaluate aerodynamic efficiency, power consumption and quality factor:

$$\overline{C_{eff}} = \frac{\overline{C_y}}{\overline{C_d}} \tag{12}$$

$$\overline{C_p} = \frac{1}{T}\int_0^T C_d(t)\frac{V(t)^3}{V_0^3}dt \tag{13}$$

$$\overline{C_q} = \frac{\overline{C_y}^{\frac{3}{2}}}{2\overline{C_p}}\sqrt{\frac{c}{a}} \tag{14}$$

where V(t) is the instantaneous airfoil velocity. The efficiency ratio represents the ratio between the effective lifting force necessary for flight and the drag force countering the wing motion, acting as a source of energy consumption. The power ratio may be assimilated to the power consumption resulting from the presence of this drag force. Finally, the Froude theory leads to the definition of a quality coefficient which estimates the relative importance of the lifting force and the power consumption. Note that in this case "a" stands for the total flapping amplitude.

4. RESULTS AND DISCUSSION

4.1 Mean analysis

The aerodynamic performances inherent to the different kinematics are evaluated using the mean coefficients described in the previous section. Figure 4 represents the mean drag, vertical (or lifting) force, efficiency ratio, power and quality coefficients for all the configurations tested. The latter is regrouped according to the value of the downstroke angle of attack αd consecutively fixed to 30°, 45° and 60°. In this way, each graph exhibits the effect of desymmetrization on the global aerodynamic performances of flapping airfoils. Precisely, the right hand symbols correspond to symmetric cases, the asymmetry being increased with decreasing upstroke angle of attack.

The overall comparison demonstrates that the mean drag and vertical force coefficients, hence the global aerodynamic force F, increase with the downstroke/ upstroke angle of attack, as denoted in previous works (Sane & Dickinson, 2001; Wu & Sun, 2004; Kurtulus, 2005). This remark is illustrated by both the enhancement of the drag and vertical force tendency curves from one graph to another (influence of α_d) and the latter's positive slope observed within a graph (influence of α_u).

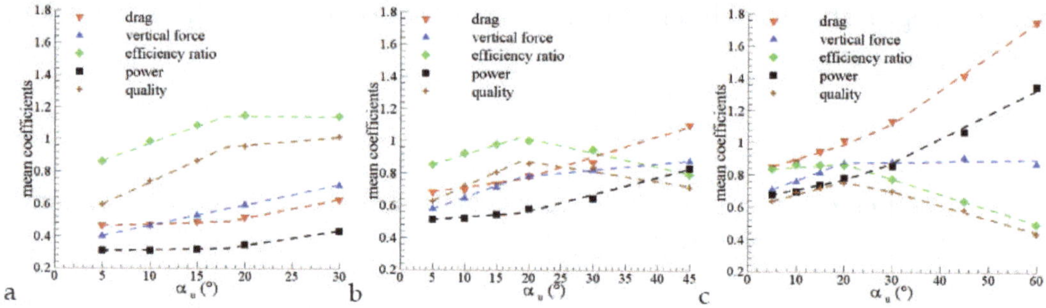

Figure 4. Mean drag, vertical force, efficiency ratio, power and quality (5*$\overline{C_q}$) **coefficients in function of α_u for the α_d =30° (a), α_d =45° (b) and α_d =60° (c) configurations**

In order to characterize the effect of asymmetry on aerodynamic performances, it is of interest to further analyse the behaviour of these tendency curves within a graph. Figure 4.a shows a linear increase of the vertical force coefficient with the upstroke angle of attack. This augmentation is similar for the drag coefficient when αu is set above 18°, but less pronounced for low αu values. The direct consequence resides in a clear change of the efficiency ratio tendency curve near α_u=18°. Comparable changes in evolution are further put into evidence for both drag and vertical force coefficients in figure 4.b. In this particular case, the behaviours of the drag and the vertical force seem opposed, resulting in a maximum efficiency ratio of approximately 1.05 at α_u≈19°. Besides these changes, Figure 4.c exhibits a quasi constant vertical force coefficient and a sharp increase in drag with increasing upstroke angle of attack. Thus, the efficiency ratio is maintained to approximately 0.85 until α_u≈20°, after which it is strongly affected and decreases to 0.498 for the normal hovering case α_d=60°, α_u=60°. As a conclusion, the common characteristic of all groups (α_d=30°, α_d=45° and α4=60°) resides in the presence of a critical upstroke angle of attack for which the tendency curve representing the efficiency ratio as a function of α_u changes behaviour and reaches a strong (maximum in some cases) value. If the cause for the existence of this critical angle of attack may arise from trivial trigonometric considerations, it is also probable that the occurence of non-linear unsteady flow structures strongly affect aerodynamic performances from one configuration to another, and may participate in the changes of behaviour previously exposed.

Furthermore, in all cases, the drag and power coefficients are characterized by comparable tendencies, still shifted by a value which increases with the upstroke angle of attack. The effect on the quality coefficient is such that the latter demonstrates analogous changes around α_u≈20°.

Table 2. Aerodynamic coefficients survey

αd	αu	$\overline{C_d}$	$\overline{C_y}$	$\overline{C_{eff}}$	$\overline{C_p}$	$\overline{C_q}$	β
30	5	0.4651	0.4018	0.864	0.310	0.119	34
30	10	0.4698	0.4647	0.989	0.309	0.148	24
30	15	0.4863	0.5300	1.090	0.321	0.173	17
30	20	0.5185	0.5966	1.151	0.348	0.191	10
30	30	0.6275	0.7190	1.146	0.433	0.203	0
45	5	0.6807	0.5802	0.852	0.512	0.125	51
45	10	0.6983	0.6444	0.923	0.520	0.144	41
45	15	0.7314	0.7149	0.977	0.543	0.161	33
45	20	0.7823	0.7826	1.000	0.580	0.172	25
45	30	0.8685	0.8217	0.946	0.644	0.167	14
45	45	1.1038	0.8801	0.797	0.832	0.143	0
60	5	0.8477	0.7080	0.835	0.676	0.127	64
60	10	0.8805	0.7598	0.863	0.695	0.138	54
60	15	0.9435	0.8134	0.862	0.735	0.144	45
60	20	1.0129	0.8706	0.860	0.782	0.150	38
60	30	1.1367	0.8811	0.775	0.855	0.140	25
60	45	1.4182	0.9087	0.641	1.070	0.117	14
60	60	1.7509	0.8713	0.498	1.350	0.087	0

4.2 Flow unsteadiness

Focusing on the spatial and temporal evolution of the vortical structures governing the flow dynamics appears as an essential step in the comprehension of the asymmetric flapping flight aerodynamic performances. In order to explain the different tendencies in the generation of lifting force and in the behaviour of the efficiency ratio from one configuration to another, the adimensional λ_2* (figure 6) and pressure C_p (figure 7) contours are displayed for the symmetric configuration α_d=45°, α_u=45° and the asymmetric configuration α_d=45°, α_u=20°. These flow properties are correlated with the time-dependent lift and drag coefficients (figure 5), putting into evidence the force generating mechanisms. The two configurations show comparable downstroke kinematics, hence comparable quasi-steady aerodynamic forces, but different upstroke kinematics. As a consequence, differences observed in the flow dynamics and in the generation of forces during the downstroke phase directly arise from the previous strokes history, especially through the wake capture phenomenon. Note that the downstroke kinematics of both cases cannot be qualified as identical since the rotation velocities are not strictly equal.

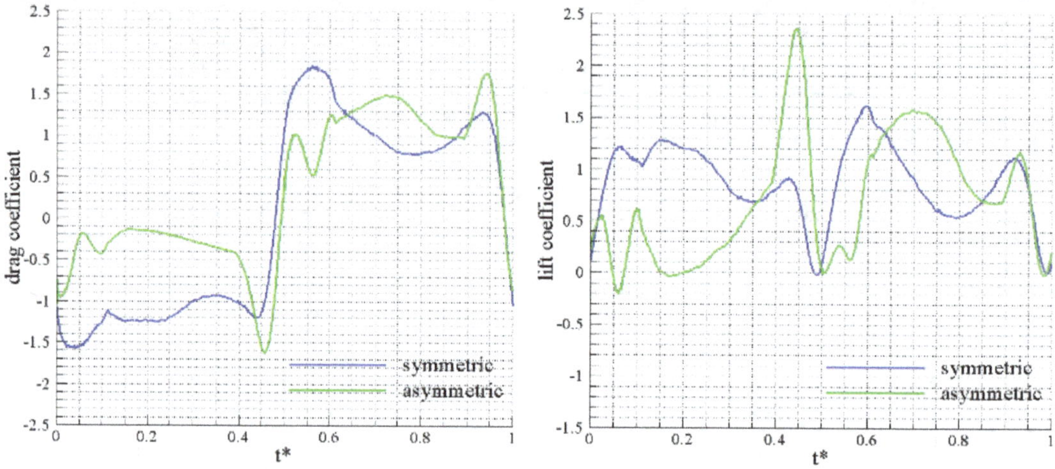

Figure 5. Drag and lift coefficient histories of the symmetric α_d=45°, α_u=45° and asymmetric α_d=45°, α_u=20° configurations

All flow properties and aerodynamic coefficients are represented at time t*, where t* corresponds to the time elapsed since the beginning of the flapping cycle, adimensionalized by the period T.

At t*=0, the airfoil begins the upstroke phase. Notwithstanding the zero translating velocity which characterizes the flapping kinematics at this instant, the drag coefficients reach a strong value (≈-1) for both configurations. Indeed, two counter-rotating vortical structures dominate the flow in the vicinity of the profile (figure 6 – t*=0), inducing the presence of a fluid jet directed towards the intrados and which acts as a drag enhancer. As previously expressed, this interaction between the airfoil and its own wake is typical of the flapping wings aerodynamics.

Between t*=0 and t*=0.1, the airfoil accelerates while still rotating to reach a constant translating speed and fixed angle of attack. The production of drag and lift here derives from the complex combination of inertial effects and vortex behaviours. Hence, significant differences can be observed between both symmetric and asymmetric configurations. On the one hand, the aerodynamic force resulting from the symmetric motion seems to primarily arise from the accelerating inertial effect, leading to the presence of a strong pressure region covering most of the airfoil intrados. On the other hand, the relative position of the airfoil and the previously shed vortices characterizing the asymmetric configuration induces a competition between strong (leading edge) and low (trailing edge) intrados pressure regions (figure 7 – t*=T/12). Therefore, at t*=0.06, the asymmetric configuration lift and drag decrease sharply whereas the symmetric ones are maintained to relatively strong values.

At t*=0.1, both cases demonstrate the formation of a LEV acting as a low pressure suction region on the extrados of the airfoil. Note that the latter is promoted by the wing/wake interaction as shown in previous studies (Birch & Dickinson, 2003). The LEV is essential when considering the symmetric motion since it dominates the vortical flow field during most of the upstroke phase. Thus, absolute values of lift and drag above 1 are reached. On the contrary, the LEV generated in the asymmetric case has a limited lifetime such that its influence is only perceptible through the occurrence of a narrow lift peak. Note that this vortical structure does not significantly affect the drag since the upstroke angle of attack is set to a low value (20°), i.e. the airfoil surface projection on the axis perpendicular to the stroke plane is weak. Moreover, one can remark that the lifetime is partially shortened by the wake capture phenomenon which leads to the LEV shedding (figure 6 – t*=T/12). Consequently, the asymmetric upstroke phase leads to a partially attached flow supported by the presence of a fluid downwash engendered by the previous strokes.

The time evolutions of the drag and lift coefficients roughly respond to the spatial and temporal evolution of the upstroke LEV until t*=0.4. At this instant, the airfoil starts to rotate while decelerating. Figure 5 shows the presence of drag and lift peaks, particularly pronounced in the asymmetric configuration for which the angle of attack changes from α_u=20° to α_d=45° with a rotating speed superior to the one encountered in the symmetric configuration. Such lift enhancements arise from the supplementary circulation generated by the combined rotating and translating motions (Kramer effect). Furthermore, whereas the rotation strengthens the Trailing Edge Vortex (TEV) formed during the latter part of the downstroke phase in the symmetric case, it is the cause for the formation of a Rotating Trailing Edge Vortex (RTEV) in the asymmetric case (figure 6 – t*=5T/12).

At t*=0.5, as observed at t*=0, the drag coefficient resulting from the symmetric motion exhibits a strong value linked to the presence of a fluid jet generated by the two shed counter rotating vortical structures. The flow behaviour induced by the asymmetric motion considerably differs since no strong separated flow yielded to the presence of a vortex dipole (figure 6 – t*=6T/12). Hence, any fluid jet tends to increase the drag coefficient which is here quasi null.

Between t*=0.5 and t*=0.6, the airfoil accelerates in the opposite direction to initiate the downstroke phase. Despite the symmetric aspect of the normal hovering case, the presence of highly unsteady non-linear structures leads to non-strictly equivalent upstroke and downstroke aerodynamic forces, as noticed by other authors on biologic cases (Van Oudheusden et al, 2008). However, global variations of upstroke and downstroke coefficients strongly resemble as arising from comparable flow behaviour. The symmetric motion engenders a rapid augmentation of lift and drag through the accelerating inertial effects. The latter is supported and maintained by the formation of a downstroke LEV on the airfoil extrados (figure 6 – t*=7T/12).

On the contrary, the influence of wake capture is reduced in the asymmetric case. Indeed, besides the action of the RTEV assimilated to a low pressure region under the intrados of the airfoil, the previously shed vorticity is not sufficient to accelerate the formation of a downstroke LEV. Consequently, the drag and lift are maintained to low values until t*=0.6. The interesting aspect here resides in the fact that the LEV develops smoothly, staying closely attached to the extrados. Thus, from t*=0.6, despite the fact that the wing kinematics are identical in both symmetric and asymmetric configurations (i.e. identical quasi-steady forces), the asymmetric motion leads to stronger aerodynamic coefficients. This observation is evidence that the wake capture phenomenon not only affects the early stages but most of the stroke resulting forces.

From t*=0.9, the wake of both configurations still exhibit different behaviour hence different drag productions as the airfoil jointly rotates and decelerates. However, the lift coefficients tend to similar time evolutions since the inertial forces conducting the Kramer effect are quasi equivalent in both configurations.

4.3 Effects of desymetrization

The correlation between the fluid dynamics and the loads experienced by the flapping airfoil for two specific symmetric and asymmetric configurations (respectively α_d=45°, α_u=45° and α_d=45°, α_u=20°) demonstrated that reducing the upstroke wake by fixing a low upstroke angle of attack leads 1) to a reduction of the aerodynamic coefficients immediately after stroke reversal and 2) to an enhancement of the latter during most of the downstroke translating phase. This characteristic partially results from the absence of significant upstroke leading edge separation, hence significant wake capture, inducing the generation of a closely attached LEV on the extrados of the airfoil. The analysis of the unsteady drag and lift coefficients obtained for all configurations tested support this observation (figure 8). Note that each graph of figure 8 regroups the flapping kinematics parameterized with a common downstroke angle of attack such that it effectively illustrates the effect of desymetrization on the production of lift and drag during the downstroke phase (the quasi steady downstroke forces being equivalent). Nevertheless, analysing the lift and drag coefficients averaged over the downstroke phase shows that both effects seem to inhibit each other. For instance, the mean drag and lift coefficients obtained during the downstroke phase of the α_d=45°, α_u=45° and the α_d=45°, α_u=20° configurations are respectively 1.09, 0.87 and 1.11, 0.89.

Besides the changes observed in the fluid dynamics, asymmetric hovering motions benefit from the combined action of both lift and drag as the effective lifting force (Wang, 2004). Indeed, for such motions, the hovering condition imposes the stroke plane to be inclined such that the drag has a vertical component and thus provides a lifting force rather than exclusively acting as a power consumer. According to the β angles calculated and listed in table 1, the resulting horizontal and vertical unsteady components are deduced from the drag and lift components and represented in figure 9. In all cases, increasing the difference α_d-α_u, i.e. the stroke plane angle, considerably enhance the resulting downstroke vertical force. For instance, the mean lifting force coefficient obtained during the downstroke phase of the α_d=45°, α_u=45° and the α_d=45°, α_u=20° configurations are respectively 0.87 and 1.28 which corresponds to an increase of approximately 47%. However, the upstroke phase appears as a harmful phase and may sometimes lead to a negative vertical force component as exhibited when the upstroke angle of attack is set to 10°. Note that such negative values can be obtained with still non-negligible angle of attack since the presence of a fluid downwash tends to reduce the effective angle of attack. Consequently, as dedicated to MAVs application, asymmetric kinematics with an upstroke angle of attack fixed below 15° should be avoided. Moreover, the quality of the resulting aerodynamic performances resides in the relative importance of the upstroke and downstroke phases.

4.4 Improving aerodynamic performances

Previous sections demonstrated that, in contrast to symmetric flapping motions, asymmetric flapping motions can result in a sustained efficiency ratio alongside a stable lifting force when the upstroke angle of attack is reduced, diminishing the wake capture phenomenon and indicating the role of drag in the vertical force component. Moreover, it is evident that the upstroke phase is detrimental, necessitating a thorough examination of the relative significance of both the upstroke and downstroke phases. This section proposes a method to enhance the aerodynamic performance of asymmetric motions by raising the upstroke translational velocity, hence diminishing the relative significance of the upstroke phase. This technique is consistent as the intrinsic properties of the prior examples remain unchanged, with the flow dynamics not influenced by the Reynolds number within this range. (Wu & Sun, 2004; Kurtulus, 2005).

Results are obtained for velocity ratios (defined as the ratio r_v between the upstroke and the downstroke translating velocity) of 1, 1.2 and 1.4. The aerodynamic coefficients are determined by adimensionalizing the upstroke and downstroke forces by means of the upstroke and the downstroke translating velocities respectively. Tables 3 and 4 illustrate the effect of increasing the upstroke velocity on two chosen asymmetric cases. It is shown that the lifting force and the efficiency can be increased in a prompter way than the power consumption such that the quality coefficient can be significantly enhanced. Indeed, one might observe that the influence on the power consumption is not monotonous (increasing between r_v=1 and r_v=1.2, decreasing between r_v=1.2 and r_v=1.4).

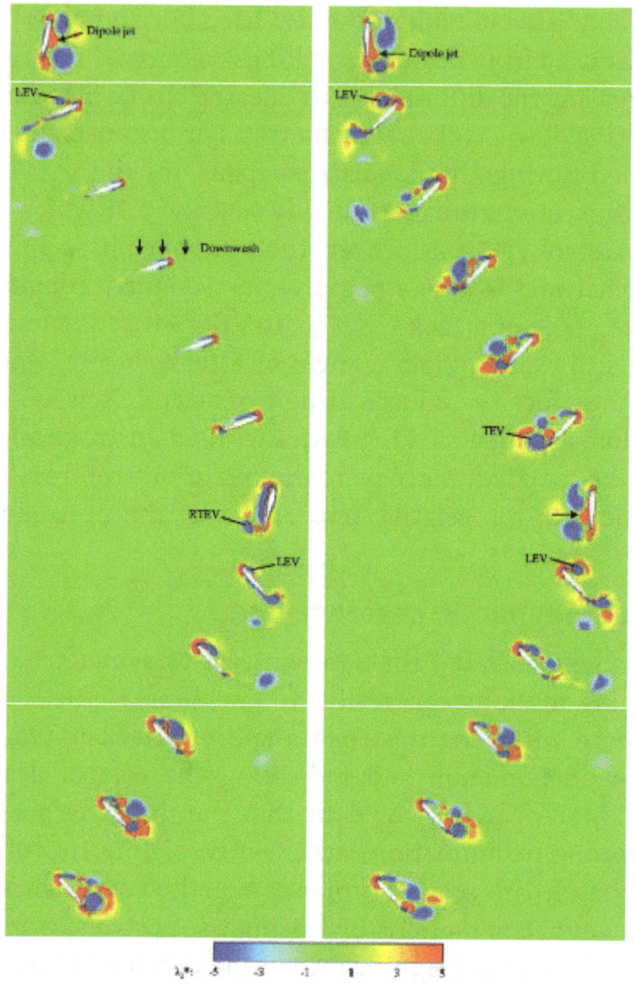

Figure 6. λ2 distribution at t*=nT/12 (n ∈ [0;11]) from top to bottom for the symmetric (right) and asymmetric (left) configurations

Table 3. Effect of velocity ratio on aerodynamic coefficients for the asymmetric configuration αd=45°, αu=15°

	rv = 1	rv = 1.2	rv = 1.4
$\overline{C_d}$	0.7314	0.7497 (+2.5%)	0.7733 (+5.7%)
$\overline{C_y}$	0.7149	0.7825 (+9.5%)	0.8308 (+16.2%)
$\overline{C_{eff}}$	0.977	1.044 (+6.9%)	1.074 (+9.9%)
$\overline{C_p}$	0.543	0.604 (+11.2%)	0.583 (+7.4%)
$\overline{C_q}$	0.161	0.165 (+2.5%)	0.187 (+16.1%)

Table 4. Effect of velocity ratio on aerodynamic coefficients for the asymmetric configuration αd=45°, αu=20°

	rv = 1	rv = 1.2	rv = 1.4
$\overline{C_d}$	0.7823	0.7848 (+0.3%)	0.8144 (+4.1%)
$\overline{C_y}$	0.7826	0.8212 (+4.9%)	0.8498 (+8.6%)
$\overline{C_{eff}}$	1.000	1.046 (+4.6%)	1.043 (+4.3%)
$\overline{C_p}$	0.580	0.628 (+8.3%)	0.599 (+3.3%)
$\overline{C_q}$	0.172	0.171 (-0.6%)	0.189 (+9.9%)

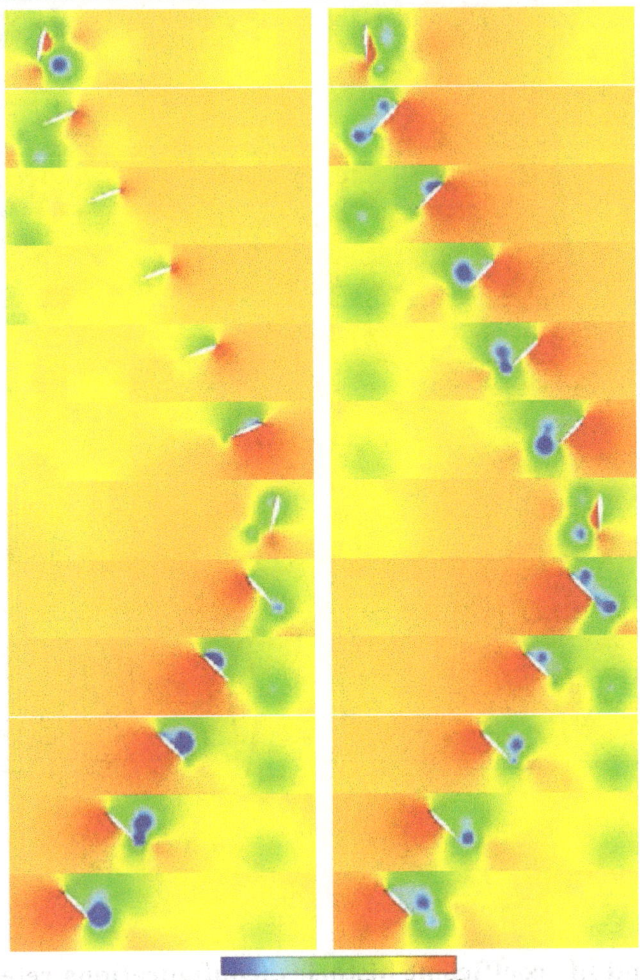

Figure 7. Adimensional pressure distribution at t*=nT/12 (n ∈ [0;11]) from top to bottom for the symmetric (right) and asymmetric (left) configurations

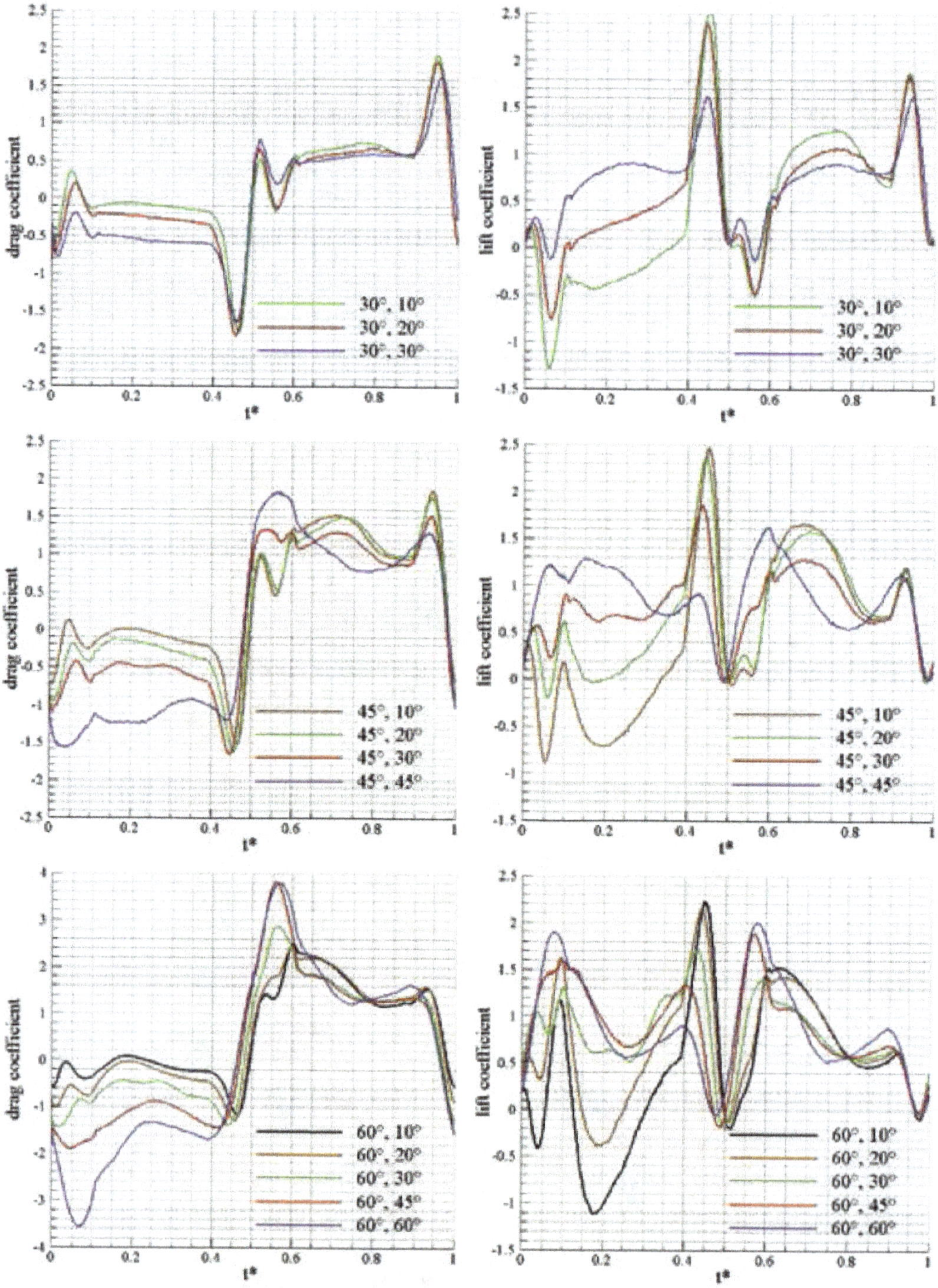

Figure 8. Drag and lift coefficient histories (configurations referenced with αd, αu)

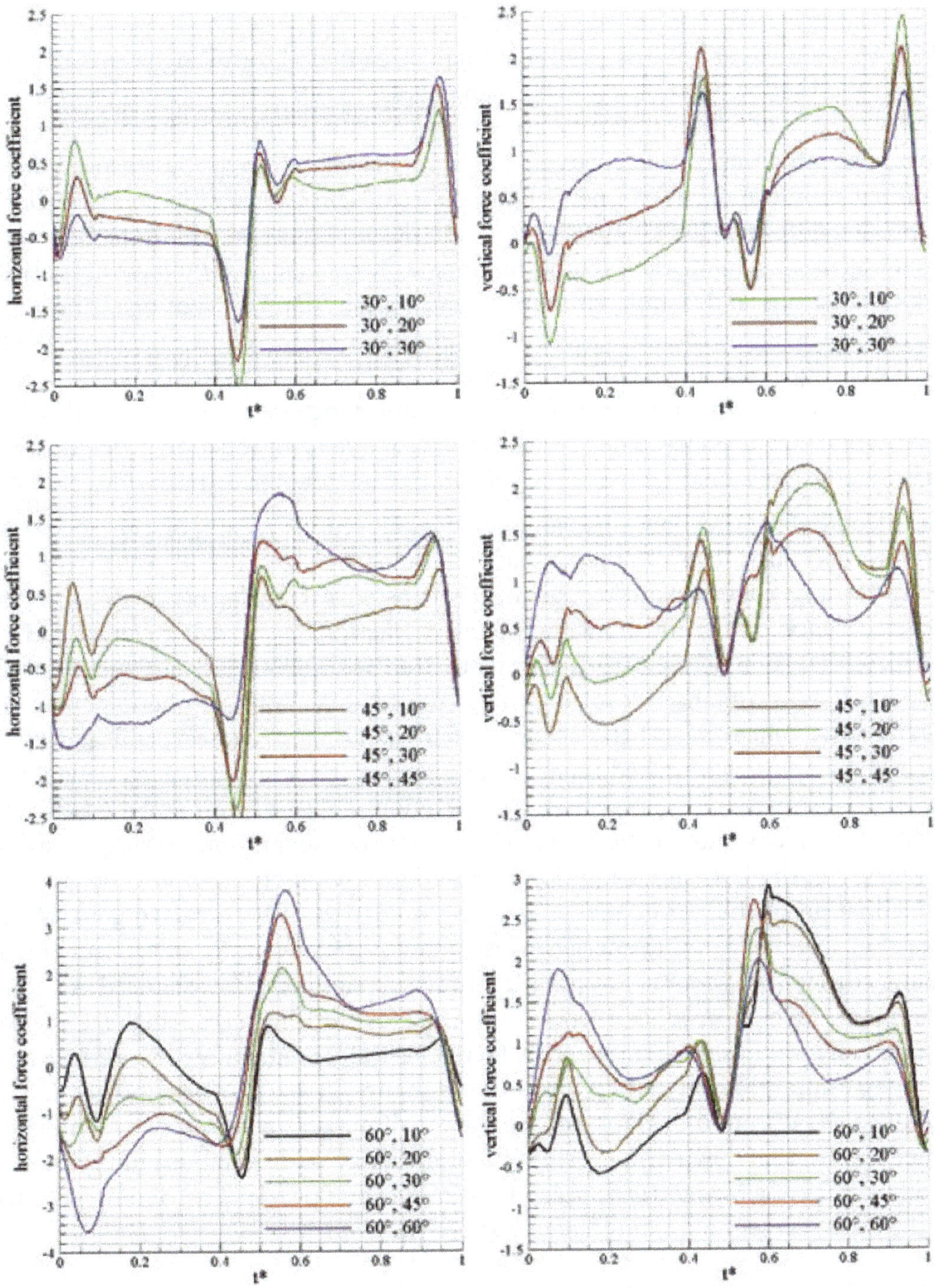

Figure 9. Horizontal and vertical force coefficient histories (configurations referenced with αd, αu)

5. CONCISE SUMMARY

Micro Air Vehicles are defined by their small size (15 cm), resulting in aerodynamic flow within the low Reynolds number range (102-104). At these Reynolds numbers, the flapping wing idea presents an alternative to traditional fixed and rotary wings, assuming improved aerodynamic performance. Prior studies focused on the examination of standard (symmetric) hovering configurations, which are regarded as the predominant motion kinematics found in the insect realm. This paper examines parameterized asymmetric flapping motions at Reynolds 1000 by two-dimensional DNS calculations and compares them to symmetric flapping motions. The initial analysis of the aerodynamic coefficients indicates that introducing asymmetry (by varying the upstroke angle of attack compared to the downstroke) typically reduces the aerodynamic force on the airfoil while improving both efficiency and quality coefficients. Furthermore, when the latter is displayed as a function of the upstroke angle of attack, two distinct tendency behaviours are denoted on both sides of $\alpha_u \approx 20°$.

In a second step, the flow unsteadiness is analysed by means of the λ_2 and the pressure contours. The symmetric case parameterized with $\alpha_d = \alpha_u = 45°$ exhibits a strong leading edge separation during the upstroke translating phase, implying the occurrence of significant wing/wake interactions at stroke reversal. On the contrary, the asymmetric case is parameterized with $\alpha_d = 45°$, $\alpha_u = 20°$ such that the upstroke flow is attached, resulting in a reduced wake capture phenomenon. The direct consequences are 1) the absence of lift peak at stroke reversal and 2) the presence of a closely attached Leading Edge Vortex inducing enhanced downstroke lift. Such observations suggest that the behaviours observed on mean coefficients might arise from the presence or not of strong separation during upstroke. Furthermore, the hovering condition imposed on asymmetric motions necessitates an inclination of the stroke plane, allowing a portion of the drag to generate lift instead of solely expending energy. Consequently, the advantage of a closely linked leading edge vortex generating both lift and drag is significantly enhanced. Although asymmetric layouts produce increased downstroke lifting force, they are marked by a detrimental upstroke phase. This factor partially contributes to a reduction in average resultant aerodynamic forces. To enhance global performance, it is vital to reduce the duration of the upstroke, hence diminishing its relative significance compared to the downstroke. This chapter offers additional insights into the aerodynamics of asymmetric flapping movements and presents intriguing potential for the advancement of high-efficiency Micro Air Vehicles.

REFERENCES

Bennett, L. (1970). Insect flight: lift and the rate of change of incidence. Science, 167, 177-179

Birch, J.M., Dickinson, M.H. (2003). The influence of wing-wake interactions on the production of aerodynamic forces in flapping flight. Journal of Experimental Biology, 206, 2257-2272

Bos, F.M., Lentink, D., Van Oudheusden, B.W., Bijl, H. (2008). Influence of wing kinematics on aerodynamic performance in hovering insect flight. Journal of Fluid Mechanics, 594, 341-368

Dickinson, M.H., Götz, K.G. (1993). Unsteady aerodynamic performance of model wings at low Reynolds numbers. Journal of Experimental Biology, 174, 45-64

Dickinson, M.H., Lehmann, F.O., Sane, S.P. (1999). Wing rotation and the aerodynamic basis of insect flight. Science, 284, 1954-1960

Ellington, C.P. (1984). The aerodynamics of hovering insect flight I–V. Philosophical Transactions of the Royal Society. Series B: Biological Sciences, 305, 1122, 1-181

Jensen, M. (1956). Biology and physics of locust flight. III. The aerodynamics of locust flight. Philosophical Transactions of the Royal Society. Series B: Biological Sciences, 239, 667, 511-552

Jeong, J., Hussain, F. (1995). On the identification of a vortex. Journal of Fluid Mechanics, 285, 69-94

Kramer, M. (1932). Die zunahme des maximalauftriebes von tragflugeln bei plotzlicher anstellwinkelvergrosserung. Z Flugtech Motorluftschiff, 23, 185-189

Kurtulus, D.F. (2005). Numerical and experimental analysis of flapping motion in hover. Application to Micro Air-Vehicles. Phd Thesis, University of Poitiers, Laboratoire d'Etudes Aérodynamiques

Maxworthy, T. (1979). Experiments on the Weis-Fogh mechanism of lift generation by insects in hovering flight Part 1. Dynamics of the 'fling'. Journal of Fluid Mechanics, 93, 47-63

Polhamus, E.C. (1971). Predictions of vortex-lift characteristics by a leading-edge suction analogy. Journal of Aircraft, 8, 4, 193 -199

Sane, S.P., Dickinson, M.H. (2001). The control of flight force by a flapping wing: lift and drag production. Journal of Experimental Biology, 204, 2607-2626

Sane, S.P. , Dickinson, M.H. (2002). The aerodynamic effects of wing rotation and a revised quasi-steady model of flapping flight. Journal of Experimental Biology, 205, 1087-1096

Walker, P.B. (1931). Growth of circulation about a wing and an apparatus for measuring fluid motion. ARC report

Wang, Z.J. (2004). The role of drag in insect hovering. Journal of Experimental Biology, 207, 4147-4155

Weis-Fogh, T. (1973). Quick estimates of flight fitness in hovering animals, including novel mechanism for lift production. Journal of Experimental Biology, 59, 169-230

Wu, J., Sun, M. (2004). Unsteady aerodynamic forces of a flapping wing. Journal of Experimental Biology, 207, 1137-1150

16

Chapter

PATH PLANNING FOR UNMANNED AERIAL VEHICLES (UAVS) IN SEARCH OPERATIONS

1. INTRODUCTION: AN OVERVIEW

An Unmanned Aerial Vehicle (UAV) is a motorized, pilotless aircraft that is operated either remotely or autonomously. Unmanned Aerial Vehicles (UAVs) are presently utilized in several military functions and several civilian applications. Approximately 32 nations are engaged in the development or production of over 250 UAV models, while 41 countries operate around 80 UAV kinds (U.S. Department of Defense, 2005). The employment of UAVs in military and civilian applications is increasing in both the short term and long term. The two fundamental methods for executing unmanned flight, remote control and autonomy, primarily depend on remote data connection and microprocessor technologies (U.S. Department of Defense, 2005). Technological advancements, which have proliferated since their inception, have significantly enhanced the capability of UAVs to tackle more complex jobs. The growing accessibility of affordable processing power will expand the limits of achievable tasks with reduced human oversight, a characteristic commonly referred to as autonomy. In numerous civil and military flight operations, the aircraft's autonomy and mobility are constrained, and the trajectory is predetermined. Based on the path, the aircraft's control task is to construct the trajectory, specifically to ascertain the necessary control maneuvers to navigate the aircraft from one location to another. In certain flight missions, the trajectory is not predetermined but must be dynamically established throughout the operation, such as in military surveillance, search and rescue missions, fire detection, and disaster response. In such cases, the objective of the UAV is to ascertain the exact location of a targeted object inside a designated area of responsibility. Typically, some ambiguous a priori information regarding the object's initial location is accessible. The UAV's trajectory cannot be predetermined before to the mission commencement, as the information may be altered during the search operation due to new reports from various sources or observations made by the UAV. This chapter addresses the issue and presents a framework for autonomous and dynamic UAV path planning in search operations. The remainder of this chapter is structured as follows: Section 2 delineates the problem broadly and specifies the particular instance addressed, section 3 reviews historical context and pertinent literature, section 4 offers a comprehensive overview of the proposed solution, sections 5 and 6 elucidate Sequential Monte Carlo (SMC) methods and their application, section 7 presents the simulation of a test case scenario along with the results obtained, and section 8 concludes the chapter with a summary.

2. PROBLEM DEFINITION

In surveillance or search and rescue operations, a UAV is designated an area of responsibility to locate a target object. The region is typically extensive, therefore minimizing the detection time to locate the object is the key parameter. In the absence of information regarding the target and the area of responsibility, the sole plan is to conduct a comprehensive and uniform search of the area. In reality, certain information typically exists that rationalizes the uneven distribution of search efforts across the entire area. An instance of this problem occurs when ambiguous information regarding the previous location of the target is accessible. This information may be integrated with assumptions on the target's velocity and movement to produce a time-dependent probability of the target's location. Another scenario occurs when sensor observations or reports from alternative sources omit certain regions of the area. Additionally, if the geographical map of the designated area is accessible, this information could be utilized to focus search efforts on regions where the target is more likely to be located. Typically, the administration of these information fragments is conducted by human operators, particularly those with extensive expertise in the domain. The operator's primary objective is to enhance the utilization of UAV resources by prioritizing and/or conducting more extensive searches in areas with a higher likelihood of locating the target. Upon the acquisition of new information during the mission, it is imperative to reiterate the procedure and adjust the trajectory as warranted. This strategy has two significant downsides. Initially, due to the mobility of both the target and the sensor (UAV), identifying high-probability locations and determining the optimal path is often a complex endeavor. The analysis of this information may exceed the capabilities of the human brain. Secondly, the time-sensitive nature of these missions renders it impractical to delegate this work to an operator. Significant time may be squandered before the human operator processes information, potentially rendering it impossible to meet time constraints, particularly when the information is subject to rapid changes or is voluminous. An enhanced method involves automating the path planning process and incorporating the analysis of target locations into the autonomous control system. To facilitate this, all pertinent information must be transmitted to the UAV, and the autonomous control system should dynamically plan and adjust the route. This study presents a simulation-based approach for UAV path planning in search and surveillance operations, utilizing ambiguous a priori information regarding the target and environment. The proposed architecture, while broadly applicable, has been specifically constructed and evaluated in a situation where a UAV seeks a mobile target traversing a known road network.

3. RELATED WORK

Work on modern search theory began in the US Navy's Antisubmarine Warfare Operations Research Group (ASWORG) in 1942 (Morse, 1982; Stone, 1989). Bernard Osgood Koopman is credited with publishing the first scientific work on the subject in 1946, Search and Screening, which was classified for ten years before it was published (Stone, 1989). He defined many of the basic search concepts and provided the probabilistic model of the optimal search for a stationary target. However, developing algorithms for optimal search plan for moving targets started in the early 1970s and when computer technology became more available. The next step in developing search planners was to consider the dynamic nature of the search process. Computer Assisted Search Planning (CASP), developed for US Coast Guard in the 1970s by Richardson is a pioneer software system for dynamic planning of search for ships and people lost at sea (Richardson & Discenza, 1980; Stone, 1983). CASP employed Monte Carlo methods to obtain the target distribution using a multi-scenario approach. The scenarios were specified by choosing three scenario types and the required parameter values for each scenario. A grid of cells was used to build a probability map from the target distribution, where each cell had a detection probability associated with it. A search plan was developed based on the probability map. Feedbacks from the search were incorporated in the probability map for future search plans, if the first search effort did not succeed. The shortage of computer power and display technique did not allow CASP to be a truly dynamic tool operating in real-time aboard aircraft. Advances in computer technology provided the possibility of developing more feasible tools, Search and Localization Tactical decision aid (SALT) was a prototype air-antisubmarine search planner system for real-time use aboard aircraft (Stone, 1989).

The problem of searching for a lost target at sea by a single autonomous sensor platform (UAV) is discussed by (Bourgault et al., 2003a). In this paper the target may be static or mobile but not evading. The paper presents a Bayesian approach to the problem and the feasibility of the method is investigated using a high fidelity UAV simulator. Bayesian analysis is a way to recursively combine the motion model of the target and the sensor measurements to calculate the updated probability distribution of the target. Time is discretized in time steps of equal length and the distribution is calculated numerically. The search algorithm chooses a strategy that minimizes the expected time to find the target or alternatively maximizes the cumulative probability of finding the target given a restricted amount of time. The paper chooses one-step lookahead, i.e. the time horizon used for optimization is one time step. Because of this myopic planning, the UAV fails to detect the target if it is outside its sensor range. A decentralized Bayesian approach is suggested to solve the same problem by coordinating multiple autonomous sensor platforms (UAVs) in (Bourgault et al., 2003b; Bourgault et al., 2004). The coordinated solution is claimed to be more efficient, however, the simulations in these papers suffer from the short time horizon as well, i.e. one-step lookahead.

The problem of path planning of UAVs in search and surveillance missions (sensor platform steering) can be considered as a sensor resource management problem and is investigated by the Information fusion community as well. Sensor management is formally described as the process of coordinating the usage of a set of sensors or measurement devices in a dynamic, uncertain environment to improve the performance of data fusion and ultimately that of perception. A brief but highly insightful review of the multi-sensor management is presented by (Xiong & Svensson, 2003). The idea of simulating the target's future movements and choosing sensor control parameters to maximize a utility function is described in (Ahlberg et al., 2004). Given a situation X_0 , all possible future situations X that are consistent with the positions in X0 at time t = 0 are generated. For each of these X's, the utility of each sensor control scheme S is calculated by simulating observations of X using scheme S. The S whose average over all X is "best" is then chosen. However, to overcome the computation complexity, the set of possible sensor schemes is kept relatively small. Simulation-based planning for allocation of sensor resources is also discussed by (Svenson & Mårtenson, 2006). For each considered sensor allocation, the possible future paths of the target are partitioned into equivalence classes. Two futures are considered equivalent with respect to a given sensor allocation if they would give rise to the same set of observations. This partitioning to equivalence classes decreases the computational complexity of the problem and speeds up the calculation process.

4. OUR APPROACH

The approach employed in this work to address the path planning problem is simulation- based. In short, this approach can be described as a method that uses simulation to approximate the future state of the target and tests alternative paths against the estimated future by running what-if simulations. These what-if simulations are conducted continuously during the mission (on-line). Utilizing information, even when it is incomplete or uncertain, is essential in constructing efficient search strategies and a system that uses all pieces of information in general performs better compared to systems not considering this information. In order to utilize this information, modeling and simulation techniques are used, which have shown to be a feasible tool handling complex and "difficult-to-analyze" systems.

The on-line simulation method for path planning in a search mission can be described as the following. The mission length is divided by a sequence of time check points, {t0, t1,...} where t0 is the start time of the mission. At time t0 the UAV chooses a default (random) path. At each time check point $t_k \in \{t_0, t_1, ..., t_n\}$, a set of simulations are started. In each simulation the state of the target for time $t \geq t_k+1$ and the effect of choosing an alternative UAV path for time $t \geq t_k+1$ are estimated. These simulations are completed during the time period $[t_k, t_k+1]$ and the results of these simulations are compared to choose the most appropriate path. At time t_k+1 the chosen path is applied and a new set of simulations are started. Observations and other received information continuously modify the estimation of the target, but this updated model is employed when the UAV reaches the next time check point. That is observations obtained in time period $[t_k, t_k+1]$ affect simulations conducted in period $[t_k+1, t_k+2]$ which determine the path of the UAV after time t_k+2.

Apart from difficulties in constructing an on-line simulation system in general, some other problems should be addressed before this method can be employed in UAV path planning. Given the state of a system, the aim of on-line simulations is to predict the future state of the system and choose a course of action that is most beneficial for the system. In a surveillance mission, the state of the system (target) is not available. Indeed, the objective of the on-line simulation in this case is to optimize the process of acquiring information. The sensor data, before the target is detected, consists mostly of "negative" information i.e. lack of sensor measurement where it was (with some probability) expected (Koch, 2004). This information should be utilized to modify our estimation of the target's location. The process of drawing conclusions from sensor data is a problem studied by the information fusion community. One powerful estimation technique used in information fusion is Sequential Monte Carlo (SMC) methods also known as Particle Filtering which is based on point mass (or particle) representation of probability densities (Arulampalam et al., 2002).

In tracking applications, SMC is an on-line simulation process, which runs in parallel with the data collection process. In on-line UAV path planning we use SMC methods to estimate the current state of the target. This estimation (particle set) which is our only picture of the reality and is updated continuously is employed in "what-if" simulations to determine how the UAV should move to collect new data as effectively as possible. This path planning algorithm consists of two parts. The first part is a main loop running in real-time in which information is collected and our picture of the state of the system is updated. The other part is a set of "what-if" simulations that are initiated and executed periodically and after reaching time check points. These simulations run faster than real-time and are executed concurrently. Comparing these simulation outputs determines the most appropriate course of action. In the two next sections we describe SMC methods briefly and explain how it is applied in the path planning framework.

5. SEQUENTIAL MONTE CARLO METHODS

In order to analyze a dynamic system using a sequence of noisy measurements, at least two models are required: First, a transition model which describes how the system changes over time and second, a sensor model which relates the noisy measurements to the state (Arulampalam et al., 2002). Usually these models are available in probabilistic form and since measurements are assumed to be available at discrete times, a discrete-time approach is convenient. In this approach the transition model $p(x_k \mid x_{k-1})$, gives the conditional probability of the state x_k given x_{k-1}. The sensor model, $p(z_k \mid x_k)$, gives the conditional probability of observation z_k, given the state x_k. We are usually interested in the conditional state of the system, given the sequence of observations $z_{1:k} = \{z_1, z_2, \ldots, z_k\}$, i.e. $p(x_k \mid z_{1:k})$. In general, this conditional probability density function, may be obtained recursively in two stages, prediction and update. The prediction is calculated before the last observation z_k is available

$$p(x_k \mid z_{1:k-1}) = \int p(x_k \mid x_{k-1}, z_{1:k-1}) p(x_{k-1} \mid z_{1:k-1}) dx_{k-1} = \int p(x_k \mid x_{k-1}) p(x_{k-1} \mid z_{1:k-1}) dx_{k-1}. \tag{1}$$

The first equality follows from $p(x_k) = \int p(x_k \mid x_{k-1})\, p(x_{k-1}) dx_{k-1}$ and the second equality is a result of the fact that the process is Markovian, i.e. given the current state, old observations have no effect on the future state (Arulampalam et al., 2002).

In the update stage the conditional probability $p(x_k \mid z_{1:k})$ is calculated using the prediction result when the latest observation z_k becomes available via Bayes' rule:

$$p(x_k \mid z_{1:k}) = \frac{p(z_k \mid x_k) p(x_k \mid z_{1:k-1})}{p(z_k \mid z_{1:k-1})} \tag{2}$$

where the denominator is calculated using

$$p(z_k \mid z_{1:k-1}) = \int p(z_k \mid x_k) p(x_x \mid z_{1:k-1}) dx_k. \tag{3}$$

If the transition model and the sensor model are linear and the process noise has a Gaussian distribution, which is a rather restrictive constraint, these calculations can be performed analytically by using Kalman Filter, otherwise some approximate method such as Particle Filtering (SMC methods) should be used (Arulampalam et al., 2002).

SMC methods are a set of simulation-based methods, which have been shown to be an appropriate tool for estimating the state of a non-linear system using a sequence of noisy measurements. Intuitively, SMC methods are simulations of how the state changes according to the transition model, and filtering the result using the sensor model. Since the system changes over time, this process is repeated in parallel with the real system when new data is received. Even if new observations are not available the prediction stage still can be used to predict the future state of the system. The procedure would be the same with the exception that since future measurements are not known yet, the update stage is not performed.

In SMC methods the probability density function of the state of the system in each time-step k is represented as a set of n points xik in the state-space and corresponding weights wik, i.e. $p_k^i = \{(x_k^i, w_k^i)\}_{i=0}^n$ where pik is particle number i in time t = k.

The simulation begins with sampling S_0 , a set of n particles, from the a prior distribution p(x0) , such that

$$S_0 = \{(x_0^i, w_0^i)\}_{i=1}^n, w_0^i = \frac{1}{n} \tag{4}$$

and number of particles in each interval [a, b] is in proportion to $\int p(x_0)dx_0$. At each iteration, particles in the set from Sk-1 are updated using the transition model, i.e. by sampling

$$p(x_k^i \mid x_{k-1}^i) \tag{5}$$

and when observations arrive the weights are recalculated using

$$w_k^i \propto w_{k-1}^i p(z_k \mid x_k^i). \tag{6}$$

Particles are resampled periodically considering their weights, i.e. they will be sampled with replacement in proportion to their weights and weights are set to $w_k^i = 1/n$. This step is necessary to replicate particles with large weights and eliminate particles with low weights and avoid degeneracy of the algorithm (Arulampalam et al., 2002).

6. APPLYING SMC METHODS TO UAV SEARCH

One natural application area of SMC methods is target tracking and surveillance. The transition model is then derived from properties of the target, terrain characteristics and other forehand information available about the target. The sensor model depends on the characteristics of the sensors and the signature of the target. Many examples of applying

SMC methods in surveillance are provided in (Doucet et al., 2001; Ristic et al., 2004). Examples of applying the methods in terrain-aided tracking are found in (Ristic et al., 2004). To demonstrate how the SMC methods work in practice, we present briefly how we have implemented them in the suggested framework. Here we assume that a single target is moving on a known road network. Some uncertain information about the initial location of the target, an approximation of its velocity and some assumption about its goal are available. The SMC methods include the following four stages: sampling, prediction, update, and resampling.

6.1 Sampling

The simulation starts with sampling $S_0 = \{(x_0^i, w_0^i)\}_{i=1}^N$ particles randomly from the a priori information $p(x_0)$, such that the number of particles on each (small) road segment is proportional to the probability of existence of the target on that road segment. Each particle is assigned a velocity randomly sampled from the distribution of the target's velocity. The weights of all particles are set equally to 1/N.

6.2 Prediction

At iteration k, particles in the set S_{k-1} are propagated forward, that is the new state of the particles are calculated using their current location, velocity and a process noise based on the transition model, $p(x_k \mid x_{k-1})$. Since the motions of the particles (vehicle) are constrained by a known road network, their state can be specified by the vector $x_k = [r_k, d_k, v_k]$, where r_k is the current road segment, d_k is the distance the particle has moved on road r_k and v_k is the instantaneous speed of the particle.

6.3 Update

As described in the previous section, the sensor model is generally described by the probabilistic model $p(z_k \mid x_k)$, where x_k is the state of the system, and z_k is the observation at time t = k. The dimensions of the state z_k are usually, but not necessarily, less than the dimensions of x_k, since the system is not completely observable. We choose to distinguish between the part of the system-state which is not under our control, i.e. the state of the target x_k and the state of the UAV (sensor), y_k. Hence the probabilistic sensor model would be $p(z_k|x_k , y_k)$. We assume that the video interpretation task is solved by some means, i.e. we have a sensor that analyzes the incoming video at a constant rate and alarm with some probability if any target is observed. That is $z_k \in$[ALARM, ¬ALARM].

Inspired by (Lichtenauer et al., 2004) we suggest the following model for sensor observations at a standard height.

$$p(\text{ALARM} \mid x_k, y_k) = \begin{cases} p_d & d \le \delta_{in} \\ p_d - \dfrac{(p_d - p_f)(d - \delta_{in})}{\delta_{out} - \delta_{in}} & \delta_{in} \le d \le \delta_{out} \\ p_f & d \ge \delta_{out} \end{cases} \quad (7)$$

In this model given the position of the target x_k and the position of the sensor y_k, the probability that the sensor indicates an ALARM, is calculated using four sensor constants. These characteristics are: detection probability (p_d), false alarm probability (p_f), inner detection range (δ_{in}), and outer detection limit (δ_{out}), see Figure 1.

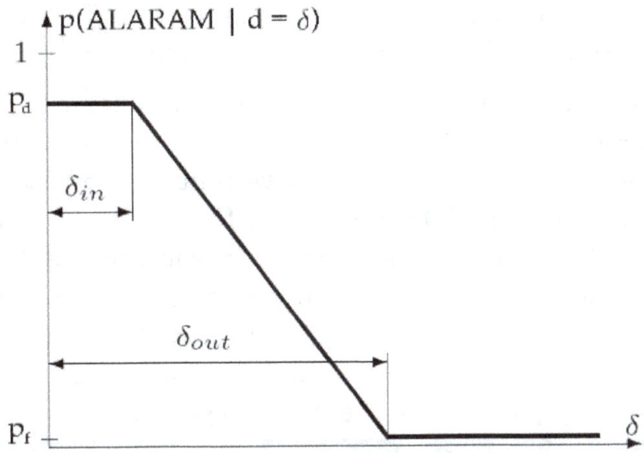

Figure 1. A graphic representation of the sensor model

After the propagation the weights of the particles are modified depending on the sensor model and current sensor observation. A sensor signal in a point increases the importance (weights) of the particles near that point. On the contrary, lack of sensor signals decreases the weights of the particles which are near the sensor. For instance if we have a perfect sensor and the UAV flies over a road segment and no sensor signal is supplied, the weights of all particles in that road segment are set to zero. After modifying the weights, they are normalized by dividing by the sum of the weights.

6.4 Resampling

A common problem with SMC methods is the degeneracy phenomenon, which refers to the fact that after many iterations, the number of particles with negligible weight increases and a large computational effort is devoted to updating particles, whose contribution to $p(x_k|z_k)$ is almost zero. One method to reduce the effect of degeneracy is to resample particles, i.e. to eliminate particles that have small weights and concentrate on particles with large weights. The resampling algorithm generates a new set of particles by sampling (with replacement) N times from the cumulative distribution function of weights of particles. The weights of these new particles are assigned the value 1/N.

7. IMPLEMENTATION AND TEST OF A SCENARIO

The suggested path planning algorithm consists of two loops, the main control loop that includes the UAV and interacts with it at each time check point to modify the path of the UAV to the best known path, and a simulation loop that estimates a picture of the reality, i.e. the probability density function of position of the target. At each time check point this picture of the reality is used in a series of what-if simulations with different possible paths of the UAV. The path that decreases the amount of uncertainty about the future is considered to be a "better" path.

Information entropy (Mackay, 2005), which is a measure of uncertainty associated with a discrete random variable $X \in \{x_1, x_2, \ldots, x_n\}$ is defined as $H(X) = -\sum_{i=1}^{n} p(x_i)\log_2 p(x_i)$.

H(X) takes only non-negative values, where H(X) = 0 indicates no uncertainty and larger values correspond to higher uncertainty. We suggest the expectation of the information entropy, E[H(X)] , as an objective function for comparing candidate UAV paths. In each step the path that decreases the expectation of the information entropy is chosen. SMC methods, which estimate the location of the target with a set of particles, provide an appropriate mechanism to estimate the expectation of the information entropy. Consider the UAV, being at point A in Figure 2(a), is facing the decision of whether to choose path ABC or ADE. The current estimate of the location of the target is shown by particles on these road segments.

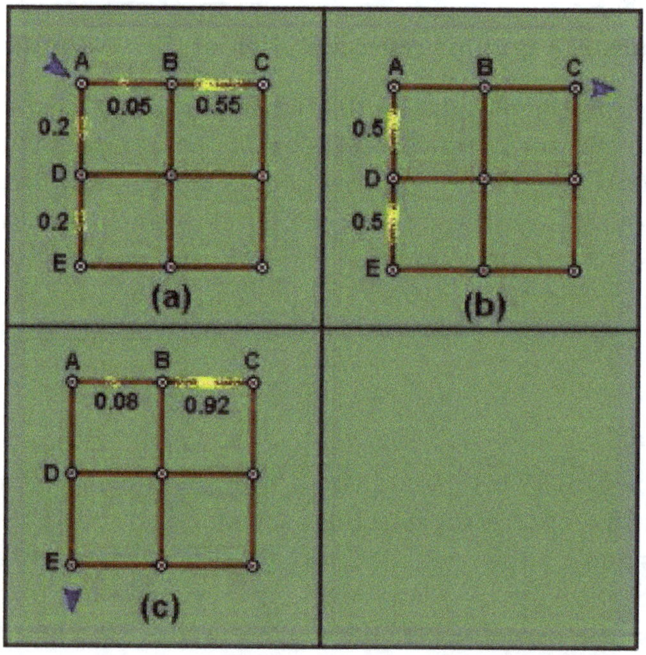

Figure 2. The impact of choosing different UAV paths on distribution of particles

Despite the fact that the total probability of finding the target on road ABC is 0.6, the most favourable path for the UAV is to choose ADE. Comparing Figures 2(b) and 2(c), shows that expectation of the information entropy by choosing path ABC is much more than choosing path ADE. See equations 8 and 9.

$$E[H(path = ABC)] = 0.p_{ABC} + (1 - p_{ABC})(-p_{AD}\log_2 p_{AD} - p_{DE}\log_2 p_{DE}) = 0.4(-0.5\log_2 0.5 - 0.5\log_2 0.5) = 0.4 \tag{8}$$

$$E[H(path = ADE)] = 0.p_{ADE} + (1 - p_{ADE})(-p_{AB}\log_2 p_{AB} - p_{BC}\log_2 p_{BC}) = 0.6(-0.08\log_2 0.08 - 0.92\log_2 0.92) = 0.24 \tag{9}$$

To evaluate the performance of the suggested method, a test scenario is designed and simulations are performed using a special purpose simulation tool, called S2-simulator, introduced in (Kamrani et al., 2006). The tool contains a "real-world" simulator including a two dimensional terrain, a target object, and a UAV that can employ different search methods, one of which is on-line simulation method as described here. The on-line simulation method employs the simulation of this "real-world" to search for the target. Clearly, the information about the location of the target in the "real-world" is not available for the UAV. However, for simplicity we assume that some part of UAV's perception from reality is exact, e.g. the map of the terrain is accurate. Terrain is modelled as a two- dimensional landscape and includes two basic elements, nodes and road segments used to build a road network.

The geography of the test scenario, as shown in Figure 3, consists of a regular road network of perpendicular crossroads. Each road segment is 15 km long and the 60 road segments make an area of responsibility that covers a square of size 75 km by 75 km. For convenience a 2D coordinate system that has the origin located at the upper left-most node, with x values increasing to the right and y values increasing downwards is introduced.

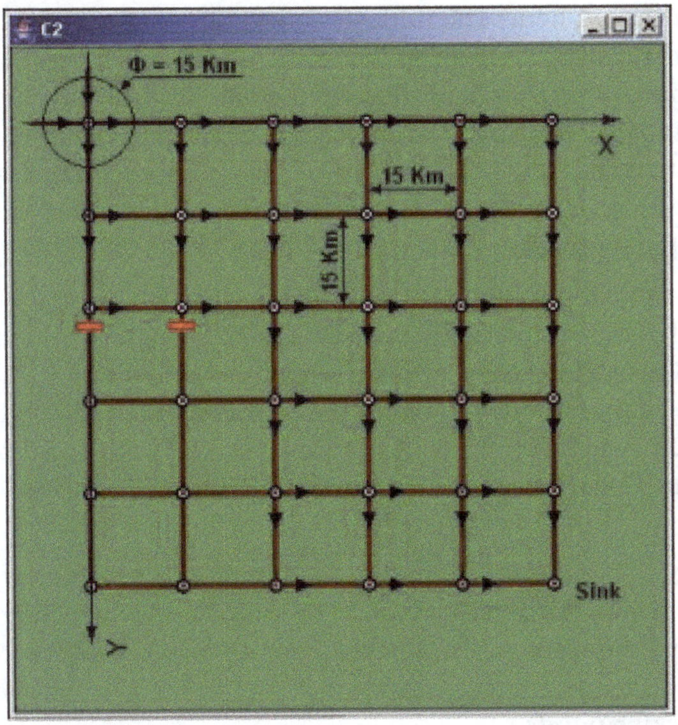

Figure 3. The target starts at the upper left-most node moving towards the sink

The target is initially located at the upper left-most node at origin. At this node and all other nodes the target moves either to east or toward south with equal probability, if any of these options are available. Hence, after passing 10 nodes and traversing

150 km the target reaches the lower right-most node and stops there. Considering these directions, the road network can be modelled as a directed acyclic graph with a source at origin and the sink located at the node in (75 km, 75 km). During the mission the road network may change as a result of the presence of friendly units, e.g. tanks may block certain roads. For example, in the scenario in Figure 3, two vertical road segments are blocked, which is shown by red bars in the figure. As seen in the figure, blocking a road segment does not affect only the blocked road segment but may change the distribution of particles in other areas. The target has a predefined path, which is unknown to the UAV. If a road in the path of the target is blocked the target chooses an alternative path. If directions to east and south are both blocked, the target stops. The average velocity of the target is 20 m/s with a velocity noise, thus it reaches its goal after approximately 7500 seconds. This velocity is unknown to the UAV.

The UAV starts its mission from a point in the III quadrant on the line y = x with a distance u0 from the origin. Velocity of the UAV is 100 m/s. A large distance between the initial location of the UAV and the area of responsibility ensures that the target has a lead over the UAV. For example if u0 = 180 km, the UAV reaches the area of responsibility after 1/2 hour. The information available to the UAV system consists of the approximate initial location and velocity of the target, i.e. it is known that the target starts from a point uniformly distributed on the roads passing origin having a maximum distance of 7.5 km and has an average velocity between 15 and 25 m/s. It is as well known, that the target chooses one of the unblocked outgoing roads (if more than one) to south or east and there is no reason to believe that the target prefers one of these outgoing roads.

In the on-line simulation every 50 seconds (time check points), a new set of 60 alternative paths are compared by simulation. The length of these simulations is 1200 seconds. These simulations are run with the maximum possible speed, which depends on the computational power of the host computer. Simulations of the real world are run by time factor 5, i.e. 5 simulated seconds are equal to 1 (wall-clock) second. All simulations are run 10 times and the presented values are the average of these results.

Figure 4 illustrates the detection time of the target as a function of the distance of the UAV to the origin. Road network and the target's paths are also depicted in the same figure for 3 different cases. Blocked road segments are marked by red bars. Blocking of road segments occurs always during the mission, but before the target reaches the blocking points.

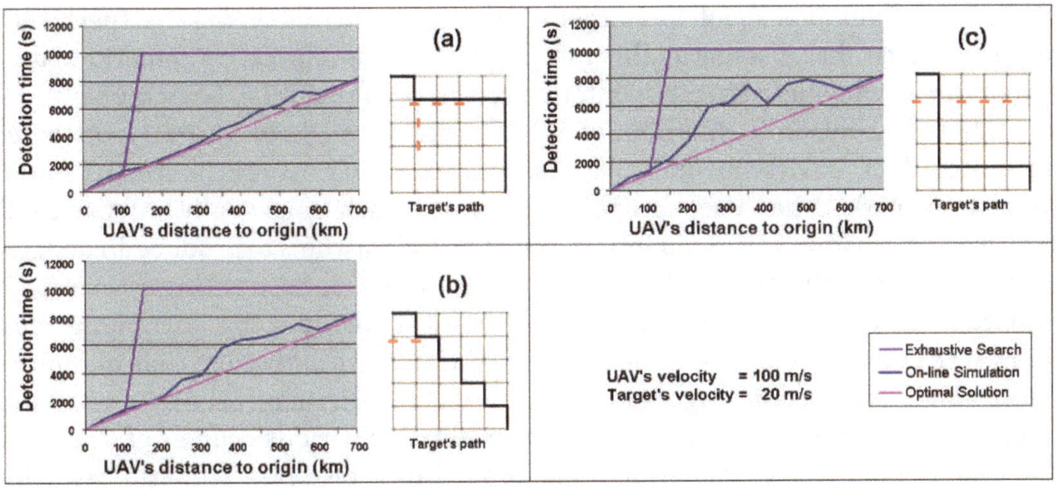

Figure 4. Detection time as a function of UAV's distance to the origin in online simulation compared with exhaustive search and the optimal solution

The horizontal axis represents the distance u0 of the UAV from the origin, ranging from 0 to a maximum of 700 km. The vertical axis represents the average time required for target detection. If the UAV does not identify the target within 10,000 seconds, the search will be terminated. The detection time for the exhaustive search approach is also presented in each instance. The exhaustive search approach entails the UAV systematically scanning the whole road network, commencing with the road segments nearest to the origin and progressing to those situated further away. Additionally, Figure 4 illustrates a lower bound for the best value of "detecting" the target. By optimal, we refer to the duration required for the UAV to reach the target when both the trajectory and velocity of the target are fully known to the UAV. The detection time escalates with the growing distance of the UAV from the origin. Nonetheless, the detection durations for the online simulation technique consistently remain below 10,000 seconds for all distances under 700 km, but the exhaustive search method is unable to detect the target when the UAV commences from a distance over 150 km. Furthermore, a comparison of the online simulation results with the "optimal" solution depicted in Figures 4-a, 4-b, and 4-c reveals that, during some intervals, the detection time approximates the optimal value, which is the duration required to reach the target when the UAV possesses complete information regarding the target and the route.

8. CONCISE SUMMARY

This chapter examines the issue of autonomous UAV path planning for search or surveillance missions, given the availability of some a priori information regarding the target and the surroundings. A search operation that leverages the available uncertain knowledge regarding the target's initial location, topography data, and plausible assumptions about the target's movement can, on average, outperform a uniform search that disregards this information. We developed a simulation-based system for employing ambiguous information in path planning. Search operations are inherently dynamic and should be adjusted throughout the mission based on sensor data, environmental changes, and external reports; hence, an online simulation method has been proposed. This approach continually integrates all available information with Sequential Monte Carlo algorithms to produce an updated representation of the probability density of the target's location. This estimation is utilized occasionally to conduct a series of what-if simulations to ascertain the most promising UAV trajectory. Among many UAV trajectories, the one that minimizes uncertainty regarding the target's location is preferred. Consequently, the anticipated information entropy serves as a metric for evaluating various courses of action for the UAV.The suggested framework was applied to a test case scenario involving a single UAV searching for a single target moving on a known road network. The performance of the method was tested by simulation, which indicated that the on-line path planning has generally a high performance. The result obtained by the on-line simulation method was compared with an exhaustive search, where the UAV searched the entire road network indiscriminately. The on-line simulation method showed significantly higher performance and detected the target in a considerably shorter time. Furthermore, the performance of the method was compared with the detection time when the UAV had the exact information about the initial location of the target, its velocity, and its path (minimum detection time). Comparison with this value indicated that the on-line simulation method in many cases achieved a "near" optimal performance in the studied scenario.

REFERENCES

Ahlberg, S. & Hörling, P. & Jöred, K. & Mårtenson, C. & Neider, G. & Schubert, J. & Sidenbladh, H. & Svenson, P. & Svensson, P. & Undén, K. & Walter, J. (2004). The IFD03 information fusion demonstrator. Proceedings of the Seventh International Conference on Information Fusion, Vol. 2, pp. 936–943, June 2004

Arulampalam, S. & Maskell, S. & Gordon, N. & Clapp, T. (2002). A tutorial on Particle Filters for on-line non-linear/non-Gaussian Bayesian tracking. IEEE Transactions on Signal Processing, Vol. 50, No. 2, pp. 174–188, February 2002

Bourgault, F. & Furukawa T. & Durrant-Whyte, H.F. (2003a). Optimal Search for a Lost Target in a Bayesian World. Proceedings of the 4th International Conference on Field and Service Robotics (FSR'03), volume 24, pp. 209–222, Lake Yamanaka, Japan, July 2003

Bourgault, F. & Furukawa, T & Durrant-Whyte, H.F. (2003b). Coordinated Decentralized Search for a Lost Target in a Bayesian World. Proceedings of the 2003 IEEE/RSJ International Conference on Intelligent Robots and Systems, Vol. 1, pp. 48–53, Las Vegas, Nevada, October 2003

Bourgault, F. & Göktogan, A. & Furukawa, T. & Durrant-Whyte H.F. (2004). Coordinated Search for a Lost Target in a Bayesian World. Advanced Robotics, Vol. 18, No. 10, pp. 979–1000

Doucet, Arnaud & de Freitas, Nando & Gordon, Neil. (2001). Sequential Monte Carlo Methods in Practice. Springer Verlag, ISBN 0-387-95146-0

Kamrani, Farzad & Lozano, Marianela Garcia & Ayani, Rassul. (2006). Path planning for UAVs using symbiotic simulation. Proceedings of the 20th annual European Simulation and Modelling Conference, ESM'2006, pp. 215–238, Toulouse, France, October 2006

Koch, Wolfgang. (2004). On 'negative' information in tracking and sensor data fusion: Discussion of selected examples. Proceedings of the Seventh International Conference on Information Fusion, Vol. 1, pp. 91–98, June 2004

Lichtenauer, J. & Reinders, M. & Hendriks, E. (2004). Influence of the Observation Likelihood Function on Particle Filtering Performance in Tracking Applications. Proceedings of the 6th International Conference on Automatic Face and Gesture Recognition. pp. 767- 772, May 2004

Mackay, David J. C. (2005). Information Theory, Inference, and Learning Algorithms. Cambridge University Press, pp. 22-46

Morse, Philip M. (1982). In Memoriam: Bernard Osgood Koopman, 1900-1981. Operations Research, Vol. 30, No. 3, May - June 1982, pp. 417–427

Richardson, Henry R. & Discenza, Joseph H. (1980). The United States Coast Guard Computer-Assisted Search Planning System (CASP), Naval Research Logistics Quarterly, Vol. 27, No. 4, March - April 1980, pp. 659–680

Ristic, Branko & Arulampalam, Sanjeev & Gordon, Neil. (2004). Beyond the Kalman Filter: Particle Filters for Tracking Applications. Artech House Series Publishers, ISBN 1- 58053-631-x

Stone, Lawrence D. (1983). The Process of Search Planning: Current Approaches and Continuing Problems. Operations Research, Vol. 31, No. 2, March – April 1983, pp. 207–233

Stone, Lawrence D. (1989). What's Happened in Search Theory since the 1975 Lanchester Prize? Operations Research, Vol. 37, No. 3, May - June 1989, pp. 501–506

Svenson, P. & Mårtenson, C. (2006). SB-Plan: Simulation-based support for resource allocation and mission planning. Proceedings of the Conference on Civil and Military Readiness (CIMI 2006), Enköping, Sweden, May 16–18, 2006

U. S. Department of Defense. (2005). Unmanned Aircraft Systems (UAS) Roadmap, 2005-2030, August 2005, pp. 47-52

Xiong, N. & Svensson, P. (2003). Multi-sensor management for information fusion: issues and approaches. Information Fusion, Vol. 3, No. 2, pp. 163–186

17 Chapter

OPTIMAL CIRCULAR FLIGHT OF MULTIPLE UNMANNED AERIAL VEHICLES FOR TARGET SURVEILLANCE IN URBAN ENVIRONMENTS

1. INTRODUCTION: AN OVERVIEW

Building upon our prior innovative formulation for the Single-Target Tracking Problem (STTP) utilizing multiple Unmanned Aerial Vehicles (UAVs) as shown in [1], we further develop our concepts for addressing the STTP with multiple UAVs in this study. The STTP to be addressed below can be articulated as follows:

Problem 2.1 Given a ground-based target and fixed-wing UAVs equipped with a camera, find an optimal strategy in a known urban environment such that the target is kept inside at least one camera's field-of-view.

Due to the importance and enormous applications of this problem, it has attracted a great deal of attention from researchers and has been studied in various directions, e.g. target identification or classification, fault-tolerant target tracking, multi-sensor target tracking, target position estimation. See [1] for a detailed literature review. However, it is our observation that most of the existing works are mainly focused on sensory data processing or fusion. In this work, we are more interested in path planning for target trackers, especially in densely populated urban domains. As far as our knowledge is concerned, there are very few works in this research area. We note that [2,3] directly address the present STTP. In particular, [3] poses the STTP as a stochastic optimisation and then proposes a real-time genetic algorithm which gradually improves an initially guessed solution.

In this present work, we develop a deterministic algorithm which is computationally much more efficient than the existing algorithms. As suggested in [1, 4], the main idea here begins with using circular motions of the fixed-wing UAVs tracking the same target. These UAVs' circular motions are highly recommended for various target tracking applications especially in urban areas, as for each UAV the maximum altitude ensures the maximum visibility and the minimum radius turn keeps the minimum distance to the target at the maximum altitude. More precisely, we first find a position, rc, over the interested target (location), rtar, which maximises the total chance of keeping the target inside the camera field of view, j, during the circular motions of the UAVs around rc. Here, all the UAVs tracking rtar keep their maximum allowed altitude and fly along the same circle C*, centred at rc with the UAVs› minimum turn radius. We assume that the probability distribution p(t) of the target location at time t is known. Note that, as the UAVs are a fixed-wing type in our present discussion, the following practical constraints are part of the STTP:

$$0 < \underline{\omega} \leq \omega_i \leq \overline{\omega} \quad \text{for} \quad i = 1, 2, \ldots, n, \tag{1}$$

where ω_i is the turning rate (magnitude) of the ith UAV, and $\underline{\omega}$ and $\overline{\omega}$ are the minimum and the maximum turning rates of UAVs, respectively. In contrast with [1], we here use ω_i as a control variable to solve the STTP problem. Note that the minimum radius turn is a function of angular velocity and bank angle. Therefore, in order to turn with the minimum radius but different speed, the angular speed must be increased or decreased by thrust, and bank angle must be adjusted appropriately so that the minimum turn radius can be maintained.

In §3.1, we first summarise how to find r_c and C^*. For more details, we direct readers to the reference [1]. Once the UAVs' flying path C^* is determined, it is natural to question about how to place more than one UAV on C^*. To answer this question, our present work is focused on the following problem:

Problem 2.2 Determine spacing Φ between the UAVs flying along C^* such that at least one UAV keeps the target inside its camera›s field-of-view all the time.

This UAV spacing problem is mathematically formulated in §3.2 and two spacing schemes are proposed thereafter. Finally, numerical examples will demonstrate the efficacy of the proposed spacing schemes in §4.

2. UAV PATH PLANNING FOR STTP

2.1 Optimal circle over target

In this section, we briefly summarise a method to find the centre r_c of an optimal circular path, along which UAVs fly and track a target at r_{tar}.

Suppose t_c is the current time and the optimal circular path is updated every t_{com} seconds. Then, assuming a known probability density function $p(r_{tar})$ of the target location during the time interval $[t_c, t_c + t_{com}]$ (or at $t_c + t_{com}/2$), one can quantify J incurred by any UAV›s travel (360-degree turn) along a circular path C using line-of-sight vectors, i.e.

$$J(C) = \int_{x \in C} \left[\int_{y \in S(x)} p(y)\, dy \right] dx, \tag{2}$$

where $S(x)$ is the set of all the ground locations within the UAV's camera's field-of-view at $x \in C$. Thus, the following optimisation problem is of our interest:

$$C^* = \underset{C}{\text{argmax}}\ J(C).$$

A sub-optimal C^*(centred at r_c with a radius of r_{uav}) may be obtained by a gradient-based search (starting with a circle whose centre is precisely over the most probable target location) combined with a random sampling scheme. Note that this search is not much computationally involved as long as a moderate number of samples is used to approximate the integration (2).

2.2 Spacing Strategy I

Once C^* is calculated, we allow more UAVs to join C^* for tracking the same target. As the introduction of more UAVs on C^* may decrease the chance of losing the target further, we are now interested in finding spacing between the UAVs so that the target is always within at least one UAV's camera's field-of-view.

Suppose the first UAV is placed on at $x_1 = r_c + [r_{uav} \cos\theta_1, r_{uav} \sin\theta_1]^3$ and other n—1 UAVs are placed on C^* according to the following sequence of spacing,

$$\Phi = \{\Phi_1, \Phi_2, \ldots, \Phi_{n-1}\},$$

where Φ_i defines the angle measured from the ith UAV to the (i+l)th UAV counterclockwise with respect to r_c. That is, for i = 1, 2, . . . , n—1 and $\Phi_n = \theta_1 - \theta_n$. We then consider the following new cost function (x), where, $\mathbf{x} = [x_1, x_2, \ldots, x_n] \subset C^*$,

$$J(\mathbf{x}) = \max_{i \in \{1,2,\ldots,n\}} \int_{y \in S_i(x_i)} p(y)\, dy, \tag{3}$$

where $S_i(x_i)$ is the set of all the ground locations within the ith UAV's camera's field-of-view at x_i. As x can be written in terms of angles, we also call (3) $J(\Theta)$, where $\Theta = \{\theta_1, \theta_2, \ldots, \theta_n\}$ and $x_i = r_c + [r_{uav} \cos\theta_i, r_{uav} \sin\theta_i]$. for i = 1, 2, . . . ,n.

As implied before, the direct optimisation to find Θ may be computationally involved as the number of UAVs on C^* increases. For this reason, we here slightly modify the problem statement as follows:

Problem 3.1 Find Θ such that the target is kept inside at least one UAV›s camera›s field-of-view with a fixed probability or probabilistic measure μ, i.e.

$$J(\Theta) \geq \mu > 0 \tag{4}$$

as UAVs fly along C^*.

To this end, we first consider $J(\Theta)$ when n = 1, i.e $\Theta = \theta_1 \overset{\text{def}}{=} \theta$. For a fixed probability distribution for the target location $p(r_{tar})$ during the time interval $[t_c, t_c + t_{com}]$, one can compute $J(\theta)$ for each $\theta \in [0, 2\pi]$or $x_1 \stackrel{\text{def}}{=} x \in C^*$. Figure 1 depicts one period of a typical $J(\theta)$. Note that, as shown in Figure 1, , which belong to $[0, 2\pi)$ for $j = 1, 2, \ldots, n_\tau$, are the angles for which the sign of $J(\theta) - \mu$ changes. The corresponding positions to $\theta_{\tau j}$ are called transition points.

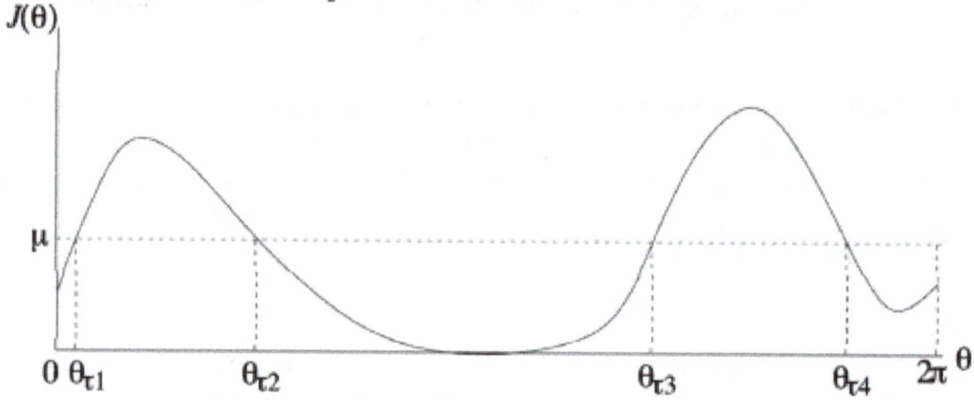

Figure 1. A typical example of $J(\theta)$ for $\theta \in [0, 2\pi]$

Our first spacing strategy is given in Table 1 Clearly, this strategy guarantees the existence of at least one UAV flying along a positive section of in which $J(\theta) \geq \mu$. Indeed, it guarantees an even better bound on average.

Table 1. Spacing Strategy I

(1)	Place the ith UAV at $x_i = r_c + [r_{uav} \cos\theta_{\tau k}, r_{uav} \sin\theta_{\tau k}]$, where $k = \mathbf{mod}(i, n_\tau)$. That is, $k = i$ if n is less than or equal to n_τ; otherwise, ith-UAV, where $i > n_\tau$, is placed at one of the already occupied transition points.
(2)	UAV's angular speeds are chosen such that the times to travel between any two consecutive transition points are all equal. UAVs keep constant angular speeds during their flights between two consecutive transition points.

Theorem 3.1 Spacing Strategy I given in Table 1 guarantees (4). Furthermore, during UAVs' 360- degree turns along C^*,

$$J(\Theta) \geq \mu + \frac{2}{n_\tau |\overline{\mathcal{I}^+}|} \int_{\theta \in \mathcal{I}^+} [J(\theta) - \mu] \, d\theta, \tag{5}$$

on average, where is the set of intervals in which $J(\theta) \geq \mu$, and $|\overline{\mathcal{I}^+}|$ is the length of the largest interval in $\mathcal{I}^+$.

Proof: As $J(\Theta) > \mu$ is straightforward to prove, we prove the second part only.

Suppose

$$\mathcal{I}^{+} = \{[\theta_{\tau 1}, \theta_{\tau 2}], [\theta_{\tau 3}, \theta_{\tau 4}], \ldots\} \tag{6}$$

and the interval lengths in $\mathcal{I}^{+}$ are l1, l3, . . ., respectively. In view of the second part of Spacing Strategy I, we set T to the time to travel between any two consecutive transition points. Then, for the example profile of $J(\theta) = J(\theta_1)$ as shown in Figure 2, if two UAVs start at θ_{τ_1} and θ_{τ_2} at t = 0 and turn 360 degrees along the circle, the corresponding $J(\Theta)$ in the time-domain can be calculated as shown in Figure 3. As seen in the figures, each positive part (peak) of $J(\theta) - \mu$ in the θ-domain appears twice with its stretched, shrunk or modified shape in the time-domain, depending on the two UAVs' locations.

More precisely, when t belongs to the interval [(k — 1)T, kT) for some k = 1, 2, . . . , n_τ , the cost is given by

$$J(\Theta) = \mu + \max_{i \in \{1,2,\ldots,n\}} [J(\theta_i) - \mu] = \mu + \max_{i \in \{1,2,\ldots,n\}} [J(\theta_{0k} + \omega_{ik} t) - \mu],$$

where ω_{ik} is the constant angular velocity of the ith UAV during the given time interval, and for i = 1, 2, ... ,n

$$\begin{aligned} \theta_i &= \frac{\theta_{\tau k}(t_{ik} + T) - \theta_{\tau(k+1)} t_{ik}}{T} + \omega_{ik} t \\ &= k\theta_{\tau k} - (k-1)\theta_{\tau(k+1)} + \omega_{ik} t \overset{\text{def}}{=} \theta_{0k} + \omega_{ik} t, \end{aligned}$$

where t_{ik} is the time when the ith UAV is at θ_{ik}.

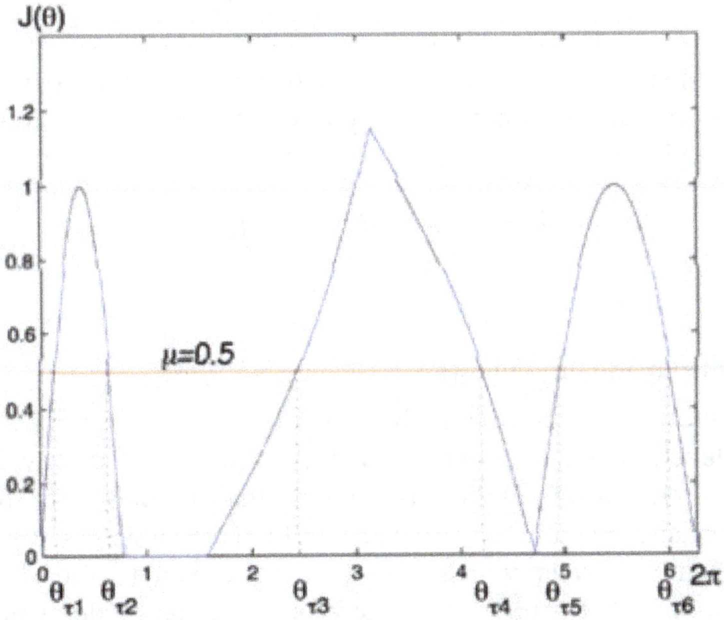

Figure 2. An example profile of $J(\theta)$ for $\theta \in [0, 2\pi]$: $\theta_{\tau j}$ (j = 1,2,..., 6) are transition points

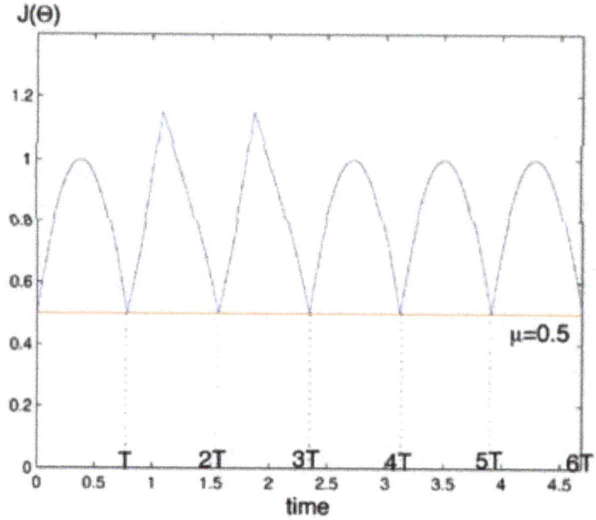

Figure 3. $J(\Theta)$ **for the** $J(\theta)$ **shown in Figure 2: T is the time for UAVs to travel between any two consecutive transition points**

Therefore, the mean value of $J(\theta)$ during UAVs› 360-degree turns can be expressed as

$$\mu + \frac{1}{n_\tau T}\sum_{k=1}^{n_\tau}\int_{(k-1)T}^{kT}\left\{\max_{i\in\{1,2,\ldots,n\}}[J(\theta_{0k}+\omega_{ik}t)-\mu]\right\}dt$$
$$\geq \mu + \frac{2}{n_\tau T}\sum_{k=1}^{n_\tau/2}\int_{2(k-1)T}^{(2k-1)T}[J(\theta_{0k}+\omega_{1k}t)-\mu]\,dt.$$

Note that n_τ is always an even number when $J(\theta)$ is continuous. Using the constant angular velocity assumption between any two consecutive transition points, one can easily check that for each k = 1, 2, . . . , $n_\tau/2$,

$$\frac{1}{T}\int_{2(k-1)T}^{(2k-1)T}[J(\theta_{0k}+\omega_{1k}t)-\mu]\,dt = \frac{1}{l_{2k-1}}\int_{\theta_{\tau(2k-1)}}^{\theta_{\tau(2k)}}[J(\theta_1)-\mu]\,d\theta_1$$
$$\geq \frac{1}{|\mathcal{I}^+|}\int_{\theta_{\tau(2k-1)}}^{\theta_{\tau(2k)}}[J(\theta_1)-\mu]\,d\theta_1.$$

Thus, the claimed inequality follows.

This theorem allows us to approximately calculate average $J(\Theta)$ (guaranteed average performance level) once C^* is given, regardless of the time T to travel between any two consecutive transition points. It is obvious that one may wish to choose μ, so that the performance level in (2.5) becomes as high as possible, but this may result in transition points between which the interval is too small. Clearly, such a small interval makes UAVs flight infeasible. Therefore, μ must be determined based on various flight constraints such as (1). In the next section, we modify the first spacing strategy to accommodate those flight constraints.

2.3 Spacing Strategy II

In this section, we take into account flight constraints on UAV's travel speed and time along C*. In view of (1.1), μ must be chosen such that

$$\underline{\omega}T \leq l_i \leq \overline{\omega}T \quad \forall i,$$

where $l_i = \theta_{T(i+1)} - \theta_{Ti}$, as defined in (6).

These flight constraints immediately imply that μ cannot be arbitrarily large. In this regard, we propose the second spacing strategy which now involves a search for a proper . The second strategy starts with setting μ to the average $J(\theta)$ over $0 \leq \theta \leq 2\pi$, i.e. $\int_0^{2\pi} J(\theta)\, d\theta/2\pi$, and check if there exist(s) an interval(s) $[\chi_{2j-1}, \chi_{2j}]$ $(j = 1, 2, \ldots, n_\chi) \subset \mathcal{I}_i^+$ in (6) such that

$$\begin{aligned} &\underline{\omega}T \leq \chi_{2j} - \chi_{2j-1} \leq \overline{\omega}T \quad \forall j = 1, 2, \ldots, n_\chi, \\ &\underline{\omega}T \leq \frac{\chi_{2j+1} - \chi_{2j}}{n-1} \leq \overline{\omega}T \quad \forall j = 1, 2, \ldots, n_\chi - 1 \text{ (if } n_\chi > 1), \\ &\underline{\omega}T \leq \frac{2\pi - (\chi_{2n_\chi} - \chi_1)}{n-1} \leq \overline{\omega}T, \end{aligned} \tag{7}$$

are satisfied. If there exist such intervals, one can increase μ and repeat the same step above. Otherwise, one should decrease μ until such intervals are found and the search is terminated.

Once the search is finished with a proper μ and its corresponding $[\chi_{2j-1}, \chi_{2j}]$, UAVs are placed on C^* at

$$x_1 = r_c + [\underline{r}_{uav} \cos \chi_1, \underline{r}_{uav} \sin \chi_1] \tag{8}$$

and if $n_\chi > 1$,

$$\begin{aligned} x_i = r_c \;+\; &\left\{ \underline{r}_{uav} \cos\left[\chi_2 + \frac{i-2}{n-1}(\chi_3 - \chi_2)\right], \right. \\ &\left. \underline{r}_{uav} \sin\left[\chi_2 + \frac{i-2}{n-1}(\chi_3 - \chi_2)\right] \right\}, \end{aligned} \tag{9}$$

otherwise

$$\begin{aligned} x_i = r_c \;+\; &\left\{ \underline{r}_{uav} \cos\left[\chi_2 + \frac{i-2}{n-1}(2\pi + \chi_1 - \chi_2)\right], \right. \\ &\left. \underline{r}_{uav} \sin\left[\chi_2 + \frac{i-2}{n-1}(2\pi + \chi_1 - \chi_2)\right] \right\} \end{aligned} \tag{10}$$

for i = 2,3, . . . ,n.

The conditions in (7) ensure that at least one UAV flies over one of the intervals in $\mathcal{I}^+$, and so $J(\theta) \geq \mu$. In fact, one can easily show that the bound obtained in Proposition 2.1 still holds.

Table 2. Spacing Strategy II

(1) Maximise μ subject to (7).
(2) Place UAVs at the locations specified by (8), (9) and (10).

Theorem 3.2 Spacing Strategy II guarantees (4) and (5).

However, finding such intervals may not be a simple task, as $\mathcal{I}^+$ is likely to be non-convex (a union of disjoint intervals). In this regard, assuming that there is a small number of intervals in , one can go through every possible combination of intervals in J^+ to find feasible $[\chi_{2j-1}, \chi_{2j}]$ (j = 1,...,n_x). In fact, we start the search with the case of n_x= 1. For example, consider an interval $[\theta_{\tau 1}, \theta_{\tau 2}] \subset \mathcal{I}^+$ and solve the following linear program for χ_1, χ_2 and T: minimise T subject to

$$\theta_{\tau 1} \leq \chi_1 \leq \theta_{\tau 2}, \quad \theta_{\tau 1} \leq \chi_2 \leq \theta_{\tau 2}, \quad \underline{\omega}T \leq \chi_2 - \chi_1 \leq \overline{\omega}T,$$
$$\underline{\omega}T \leq \frac{2\pi - (\chi_2 - \chi_1)}{n-1} \leq \overline{\omega}T.$$

If no feasible solution is found for each interval in , is now increased to 2 and two intervals are chosen from $\mathcal{I}^+$. For example, we choose , and solve the following linear program for $\chi_1, \chi_2, \chi_3, \chi_4$ and T: minimise T subject to

$$\theta_{\tau 1} \leq \chi_1 \leq \theta_{\tau 2}, \quad \theta_{\tau 1} \leq \chi_2 \leq \theta_{\tau 2}, \quad \theta_{\tau 3} \leq \chi_3 \leq \theta_{\tau 4}, \quad \theta_{\tau 3} \leq \chi_4 \leq \theta_{\tau 4},$$
$$\underline{\omega}T \leq \chi_2 - \chi_1 \leq \overline{\omega}T, \quad \underline{\omega}T \leq \chi_4 - \chi_3 \leq \overline{\omega}T,$$
$$\underline{\omega}T \leq \frac{\chi_3 - \chi_2}{n-1} \leq \overline{\omega}T, \quad \underline{\omega}T \leq \frac{2\pi - (\chi_4 - \chi_1)}{n-1} \leq \overline{\omega}T.$$

When this trial is still not successful with different two intervals in $\mathcal{I}^+$, n_χ is increased (up to $n_\tau/2$) and gives rise to similar linear programs as above. If no feasible solution is found for each n_χ, must be decreased to relax the constraints.

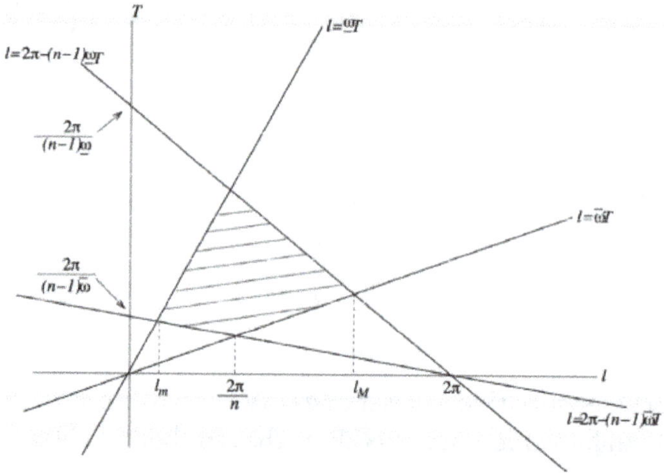

Figure 4. The polygon (dashed area) formed by the constraints when $n_\chi = 1$

As a matter of fact, the case of $n_\chi = 1$ allows a more efficient way of obtaining the solution.

First note that the constraints in the associated linear program can be written as

$$\underline{\omega}T \leq l \leq \overline{\omega}T \quad \text{and} \quad \underline{\omega}T \leq \frac{2\pi - l}{n-1} \leq \overline{\omega}T,$$

where $l = \chi_2 - \chi_1$. These two constraints in l and T form a polygon in two-dimensional space, as shown in Figure 4. From the figure, one can read that the minimum and maximum l possible in the polygon are

$$l_m = \frac{2\pi}{1 + (n-1)(\overline{\omega}/\underline{\omega})} \quad \text{and} \quad l_M = \frac{2\pi}{1 + (n-1)(\underline{\omega}/\overline{\omega})}.$$

Therefore, in order for some to be feasible, the corresponding $\mathcal{I}^+$ must contain an interval whose length is in between l_m and l_M. If there exists such an interval(s) in $\mathcal{I}^+$ whose lengths are l_1, l_2, . . ., one can then read the corresponding T_1, T_2, . . . from the figure, i.e.

$$T_j = \begin{cases} (2\pi - l_j)/(n\overline{\omega} - \overline{\omega}) & \text{if} \quad l_m \leq l_j \leq 2\pi/n, \\ l_j/\overline{\omega} & \text{if} \quad 2\pi/n < l_j \leq l_M. \end{cases}$$

Thus, the solution $[\chi_1, \chi_2]$ is the interval in $\mathcal{I}^+$ whose length l* is such that Tj is minimised over all j. One can repeat finding $[\chi_1, \chi_2]$ with an increased to improve the solution.

3. NUMERICAL TESTS

To demonstrate the performance of the suggested algorithm (Spacing Strategy II), we randomly generate urban-area maps and model a ground target as a random jump process. As the employed strategy directs, each UAV flies along a designed circular path (consisting of several segments which are defined by separation angle(s)) at a different speed, but at a constant speed along each segment. In our tests, UAVs' minimum turn radius is assumed to be equal to 200 m, and their flight altitude is fixed to 100 m. The corner velocity, i.e. the velocity for a minimum radius turn at a maximum rate, is 30 m/s (approximately 100 km/h). Therefore, the maximum angular rate ($\overline{\omega}$) is given by 0.15 rad/s. The minimum angular velocity ($\underline{\omega}$) is assumed to be equal to the half of the maximum rate, i.e. 0.075 rad/s. The number of UAVs is two and they are of the same type.

One test example is shown in Figure 5. In the figure, two UAVs fly along the optimised circular path (thick solid line) in a counterclockwise direction, with an initial separation of 130 degrees. The starting position of each UAV is indicated by a triangle. The ground moving target (denoted by a red star) jumps to a random place every 1 s. Note that the target's operational area is densely populated by buildings, and thus tracking the target is very challenging. The simulation shows that, as expected, the two UAVs swap their starting positions and reach there at the same time by changing their angular velocities. Figure 6 shows $J(\theta)$ (dotted line; before introducing two UAVs), $J(\Theta)$ (solid line; after introducing two UAVs with the separation angle suggested by our spacing strategy), and the corresponding lower bound μ. The actual cost slightly violates the lower bound at 27s. This violation is caused by the discretisation of $J(\theta)$ when numerically obtaining separation angles. A finer discretisation may alleviate this problem, but this slight violation does not make any significant physical difference in tracking the target. Figure 7 shows when the target is captured by the cameras. During one complete 360-degree turn along the circular path, the joint visibility is guaranteed during more than 69% of the total flight time. That is, the target is outside the two cameras' field-of-views for about 16s and the longest continuous time during which the target is lost is about 6 s.

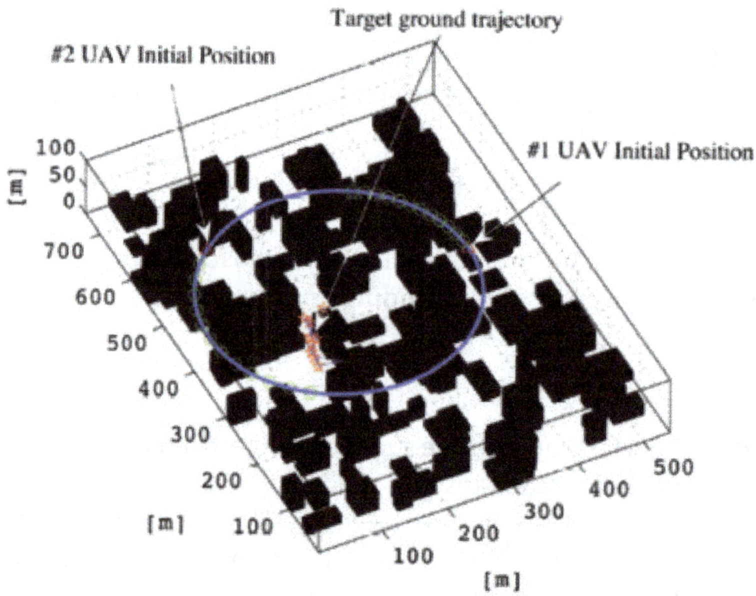

Figure 5. Two UAVs tracking a ground moving target in a dense urban area

A 1000-Monte-Carlo random simulation is also performed. This simulation is done using a personal computer (Windows-XP Professional version 2002, Intel Core 2 Quad CPU, 2.66GHz, 2.75GB of RAM, MATLAB 7.5 (R2007b)). The mean time spent to calculate separation angles for each case is 0.4963 s. In Figure 8, our proposed spacing strategy is compared with a simple separation strategy in which two UAVs are initially separated by 180 degrees. Figure 8 depicts, for both strategies, the number of cases versus the per cent of visible time during one 360-degree turn along the circular path associated with each case. The mean visibility is around 56% for both strategies. However, the variance corresponding to our spacing strategy is slightly better (smaller) than the simple one: our spacing strategy yields about 23% and the simple one about 27%. Furthermore, observe that the two distributions exhibit significant divergence: the distribution for the simple technique is nearly uniform, while that for our proposed spacing strategy is right-skewed. This indicates that our separation strategy manages complex situations significantly more effectively than the simplistic approach.

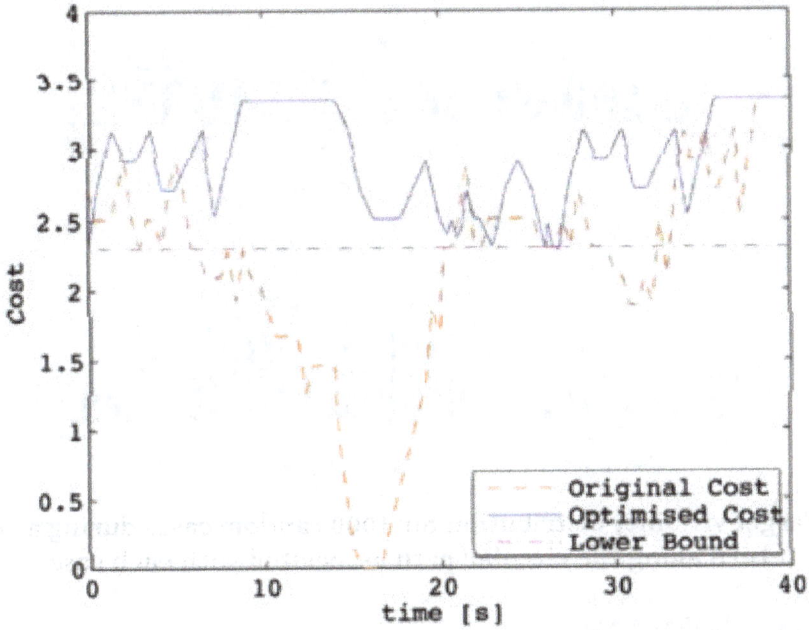

Figure 6. Original cost $J(\theta)$, optimised cost $J(\Theta)$ and lower bound

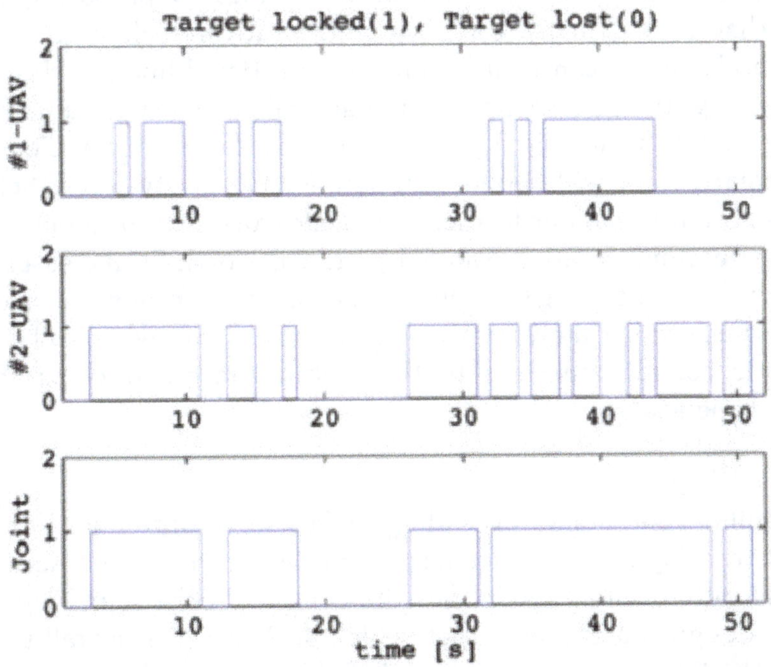

Figure 7. Target visibility during a 360-degree turn along the circular path shown in Figure 5

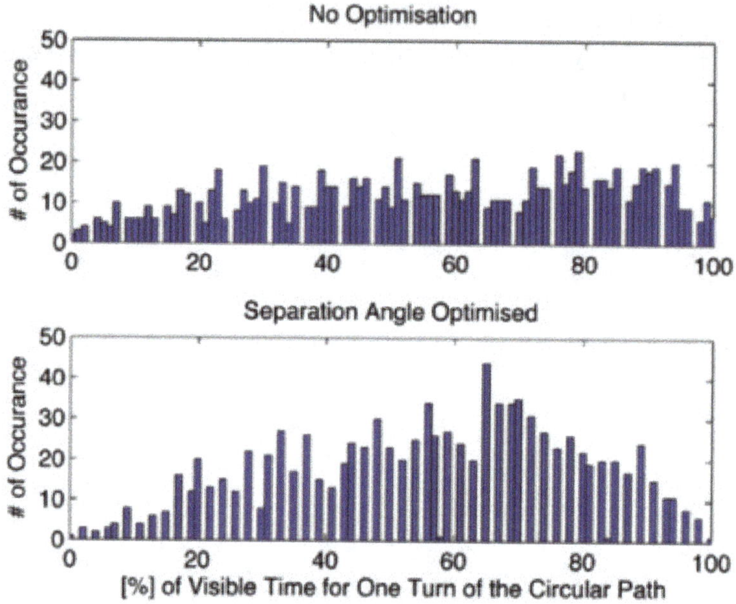

Figure 8. Target visibility distribution for 1000 random cases during a 360-degree turn along the circular path associated with each case

4. CONCISE SUMMARY

In this study, we expanded upon our prior findings by proposing two spacing techniques that enable many UAVs to collaboratively follow a target with a guaranteed probabilistic constraint. Unlike conventional line search methods, the suggested spacing strategies exhibit computational complexity that is independent of the number of UAVs pursuing the same goal, hence facilitating implementation on UAVs, particularly those with limited computational capacity. Nevertheless, certain issues require attention in our forthcoming endeavors.. For example, suppose that a target moves beyond the area covered by a circular path C_1, and so a new circular path C_2 and the corresponding spacing are calculated. Then, one needs to consider a transition strategy which allows UAVs to move from C_1 to C_2 while still guaranteeing a probabilistic bound Extensions to the multiple target tracking scenario are undoubtedly beneficial.

5. AT A GLANCE

This study builds upon our prior findings, wherein we introduced an innovative single-target tracking technique for fixed-wing UAVs (Unmanned Aerial Vehicles). The prior algorithm initially identifies the center of a circular flight trajectory, r_c, above the designated ground target, which optimizes the overall probability of maintaining the target inside the camera's field of vision of UAVs, j, as the UAVs navigate along the circular path. All UAVs maintain their maximum permitted altitude and navigate in a circular path centered at rc, utilizing the shortest possible turn radius for UAVs. According to [1,4], these circular flights are strongly endorsed for diverse target tracking applications, particularly in urban environments, as the maximum altitude flight for each UAV guarantees optimal visibility, while the minimum radius turn maintains the closest distance to the target at that altitude. Given a known probability distribution for the target position, one can estimate j, incurred by the flight of a single UAV along an arbitrary circle, utilizing line-of-sight vectors. Based on this insight, the center of an ideal circle C* can be determined among several feasible options through a gradient-based search integrated with random sampling, as proposed in [1]. The ideal circle C* is subsequently utilized by the other UAVs for the joint tracking of the same target. The inclusion of numerous UAVs may further minimize j, hence necessitating the consideration of ideal spacing between the UAVs. A conventional line search strategy is proposed for this optimal spacing issue in [1]. Nonetheless, it is predictable that the computational complexity of this search strategy may increase unfavorably with the rise in the number of UAVs. This study proposes a solution for the ostensibly intricate optimal spacing issue. Rather than relying on laborious search methods, we have devised the following approach, which is far more computationally efficient. Initially, we compute the distribution j(x), where x is an element of C*, representing the probability of capturing the target by a single camera along C*. Secondly, utilizing the distribution function j(x), determine the separation angles between UAVs to ensure that the target is consistently monitored by at least one UAV with a guaranteed probability measure. The guaranteed probabilistic measure is selected considering practical constraints, such as required tracking precision and the minimum and maximum speeds of UAVs. The efficacy of our proposed spacing system is validated using numerical simulations.

REFERENCES

J. Kim and Y. Kim. Moving ground target tracking in dense obstacle areas using UAVs, In Proceedings of the 17th IFAC World Congress, July 2008. [1]

Mark A. Peot, Thomas W. Altshuler, Arlen Breiholz, Richard A. Bueker, Kenneth W. Fertig, Aaron T. Hawkins and Sudhakar Reddy. Planning sensing actions for UAVs in urban domains, In Proceedings of the SPIE, October 2005. [2]

Vitaly Shaferman and Tal Shima. Unmanned aerial vehicles cooperative tracking of moving ground target in urban environments, Journal of Guidance, Control, and Dynamics, Vol. 31, No. 5, September-October 2008, pp. 1360-1371. [3]

Rafi, S. Khan, K. Shafiq and M. Shah. Autonomous target following by unmanned aerial vehicles, In Proceedings of the SPIE, May 2006. [4]

18 Chapter

STIFFNESS-FORCE FEEDBACK IN UNMANNED AERIAL VEHICLE (UAV) TELEOPERATION

1. INTRODUCTION: AN OVERVIEW

The operator and vehicle are physically isolated during the tele-operation of an uninhabited aerial vehicle (UAV). Poor situation awareness is frequently the result of the limited sources of environmental information, such as the absence of a comprehensive outside view, noise, vehicle motions, and vibrations. A visual interface on a ground station is typically employed in tele-operation to provide a navigation display and an external visual that is generated by a camera affixed on the vehicle. However, the visual information is typically insufficient due to a restricted field of view, which results in a teleoperation that is frequently inefficient and may even be hazardous. (Elhaij et al., 2001; McCarley & Wickens, 2005).

Previous research efforts indicated that a haptic interface, using force feedback via a haptic control device, can be used to complement the visual interface. Haptic feedback provides information about the environment through the sense of touch (Lam et al., 2004; Lam et al., 2007). Indeed, the multi-sensory haptic interface improved operator performance and significantly reduced the number of collisions, leading to an overall highly increased level of safety (Lam et al., 2007). Operator control activity and workload, however, increased as well. This may be attributed to the incompatibility between the haptic interface and the operator's intentions, based on her or his internal representation of the environment.

High operator control activity and workload particularly occur in situations where the UAV is surrounded by many obstacles. Although force feedback would indeed help the operator to avoid a collision with one obstacle, through deflecting the control device away from the direction of the obstacle, it may also direct the UAV towards another obstacle located at the other side. Whereas the haptic interface does not know about the other closely-located obstacle, the operator might already have located it from the visual interface, and would prefer to adjust the direction of motion only "just enough" to avoid collision.

The operator and vehicle are physically isolated during the tele-operation of an uninhabited aerial vehicle (UAV). Poor situation awareness is frequently the result of the limited sources of environmental information, such as the absence of a comprehensive outside view, noise, vehicle motions, and vibrations. A visual interface on a ground station is typically employed in tele-operation to provide a navigation display and an external visual that is generated by a camera affixed on the vehicle. However, the visual information is typically insufficient due to a restricted field of view, which results in a teleoperation that is frequently inefficient and may even be hazardous.

2. HAPTIC FEEDBACK

The use of haptic feedback through a control device allows the operator to perceive tactile cues through the sense of touch. The tactile information can represent various physical or mechanical properties, such as temperature level and relative distance (Elhaij et al., 2001). For collision avoidance in the tele-operation of a moving vehicle, it would be compatible with the operator's internal representation when the tactile cues would represent "repulsive forces" exerted from obstacles located near-by in the environment. These "repulsive forces" can be generated by an artificial force field (AFF) (Boschloo et al., 2004; Krogh, 1984).

The purpose of the artificial force field is to transform the UAV position and velocity relative to an obstacle, to a certain "value" that can be fed back to a haptic control device, yielding an impedance on the operator's control deflections. Fig. 1 shows a schematic representation of haptic feedback from a haptic control device, in our study a side-stick, to the human operator.

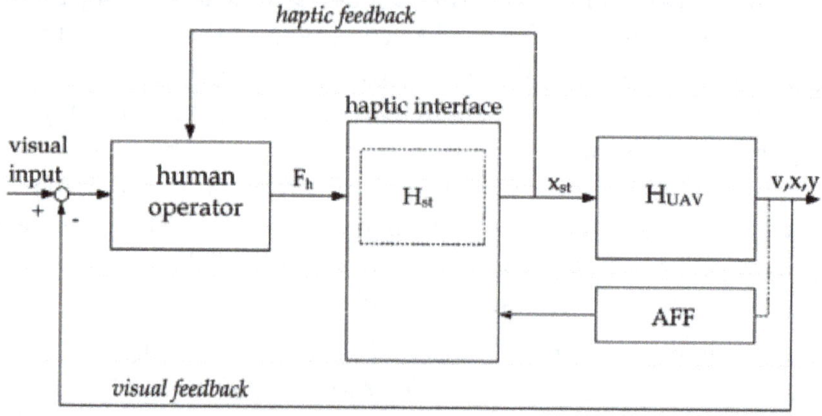

Figure 1. Schematic representation of a human-in-the-loop system with visual and haptic feedback

The impedance of the haptic control device consists of the stick mass-spring-damper dynamics Hst, depicted in Eq. (1), and the extra impedance from the output of the AFF.

$$H_{st}(s) = \frac{1}{ms^2 + bs + k}, \quad (1)$$

with m, b and k the stick's mass, damping coefficient and spring constant, respectively. The generated value of the AFF can be fed back in many different ways to the haptic device. In the following sections, three alternatives will be discussed.

2.1 Force feedback

In force feedback, a force offset F_f is applied to the control manipulator to guide the operator, see Fig. 2. The total force that the operator perceives is the sum of the reaction force from the stick dynamics, F_{st}, and the external force offset. Assume that the stick is displaced to a certain position, x_{st}, then the force exerted by the operator, i.e., the force on the hand, F_h, is written as:

$$F_h(x_{st}, i) = F_{st}(x_{st}) + F_f(i), \quad (2)$$

$$F_h(x_{st}, i) = kx_{st} + F_f(i), \quad (3)$$

with $F_{st}(x_{st})$ and $F_f(i)$ the reaction force from the control device as function of the stick deflection x_{st}, and the external force offset as function of haptic feedback information i, respectively. Note that, with haptic feedback, the stick deflection xst acts as the input to the system to be controlled, in our case the UAV helicopter.

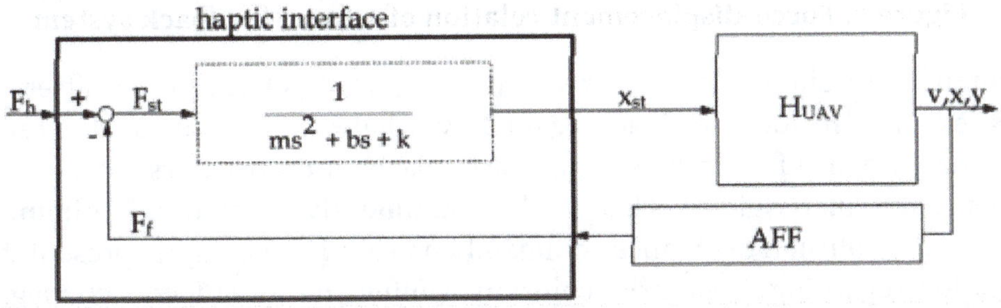

Figure 2. Schematic representation of the force feedback algorithm

Fig. 3 shows that due to the force offset, the stick will have a non-zero neutral position (A), i.e., the position where the stick is in equilibrium in the absence of external forces. When the stick is released, $F_h(x_{st}, i)=0$, repulsive forces can still exist and the stick indeed actively deflects away from the direction with a possible collision: $x_{st}=-F_f(i)/k$. This active deflection with hands-off can be considered as an "autonomous collision avoidance" function. In fact, the force feedback can be regarded to yield a "commanded" stick deflection that the operator should follow as good as possible. That is, when yielding to the forces applied on the hand, the operator would deflect the stick in a way that satisfies the collision avoidance function. Because a control device is generally limited in its deflections, there is also a limit to the amplitude of the force feedback, i.e., a natural constraint to the force feedback gain.

In previous studies, Lam et al. found that the amplitudes of stick motions due to neutral position changes may contribute to workload (Lam et al., 2007; Lam et al., 2008). In particular when flying through a narrow corridor, or when moving along multiple smaller and closely-spaced obstacles, these force offsets may vary continuously. The operator is not able to follow the force feedback accurately and overshoots and control oscillations occur.

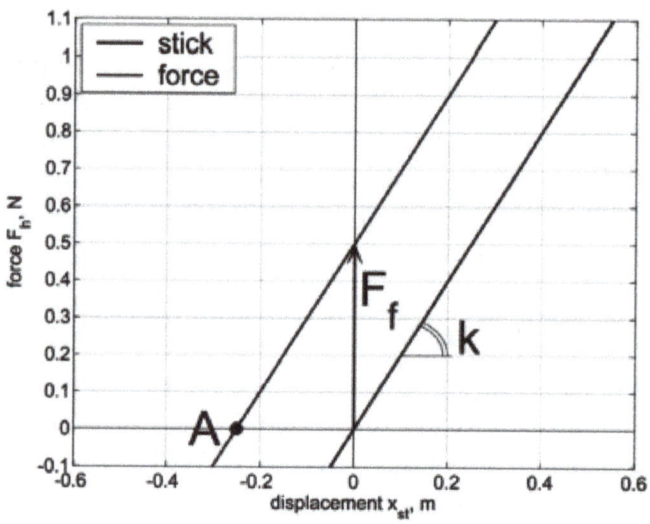

Figure 3. Force-displacement relation of a force feedback system

One potential resolution is to decrease the amplitude of the force offset, which entails reducing the force feedback's gain. Nevertheless, this could lead to repulsive forces that are insufficiently strong, particularly for operators who employ a high neuromuscular rigidity. This would consequently substantially diminish the efficacy of the collision avoidance command and restrict the haptic presentation of information regarding a potential collision. Tuning the force feedback magnitude necessitates a compromise between the haptic interface's efficacy and the operator's workload.

2.2 Stiffness feedback

Stiffness feedback involves addition of an extra spring load, $k_s(i)$ as a function of haptic feedback information i, to the nominal stick dynamics' spring constant k, see Fig. 4. Rather than a force offset, the stick becomes stiffer in the presence of an obstacle. This extra stiffness creates an impedance, resulting in an additional force that is contingent upon the operator's deflection of the stick. The stick will not actively deflect away from a potential collision when it is released, as is the case when force feedback is employed. Rather, it will revert to the neutral position. Therefore, autonomous collision avoidance cannot be implemented solely through rigidity feedback. The total force that an operator perceives in this situation can be written as:

$$F_h(x_{st},i) = F_{st}(x_{st}) + F_s(x_{st},i), \tag{4}$$

$$F_h(x_{st},i) = kx_{st} + k_s(i)\, x_{st}. \tag{5}$$

with $F_s(x_{st}, i)$ the force due to an extra spring load $k_s(i)$. The slope of the force-excursion relation increases as a result of the additional spring load, as illustrated in Fig. 5. Therefore, a zero displacement results in a zero repulsive force, as the line rotates around the origin.

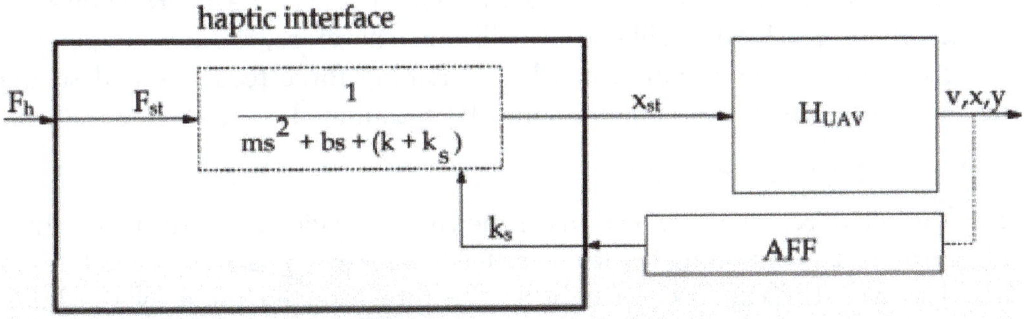

Figure 4. Schematic representation of the stiffness feedback algorithm

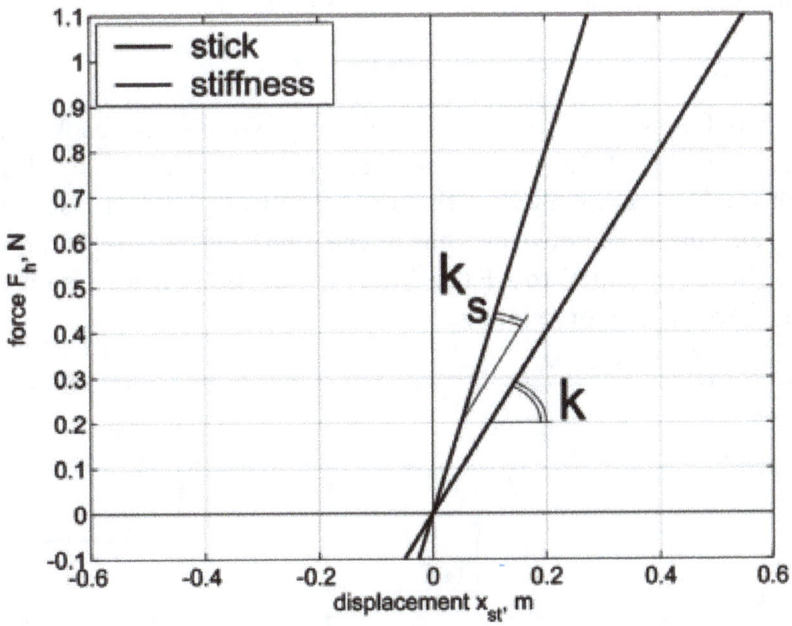

Figure 5. Force-displacement relation of a stiffness feedback system

Human operators may not perceive adequate information when employing only small stick displacements, as small displacements generate relatively small repulsive forces. The operator will not receive any haptic feedback when the stick displacement is zero. In this scenario, the rigidity feedback would have been so significant that the stick could not be moved, as when the UAV would have moved in close proximity to the obstacle, such as during a drift caused by the wind. Consequently, it is not preferable to provide solely feedback regarding stiffness. In order to prevent the introduction of control problems, the haptic device should be able to actively deflect the stick away from a potential collision, preferably with only a small overshoot from the zero displacement. However, it could be advantageous if the haptic device could generate substantial resistance when the stick is deflected in the incorrect direction, i.e., when there is a high likelihood of collision. Therefore, the two properties of haptic feedback can be accomplished by combining force feedback and stiffness feedback. Subsequently, this combination will be examined.

2.3 Stiffness-force feedback

The lack of active repulsive deflections of the control device with stiffness feedback can be resolved by combining it with force feedback, see Fig. 6. In this combination, referred to as "stiffness-force" feedback, the total exerted force by the human operator is defined as:

$$F_h(x_{st},i) = F_{st}(x_{st}) + F_s(x_{st},i) + F_f(i), \tag{6}$$

$$F_h(x_{st},i) = kx_{st} + k_s(i)\, x_{st} + F_f(i). \tag{7}$$

Fig. 7 shows the force-excursion relation of stiffness-force feedback. A property of this configuration is that due to the increase in stiffness, the desired offset of the neutral position, (A), commanded by the force feedback will decrease, which results in (B). It can also be seen, however, that in the hands-off case ($F_h(x_{st}, i)=0$) the stick deflection x_{st} becomes equal to - $F_f(i)/(k+k_s(i))$, i.e., the stiffness feedback actually reduces the effects of the force feedback.

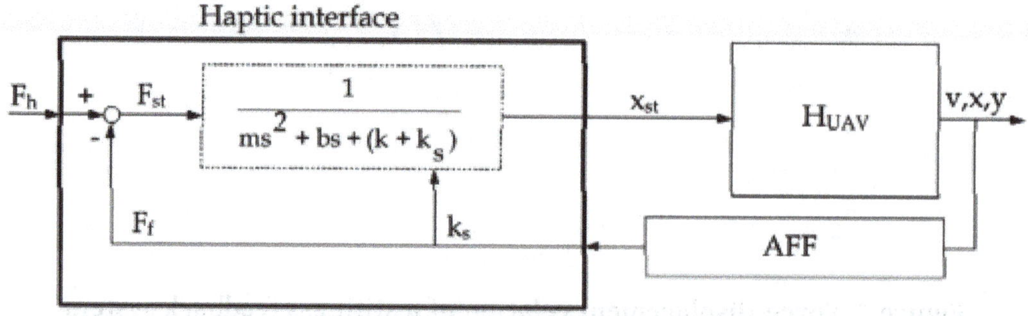

Figure 6. Schematic representation of the stiffness-force feedback algorithm

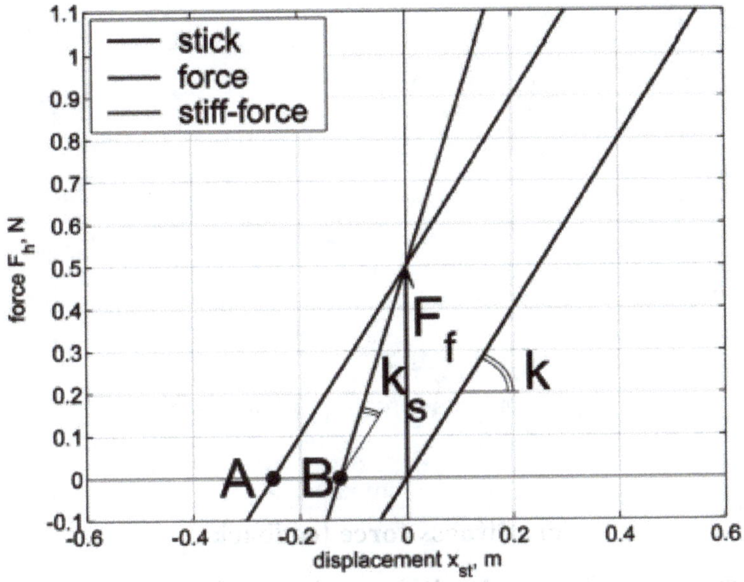

Figure 7. Force-displacement relation of a stiffness-force feedback system

2.4 Stability analysis

As mentioned in the introduction, force feedback may cause the stick to overshoot the neutral position. Operators sometimes need to counteract the active motions, particularly when obstacles are present at the other side. This indicates that active stick motion, a property of force feedback, could be a source for undesired oscillations. This section will consider the stability of force feedback, stiffness feedback and stiffness-force feedback at the hand of computer simulations, using the closed-loop system from Fig. 1. The human operator model contained a neuromuscular model for the lateral direction from (Lam et al., 2005). The stick dynamics were described as:

$$H_{st}(s) = \frac{1}{0.015s^2 + 0.3s + 3}. \tag{8}$$

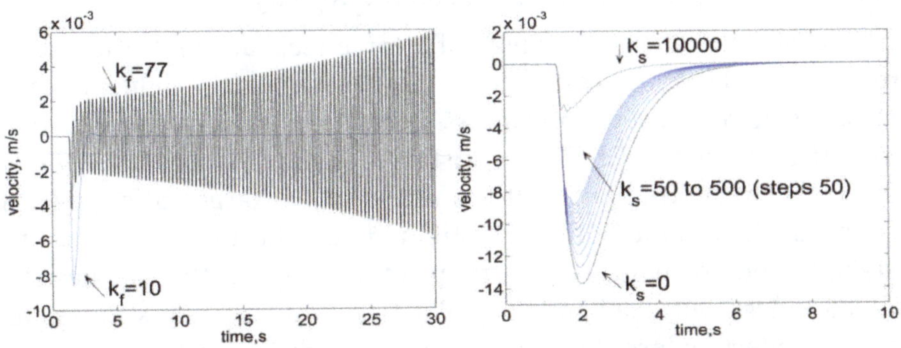

a) force feedback **b) stiffness feedback**

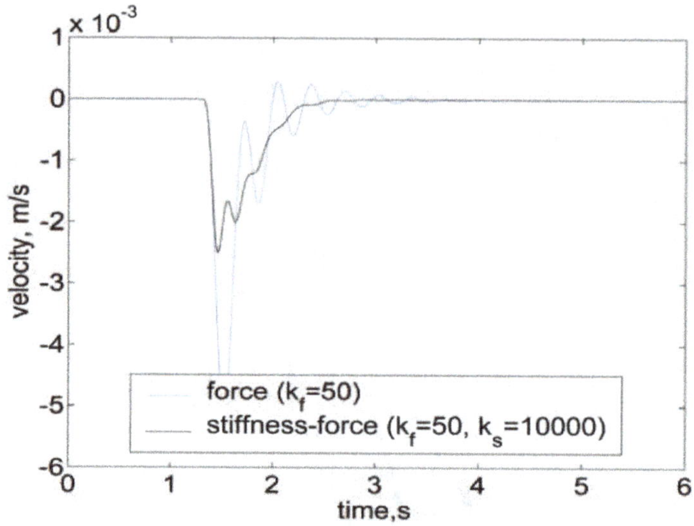

c) stiffness-force feedback

Figure 8. Step responses of the different haptic feedback configurations

The UAV helicopter was assumed to be control-augmented, with dynamics equivalent to a first order relation between stick deflection and UAV velocity (Voorsluijs et al., 2004):

$$H_{UAV}(s) = \frac{2}{s+2}. \tag{9}$$

The function of the AFF was to generate a value, based on relative position and velocity toward an obstacle, which could be used as input to the haptic device. Since this theoretical investigation only focuses on the behavior of the closed-loop system response to the different feedback of the AFF signal, the actual (non-linear) algorithm of AFF was left out and replaced by a gain for direct feedback of the velocity. For force feedback and stiffness feedback, the gains were k_f and k_s, respectively. It was assumed that the AFF primarily depended on the velocity.

Figs. 8(a) and 8(b) show step responses of a force feedback and stiffness feedback for increasing k_f and k_s gains, respectively. Fig. 8(c) shows the influence of stiffness feedback on a step response with stiffness-force feedback. It can be seen that a step response with force feedback becomes unstable for large force gain $k_f \geq 77$, whereas the stiffness feedback will not result in instability. For large stiffness gains k_s the stick dynamics will become "infinitely" stiff and no deflection will occur at all. The better stability and no change of neutral position suggests that stiffness feedback would result in less workload and control activity. Additionally, stiffness feedback has a reducing effect on force feedback, as was already illustrated in Fig. 7.

3. TUNING OF THE STIFFNESS FEEDBACK

Both force feedback and stiffness feedback will cause the human operator to exert a larger force to keep the same deflection. However, the two configurations have a different contribution to the repulsive force. The force feedback is directly mapped as a repulsive force, whereas the stiffness feedback is only an increase of the spring constant. Hence, when a same value from the AFF is used to generate external force or external spring constant, the latter will result in a smaller repulsive force, which also depends on the stick deflection.

Because the side stick in the experiment uses moments and angles instead of forces and displacements, we will refer to moments and angles from this point. When considering a hand that is exerting a certain moment Mh on the haptic device the following holds:

$$M_h(\theta_{st}, i) = M_{st}(\theta_{st}) + M_{ext}(\theta_{st}, i), \tag{10}$$

$$M_h(\theta_{st}, i) = \underbrace{I\ddot{\theta}_{st} + B\dot{\theta}_{st} + K\theta_{st}}_{\text{stick dynamics}} + \underbrace{M_f(i) + M_s(\theta_{st}, i)}_{\text{haptic dynamics}}. \tag{11}$$

with I, B, and K the stick moment of inertia, damping coefficient, and spring constant, respectively. θ_{st} is the stick rotation, $M_{st}(\theta_{st})$ is the reaction moment from the stick dynamics, and $M_{ext}(i)$ is the repulsive moment from the haptic device. $M_{ext}(i)$ may consist of a moment offset $M_f(i)$ from force feedback and/or $M_s(i)$ from the stiffness feedback, due to extra spring constant.

To compare force feedback and stiffness feedback, it is important to tune the stiffness feedback in such a way that both systems will generate approximately the same level of repulsive force. The strategy adopted in this chapter is to tune the maximum repulsive moment based on the so-called Parametric Risk Field (PRF), an AFF adopted from (Boschloo et al., 2004).

For a maximum risk value of 1 (the limit of the Parametric Risk Field) and a force gain k_f=1.5, the maximum repulsive moment ($M_{ext}(\theta_{st}, i)$) the force feedback can generate is 1.5Nm. For tuning the stiffness gain, the maximum repulsive moment provided by stiffness feedback should also be 1.5 Nm. Consider the contribution of stiffness to the haptic dynamics:

$$M_s(\theta_{st}, i) = K_s(i)\theta_{st}. \tag{12}$$

The maximum repulsive moment from stiffness feedback depends on the extra spring constant $K_s(i)$ and the maximum stick deflection, which was $\theta_{st(long)}$ = 0.35 rad and $\theta_{st(lat)}$ = 0.4 rad in the longitudinal and lateral direction, respectively. Hence, for $M_{ext}(\theta_{st}, i) = M_s(\theta_{st}, i)$ = 1.5 Nm the stiffness gain for the longitudinal and lateral direction is:

$$K_{s(long)} = \frac{1.5}{0.35} = 4.29 \quad Nm/rad, \tag{13}$$

$$K_{s(lat)} = \frac{1.5}{0.40} = 3.75 \quad Nm/rad. \tag{14}$$

Pure stiffness feedback, however, has some drawbacks as mentioned in Section 2.2. First, the repulsive force generated by the stiffness feedback only equals the force feedback when the stick is maximally deflected. Second, for small deflections, small repulsive forces are generated that are very hard to sense. Third, in case of drifting toward an obstacle due to wind the operator would not feel any feedback (assuming there is no controller, rejecting the drift), because the stick is not deflected into the direction of the obstacle. But the stiffness would also be very high, making it difficult for the operator to deflect the stick away from the direction of a potential collision.

It is, therefore, recommended to use stiffness-force feedback with a small part of force feedback in order to provide some feedback when small or no deflections are given, while the UAV is in a close vicinity of an obstacle. Through trial and error, it was decided to use 20% and 40% of the force offset in the longitudinal and lateral direction, respectively. Now, for a maximal deflection, the total external moment can exceed 1.5 Nm due to the partial addition of the force feedback. Since operators are able to release the stick, resulting in a rapid decrease of the repulsive force to avoid large overshoot, it is unlikely that the higher maximum external moment would lead to larger workload and control activity.

4. EXPERIMENT

A human-in-the-loop experiment was conducted to investigate the effects of the various force feedback and stiffness-force feedback alternatives on human operator performance, collision avoidance, control activity and workload.

4.1 Subjects and instructions

Eight subjects with no flight experience participated in the experiment. Their main task was to fly from waypoint to waypoint as accurately as possible in an obstacle-laden environment. The experiment simulated a reconnaissance task in a hazardous environment. The waypoints were represented by smoke plumes, strategically positioned near particular obstacles in the environment. Note that subjects were not provided with information whether or not, or when a collision occurred. After each run, a subject was asked to rate the workload she or he experienced using the NASA TLX rating scale (Hart & Staveland, 1988).

4. 2 Apparatus

The experiment was conducted in a fixed-base simulator. Subjects were seated on an aircraft chair in front of an 18 inch screen, presenting a navigation display (Fig. 9(a)), and looking forward towards a large white wall, on which the outside view from a hypothetical onboard camera was projected (Fig. 9(b)).

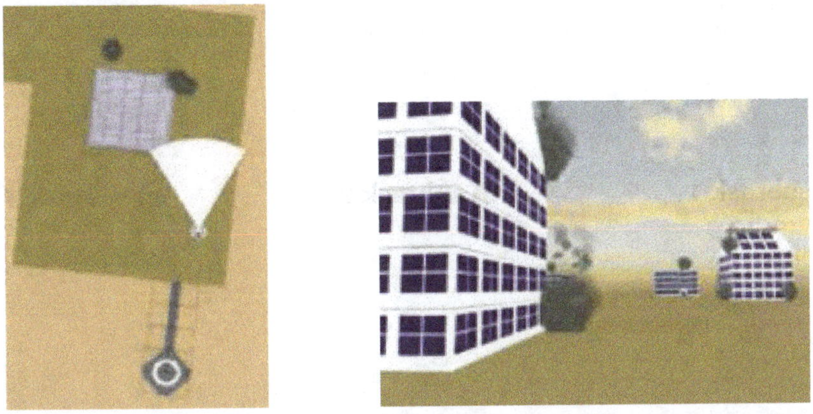

a)2-dimensional navigation display b) 3-dimensional camera display

Figure 9. Experimental navigation and camera displays

On the right side of the aircraft chair, an electro-hydraulic side stick was used as the haptic control device. The mass-spring-damper stick dynamics were simulated with the inertia I =0.01 kgm^2, damping coefficient B = 0.2 Nms/rad and spring constant K = 2 Nm/rad. The UAV was simulated by a control-augmented helicopter model, Eq. (9), with a maximum velocity of 5 m/s and maximum acceleration of 1 m/s^2. The altitude was kept constant by the control augmentation.

4.3 Independent variables

Two independent variables were used in the experiment: three levels of haptic feedback (HF) and six levels of subtask (ST).

The three haptic feedback conditions were:

Haptic off (HFoff): Subjects would only feel the side stick simulated mass-spring-damper dynamics;

Force (HF1): Subjects would feel external forces due to force offset generated by the PRF (Boschloo et al., 2004); and

Stiffness-force (HF2): Subjects would feel external forces, due to external spring constant and partial force offset, also based on the PRF.

The six subtasks are listed below, with the item number corresponding to the subtask number. The subtasks are illustrated in Fig. 10, with an arrow showing the flight direction and the red stars indicating the waypoint location, i.e., the smoke plumes. Note that a secondary purpose of the smoke plumes was to reduce the visibility of the obstacle boundaries.

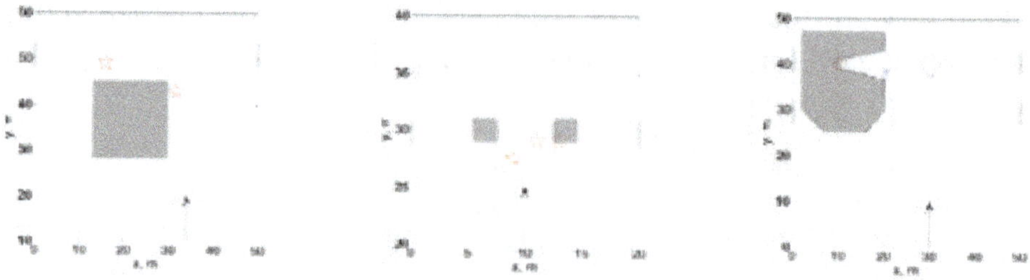

a) subtask 1 b) subtask 2 c) subtask 3

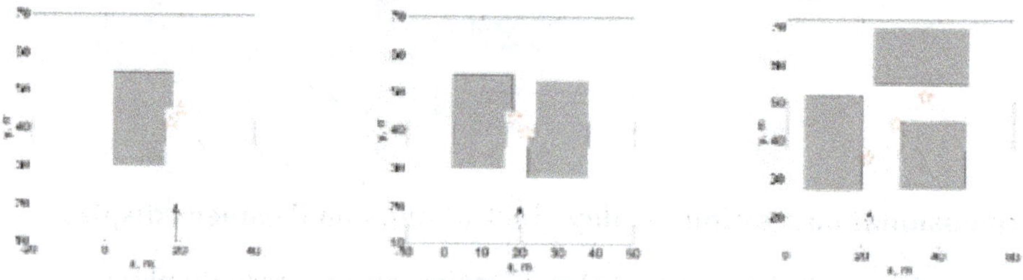

a) subtask 4 b) subtask 5 c) subtask 6

Figure 10. Top-down view of the six experimental subtasks

Subtask 1 In this subtask, the helicopter had to make a 90 degrees turn around a building, Fig. 10(a). Before the turn, the UAV had to approach the first waypoint, forcing the UAV to fly closely to the corner. The second waypoint was located after the corner in order to force subjects to make a turn as sharp as possible.

Subtask 2 In this subtask, the helicopter was to fly through a narrow gate of which the posts served as two closely-spaced, small obstacles. Fig. 10(b) shows a cross section of the gate. The smoke plume positioned between the posts was meant to reduce the visibility of the posts, making it more difficult to fly through the gate, rather than that it should serve as a waypoint.

Subtask 3 This subtask demanded a special task during hover. Once the helicopter had reached the (blue) diamond, it should hover backward toward the building until the operator could see a certain stop sign that was below the flight altitude and fixed in the world (blue asterisk), Fig. 10(c). Here, the camera would not point in the direction of motion and it was expected that haptic feedback would become very useful. Smoke on top of the building was placed to reduce the visibility of the edges of the building, leading to difficult estimation of the building edges, rather than to serve as waypoint.

Subtask 4 This subtask consisted of a building with a discrete change in the shape of the wall. The smoke was located before the discrete change of the wall and forced the UAV to approach the wall followed by an escape maneuver to avoid collision with the extension of the wall, Fig. 10(d). The second smoke plume was placed in order to force the UAV to stay close to the building during the escape maneuver.

Subtask 5 In this subtask, two buildings with discrete changes in opposite directions might lead to oscillatory behaviour in the stick and might cause considerable control difficulties. The first location of the smoke plume would force the UAV to make a sharp turn, whereas the second smoke plume would force the UAV to make an escape maneuver, Fig. 10(e).

Subtask 6 In this subtask, the turn radius with haptic feedback would be limited due to the obstacles in front and at the left side. It was expected that this subtask would lead to control difficulties, when approaching with high speed. The first smoke plume would force the UAV to approach the side of a building, whereas the second smoke plume would force the UAV to make a quick turn to fly closely along the corner of the building. The third smoke plume would force the UAV to approach another wall after the turn, Fig. 10(f).

4.4 Trajectory

In general, the trajectories (scenarios) were the same as was used in (Lam et al., 2007). The difference was the absence of a reference path in this experiment. For navigation, the subtask areas had a darker background colour on the navigation display (Fig. 9(a)).

The trajectories consisted of three sectors, which were in turn clusters of the six subtasks in a different order. To prevent boredom, 6 different trajectories were designed. Each trajectory contained a different order of the sectors. Each trajectory was flown once for each condition. Hence, each subject had to fly 6x3=18 runs; each run took approximately 5 min.

4.5 Dependent measure

The performance of collision avoidance was expressed by the number of collisions (cnt). The level of safety was expressed by the minimum allowable distance toward an obstacle (Dmin). The performance of tele-operation was expressed by the minimum distance from the smoke location (Dsmoke). Speed-related measures were expressed by the average of the total speed ($\overline{v}$) and the standard deviation of the total speed (σ_v).

The standard deviations of the total stick deflection $\sigma_{\delta tot}$ and the total exerted moment by the hand (σ_{Mh}) represented control activity. Haptic activity was represented by the standard deviation of the total external moment by the haptic device (σ_{Mext}). Workload was measured using the NASA TLX rating scale, after each measurement run.

4.6 Procedure

Each subject flew 6 runs for each haptic configuration. Before the actual experiment, subjects got the opportunity to get familiar with the three haptic configurations by training runs. After each experiment run, subjects were asked to rate their workload using the NASA TLX rating scale. After the experiment, subjects were asked to complete a questionnaire, regarding their experience with the three haptic feedback configurations.

4.7 Results and discussion

The main results are discussed in this section. A full-factorial ANOVA was applied. The means and 95% confidence intervals of the dependent measures are shown in Fig. 12.

4.7.1 Collisions

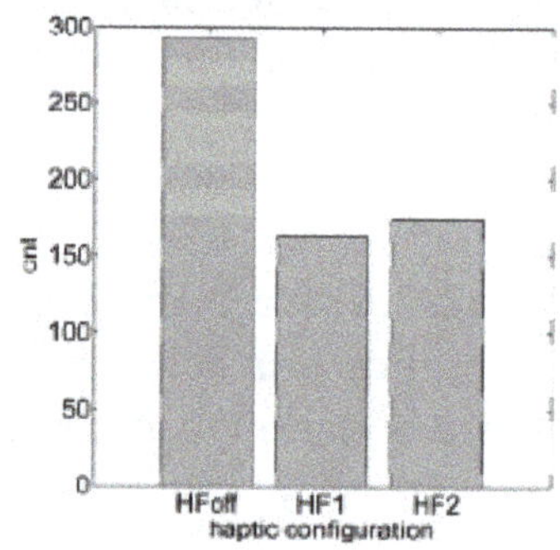

Figure 11. Total number of collisions (data of all subjects)

The total number of collisions, illustrated in Fig. 11, was highest when flying without haptic feedback, which was expected. The absence of haptic feedback, HFoff, led to a lack of situation awareness, especially when the visual information was reduced due to the smoke plumes. The force feedback, HF1, resulted in slightly less collisions than the stiffness-force feedback, HF2.

4.7.2 Approach performance

Performance expressed in terms of the distance from the smoke plumes (the waypoints) is shown in Fig. 12(a). Subtasks 2 and 3 were not included, because here the smoke plumes were placed with the objective to reduce the visual information of the obstacle boundaries, rather than to serve as a waypoint. Fig. 12(a) shows that HF1 resulted in the largest distance from the waypoints, whereas no haptic feedback resulted in the smallest distance from the waypoints, a highly-significant effect (HF: $F2,14 = 8.129$, $p \leq 0.01$). A post-hoc analysis revealed that the stiffness-force feedback lies in-between the force feedback and no haptic feedback conditions (Student-Newman Keuls (SNK), $\alpha = 0.05$). Regarding the effect of subtask, in subtasks 1 and 4, subjects were able to fly closely to the smoke. In subtasks 5 and 6, with objects surrounding the vehicle, subjects were not able to closely approach the waypoints, due to the difficult escape maneuver the close approach would cause afterwards, a highly-significant effect (ST: $F3,21 = 10.024$, $p \leq 0.01$).

4.7.3 Control activity

Fig. 12(b) indicates no significant effect of HF on the stick deflections. In subtask 3, subjects had to fly backward and forward, resulting in the largest standard deviation of the stick deflection. In subtask 2, subjects only had to fly straight through a passage, resulting in the smallest standard deviation of the stick deflection, a highly-significant effect (ST: $F_{5,35}$ = 29.343, $p \leq 0.01$). Fig. 12(c) shows that without haptic feedback the standard deviation of the measured moment was significantly smallest (HF: $F_{2,14}$ = 8.173, $p \leq 0.01$). Post-hoc analysis (SNK, α =0.05) revealed that the difference between force and stiffness-force feedback was not significant. Similar to the stick deflection, subtask 3 and subtask 2 resulted in the largest and smallest variation of the measured moment, respectively (ST: $F_{5,35} = 33.592$, $p \leq 0.01$).

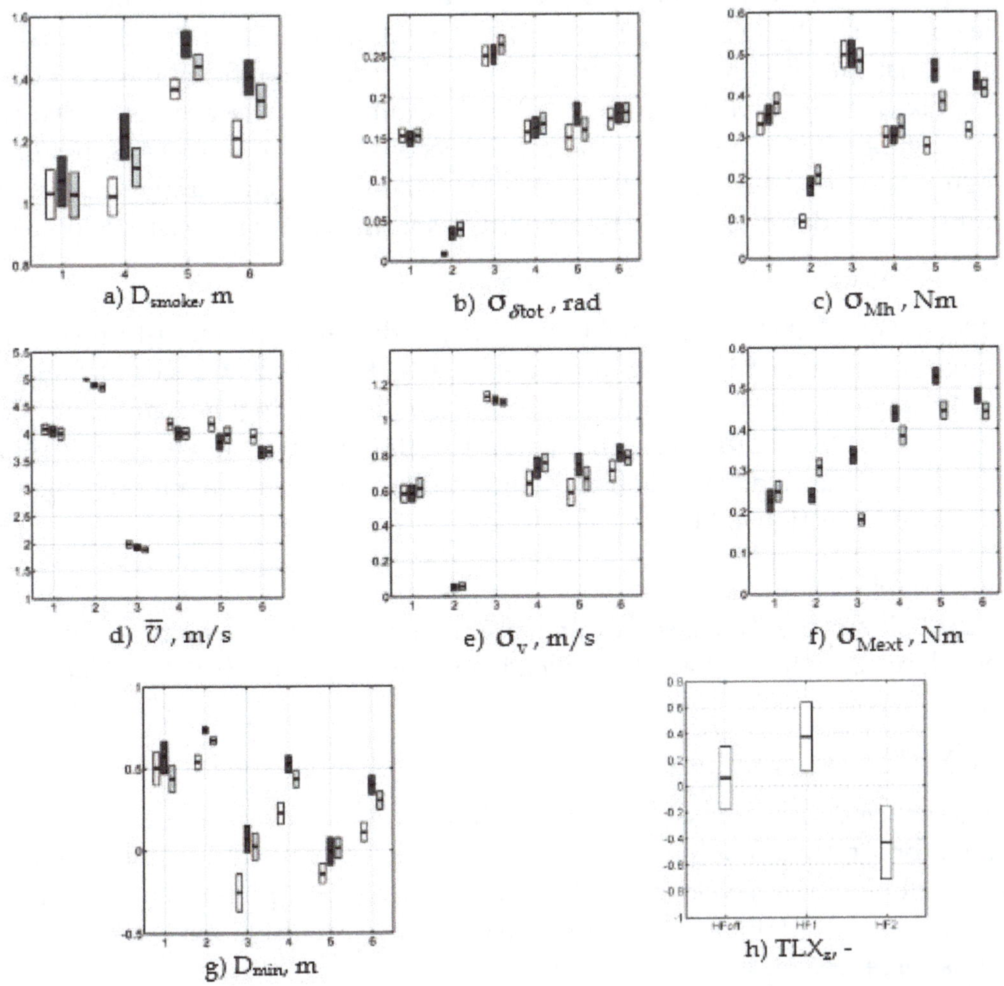

Figure 12. The means and 95% confidence intervals of the main dependent measures. The numbers 1 to 6 on the horizontal axis correspond to the subtask numbers. In all subfigures except (h), the white, dark gray, and light gray bars represent the haptic feedback conditions HFoff, HF1, and HF2, respectively

4.7.4 Haptic activity

Fig. 12(f) indicates that the standard deviation of the external moment was highly-significantly larger for the force feedback condition (HF: $F_{1,7}$ = 39.700, p ≤ 0.01). Obviously, the condition without haptic feedback was not considered. Subtasks 4, 5 and 6, all demanding a difficult approach and escape maneuver, resulted in large variations of the external moment (ST: $F_{5,35}$ = 28.255, p ≤ 0.01). In subtask 2, the external moment was larger for the stiffness-force feedback; this may be due to the partial addition of force feedback as discussed in the theoretical analysis section.

4.7.5 Speed

The average of the UAV velocity, a measure of tele-operation efficiency, was highly-significantly larger without haptic feedback, due to the lack of the repulsive forces in the opposite flight direction (HF: $F_{2,14}$ = 7.338, p ≤ 0.01), Fig. 12(d). Post-hoc analysis revealed that the difference between HF_1 and HF_2 was not significant (SNK, α = 0.05). In subtask 2 (short and relatively easy) and 3 (difficult), the average speed was highly-significantly largest and smallest, respectively (ST: $F_{5,35}$ = 59.978, p ≤ 0.01). Fig. 12(e) shows that subtasks 2 and 3 resulted in a highly-significantly smallest and largest standard deviation of the total speed, respectively (ST: $F_{5,35}$ = 24.635, p ≤ 0.01).

4.7.7 Minimum distance

Fig. 12(g) shows that HFoff resulted in the smallest distance, whereas HF1 resulted in the largest distance, a highly-significant effect (HF: $F_{2,14}$ = 24.209, p ≤ 0.01). Subtasks 3 and 5 resulted in the smallest distance from a building, a highly-significant effect (ST: $F_{5,35}$ = 19.058, p ≤ 0.01). Note that a negative distance was due to collisions and the smallest distance due to HFoff corresponded to the largest amount of collisions as discussed in Section 4.7.1.

4.7.7 Workload

Fig. 12(h) shows the TLX z-scores and indicates that stiffness-force feedback, as expected, resulted in the lowest workload, a significant effect (HF: $F_{2,14}$ = 5.888, p = 0.014).

5. CONCISE SUMMARY

It can be concluded that, when considering the approach performance and workload, stiffness-force feedback outperforms force feedback with equivalent control activity, level of safety and UAV velocities. The improved approach performance and lower workload can be attributed to the relatively lower repulsive force activity with stiffness-force feedback.

The additional stiffness changes the dynamics of the haptic interface, which influences the interactions with the operator. A stiffer side stick would make it more difficult for the operator to overrule the force feedback. Hence, stiffness feedback in combination with force feedback will not only add extra repulsive force, but it may also serve as a haptic feedback design parameter to establish the level of operator autonomy. In order to achieve this, however, the reducing effect of stiffness feedback on force feedback should be eliminated. The first steps towards this new concept, provisionally called "force-stiffness feedback" have proven to be successful (Abbink & Mulder, 2008).

REFERENCES

Abbink, D. A. & Mulder, M. (2008), Exploring the Dimensions of Haptic Feedback Support in Manual Control, Journal of Computing and Information Science Engineering, accepted

Boschloo, H. W.; Lam, T. M.; Mulder, M. & Van Paassen, M. M. (2004), Collision Avoidance for a Remotely-Operated Helicopter Using Haptic Feedback, Proceedings of the IEEE International Conference on Systems, Man and Cybernetics, The Hague, The Netherlands, 10-13 October, pp. 229-235

Elhaij, I.; Xi, N.; Fung, W. K.; Liu, Y. H.; Li, W. J.; Kaga, T. & Fukuda, T., (2001). Haptic Information in Internet-Based Tele-operation, IEEE/ASME Transactions on Mechatronics, Vol. 6, No. 3, pp. 295-304

Hart, S. G. & Staveland, L. E. (1988), Development of NASA-TLX (Task Load Index): Results of Empirical and Theoretical Research, Hancock, Human Mental Workload, P. A. and Meshkati, N., pp. 139-183, Elsevier Science Publishers (North-Holland)

Krogh, B. H. (1984), A Generalized Potential Field Approach to Obstacle Avoidance Control, Society of Manufacturing Engineers Technical Paper, No. MS84-484.

Lam, T. M.; Boschloo, H. W.; Mulder, M.; Van Paassen, M. M. & Van der Helm, F. C. T. (2004), Effect of Haptic Feedback in a Trajectory Following Task with an Unmanned Aerial Vehicle, Proceedings of IEEE International Conference on Systems, Man and Cybernetics, The Hague, The Netherlands, 10-13 October, pp. 2500-2506

Lam, T. M.; Mulder, M.; Van Paassen, M. M. & Van der Helm, F. C. T. (2005), Tele-operating a UAV Using Haptics - Modeling the Neuromuscular System, Proceedings of the IEEE International Conference on Systems, Man and Cybernetics, Hawaii, USA, 10-12 October, pp. 2695-2700

Lam, T. M.; Mulder, M. & Van Paassen, M. M. (2007), Haptic Interface for UAV Collision Avoidance, International Journal of Aviation Psychology, Vol. 17, No. 2, pp. 167-195

Lam, T. M.; Mulder, M. & Van Paassen, M. M. (2008), Haptic Feedback in UAV Tele-operation with Time Delay, AIAA Journal of Guidance, Control, and Dynamics, Vol. 31, No. 6, pp. 1728-1739

McCarley, J. S. & Wickens, C. D. (2005), Human Factors Implications of UAVs in the National Airspace, Technical report Aviation Human Factors Division, AHFD-05- 05/FAA-05-01, pp. 1-5, Savoy, Illinois

Voorsluijs, G. M.; Bennani, S. & Scherer, C. W. (2004), Linear and Parameter-dependent Robust Control Techniques Applied to a Helicopter UAV, Proceedings of the 38-th AIAA Guidance, Navigation and Control Conference, Providence (RI), August 16-19, AIAA paper 2004-4909.

INDEX

A

Activation 314

Actuator 247, 267, 284

Aerial 1, 89, 91, 357

Aircraft 137-138, 141, 144-145, 167, 251-252, 254-255, 257-258, 261-264, 267, 270-275, 357

Airflow 96

Airfoil 342, 346-348

Airframe 227-228

Airspeed 157

Algorithm 25, 102, 104, 150, 154, 156

Altitude 158, 374, 385

Amplifier 285

Amplitude 207

Angular 42, 45, 118, 145, 181-182, 187, 228, 243-244, 247, 318, 321

Anti-phase 204, 209-211

Asymmetric 338, 346-349, 354

Augmented 325

Automation 30

Avionics 258-259, 261

B

Bandwidth 282, 288

Binary 46

Bluetooth 3-5, 8-9, 11-12

C

Cables 286, 293

Camera 16, 19, 175-180, 183-189, 195, 198-202, 375, 397

Ceramic 302

Collision 389-390, 396

Commands 65, 177, 180

Component 208, 269

Computer 15-16, 104, 259, 284, 359

Connect 63

Constant 41, 241, 319-320, 395, 397

Controller 40, 49, 79, 99, 118-119, 123, 134, 137-138, 141, 143-144, 146-149, 151, 154, 156, 158, 162, 164-166, 169-170, 180, 195, 225-226, 231, 253-254, 257, 267-268, 271, 321, 327

Crystal 306

Cycles 221

D

Derivative 68, 139

Design 35-36, 44, 46, 56, 119, 137, 139-141, 150, 169-171, 173, 236, 321

Detection 217, 364, 369

Detector 217, 285, 288

Device 284-285, 388, 392

Dimensions 229

Distance 25, 145, 149, 166, 253, 269, 329-331, 368-369, 400, 402

Downstroke 204, 212, 339, 347-349, 354

Dynamics 122-124, 126-127, 131, 134, 174, 237, 239-240, 312, 316, 319-320

E

Electronic 51, 280

Encoding 152

Engine 265-267

Experiment 271-272, 396, 399

F

Feedback 40, 118-119, 134, 137, 172, 387-396, 400, 402-403

Filters 45, 291

Fitness 102-103, 154, 156-157, 159, 161-163

Flapping 337-339, 343, 346, 354-355

Flightgear 195

Flowchart 220

Flying 57

Forces 92, 99, 121, 208-211, 348, 389

Framework 65

Frequency 3, 43-44, 51, 141, 214, 218, 220-221, 279-283, 287-289, 294, 305-309

Function 9, 37, 42, 45, 48, 102-103, 314, 389

Fuselage 99

G

Generation 64, 155, 162

Geometry 253

Gimbal 178

H

Haptic 387-390, 392, 395, 397, 399-400, 402-403

Helicopter 15-16, 28, 64, 85, 91, 94, 96, 99, 108, 113, 115, 398

Heuristic 91

Hierarchy 7

Hindwing 204-205, 207, 209, 211-213

Hovercraft 38-41, 43, 52, 55

I

Inner-loop 257, 271

In-phase 204, 210

Intrados 346

Isolation 280, 302

K

Kinematic 119, 206-207

L

Leader 230, 241-242, 253, 258, 272, 315, 323-325, 327-328, 332

Linear 118, 131, 140, 142, 145, 262, 380

M

Matrix 34, 120, 181-182, 184-185, 187, 199-200, 313

Microwave 290-291, 308
Mission 61-62, 64-65, 84, 108, 360, 368
Multi-hop 328
Mutation 102, 104, 150, 155

N

Network 1-3, 8, 11, 143, 312, 328-329, 331, 368
Nonlinear 141-142, 172, 255, 276-277, 311, 335

O

Object 17, 19, 21, 23
Obstacle 13-14, 17, 28-29, 387, 396
On-board 270
Operations 35, 357, 371
Operator 175, 177, 179, 195, 358, 387-390, 396, 403
Optical 188, 293-295, 303, 306-307
Oscillator 279, 281, 283, 287, 290, 294, 296-297, 300, 302, 306
Outer-loop 253-254
Output 45, 138, 285, 294

P

Pan-tilt 176, 199
Parameters 100, 102-106, 144, 151, 154, 167, 170, 267, 269, 332
Position 79, 108, 113, 118, 132, 146, 167, 186, 188-189, 195, 241, 243-244, 254, 312, 316, 389
Pressure 155, 260, 341-342
Propeller 231

Q

Quadrotor 175, 179-183, 188, 195, 312

R

Real-time 3, 11, 269
Recovery 137, 144, 146, 149, 169
Regulator 39-40, 219
Robotics 29-30, 115
Rotation 82, 93, 185, 187, 189, 199-200, 229, 238
Rotorcraft 29, 226
Routing 34, 328-329

S

Sensor 14, 16, 18-19, 23, 28, 217-219, 359-362, 364-365
Set-point 40
Setside 40
Shifter 292
Sidebands 283
Sideways 40
Signal 40, 47, 118, 139, 158, 260-261, 289, 294
Simulation 84-85, 106, 153, 164, 195, 241, 268-269, 360-361, 367, 369-370
Software 35, 65, 220, 261, 269-270
Spacecraft 251
Stiffness 390, 392, 394-396, 403
Strategy 146, 230, 376, 379, 382
Stroke 205, 207, 339, 348, 354
Structure 150-152, 155-156
Switching 138, 261
Symmetric 346-347, 354

T

Techniques 140, 142, 225, 287
Testing 164, 169, 261
Throttle 156-157, 266
Thrust 40, 97, 208-209, 213, 266, 320
Time-scale 122-123
Toolbox 35
Torque 96
Touchdown 146
Tracking 23, 135-136, 179, 226, 233, 274, 316, 318, 320-321, 332, 371, 373, 385
Traffic 3
Trajectory 62, 64, 71, 73, 76, 83, 86, 88, 145, 148, 167, 195, 207, 357, 399
Transition 362, 378
Turbine 258

U

Unmanned 29, 89, 115, 217, 357
Upstroke 204, 339, 344, 347-349, 354

V

Variable 153, 236-237, 239-240
Vector 77, 79, 120, 182, 238, 263, 282, 313, 317
Vehicle 22, 40, 65, 79-80, 84, 118, 198, 226, 387-388
Velocity 62, 64, 73-76, 79-80, 86, 88, 99, 106, 108-109, 113, 118, 131, 134-135, 181, 186-187, 241-244, 247, 254, 316-318, 320-321, 324-325, 342, 349, 364, 368, 381, 394
Vibration 211, 279-281, 283-287, 290-295, 297-299, 301-306, 308
Virtual 12, 124, 131, 319, 321, 325
Visual 32, 41, 59, 387-388
Voltage 219, 288

W

Waypoints 63, 400
Wing-wing 205, 210
Wireless 1-5, 11-12, 328
Workload 396, 402

Z

Z-axis 295, 299-300, 305
Z-value 26

www.ingramcontent.com/pod-product-compliance
Lightning Source LLC
LaVergne TN
LVHW062215200926
848484LV00006B/1400